ANNALS OF THE NEW YORK ACADEMY OF SCIENCES

Volume 303

ELECTRICAL PROPERTIES OF BIOLOGICAL POLYMERS, WATER, AND MEMBRANES

Edited by Shiro Takashima and Harvey M. Fishman

The New York Academy of Sciences
New York, New York
1977

Library of Congress Cataloging in Publication Data

Main entry under title:

Electrical properties of biological polymers, water, and membranes.

(Annals of The New York Academy of Sciences; v. 303)
"This series of papers is the result of a conference ... held by The New York Academy of Sciences on January 26, 27, 28, 1977."
Includes index.
1. Electrophysiology—Congresses. 2. Polymers and polymerization—Electric properties—Congresses. 3. Membranes (Biology)—Electric properties—Congresses. 4. Water—Electric properties—Congresses. I. Takashima, Shiro, 1923- II. Fishman, Harvey M. III. New York Academy of Sciences. IV. Series: New York Academy of Sciences. Annals; v. 303. [DNLM: 1. Biophysics—Congresses. 2. Membranes, Artificial—Congresses. 3. Polymers—Congresses. 4. Water—Congresses. 5. Electrophysiology—Congresses. W1 AN626Y1 v. 303 1977 / QT34E33 1977]
Q11.N5 vol. 303 [QH517] 508'.1s [574.1'917] 77-16565
ISBN 0-89072-049-5

SP
Printed in the United States of America
ISBN 0-89072-049-5

ANNALS OF THE NEW YORK ACADEMY OF SCIENCES

VOLUME 303

December 30, 1977

ELECTRICAL PROPERTIES OF BIOLOGICAL POLYMERS, WATER, AND MEMBRANES*

Editors and Conference Organizers
Shiro Takashima and Harvey M. Fishman

Honorary Chairpersons
Kenneth S. Cole and Rita Guttman

CONTENTS

Opening Remarks. *By* KENNETH S. COLE 1

Part I. Electrical and Physical Properties of Biopolymers and Water-Biopolymer Interaction

Intermolecular Potential for Water and the Hydration of Proteins. *By* HAROLD A. SCHERAGA 2

On the Nature of Liquid Water. *By* JULIAN H. GIBBS 10

Discussion. W. H. STOCKMAYER, *Moderator* 20

Direct Solution of the Poisson Equation for Biomolecules of Arbitrary Shape, Polarizability Density, and Charge Distribution. *By* WILLIAM H. ORTTUNG 22

Field Fluctuation of Ionic Solutions and its Biological Significance. *By* FUMIO OOSAWA 38

Real-Time Measurement of Dielectric Relaxation of Biomolecules: Kinetics of a Protein-Ligand Binding Reaction. *By* WALTER SCHEIDER 47

Discussion. M. MANDEL, *Moderator* 57

Part II. Theoretical and Technical Aspects of Dielectric Relaxation

Dielectric Theory and Properties of DNA in Solution. *By* ROBERT H. COLE . 59

Dielectric Properties of Charged Linear Macromolecules with Particular Reference to DNA. *By* M. MANDEL 74

Discussion. E. H. GRANT AND FUMIO OOSAWA, *Moderators* 88

Subsecond Measurement of Dielectric Dispersion. *By* YUZURU HUSIMI, HARUKI NAKAMURA, AND AKIYOSHI WADA 90

Dielectric Dispersion of Polyelectrolytes Due to Ion Fluctuation. *By* AKIRA MINAKATA 107

Discussion. ROBERT H. COLE, *Moderator* 121

*This series of papers is the result of a conference entitled Electrical Properties of Biological Polymers, Water, and Membranes, held by The New York Academy of Sciences on January 26, 27, and 28, 1977.

Part III. Dielectric Properties of Biological Macromolecules

Aggregation and Conformational Changes in Solution Studied by Dielectric Absorption of Polypeptides and Substituted Polysaccharides. *By* EVE MARCHAL 123

Dielectric Properties of Normal and Abnormal Lipoproteins in Aqueous Solution. *By* C. G. ESSEX, F. H. GRANT, R. J. SHEPPARD, G. P. SOUTH, M. S. SYMONDS, G. L. MILLS, AND JOAN SLACK 142

Discussion. GERHARD SCHWARZ AND CHESTER T. O'KONSKI, *Moderators* ... 156

Part IV. Dielectric Behavior of Biopolymers in Strong Fields

Dynamics of Superhelical DNA and its Complexes with Ethidium Bromide from Electro-Optic Relaxation Measurements. *By* ARTHUR E. PRITCHARD AND CHESTER T. O'KONSKI 159

Electric Dichroism and Birefringence of DNA, Chromatin, and Their Complexes with Cationic Dyes. The Structure of Chromatin. *By* CLAUDE HOUSSIER, JOSÉ BONTEMPS, XAVIER EMONDS-ALT, AND EUGÈNE FREDERICQ 170

Chemical Transitions of Biopolymers Induced by an Electric Field and Their Effects in Dielectrics and Birefringence. *By* GERHARD SCHWARZ 190

Field Interaction with Biological Matter. *By* H. P. SCHWAN 198

Discussion. J. L. ONCLEY AND A. WADA, *Moderators* 214

Part V. Chemical Probes and Transport Phenomena in Biomembranes

Mechanisms of Rapid Optical Changes of Potential Sensitive Dyes. *By* A. S. WAGGONER AND A. GRINVALD 217

Discussion. HANS MEVES, *Moderator* 242

Part VI. Electrical and Transport Phenomena of Artificial Membranes

Studies of the Physical Chemistry of Planar Bilayer Membranes Using High-Precision Measurements of Specific Capacitance. *By* STEPHEN H. WHITE 243

Electrical Characteristics of Ion Transport in Lipid Bilayer Membranes. *By* GABOR SZABO 266

Channels Across Black Lipid Membranes. *By* MOISÉS EISENBERG, MICHAEL E. KLEINBERG, AND JOEL H. SHAPER 281

Discussion. ROBERT E. TAYLOR, *Moderator* 292

Discussion Paper: Electrical Properties of Lipid Bilayers in Nonaqueous Media. *By* PAUL KRONICK 295

Part VII. Membrane Phenomena Related to Electrical Impedance

Electrical Impedance of Excitable Membranes. *By* ROBERT E. TAYLOR 298

Investigation of Voltage-Dependent Membrane Capacity of Squid Giant Axons. *By* SHIRO TAKASHIMA AND R. YANTORNO 306

Activation, Inactivation, and Chemical Blockage of the Gating Current in Squid Giant Axons. *By* HANS MEVES 322

Discussion. ROBERT S. EISENBERG, *Moderator* 339

Measurement, Modeling, and Analysis of the Linear Electrical Properties of Cells. *By* ROBERT S. EISENBERG, R. T. MATHIAS, AND J. S. RAE 342

Ion Movements and Kinetics in Squid Axon. I. Complex Admittance. *By* D. POUSSART, L. E. MOORE, AND H. M. FISHMAN 355

Discussion. A. A. VERVEEN, *Moderator* 380

Part VIII. Membrane Phenomena Related to Noise

Noise Analysis and Channel Properties. *By* YI-DER CHEN 382

Some Problems in Membrane Noise Research. *By* J. DE GOEDE, M. W. VONK, R. J. VAN DEN BERG, H. VAN RIJN, AND A. A. VERVEEN 389

Ion Movements and Kinetics in Squid Axon. II. Spontaneous Electrical Fluctuations. *By* H. M. FISHMAN, L. E. MOORE, AND D. POUSSART 399

Discussion. RITA GUTTMAN AND H. M. FISHMAN, *Moderators* 424

Closing Remarks. *By* RITA GUTTMAN 429

Financial assistance was received from:

- THE BIOLOGICAL SCIENCES PROGRAM, CHEMICAL AND BIOLOGICAL SCIENCES DIVISION, UNITED STATES ARMY RESEARCH OFFICE, DEPARTMENT OF THE ARMY
- THE NATIONAL INSTITUTE OF GENERAL MEDICAL SCIENCES
- TOWER INTERNATIONAL

OPENING REMARKS*

Kenneth S. Cole

Laboratory of Biophysics
National Institutes of Health
Bethesda, Maryland 20014

This is a truly remarkable program that you have in the three days ahead. On the statistics, you have seven foreign countries and eleven states represented. But their coverage is even more spectacular—from liquid water and hydration through bio- and other polymers to ions crossing biological membranes.

I truly wish that I could be with you in fact, instead of only as an absentee honorary chairperson. It would be so much more fun to hear what is happening, to see old friends and colleagues and meet new ones, than to have to wait for it all in cold print.

Some years ago I was proposed for a seminar at Bell Laboratories in New York, but no one knew that I could talk. So the microphone of their endless iron wire "Voice Mirror" was shoved at me. "Say something." I was a little surprised at the playback, but even more bothered. I kept worrying, until I recognized the voice of my brother and announced my great discovery, to which our guide only commented, "That's what everyone says!" Some of you may notice a resemblance, too!

Your speakers on these three days go far into the unknown of my fifty-year obsession with cell membranes. But these are not "directed" or "mission-oriented" research, except by the demands of funding.

Last Columbus Day, in Washington, Lord Todd, the new President of the Royal Society, London, gave a wonderful example. "When Columbus left Spain he had no idea where he was going, on the way he had no idea how he would get there, when he landed he had no idea where he was, and when he got back he had no idea where he'd been."

Don't we call that basic research?

Although I've been accused of starting a new era of neurophysiology and, later, of creating a revolution in electrobiology, I've also been lucky enough to see the many and often bewildering subsequent steps. Nonetheless, I still worry about dielectric loss as we usually see it in biological membranes, and about the gory details of how ions get through them. But I'm not completely convinced that some of our data may not be results of less than adequate instrumentation. So I plead that you be sure your facts are the facts you think they are.

And I wish you well.

*Read by Robert H. Cole.

INTERMOLECULAR POTENTIAL FOR WATER AND THE HYDRATION OF PROTEINS*

Harold A. Scheraga

Department of Chemistry
Cornell University
Ithaca, New York 14853

INTRODUCTION

Since water plays an important role in the determination of how proteins acquire their native structure and then interact with other molecules, it is necessary to understand the nature of the interactions between water and the functional groups of proteins. This is a problem that we have considered on two previous occasions in these Annals,[1,2] and the present discussion is devoted to recent advances in this field. Further details on water and aqueous solutions may be found in a number of treatises.[3-6]

At present, a shell model[7,8] is used to treat hydration in conformational energy calculations on proteins. In this model, a layer of water, having the thickness of one water molecule, is assumed to exist around each functional group. As the conformation of the protein is altered (to minimize its conformational energy), water must be eliminated from the hydration layers whenever they overlap. The free-energy change accompanying this "dehydration" depends on the amount of water eliminated from the first layers (which, in the computations, depends on the degree of overlap) and on the free energy of hydration of the various groups. Such a model leads to a tendency for polar groups to lie on the surface of proteins, in contact with water, and for nonpolar groups to lie in the nonpolar interior, out of contact with water. The free energies of hydration of the various functional groups were obtained[7,8] from a variety of physical chemical data, but experience with these computations has demonstrated a need to obtain more accurate values of the free energies of hydration.

For this purpose, considerable attention has been focused on the structure and thermodynamic properties of water and of aqueous solutions of polar and nonpolar solutes, which can serve as models for understanding the hydration of the functional groups of proteins. We shall, therefore, consider some of the recent developments in the theory of water and aqueous solutions. It is hoped that such studies will provide a more accurate description of hydration that can then be used to improve the shell model, whose simplicity is ideal for conformational energy calculations on proteins.

*Supported by research grants from the National Science Foundation (PCM-08691) and from the National Institute of General Medical Sciences of the National Institutes of Health, U.S. Public Health Service (GM-14312).

THEORETICAL APPROACHES TO THE STUDY OF AQUEOUS SOLUTIONS

Essentially three different types of theoretical approaches have been taken to study the properties of aqueous solutions. In the first of these, one formulates a model and then carries out a statistical mechanical treatment of it. The cluster model of Némethy and Scheraga,[9-12] with its recent improvements by Hagler *et al.*,[13] Lentz *et al.*,[14] and Owicki *et al.*,[15] is an example of this approach. In the second, one assumes a knowledge of the pair interaction potential between water molecules and solves approximate integral equations for observable quantities,[16,17] as in the general theory of liquids. In the third approach, one also assumes a knowledge of the pair potential energy function and carries out a molecular dynamics[18-22] or Monte Carlo[23-31] calculation. Currently, with the availability of reasonable pair interaction potentials for water, most efforts are being devoted to the second and third approaches.

PAIR POTENTIALS

In recent years, a number of empirical pair potentials have been used to treat water · · · water interactions. These include the Weissmann-Blum,[32] Ben-Naim-Stillinger,[18,33] Analytical Fit to Hartree-Fock,[34,35] ST2,[20] Shipman-Scheraga,[36] and EPEN (Empirical Potential using Electrons and Nuclei)[37-39] potentials, all of which were discussed by Owicki *et al.*,[40] and the configuration-interaction potential of Matsuoka *et al.*[41]

An advantage of the EPEN potential is that it is applicable not only to water but also to saturated and unsaturated organic molecules, which can be assembled from molecular fragments based on carbon, oxygen, and nitrogen atoms, respectively; the formation of molecules from molecular fragments is an essential feature of the EPEN formalism. The EPEN potential has been applied to a variety of problems, e.g. to studies of the structure and energetics of crystals,[37,38] to calculations of dipole moments and barriers to internal rotation,[37-39] and to a study of the structure, energetics, and dynamics of small (hydrogen-bonded) water clusters.[40] In attempts to apply EPEN to Monte Carlo calculations on liquid water, it was found that nonhydrogen-bonded configurations of a water dimer were not sufficiently repulsive. The EPEN formalism was therefore modified[42,43] to rectify this deficiency, and also to extend the original treatment of saturated molecules to unsaturated ones.

As EPEN was used originally,[37-39] a molecule was constructed from molecular fragments consisting of a single heavy (nonhydrogen) atom nucleus, any hydrogen atom nuclei bonded to this nucleus, and point charges representing the bonding and lone-pair electrons associated with this heavy-atom fragment. There were three types of interactions to be evaluated when using EPEN: (1) overlap repulsion between electrons, of the form $A \exp(-Br)$; (2) coulombic interactions between all charge centers (electrons and nuclei), of the form $q_i q_j / r$, where q_i and q_j are the charges at two points separated by a distance r, and (3) dispersion and other attraction terms resulting from the interaction of fragments (this was taken to be effective between heavy atom nuclei only, and was of the form C/r^6). In the modifications that were introduced later,[42,43] the centers of the C/r^6-attractive terms were shifted from the heavy atom nuclei to the electrons, and a specific function for intramolecular rotations in unsaturated molecules where π electrons are involved was added.

An important aspect of EPEN is that it is parameterized on *experimental* data such as dipole moments, lattice constants, lattice energies, barriers to rotation, and so on. Such data provide information about a pair potential in the long-range attractive region and near the minimum. However, no experimental data are available to specify the shape of the potential in the repulsive region, and it is important to know the potential in the repulsive region, as well, for calculations on liquids. Therefore, resort has been had to *ab initio* calculations to determine the repulsive part of the EPEN potential. For water · · · water interactions, the potential of Matsuoka *et al.*[41] has been used.[42,43] For interactions of water with small molecules such as methane, methanol, and others, similar *ab initio* calculations are being carried out.[44] For the methane · · · water interaction, Owicki and Scheraga[29] used an empirical fit to the *ab initio* results of Ungemach and Schaefer[45] to treat aqueous solutions of methane (see below). Preliminary results from the use of the modified EPEN potential in a Monte Carlo treatment of water[46] indicate that previous deficiencies in the repulsive region of the potential have been corrected.

An empirical approach for treating long-range interactions (including solvent effects) in proteins has been developed by Tanaka and Scheraga.[47] It is based on the frequencies of contacts of pairs of amino acid residues in known crystal structures of a large number of proteins.

Treatment of Liquid Water

Using molecular dynamics techniques, Rahman and Stillinger[18-22] have carried out a series of calculations on water and aqueous solutions, obtaining considerable insight into the properties of these systems. Similarly, interesting results have been obtained by Monte Carlo methods.[23-31]

Owicki and Scheraga[28] carried out a Monte Carlo simulation of liquid water in the isothermal-isobaric ensemble at 298°K and atmospheric pressure. Since the difficulties with the EPEN potential, referred to above, had not yet been rectified at the time that the Monte Carlo calculations were carried out,[28] the configuration-interaction potential of Matsuoka *et al.*[41] was used. Lie *et al.*[27] carried out similar computations in the isothermal-isovolumic ensemble at 298°K, with the volume fixed at its low-pressure value (for $P \sim 1$ atm). The calculated internal energies were −7.1 and −6.5 kcal/mole in the T-P-N and T-V-N ensembles, respectively, compared to an experimental value of −8.1 kcal/mole. For the calculated heat capacity, $C_p = 21$ and $C_v = 14$ cal/mole deg in the T-P-N and T-V-N ensembles, respectively, compared to an experimental value of $C_p = 18$ cal/mole deg; however, because of large errors in the computed values, the agreement is really not as satisfactory as it appears. The molar volume is overestimated in the T-P-N calculation; the value of the pressure in the T-V-N calculation was correspondingly in error. The preliminary calculations with the improved EPEN potential (referred to above) appear to be giving better values of the molar volume in the T-P-N ensemble.

There are as yet serious practical difficulties in computing free energies by the Monte Carlo technique,[28,29,31,48,49] which precluded the attempt to compute free energies in the work cited above. Whereas it is possible to compute the average potential energy, $\overline{U}$, efficiently in a Monte Carlo calculation, it is very difficult to obtain the average value of exp (U/kT), which is essentially the free energy. This difficulty arises because the probability distribution function for U has its maximum in the region where $\overline{U}$ appears, but is close to zero in the region where large values of exp (U/kT) occur; thus, the Monte Carlo technique rarely samples the region of U where the value of exp (U/kT) is large. However, some attempts have been made[31] to circumvent this problem by using non-Boltzmann sampling[48,49] in a computation of the free energy of a Lennard-Jones fluid containing a hard-sphere solute.

Radial distribution functions and energy distribution functions were also computed.[28] It was found that there is a broad, smooth distribution of hydrogen-bond energy in this model for liquid water, rather than relatively discrete sets of bonded and unbonded energies. Stillinger and Rahman[19] reached similar conclusions by a somewhat different route in their molecular dynamics study of a water model.

Monte Carlo Treatment of Dilute Aqueous Methane

The Monte Carlo simulation described above was also carried out[29] for a dilute aqueous solution of methane in the isothermal-isobaric ensemble at 298°K and atmospheric pressure. Again, because the EPEN potential had not yet been reparameterized at the time that the Monte Carlo calculations were carried out,[29] the potential used was that of Ungemach and Schaefer.[45]

Recognizing the large standard errors, the calculated partial molar energy and volume of the solute agree with the experimental values (a calculated partial molar energy of -11 ± 15 kcal/mole compared to an experimental value of -2.6 kcal/mole, and a calculated partial molar volume of 25 ± 34 cm^3/mole compared to an experimental value of 37 cm^3/mole). It was not possible to compute the partial molar free energy by the Monte Carlo technique because of the difficulty cited in the previous section.

Probably the most important single observation from these computations is that the first hydration shell is broad, extending from a C · · · O distance of ~3.1 to 6.0 Å, and contains ~23 water molecules. For comparison, in the computation on pure water, cited above, the first hydration shell extends from an O · · · O distance of ~2.4 to 3.6 Å and contains ~4 water molecules. Further, computed energy probability distribution functions indicate that the environments of water molecules in the first hydration shell around the methane molecule are characterized by lower, more sharply distributed energies than are the environments of water molecules far from the methane (i.e., in the bulk). These results support the concept[9-12,50,51] that nonpolar solutes increase the degree of hydrogen bonding or structure in their hydration shells. Such ordering is not strictly equivalent to the formation of a clathrate cage, but strong resemblances do exist—particularly between the calculated coordination number of methane (viz. ~23) and the sizes of the cages in the methane clathrate (viz. 20 and 24).[52] This ordering accounts for the thermodynamic properties of aqueous solutions of hydrocarbons,[10,50,51] and such systems have been used as a model to compute the thermodynamic properties of the hydrophobic bond.[1,2,11]

This type of investigation, i.e., the use of pair interaction potential functions and Monte Carlo procedures to treat aqueous solutions, is being extended not only to improve the treatment of dilute aqueous solutions of methane, but also to consider dilute solutions of polar molecules such as methanol (44), methyl amine, and so on, and to treat interactions between (polar and nonpolar) solutes, in water. Thus far, we have information (from older, model treatments) only about hydrophobic bonding, which we consider next.

Hydrophobic Bonding

Early model approaches, based on ideas of Frank and Evans[50] and Kauzmann[51] that hydrocarbons increase the "degree of ice-likeness" in the surrounding water and that association of nonpolar groups in water reverses this process, provided theoretical thermodynamic parameters for formation of hydrophobic bonds.[11] A variety of experiments were used to verify and apply these parameters. These include the role of hydrophobic bonds in hydrophobic chromatography (and the effect of urea thereon), in the helix-coil transition of polyamino acids, in stabilizing carboxylic acid dimers and other binary complexes in water, in stabilizing hydrogen bonds, in stabilizing proteins, in protein association, and in determining volume changes in hydrocarbon-water systems; the effect on hydrophobic bond strength of substituting D_2O for H_2O was also considered. All of these were reviewed previously.[1,2] Other applications include the effect of hydrophobic bonds on micelle stability,[53,54] on the interactions between poly(L-alanine) α-helices in water,[55] and on the helix-coil transition of poly(L-valine) in water.[56]

Similar studies were carried out in other laboratories, notably those of Ben-Naim, Franks, Tanford, Edelhoch, Hvidt, and Kauzmann. The main objective of the studies of Ben-Naim[57-60] was to find a simple measure of the strength of the hydrophobic interaction between two nonpolar solutes in water. He related the change in free energy for bringing two simple solutes from infinite separation to some close distance, in water, to experimental quantities, and determined its dependence on temperature, pressure, and the addition of various solutes. He also examined its variation when H_2O was replaced by D_2O, ethanol, and dioxane, respectively. Franks carried out similar studies with similar objectives;[61-74] a comprehensive discussion of the work of Franks is provided in Ref. 75. He also treated aqueous solutions of alcohols, micelle-forming solutes, volumetric behavior of nonpolar solutes in water, and the effect of urea on hydrophobic bonding, and used a variety of experimental techniques such as solution calorimetry and neutron scattering to study these systems. Tanford[76] and Edelhoch and Osborne[77] have also considered experimental data involving the interactions of nonpolar solutes in water. Bøje and Hvidt[78,79] and Kauzmann[80,81] have paid considerable attention to volume effects in hydrocarbon-water and related systems. In particular, Kauzmann[81] has questioned the ability of theories based on models to account for the effects of pressure on water, and has focused attention on the more recent approaches to liquid water based on the use of a water · · · water interaction potential function. The latter view seems to be the one that is now held by workers interested in elucidating the structure of water and aqueous solutions. All of the approaches to hydrophobic bonding, mentioned above, have similar objectives, are based on similar kinds of experimental data, apply similar models to account for the observations, and are all in agreement on the essential features of the phenomena involved.

However, this earlier reliance on models is now giving way, as indicated above, to techniques such as molecular dynamics and Monte Carlo calculations (and to statistical mechanical procedures applied in general in the theory of liquids), all of which make use of empirical pair interaction potentials. It is hoped that these theoretical approaches (not only for nonpolar solutes but for polar ones as well) will provide the necessary data for proper parameterization of a shell model (or some improvement thereof) to treat solvation properly in conformational energy calculations on proteins.

References

1. Scheraga, H. A. 1965. Annals N.Y. Acad. Sci. **125:** 253.
2. Hagler, A. T., H. A. Scheraga & G. Némethy. 1973. Annals N.Y. Acad. Sci. **204:** 51.
3. Franks, F. 1972. Water, A Comprehensive Treatise. Vol. 1. Plenum Press. New York, N.Y. Also, 1973, Vol. 2; 1973, Vol. 3; 1975, Vol. 4; 1975, Vol. 5.
4. Horne, R. A. 1972. Water and Aqueous Solutions. Wiley-Interscience. New York, N.Y.
5. Luck, W. A. P. 1974. Structure of Water and Aqueous Solutions. Verlag Chemie. Weinheim, Germany.
6. Ben-Naim, A. 1974. Water and Aqueous Solutions. Plenum Press. New York, N.Y.
7. Gibson, K. D. & H. A. Scheraga. 1967. Proc. Nat. Acad. Sci. USA **58:** 420.
8. Hopfinger, A. J. 1971. Macromolecules **4:** 731.
9. Némethy, G. & H. A. Scheraga. 1962. J. Chem. Phys. **36:** 3382.
10. Némethy, G. & H. A. Scheraga. 1962. J. Chem. Phys. **36:** 3401.
11. Némethy, G. & H. A. Scheraga. 1962. J. Phys. Chem. **66:** 1773.
12. Némethy, G. & H. A. Scheraga. 1964. J. Chem. Phys. **41:** 680.
13. Hagler, A. T., H. A. Scheraga & G. Nemethy. 1972. J. Phys. Chem. **76:** 3229.
14. Lentz, B. R., A. T. Hagler & H. A. Scheraga. 1974. J. Phys. Chem. **78:** 1531.
15. Owicki, J. C., B. R. Lentz, A. T. Hagler & H. A. Scheraga. 1975. J. Phys. Chem. **79:** 2352.
16. Andersen, H. C. 1974. J. Chem. Phys. **61:** 4985.
17. Lowden, L. J. & D. Chandler. 1974. J. Chem. Phys. **61:** 5228.
18. Rahman, A. & F. H. Stillinger. 1971. J. Chem. Phys. **55:** 3336.
19. Stillinger, F. H. & A. Rahman. 1972. J. Chem. Phys. **57:** 1281.
20. Stillinger, F. H. & A. Rahman. 1974. J. Chem. Phys. **60:** 1545.
21. Stillinger, F. H. & A. Rahman. 1974. J. Chem. Phys. **61:** 4973.
22. Rahman, A., F. H. Stillinger & H. L. Lemberg. 1975. J. Chem. Phys. **63:** 5223.
23. Barker, J. A. & R. O. Watts. 1969. Chem. Phys. Lett. **3:** 144.
24. Watts, R. O. 1974. Mol. Phys. **28:** 1069.
25. Sarkisov, G. N., V. G. Dashevsky & G. G. Malenkov. 1974. Mol. Phys. **27:** 1249.
26. Lie, G. C. & E. Clementi. 1975. J. Chem. Phys. **62:** 2195.
27. Lie, G. C., E. Clementi & M. Yoshimine. 1976. J. Chem. Phys. **64:** 2314.
28. Owicki, J. C. & H. A. Scheraga. 1977. J. Am. Chem. Soc. In press.
29. Owicki, J. C. & H. A. Scheraga. 1977. J. Am. Chem. Soc. In press.
30. Owicki, J. C. & H. A. Scheraga. 1977. Chem. Phys. Lett. **47:** 600.
31. Owicki, J. C. & H. A. Scheraga. 1977. J. Phys. Chem. To be submitted.

32. Weissmann, M. & L. Blum. 1968. Trans. Faraday Soc. **64:** 2605.
33. Ben-Naim, A. & F. H. Stillinger, Jr. 1972. *In* Structure and Transport Processes in Water and Aqueous Solutions. R. A. Horne, Ed.: 295. John Wiley & Sons. New York, N.Y.
34. Popkie, H., H. Kistenmacher & E. Clementi. 1973. J. Chem. Phys. **59:** 1325.
35. Kistenmacher, H., G. C. Lie, H. Popkie & E. Clementi. 1974. J. Chem. Phys. **61:** 546.
36. Shipman, L. L. & H. A. Scheraga. 1974. J. Phys. Chem. **78:** 909.
37. Shipman, L. L., A. W. Burgess & H. A. Scheraga. 1975. Proc. Nat. Acad. Sci. USA **72:** 543.
38. Burgess, A. W., L. L. Shipman & H. A. Scheraga. 1975. Proc. Nat. Acad. Sci. USA **72:** 854.
39. Burgess, A. W., L. L. Shipman, R. A. Nemenoff & H. A. Scheraga. 1976. J. Am. Chem. Soc. **98:** 23.
40. Owicki, J. C., L. L. Shipman & H. A. Scheraga. 1975. J. Phys. Chem. **79:** 1794.
41. Matsuoka, O., E. Clementi & M. Yoshimine. 1976. J. Chem. Phys. **64:** 1351.
42. Snir, J., R. A. Nemenoff & H. A. Scheraga, J. Phys. Chem. To be submitted for publication.
43. Nemenoff, R. A., J. Snir & H. A. Scheraga, J. Phys. Chem. To be submitted for publication.
44. Wertz, D. H., C. Tosi, E. Clementi & H. A. Scheraga, J. Chem. Phys. To be submitted for publication.
45. Ungemach, S. R. & H. F. Schaefer III. 1974. J. Am. Chem. Soc. **96:** 7898.
46. Owicki, J. C., R. A. Nemenoff, J. Snir & H. A. Scheraga, J. Phys. Chem. To be submitted for publication.
47. Tanaka, S. & H. A. Scheraga. 1976. Macromolecules **9:** 945.
48. Torrie, G. M. & J. P. Valleau. 1974. Chem. Phys. Lett. **28:** 578.
49. Torrie, G. M. 1975. Statistical Mechanics of Small Systems. Doctoral Thesis. University of Toronto, Toronto, Canada.
50. Frank, H. S. & M. W. Evans. 1945. J. Chem. Phys. **13:** 507.
51. Kauzmann, W. 1959. Adv. Protein Chem. **14:** 1.
52. Davidson, D. W. 1973. Ch. 3 in Vol. 2 of Ref. 3.
53. Poland, D. & H. A. Scheraga. 1965. J. Phys. Chem. **69:** 2431, 4425.
54. Poland, D. & H. A. Scheraga. 1966. J. Colloid Interface Science **21:** 273.
55. Silverman, D. N. & H. A. Scheraga. 1972. Arch. Biochem. Biophys. **153:** 449.
56. Go, M., F. T. Hesselink, N. Go & H. A. Scheraga. 1974. Macromolecules **7:** 459.
57. Ben-Naim, A. 1974. Ch. 8 of Ref. 6.
58. Ben-Naim, A. & M. Yaacobi. 1975. J. Phys. Chem. **79:** 1263.
59. Ben-Naim, A. 1975. J. Phys. Chem. **79:** 1268.
60. Ben-Naim, A. 1975. Biopolymers **14:** 1337.
61. Franks, F. 1965. Annals N.Y. Acad. Sci. **125:** 277.
62. Eagland, D. & F. Franks. 1965. Trans. Faraday Soc. **61:** 2468.
63. Franks, F. & H. T. Smith. 1967. Trans. Faraday Soc. **63:** 2586.
64. Franks, F., M. J. Quickenden, J. R. Ravenhill & H. T. Smith. 1968. J. Phys. Chem. **72:** 2668.
65. Franks, F. & B. Watson. 1967. Trans. Faraday Soc. **63:** 329.
66. Franks, F. & H. T. Smith. 1968. Trans. Faraday Soc. **64:** 2962.
67. Frank, H. S. & F. Franks. 1968. J. Chem. Phys. **48:** 4746.
68. Franks, F. & B. Watson. 1969. Trans. Faraday Soc. **65:** 2339.
69. Franks, F., M. A. J. Quickenden, D. S. Reid & B. Watson. 1970. Trans. Faraday Soc. **66:** 582.
70. Franks, F., J. Ravenhill, P. A. Egelstaff & D. I. Page. 1970. Proc. Roy. Soc. London **A319:** 189.
71. Franks, F., J. R. Ravenhill & D. S. Reid. 1972. J. Solution Chem. **1:** 3.
72. Ablett, S., M. D. Barratt & F. Franks. 1975. J. Solution Chem. **4:** 797.
73. Franks, F., M. Pedley & D. S. Reid. 1976. J. Chem. Soc., Faraday Trans. I. **72:** 359.
74. Clark, A. H., F. Franks, M. D. Pedley & D. S. Reid. 1977. J. Chem. Soc., Faraday Trans. I. **73:** 290.
75. Franks, F. 1975. Vol. 4, Ch. 1 of Ref. 3.

76. TANFORD, C. 1973. The Hydrophobic Effect—Formation of Micelles and Biological Membranes. John Wiley & Sons. New York, N.Y.
77. EDELHOCH, H., & J. C. OSBORNE, JR. 1976. Adv. Protein Chem. **30:** 183.
78. BØJE, L. & A. HVIDT. 1971. J. Chem. Thermodynamics **3:** 663.
79. BØJE, L. & A. HVIDT. 1972. Biopolymers **11:** 2357.
80. KAUZMANN, W. 1974. Proc. 4th Int. Conf. High Pressure: 619.
81. KAUZMANN, W. 1976. Colloques internationaux du C.N.R.S. No. 246—L'Eau et Les Systèmes Biologiques: 63. (Organisé par A. Alfsen et A. J. Berteaud). Éditions du Centre National de la Recherche Scientifique, 15 quai Anatole-France, Paris 75700, France.

ON THE NATURE OF LIQUID WATER

Julian H. Gibbs

Department of Chemistry
Brown University
Providence, Rhode Island 02912

Introduction

As a basis for discussion in the remainder of this conference of the electric properties of aqueous systems of biological significance, some conception is needed of the nature of individual water molecules, the forces between them, and the multimolecular aggregates of which bulk water is composed. It is the last of these topics with which this paper is concerned, the others having been discussed by Professor Scheraga. This last topic is one over which there remains considerable disagreement. In a conference not primarily devoted to water *per se*, one cannot hope to review the many positions that have been taken. On the other hand, it would be clearly inappropriate to usurp this occasion for the detailed presentation of a highly technical paper on one narrow aspect of the subject. What follows is accordingly neither a scholarly review nor a research paper, but rather, (1) a somewhat personal view regarding the limits one may impose on the range of possibly valid models of liquid water, and (2) a suggestion for a qualitative microscopic interpretation of the so-called "hydrophobic effect" in aqueous solutions and suspensions. Neither of these will be buttressed with mathematical sophistication here.

Any theory for the "structure" and properties of liquid water must satisfy two somewhat divergent criteria. On the one hand, it must account for the differences between the properties of water and those of other liquids; on the other hand, its interpretation of the fundamental differences between the liquid state and the crystal and vapor states must be applicable to all liquids (Occam's razor). Thus some properties of water are unique, or at least peculiar, but others are not. Some observations must be made concerning a few of these peculiarities before attention is turned to the possibilities for theoretical interpretation. The peculiarities to be discussed here involve the thermal expansion coefficient, the pressure dependence of transport properties and the insolubility of hydrocarbons ("hydrophobic effect"). Other peculiarities have been discussed on another occasion[1] in terms of the same model as that to be invoked today.

Peculiarity of the Thermal Expansion Coefficient Extends over the Whole Liquid Range

Some of the strange properties of water are often felt to be strange only close to the freezing point. Specifically, the circumstance that the thermal expansion coefficient is negative only between 0°C and 4°C (at 1 atm) suggests peculiar behavior only near the freezing point. If, however, one recognizes that thermal expansion is a consequence of two different phenomena, an altered view emerges.

Thermal expansion arises both from enlargements of amplitudes of vibrations of particles around positions of minimum potential energy associated with any given configuration (defined for water, for example, as any given topology of hydrogen bonds) and also from changes in the relative contributions of various configurations to the time-averaged "structure" of the liquid. For many liquids the existence of these as disparate factors can be inferred from the discontinuous jump in the thermal expansion coefficient at the glass transition temperature. The approximate constancy of the thermal expansion coefficient in the glassy state suggests that the volume (V) vs. temperature (T) curve for the glassy state may be extrapolated above the glass transition to provide an estimate of the contribution of changes in vibrational amplitudes to the thermal expansion in the liquid region. Subtraction of this from the total V vs. T-curve then yields the contribution made by configurational changes (V_c vs. T). In the case of water, the unavailability of thermal expansion coefficient data for a glassy state requires that we utilize instead the thermal expansion coefficient of ice. (Thermal expansion coefficients of glassy substances run closely parallel to those of the corresponding crystalline materials.)

With this procedure, the resulting curve for V_c vs. T is found[2] to have a slope that is negative over approximately half of the liquid range and small over all of it. Thus the temperature range of this anomaly does not fail to coincide with those of the other anomalous properties of water. With this observation one may hope to find a common molecular basis for these various anomalies.

Anomalous Pressure Coefficient of Viscosity Related to the Anomalous Thermal Expansion Coefficient

Again we can refer to the concepts relating to the glass transition to gain clarification of peculiarities in the behavior of water. Both experimental and theoretical studies of glass-forming liquids have indicated that the sluggishness of transport in a supercooled liquid near its glass transition temperature, detectable for example in its viscosity, is correlated with smallness in configurational entropy in this T regime. Indeed, it was proposed[3] some years ago that the viscosity of a liquid not too far from its glass temperature can be approximately described by the equation

$$\eta = A e^{B/TS_c} \tag{1}$$

where A and B are constants and $S_c(T)$ is the configurational entropy at absolute temperature T. This equation can be approximated by the familiar empirical equation of Fulcher, Vogel, Tammann, Cole, and Williams-Landel-Ferry,

$$\eta = A e^{c/(T - T_0)} \tag{2}$$

provided that T_0 is identified with the second-order-transition temperature T_2 of the configurational entropy theory[4] for the glass transition. (The identity $T_0 = T_2$ is well substantiated in the cases of those few substances for which both the relevant viscosity data and the relevant calorimetric data in the supercooled liquid state are available).

If, then, the viscosity is a monotonically decreasing function of S_c, the sign of the pressure coefficient of viscosity can be related to that of the thermal expansion coefficient through the Maxwell relation

$$(\delta S_c/\delta P)_T = -(\delta V_c/\delta T)_P \tag{3}$$

The fact that liquid water has the unorthodox (negative) sign of $(\delta V_c/\delta T)_P$ over the lower half of its liquid range implies, then, that the pressure coefficient of viscosity will be similarly anomalous in sign in this range. This is precisely the case.[5]

Thus, whatever the molecular mechanism may be for these peculiarities of water, it is likely that it is the same for each.

Anomalous Temperature Coefficient of the Water Solubility of Nonpolar Substances can be Related to the Anomalous Thermal Expansion Coefficient

To the extent that one may ignore the energies of interaction between methane (for example) and water, relative to interactions between water molecules alone (hydrogen bonds and dipole-dipole interactions), one may say that introducing a methane molecule into liquid water is roughly akin to introducing nothing at all (i.e., a hole) into the liquid structure. Thus it is reasonable to inspect a Maxwell relation for $(\delta N_2/\delta T)_{\mu 2}$ that is somewhat analogous to that discussed in the previous section for $(\delta V/\delta T)_P$, N_2 being the number of moles of methane dissolved in an aqueous phase subject to a chemical potential μ_2 of methane. Such a relation is readily found via Legendre transformation of the Helmholtz free energy relative to N_2 to yield the state function $G - \mu_2 N_2$ or the exact differential

$$dG - \mu_2 dN_2 - N_2 d\mu_2 = V\,dp - S dT + \mu_1 dN_1 - N_2 d\mu_2$$

from which it follows that

$$(\delta N_2/\delta T)_{N_1,P,\mu_2} = (\delta S/\delta \mu_2)_{N_1,P,T} \tag{4}$$

the subscript 1 referring to water.

It is illuminating to reexpress both Equation 4, for the temperature dependence of the equilibrium concentration of methane, and Equation 3, for the equilibrium concentration of holes:

$$(\delta V_c/\delta T)_{N_1,P} = -(\delta S_c/\delta V)_{N_1,T}(\delta V/\delta P)_{N_1,T} \tag{3a}$$

and

$$(\delta N_2/\delta T)_{N_1,P,\mu_2} = (\delta S/\delta N_2)_{N_1,P,T}(\delta N_2/\delta \mu_2)_{N_1,P,T} \tag{4a}$$

In (**3a**), $(\delta S_c/\delta V)_{N_1,T}$ is seen to be negative, over the lower half of the liquid range, by virtue of the experimentally observed negative sign of the thermal expansion coefficient, in this regime, and the necessarily negative sign of $(\delta V/\delta P)_{N_1,T}$.

A parallel analysis of the signs in (**4a**) is not quite relevant, inasmuch as $(\delta N_2/\delta T)_{N_1,P,\mu_2}$ is not the quantity normally measured. The quantity $(\delta N_2/\delta T)_{N_1,P}$ is essentially the conventional temperature coefficient of solubility. It can be expressed in terms of $(\delta N_2/\delta T)_{N_1,P,\mu_2}$ by

$$(\delta N_2/\delta T)_{N_1,P} = (\delta N_2/\delta T)_{N_1,P,\mu_2} + (\delta N_2/\delta \mu_2)_{N_1,T,P}(\delta \mu_2/\delta T)_{N_1,P}$$

which, on substitution of (**4a**) becomes

$$(\delta N_2/\delta T)_{N_1,P} = (\delta N_2/\delta \mu_2)_{N_1,T,P}[(\delta S/\delta N_2)_{N_1,T,P} + (\delta \mu_2/\delta T)_{N_1,P}] \qquad \textbf{(4b)}$$

In this equation the quantity in square brackets must be negative, in the case of aqueous solutions of many hydrocarbons, because $(\delta N_2/\delta T)_{N_1,P}$ is known experimentally to be negative, over the lower half of the liquid range of water, and $(\delta N_2/\delta \mu_2)_{N_1,T,P}$ is required to be positive by a thermodynamic stability condition (LeChatelier's principle). Although the negative sign of $(\delta \mu_2/\delta T)_{N_1,P}$ might, at first sight, seem to offer an explanation for this, it must be noted that the peculiarity of aqueous solutions cannot be attributed to this term, which is independent of the nature of the solvent (inasmuch as it can be ascribed to the coexisting hydrocarbon phase, the μ_2 of which must equal that in the aqueous phase). In "normal" cases, for which the temperature coefficient of solubility is positive, $(\delta S/\delta N_2)_{N_1,T,P}$ must be positive and of greater absolute magnitude than $(\delta \mu_2/\delta T)_{N_1,P}$. The anomaly, in the case of aqueous solutions of "hydrophobic" substances, must reside in $(\delta S/\delta N_2)_{N_1,T,P}$. Indeed, this quantity, the partial molal entropy, is found experimentally to be anomalous.[6] When its "cratic" component (that attributable to ideal mixing) is subtracted, the remaining "unitary" component is seen to be negative.[7]

Considerations of the Nature of Melting and Boiling Indicate that Models of the Liquid State must Probably be "Random Gel" Models

The considerations presented so far suggest that the anomalies of liquid water discussed here have a common origin.

This reduction in the number of independent anomalies requiring interpretation must now be accompanied by a reduction in the number of models at our disposal for the task. It has been argued elsewhere[1] that comparison of the results yielded by various theoretical models for water with the experimentally observed properties of liquid water alone may not by itself suffice for the discrimination among the models being so tested. The number of physically recognizable but quantitatively unknown parameters that must appear in any model that is sufficiently detailed to be quantitatively realistic for this complex problem is sufficiently large that adjustment of their values will probably permit a number of disparate models to fit the experimental data within the precision limit requirements that one may reasonably impose upon even the essentially correct model. Accordingly, it has been argued that one must enlarge one's ambitions and attempt to account with any proposed model not only for the properties of water in its liquid range but also for the phase transitions (melting and boiling), which abruptly bound this liquid range. Any valid model for a liquid must be compatible with the concept that it is the consequence of highly cooperative, discontinuous changes attendant upon heating and cooling from the crystalline and gaseous states respectively.

Such a program is a very ambitious one, not likely to be completed in a fully satisfactory way in our lifetimes. Nevertheless, such considerations do permit certain qualitative statements to be made with some confidence even now. Experimentally observed similarities among members of a particular set of phase transitions suggest that certain conclusions that can be drawn from study of one member may be applicable to other members. Specifically, one observes such similarities between the gas-liquid system and the ferromagnet as (1) the existence of a critical (or Curie) temperature at which the otherwise first-order phase transition becomes second-order and above which it no longer exists; (2) the similarity of the values observed for the analogous "critical exponents" (the powers of $[1/(T - T_c)]$ according to which various properties, such as compressibility for the gas-liquid system and magnetic susceptibility for the magnet, diverge at the critical temperature T_c); and, (3) the existence under certain conditions of hysteresis in analogous properties in these transitions. These similarities argue that, to the extent that the Ising model is a reasonable model for a magnet, the lattice-gas model (which was shown by Lee and Yang[8] to be mathematically isomorphous with the Ising model) may be used for those properties of the gas-liquid system that are similar to those of the magnet. The clear conclusions are thus obtained that (1) the gas-to-liquid transition involves the onset of a form of *long-range* particle order (ferromagnetic "domains" are of *macroscopic* size), and (2) the order thus obtained is highly imperfect, the degree of imperfection depending upon the temperature.[1,9]

What is suggested, then, is that the liquid phase may be viewed as a random, internally mobile, network (or "gel") of intermolecular bonds (van der Waals bonds in the case of simple liquids, hydrogen bonds in the case of water), this network being of macroscopic size.[1,9] Such a structure is the analog of the structure of a ferromagnetic domain, which is spanned by a network of spins, all of which are similarly oriented in the sense that any two neighbors are "bonded" by virtue of close parallelism of orientation (exact parallelism in the Ising model). The order is imperfect in that nonconforming spins exist within the domain; not part of the "spongy" network ("gel") of conforming spins, they are nonetheless occluded by it. In the other of the two phases that coexist below the Curie temperature (only two are allowed in the Ising model), it is spins of the opposite orientation that comprise a domain-spanning "gel." Since, in the lattice-gas version of the Ising model, the two possible states of a given lattice site are occupation by a particle and vacancy (analogous to spin up or spin down in the magnetic Ising model), the gaseous state that coexists with the liquid below the critical point is viewed (in the lattice-gas model) as a gel of "holes" (vacant sites), within which there are occluded nonconforming sites containing particles. This low-density structure is just the inverse of the higher density liquid phase (the well-known particle-hole symmetry of the lattice-gas model).

The term "random gel," as applied to the liquid state, must not be misinterpreted as implying a single persistent configuration any more than it does in the case of the spin system. It is intended only to convey the concept that, if a photograph were taken of the liquid structure with a sufficiently short time-exposure (presumably less than 10^{-11} seconds in the case of water), a particular spanning-gel topology would be observed. If another such photograph were taken later, thc structure would again be seen to be held together by a spanning-gel of bonds, but the gel topology would be seen to have been altered during the interim, some of the bonds having been reshuffled. At any intervening instant of time there would also be a gel, the multiplicity of long paths of bonded particles being so great that the breaking of some bonds in a given time interval does not break the gel apart.

Although the Ising or lattice-gas model has been useful as the only model for cooperatively interacting particles upon which it has been possible to perform highly accurate calculations for systems in more than one dimension, the fact remains that it is surely not, in the case of the gas-liquid system, an adequate model for the inspection of properties other than the aforementioned purely qualitative ones. Confidence in the validity of these qualitative conclusions can be drawn from the fact that they can also be inferred, albeit less rigorously, without the use of the lattice-gas model. Indeed, the chairman of this session noted the analogy between vapor condensation and the sol-gel transition exhibited in the chemical polymerization of polyfunctional monomers more than thirty years ago.[10] One of the appealing features of this analogy is the simple physical interpretation it provides of the cooperativity responsible for the sharpness (first-order phase transition property) of the condensation transition: the larger a multi-molecular cluster (or branched polymer), the greater its number of unbonded attachment points (or functional groups) onto which it can absorb more molecules.

The crudeness of the lattice-gas model, even as adapted to allow for the rotational degrees of freedom of water molecules,[11-13] and the lack of rigor in the theories that avoid its use indicate that the conclusions concerning the nature of the liquid state that can be drawn from theoretical considerations of the nature of vapor condensation are limited to the qualitative ones that have just been drawn. These merely place limits on the range of plausible models for the liquid state. Nevertheless, the gel (macroscopic cluster) feature eliminates or at least indicates the need for modification of those models for liquid water that have received greatest attention. "Significant structures," "icebergs," "flickering clusters," and indeed any constructs based on the concept of a "mixture" of multi-molecular aggregates of qualitatively varied type and/or of microscopic size cannot be the essence of the difference between a liquid and its vapor—if they exist at all.

The random gel concept gains further support from considerations of the nature of crystal melting. Such considerations are limited in that they cannot be based on the somewhat tractable Ising or lattice-gas model, inasmuch as the simple lattice-gas model cannot display a liquid-crystal transition. (It imposes crystal-like order *a priori* on whatever phases it might exhibit; this circumstance and the fact that the order parameter of the single transition it displays is the density are the reasons for identifying its transition as a gas-liquid transition). Nevertheless, it has been possible to make the following argument.[1] Given the proposition that the essential cooperative process involved in vapor condensation is the cascade-wise formation of a gel in the sense originally treated by our chairman—i.e., chain-branching to "infinity"—the only bond-forming process that remains to provide the cooperative cascade responsible for freezing is one involving the closing of rings. As viewed in the direction of melting, the breakup of an orderly connected array of rings of specified size (the crystal), even though unaccompanied by the disruption of the network (or "random gel") thus formed, is sufficiently cooperative to constitute a first-order phase transition. This audience, doubtless familiar with the theory of the helix-coil transition both in polypeptides and in polynucleotides,[14-16] will appreciate the basis for this speculation. The fact that the breakup of connected rings that constitutes the helix-coil transformation, although highly cooperative, falls short of producing a true first-order phase transition is no argument against the proposition that such a process in three dimensions would constitute a true phase transition, for Landau and Lifschitz[17] have shown that no physical system that is not of infinite extent in more than one dimension can display a true phase transition. Some evidence has in fact been presented to the effect that such a process in three dimensions does indeed constitute a first-order phase transition.[1] Furthermore, an argument has also been presented to the effect that the melting process is essentially confined to this ring-breaking process. This argument also indicates that at sufficiently low densities (pressures) the ring-breaking process should become accompanied by the tree-breaking process; that is, the two cooperative processes are expected to coalesce. Thus an interpretation of the triple point and of sublimation is afforded by these views of melting and boiling.[1]

This interpretation of crystal melting clearly leads to a physical picture of the liquid state that is qualitatively similar to that inferred from interpretation of vapor condensation. A major question left unresolved by both of these interpretations, incomplete as they are, pertains to the distribution of rings in the liquid (gel) state. It should be clearly understood that the identification of condensation with a cascade of chain branching and of freezing with a cascade of formation of connected rings (hexagonal in ice) does not mean that there are no ring closures in each liquid network. Just as small treelike clusters (and perhaps rings) exist in the vapor not-too-far above the boiling point, so also rings exist as a feature of every bond topology in the liquid. With regard to the analogy between melting and the helix-coil transition, it should be recognized that the treatment, in the theories, of the random coil as a structure without intrachain ring-closing hydrogen bonds is an approximation. The transition from helix to random coil is one in which the ordered one-dimensional array of connected rings (all the non-short-circuited members of which are of the same size) of the helical polypeptide or polynucleotide is disrupted and replaced with a structure in which the (non-short-circuited) rings are of random size and location; the number of ring-closing hydrogen bonds is decreased significantly in the transition but surely does not drop to zero.

This issue of the number of nonshort-circuited rings and of their size distribution in the liquid is an important one that cannot be resolved at present. Rahman and Stillinger[18,19] have obtained evidence pertaining to this from their computer simulation of the molecular dynamics of molecules that interact pair-wise via a potential that has the appropriate symmetry for water molecules. Evidence from real water can be gleaned from consideration of the conclusion of King and Barletta[20] that the average number of hydrogen bonds with which a typical water molecule is associated at 0°C is three (this result having been obtained from analysis of oxygen-isotope fractionation). In a pure "tree" structure (one containing no rings), the average water molecule is involved in two hydrogen bonds if the tree is of infiinte size. Accordingly, the value 3 implies a considerable degree of ring closing in the liquid at its freezing point.

It should be noted that the "structure" of liquid water that is provided by the computer simulation of molecular dynamics falls within the range of models permitted by the considerations of melting and boiling that have been qualitatively reviewed here today.

Possible Relationship of Hydrophobic Effect to Elasticity of the Water Network

The random gel model, qualitative though it is at the present time, nonetheless provides a suggestion concerning the probable nature of the hydrophobic effect. Chemically bonded polymer gels (e.g., synthetic elastomers), when immersed in a liquid composed of small molecules, exert a resistance to the swelling pressure of these small molecules. In these cases, in which the chemical bonds of the polymer gel are too strong to be broken by the admission of small molecules, the resistance to swelling is well understood. For the case of synthetic elastomers, in which the contour lengths of chains between cross-link points is not inappropriately short, this understanding has been quantified in the theory of Flory and Rehner.[21] The expansion of the gel structure enforced by the entrance of the small molecules constitutes a straightening of polymer chains between points of cross-linkage. Inasmuch as elongated chain configurations are fewer in number than the folded ones that yield the unswollen equilibrium (mean square) end-to-end length, this straightening is achieved at the price of a decrease in entropy. The resistance to swelling is basically an entropy effect identical to that responsible for the elastic resistance to mechanical stretching.

A qualitatively similar elastic effect must resist the "swelling" of water by hydrocarbons. In contrast to molecules with hydrogen-bonding capabilities (e.g. methanol), which can be enlisted in the chains of the gel network itself and thereby be absorbed in the aqueous phase without limit (infinite miscibility), "hydrophobic" substances, which can enter only spaces between the chains of the network, find their entrance resisted by the elasticity of each network structure (topology of hydrogen bonds) that contributes to the time-averaged water structure. The experimental fact that the heat of mixing is small (and probably even negative) shows that this elastic effect is not significantly relieved by a shifting of the equilibrium among hydrogen-bonded topological structures in the direction of structures containing fewer hydrogen bonds. Increase of temperature, however, achieves the latter and contributes to the observed large positive partial molal heat capacity of mixing.

A quantitative theory for the hydrophobic bond cannot, however, be based on an exact parallel to the Flory-Rehner theory for polymer gels. In the first place, the partition function in the case of aqueous solutions involves a sum over all topologies of approximately the optimum number (at each temperature) of hydrogen bonds. In the second place, the degree of cross-linking indicated by the cited results of King and Barletta, though less than that usually supposed for water, is much too large to lend utility to such concepts of the Flory-Rehner theory as that of chains between cross-link points (almost every water molecule is a cross-link point). Indeed, this tightness of the water gel, as contrasted with that of synthetic elastomers, is what is responsible for its high degree of inelasticity and therefore for the low water-solubility of hydrocarbons. The low solubility, moreover, indicates that the assumption of an "affine" transformation of cross-link points could not be taken as valid, even if the concept of chains between cross-links could be.

For these reasons, the "swelling" of water by hydrocarbons will best be treated in terms of the forced expansion of (previously collapsed) rings, as opposed to stretching of (previously folded) chains as in the Flory-Rehner theory.

In any event, the negative sign of the unitary entropy of mixing of hydrocarbons in water, as manifest in the peculiar temperature coefficient of solubility, has a parallel in the negative "unitary" entropy of swelling of chemically bonded polymer gels.

Discussion

At the outset, the point was made that any interpretation of the properties of water would have to possess some elements that are general (applicable to all liquids) and some that are peculiar to water. The interpretations of crystal melting and vapor condensation that have been present are general, although quantitative applications of each are most readily perceived[1,9] in the case of freezing and boiling of liquid water, for which the assumption (common to both Ising and Stockmayer models) of a small number of discrete bonded states per molecule is most appropriate (e.g. the discrete number of ways a water molecule may possess 0, 1, 2, 3, or 4 hydrogen bonds).

The smallness of the maximum bonded-coordination number (4 for water), coupled with the gel concept, implies that 3- and probably 4-membered rings are not importantly represented in liquid water. The value 3 (of King and Barletta) for the average bonded-coordination number (not to be confused with the number of nearest-neighboring oxygen atoms as obtained from x-ray diffraction, which cannot distinguish between hydrogen-bonded and non-bonded oxygen pairs as a consequence of the similarities in the distances involved[1]) at 0°C. implies that many of the nonshort-circuited rings in liquid water are far larger than 6-membered. This is the second of the three propositions on which the present interpretation of various properties of liquid water is based (the first being the gel feature common to all liquids).

The third proposition relates to the strength of hydrogen bonds as contrasted, for example, with the van der Waals bonds holding the liquid argon "gel" together. This strength implies that solutes such as hydrocarbons, which are not themselves capable of hydrogen bonding, cannot enter the aqueous gel network but can at best enter its larger interstices, forcing the surrounding aqueous rings into a less collapsed configuration.

This elastic effect, discussed briefly today in the context of the "hydrophobic effect" (hydrocarbon insolubility), should also contribute to the observed anomalies in the thermal expansion and pressure dependence of transport properties. As has been demonstrated, all of these anomalies are consequences of entropy-control, and the temperature ranges over which these anomalies persist (ca 0–50°C) are similar. The elastic contribution to the thermal contraction can be recognized, for example, in the fact that chemically bonded elastomeric structures (e.g. rubber) contract on heating. These properties, and indeed also the temperature-dependence of the hydrophobic effect, are presumably also strongly influenced by the decrease in hydrogen-bonding with increasing temperature, inasmuch as this decrease increases the average nonshort-circuited-ring size in the gel and thereby decreases the magnitude of elastic effects.

References

1. Gibbs, J., C. Cohen, P. D. Fleming, III & H. Porosoff. 1973. J. Sol. Chem. **2:** 277.
2. Gibbs, J. H., S. Van Kirk & J. Roe. Unpublished.
3. Adam, G. & J. H. Gibbs. 1965. J. Chem. Phys. **43:** 139.
4. Gibbs, J. H. & E. A. DiMarzio. 1958. J. Chem. Phys. **28:** 373.
5. Bett, K. E. & J. B. Cappi. 1965. Nature **207:** 620.
6. Frank, H. S. & M. W. Evans. 1945. J. Chem. Phys. **13:** 507.
7. Kauzmann, W. 1959. Adv. Protein Chem. **14:** 1.
8. Lee, T. D. & C. N. Yang. 1952. Phys. Rev. **87:** 410.
9. Cohen, C., J. H. Gibbs & P. D. Flemming, III. 1973. J. Chem. Phys. **59:** 5511.
10. Stockmayer, W. H. 1943. J. Chem. Phys. **11:** 45.
11. Bell, G. M. 1972. J. Phys. C; Solid State Phys. **5:** 889.
12. Fleming, P. D., III & J. H. Gibbs. 1974. J. Stat. Phys. **10:** 157.
13. Fleming, P. D., III & J. H. Gibbs. 1974. J. Stat. Phys. **10:** 351.
14. Gibbs, J. H. & E. A. DiMarzio. 1959. J. Chem. Phys. **30:** 271.
15. Zimm, B. H. & J. K. Bragg. 1959. J. Chem. Phys. **31:** 526.
16. Poland, D. & H. A. Scheraga. 1970. Theory of Helix-Coil Transitions in Biopolymers. Academic Press. New York, N.Y.
17. Landau, L. & E. Lifschitz. 1958. Statistical Physics: 232. Oxford Press. London, England.
18. Rahman, A. & F. Stillinger. 1971. J. Chem. Phys. **55:** 3336.; 1972. **57:** 1281.
19. Rahman, A. & F. Stillinger. 1972. J. Chem. Phys. **57:** 1281.
20. King, W. T. & R. E. Barletta. 1977. J. Chem. Phys. In press.
21. Flory, P. J. & J. Rehner. 1943. J. Chem. Phys. **11:** 521.

DISCUSSION

W. H. Stockmayer, *Moderator*

Department of Chemistry
Dartmouth College
Dartmouth, New Hampshire 03755

DR. D. BEVERIDGE (*Hunter College, CUNY, New York, N.Y.*): Professor Scheraga, I am most interested in your distribution of binding energy. Mr. Swaminathan and I carried out the identical calculation with the same result. The fact that one gets a continuous distribution is most interesting, and I would like to know if you would care to comment on that with regard to the flickering cluster model, should it be representable in those terms, and the effect of the assumption of pairwise additivity on that result.

DR. SCHERAGA: There are essentially three different approaches to water. The treatment by Nemethy and me represents the model approach where one formulates ad hoc models and then applies statistical mechanics to them. That approach was based on the flickering cluster concept that Henry Frank introduced. The second approach is that of the general theory of liquids using a potential. Finally, there are the molecular dynamics and Monte Carlo approaches based on potentials, which several people have adopted. I interpret your question on flickering clusters in the context of the first type of approach. We started by assuming the existence of clusters of various sizes in the presence of monomeric unhydrogen-bonded water molecules, and determined the average cluster size as a function of temperature. These Monte Carlo calculations suggest that the concept of clusters imbedded among nonbonded water molecules is no longer valid and that, whether one looks at the distribution of pair-energies or the binding energy of any given water molecule with the whole surroundings, one gets a continuous distribution, as you indicated. The continuous distribution is incompatible with the mixture model that was used in our treatment of flickering clusters. A few years later, with Barry Lentz, we introduced the possibility of rings into the mixture model and found that the most probable ring size is about five or six in liquid water.

DR. FANDOR (*Fairleigh Dickinson University, Rutherford, N.J.*): This is to Dr. Gibbs. What is the nature of the long-range forces that you were referring to in water?

DR. GIBBS: I do not think I referred to any long-range forces. I was presenting the argument totally in terms of hydrogen bonds. Dipole-dipole interactions are longer range.

DR. FANDOR: I thought you had an analogy of magnets and magnetic field.

DR. GIBBS: In the Ising model for the magnet, you use only short-range interactions between the spins, interaction between nearest neighbors, at least in the simplest Ising models.

DR. FANDOR: So then there are not any forces or any further reaching than the hydrogen bonding?

DR. GIBBS: They exist, but in the approximate model that would give you the phase transition at all, you do not need them.

DR. SCHERAGA: I was a little confused by your statements. Isn't the structure of a liquid also determined by short-range interactions, the repulsive forces? Are you saying that that is not true? And that it is the long-range interactions that determine structure?

DR. GIBBS: I am not saying that the repulsive forces can be ignored. One knows that the simple perturbation theory approach (i.e., Barker-Henderson approach) worked well with the simple liquids; in it you can say that although you cannot have a liquid at all if you do not have attractive forces, once you have got the liquid, you can ignore the attractive forces. As far as the detailed geometry of the liquid is concerned, it works very well for simple liquids. We know that it fails badly for water, but we must not go over to the other side of the fence totally and say that it is only the attractive forces that determine the structure of water.

DR. SCHERAGA: I would say that slightly differently. In water, it is the repulsive forces that determine the structure, but the overall total energy in the system gets a significant contribution from the longer range interaction, the attractive forces.

DR. STOCKMAYER: I would like to say that the important thing is the orientational dependence, rather than whether it is attractive or repulsive.

DR. GIBBS: Exactly.

DR. MANDEL: I have some difficulties with the analogy between the gel and water. I wonder whether the difference may not be within that of topology.

DR. GIBBS: Professor Mandel is concerned with whether the bonds in water are reshuffling all the time. I suspect that it may well be crucial for the partial molal heat capacity of mixing, and it certainly contributes to the peculiar temperature dependence. In the case of hydrophobic effect, the situation is simpler. Hydrocarbons cause a negative enthalpy change, so you hardly break hydrogen bonds. Surely, hydrogen bonds are reshuffling among topologies. But all of them have essentially the same number of hydrogen bonds in them.

DR. SCHERAGA: The negative enthalpy is built into our flickering model. It is due to the fact that you are making hydrogen bonds and ordering the clathrate structure, thereby decreasing entropy. One can easily reconcile this with a model for the solubility of hydrocarbon in water, and the entropy effect dominates over the enthalpy effect, as H. Frank pointed out.

DR. GIBBS: All I am saying is that you do not need to form more hydrogen bonds when you put in the hydrocarbon in order to get these negative entropy effects. You have it even without the increase in hydrogen bonding.

DR. SCHERAGA: Where do you get the negative entropy, if you are not ordering the system?

DR. GIBBS: The elastic effect, for exactly the same reasons as in the swelling of a polymer gel.

DIRECT SOLUTION OF THE POISSON EQUATION FOR BIOMOLECULES OF ARBITRARY SHAPE, POLARIZABILITY DENSITY, AND CHARGE DISTRIBUTION

William H. Orttung

Department of Chemistry
University of California
Riverside, California 92502

INTRODUCTION

Classical electrostatics has been used with considerable success to explain the dielectric, optical, and energetic properties of molecular systems. Three general approaches have been used for problems where polarizability effects are important.

In the first, the molecule and its environment are represented by dielectric media that may contain embedded charges or dipoles. Polarization of the dielectric by a field corresponds to the nuclear and electronic displacements.[1,2]

In the second, the diffuse polarizability of the electron cloud is replaced by a set of point polarizabilities, usually situated at atomic centers. The through-space interactions of the induced point dipoles then explain the molecular properties.[3]

In the third, the bonds of a molecule are assigned polarizability tensors that are often assumed additive for the purpose of predicting molecular properties. It is somewhat surprising that this approach has been found useful, since it neglects the important through-space interactions.[4]

In spite of important successes, there are some serious limitations. Solutions by the first method have been achieved only for simple shapes and uniform dielectric constants. The second method becomes uncertain when the diffuse charge distributions overlap significantly, or when there is significant delocalization of charge, as in double bonds.

The classical approach would be expected to apply more rigorously to larger systems such as proteins. However, considerable success has also been achieved at the atomic level, where the neglect of dynamic and quantum effects may be expected to cause errors. The problem of colliding rare gas atoms in an electric field has recently been treated with considerable success by the classical electrostatic formalism,[5] and quantum corrections have been considered.[6]

While a general examination of the validity and limits of the classical model at the atomic level would be of interest, it is the purpose of the present paper to show that it is now possible to obtain more general solutions by the first method described above. Over the last ten to fifteen years, structural engineers have developed a new technique, the finite element method, for obtaining solutions to partial differential equations with general boundary conditions, material properties, and source distribution.[7] The method first came to my attention two years ago.[8]

In the following sections, the Poisson equation will be solved for two types of molecular problem to illustrate the potential usefulness of the finite element method for molecular situations. The first type deals with the polarization of molecules in a vacuum subjected to an applied field; the second deals with the reaction field of a dielectric environment induced by charge and dipole distributions within a cavity of molecular size in the dielectric. In addition to further studies on these two types of problem, other important areas for future investigation include solvent polarization effects (around nonspherical or nonuniform molecules), solvent shifts of spectral lines, crystal optical phenomena, topics in physical organic chemistry (substituent effects, etc.), ionic distribution problems in protein systems, and so forth.

Basic Theoretical Considerations

Electrostatics

The Coulomb Law of force between point charges, in electrostatic c.g.s. units, can be transformed into the Poisson partial differential equation,

$$\frac{\partial}{\partial x_\alpha}\left[\epsilon_{\alpha\beta}\,\frac{\partial\phi}{\partial x_\beta}\right] = -4\pi\rho, \tag{1}$$

where ϕ is the electrostatic potential, ρ is the free charge density, $\epsilon_{\alpha\beta}$ is the dielectric constant, and x_α is a cartesian coordinate component. A symbol with one (or two) Greek subscripts is a vector (or tensor). If the same Greek subscript appears twice in a symbol or product of symbols, summation over that index is implied. Since only isotropic dielectric constants will be invoked in this paper, we replace $\epsilon_{\alpha\beta}$ by $\epsilon\delta_{\alpha\beta}$, where $\delta_{\alpha\beta}$ is the Kronecker delta (unit diagonal matrix). Once the potential ϕ has been obtained by solving Equation 1, other electrostatic quantities may be deduced from it, as indicated below.

The distinction between free and bound charges in Equation 1 is valid when the smallest element of volume contains many atoms or molecules. Polarization of the bound charges of a medium occurs in an applied field when the dielectric constant is greater than unity. If P_α is the polarization (dipole moment per unit volume, or induced dipole density) of the bound charges, and E_α is the electric field inducing the polarization, then

$$E_\alpha = -\frac{\partial\phi}{\partial x_\alpha}, \qquad \text{and} \qquad P_\alpha = \frac{1}{4\pi}(\epsilon - 1)E_\alpha = \chi E_\alpha. \tag{2a, b}$$

The last equality defines the dielectric susceptibility χ, which may also be called the polarizability density.[5] The integral of P_α over a volume V is the dipole moment M_α of V. If P_α arises from a uniform applied field $E_{0\alpha}$, then the polarizability $\alpha_{\alpha\beta}$ of the volume is defined by the equation

$$M_\alpha = \alpha_{\alpha\beta}E_{0\beta}. \tag{3}$$

If the classical electrostatic formalism is to be used for volume elements within an atom or bond, the idea of a polarization of bound charges must be modified to the idea of a (time-averaged) shift of the initial charge density upon application of the field. The incremental interaction of the shifted charge elements might then be capable of description by the classical formalism. The viewpoint just expressed seems reasonable for a complex atom or molecule, but obviously fails in the limiting case of the hydrogen atom, since one electron cannot interact with itself at a distance. The purpose of this paper is not so much to resolve the question raised in this paragraph as it is to proceed empirically in cases of this type to see whether such an effort might be worth undertaking.

Variation Method for the Poisson Equation

The Poisson equation can be converted into an integral form that is stationary to variations of ϕ:

$$\delta \int_V \left(\frac{1}{2} \epsilon_{\alpha\beta} \frac{\partial \phi}{\partial x_\beta} \frac{\partial \phi}{\partial x_\beta} - 4\pi\rho\phi \right) dv - \int_S \epsilon_{\alpha\beta} \frac{\partial \phi}{\partial x_\beta} \delta\phi \, dS_\alpha = 0. \qquad \textbf{(4)}$$

Here δ acts on ϕ to cause a variation $\delta\phi$, and S is the surface enclosing volume V. In all cases of current interest, the surface integral disappears, either because the potential is specified on the boundary ($\delta\phi = 0$ on S), or because the normal derivative, $D_n = -\epsilon_{\alpha\beta}(\partial\phi/\partial x_\beta)dS_\alpha/dS$, is zero on S. Conversely, Equation 1 is the condition that ϕ satisfy Equation 4.

A solution can be obtained from Equation 4 by the Rayleigh-Ritz variation method. Suppose that a trial function of the form,

$$\phi = \sum_{i=1}^{n} c_i \phi_i, \qquad \textbf{(5)}$$

involving known functions ϕ_i and unknown coefficients c_i, is substituted in Equation 4. Then, denoting the volume integral of Equation 4 by I_v, the c_i can be determined by solving the equations

$$\partial I_v / \partial c_i = 0, \qquad i = 1, \ldots, n. \qquad \textbf{(6)}$$

These equations are of the form,

$$\sum_{j=1}^{n} a_{ij} c_j = b_i, \qquad i = 1, \ldots, n, \qquad \textbf{(7)}$$

where

$$a_{ij} = \int_V \epsilon_{\alpha\beta} \frac{\partial \phi_i}{\partial x_\alpha} \frac{\partial \phi_j}{\partial x_\beta} dv; \qquad b_i = 4\pi \int_V \rho\phi_i dv. \qquad \textbf{(8a, b)}$$

The ϕ_i of Equation 5 usually extend over the volume V and are required to satisfy any boundary conditions that are imposed.

The Finite Element Method

For present purposes, the finite element method can be regarded as an extension of the variation method. In the finite element method, space is divided into small elements, and the ϕ_i are usually taken as simple polynomials defined within each element. The c_i coefficients are typically taken as the values of ϕ at particular nodal points on the surface or interior of the elements. The ϕ_i, referred to as "shape" or "interpolating" functions, can be defined to equal unity at the ith node and zero at all other nodes of the element.

Among the many advantages of the finite element approach, we mention here the direct and simple significance of the c_i, and the fact that $a_{ij} = 0$ unless nodes i and j are associated with the same element. The latter property causes the equations to be "banded," and therefore much easier to solve. Since the integrations of Equations 8a and 8b are carried out numerically, it is possible to fit elements to boundaries of any shape, and to use arbitrary dielectric constant and charge density within each element. This power is not so readily available in older techniques, such as the finite difference method.

Calculation Methods

An extensive literature has evolved on engineering applications of the finite element method,[9] but only a few references have appeared so far in the chemical literature. Most of the latter deal with the one-dimensional Schroedinger equation. Since the engineering applications are quite different from anticipated molecular problems, only the basic results and general ideas from key sources[7,10,11] have been used in the present project. The computer programs have been written in FORTRAN IV for use on an IBM 360/50 computer.

Preliminary calculations were carried out with simple elements such as triangles and axis-aligned rectangles with linear or quadratic interpolating functions. The element integrations, Equations 8a and 8b, and the polarization vector integral (PVI), were evaluated analytically for uniform ϵ and ρ. This approach was extended to cylindrical coordinates with axial symmetry (CCAS) so that a limited class of three-dimensional problems could be explored relatively easily.

The equations are solved by the Cholesky method of Gaussian elimination. An efficient and compact algorithm[12] was adapted for this purpose. The equations are well-conditioned, and iteration on the residues is not required. In addition, a computer system for the plotting of potential contours was developed. This system allows the calculated potential to be plotted element by element, and to be compared with plots of the true solution contours when known. This capability has been essential, both for error detection and for rapid assessment of the quality of the solutions.

Gaussian numerical integration was then introduced, along with cubic interpolating functions. "Isoparametric" elements, allowing general vertex positions and curved edges, were also added.[7] A new system was then developed for three-dimensional calculations. Only the 8-vertex "brick" element has been implemented so far, although provision has been made to include the 4- to 7-vertex "degenerate" elements when they are needed. Elements of mixed degree (linear, quadratic, cubic interpolating functions) are integrated by the highest degree present and then adjusted in the equation assembly step. This procedure sacrifices some efficiency for increased flexibility. In larger problems it is often advantageous to store guassian integrand factors on tape for repeat calculations if only the dielectric function is altered, as often happens. (See the discussion of scaling in the first illustrative calculation below.)

For a particular calculation, it is necessary to number the nodes, and to associate each node with its local position in one or more elements. A minimum number of nodal coordinates must also be specified to uniquely define the geometry. The indexing and geometric operations comprise the mesh generation step. The free nodes of unknown potential are numbered first, followed by the fixed nodes occurring on some of the boundaries. The bandwidth of the assembled equations is sensitive to the method used for indexing the nodes. Experience suggests that a systematic layer-by-layer numbering scheme is practical and efficient in most of the cases considered so far.

Although methods were developed so that related problems could be generated from each other with a minimum of repetition, minor changes of the mesh and other simple changes (such as altering the axis orientation of the applied field) often required extensive reindexing, particularly for three-dimensional examples. To overcome this potentially crippling obstacle, an automated mesh generation system was developed. Each vertex and edge is represented by a symbol on a topological map. The computer then automatically indexes the mesh for the desired calculation. With this system, it became very easy to specify the mesh, to study the effect of small changes, and to alter the problem in simple ways.

Further comments on the calculation methods will be made as needed in discussing particular calculations in the following section.

Illustrative Calculations

Dielectric Sphere in a Uniform Field

This problem was used to gain experience in the design of finite element meshes for molecular calculations and to investigate variable polarizability densities. The response of a sphere of uniform dielectric constant ϵ to a uniform applied field E_0 in vacuum is a well-known analytical result.[1] The uniform field E and polarizaton P in the sphere are

$$E = \frac{3}{\epsilon + 2} E_0 \quad \text{and} \quad P = \frac{3}{4\pi} \frac{\epsilon - 1}{\epsilon + 2} E_0, \qquad \textbf{(9a, b)}$$

and the moment of the sphere is $M = PV = \alpha E_0$. For radius r_m, $V = 4\pi r_m{}^3/3$, so that

$$\alpha = r_m{}^3 \frac{\epsilon - 1}{\epsilon + 2}. \qquad \textbf{(10)}$$

For calculating M to 0.1% accuracy with uniform ϵ inside the sphere, the element geometry shown in FIGURE 1 was found to be minimally adequate and was the prototype for the two-sphere calculations described below. For this calculation, the elements shown in FIGURE 1 all have quadratic interpolating functions (nodes at the vertex and mid-edge positions). The curvature of the element edges on the circle of unit radius is quadratically determined by the mid-edge coordinates. FIGURE 1 shows only the inner region; an outer region extending to 10 units along each axis is not shown. Beyond a connecting layer of quadratic elements from 2 to 3.5 units, the outer region consists of two layers of simple (axis-aligned) linear rectangles. When the elements between 1 and 3.5 units were converted to cubic (two nodes per edge), M was calculated to 0.01% for uniform ϵ inside the sphere.

For these calculations, the nodal potentials along the outer boundaries, $z = 10$ and $r = 10$, were assigned the uniform applied field values, $\phi_0 = -zE_0$. E_0 was usually assigned unit magnitude for convenience.

Nodes along $z = 0$ were assigned $\phi = 0$. Since the problem has plus-minus symmetry for $\pm z$, there is no reason to include the lower half of the sphere in the calculation. Similarly, along the symmetry axis ($r = 0$), the normal derivative of ϕ

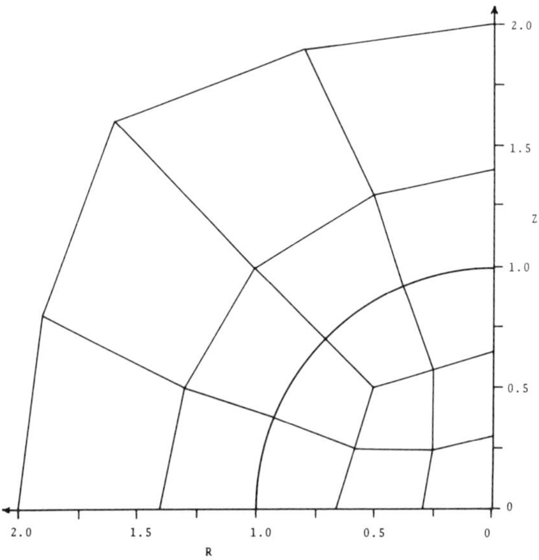

FIGURE 1. Element geometry in a two-dimensional section of a hemisphere of unit radius in cylindrical coordinates. Only the inner elements are shown.

is zero, and the surface integral (Equation 4) disappears, so that these nodes are free even though they are on the boundary. This paragraph illustrates two ways that symmetry can be exploited in these calculations. The second example applies to a reflection plane as well as to a symmetry axis.

The mesh of FIGURE 1 defines a sphere of unit radius. This choice was motivated by scaling considerations. If all dimensions are multiplied by a scale factor, $s(\ell' = s\ell)$; then $E' = E/s$, $P' = P/s$, $M' = s^2M$, and $\alpha' = s^3\alpha$. These relations are independent of the shape and dielectric function of the object. Suppose, for example, that a series of rare gases is to be studied. Then the mesh of FIGURE 1 would serve in each case, although different ϵ functions would be used in each calculation, and the results would also be scaled differently ($s = r_m$) in each case.

Variable ϵ functions were then tried, analytic results not being available. Calculations referring to the argon atom will be discussed. If $\alpha = 1.642$ Å^3 [13] and $r_m = 1.93$ Å (half of the pair separation at the minimum of the intermolecular potential),[14] then Equation 10 gives $\epsilon = 1.88$ as the effective dielectric constant if the atom is considered to be a uniform sphere. Although any reasonable variable function can be used in the calculations, the simplest choice for exploratory purposes is a linear function of the form,

$$\epsilon = 1 + 4\pi x_0(1 - r/r_0), r \leq r_0. \quad \textbf{(11)}$$

(For this discussion, r,θ,ϕ are spherical polar coordinates.) Since the parts of the atom interact with each other in the applied field, we cannot simply integrate $\chi = (\epsilon - 1)/4\pi$, the polarizability density, over the atom to obtain α. Instead, χ_0 was found by successive approximations, each involving a calculation of $M = \int P dv$, until the result was the same as the M from Equation 10 with $\epsilon = 1.88$ (with $E_0 = 1$). r_0 was taken as 1.23 units, and χ_0 was found to be 1.76.

To investigate the form of the function $P(r,\theta)/E_0$ calculated from the $\chi(r)$ of Equation 11, the mesh shown in FIGURE 2 was set up. All elements were quadratic, and the outer region was the same as for FIGURE 1. The z-component of the polarization vector integral (PVI) was calculated for each cylindrical (ring-shaped) element for $r \leq 1.25$. When the element results for each radial shell except the inner two were plotted versus the mean angle $\overline{\theta}$ of the element, an angular dependence of the type, $P_2(\cos\theta) = \frac{1}{2}(3\cos^2\theta - 1)$, was found, with an amplitude of the form, $P(r)[1 + ar + br^2]$, a and b being positive. $P_2(\cos\theta)$ is the well-known second Legendre polynomial. This angular dependence is not surpris-

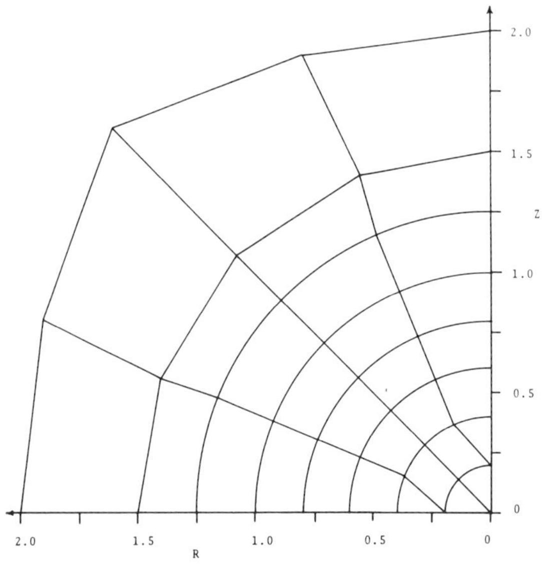

FIGURE 2. Element geometry, as in FIGURE 1, but the elements out to a radius of 1.25 units are arranged in a shell and sector structure.

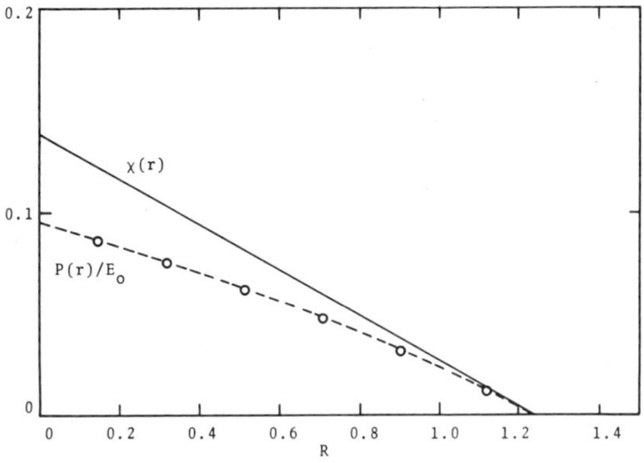

FIGURE 3. Angularly averaged polarization (dipole moment density) per unit applied field, $P(r)/E_0$, calculated for the shells of FIGURE 2 using the polarizability density $\chi(r)$ of Equation 11, and plotted against the weighted mean of r for each shell.

ing, since $\chi(r)$ from Equation 11 weights the polarizability density toward the center of the sphere, and the z-field of the induced moment of the central region should follow a $P_2(\cos\,\theta)$ dependence in the outer region.

The PVI's of the elements in each radial shell were then added, divided by the volume of the shell, and plotted versus $\bar{r}$, the mean radius of the shell (weighted by r^2), for a preliminary examination. The result, shown in FIGURE 3, indicates that the angularly averaged dipole moment density per unit applied field, $P(r)/E_0$, approaches $\chi(r)$, the polarizability density, at large r (as expected, since $P_2(\cos\,\theta)$ averages to zero over all directions), but the interaction effects at small r do not average out, and reduce $P(r)/E_0$ relative to $\chi(r)$. It is interesting to compare this result with that for the uniform sphere, for which, according to Equations 2 and 9, P/E_0 differs from χ by the factor $3/(\epsilon + 2)$, which is 0.77 if $\epsilon = 1.88$.

The preceding calculation draws attention to the important cancellation effects that occur in the uniform sphere in a uniform field. The calculation is also of considerable interest with respect to the work of Oxtoby and Gelbart.[5,6] It will be of interest to repeat the above calculation with a $\chi(r)$ deduced from their quantum-mechanically derived $P(r,\theta)/E_0$. The possibility that $\chi(r)$ is anisotropic is considered in the DISCUSSION section below.

Nonspherical Dielectric in a Uniform Field

The sphere mesh of FIGURE 1 can be mapped into various nonspherical shapes for illustrative calculations. For example, the equation of the sphere in cylindrical coordinates can be written more generally as

$$(z/a)^n + (r/b)^n = 1. \qquad \textbf{(12)}$$

For a unit sphere, $n = 2$ and $a = b = 1$. For prolate or oblate spheroids (ellipsoids of revolution), $n = 2$ and a/b is greater or less then unity. For $n > 2$, the sphere is distorted toward the right cylinder ($n = \infty$); and for $n < 2$, the distortion is toward the cone ($n = 1$).

TABLE 1

AXIAL POLARIZABILITIES OF NONSPHERICAL DIELECTRICS OF EQUAL VOLUME

Calculation	a/b	n	α/α_{sphere}
1. Prolate spheroid	1.953	2	1.118
2. Oblate spheroid	0.512	2	0.887
3. Distorted sphere	1.0	3	1.011
4. Right cylinder	1.0	∞	1.029

For spheroids, the mapping of the element mesh is achieved simply by applying scale factors a and b to the inner z and r coordinates of FIGURE 1. The volume is held constant by requiring that $ab^2 = 1$. For other distortions, such as $n = 3$, $a = b = 1$, the choice of mapping function is not unique, but several simple and practical forms are easily constructed.

The results of four calculations of the polarizability along the symmetry axis are summarized in TABLE 1. A uniform $\epsilon = 1.88$ was used inside the distorted sphere in each case ($\epsilon = 1$ outside). The prolate and oblate spheroid results are within 0.1% of the analytic results.[15] In Case 3, the volume was obtained as part of the calculation. It is interesting that the spheroids differ from the sphere by about ±12%, but the distorted sphere and right circular cylinder differ by only 1% and 3%. This result suggests that the sphere may be a good approximation for certain molecular shapes, but not for others. The question is apparently decided by the way a given distortion affects the cancellations occurring in the sphere.

Further calculations along these lines should be of interest for the interpretation of protein dielectric properties and optical anisotropies. Polarizabilities perpendicular to the symmetry axis, or for molecules without an axis of symmetry, must be obtained by calculations with three-dimensional meshes.

Overlapping Dielectric Spheres in a Uniform Field

If two identical dielectric spheres (or atoms) are represented by isotropic point polarizabilities α at their centers, and if the centers are separated by distance R, then the polarizability of the pair for an applied field parallel or perpendicular to the line of centers is

$$\alpha_{\parallel}/2\alpha = \frac{1}{1 - 2\alpha/R^3} \quad \text{or} \quad \alpha_{\perp}/2\alpha = \frac{1}{1 + \alpha/R^3} \qquad \textbf{(13a, b)}$$

In addition, the mean polarizability, $\overline{\alpha} = (\alpha_{\parallel} + 2\alpha_{\perp})/3$, for this model is given by

$$\overline{\alpha}/2\alpha = \frac{1 - \alpha/R^3}{(1 - 2\alpha/R^3)(1 + \alpha/R^3)} = 1 + 2(\alpha/R^3)^2 + \ldots \qquad \textbf{(14)}$$

These results are consequences of the through-space interaction of the induced dipoles.[3]

The above model, in its simplest form, has been found to be inadequate for interactions between rare gas atoms at pressures where binary collisions are important, because it does not recognize the overlap of the charge and polarizability densities.[13,16,17] It has also been shown recently that the classical electrostatic approach can be applied to the diffuse overlapping polarizabilities of an atom pair.[5,6] Analytic results in closed form are not available for this problem, and it is the purpose of the following illustrative calculation to show that complete solutions of the classical electrostatic problem are possible by the finite element approach.

Starting from the mesh of FIGURE 1, meshes for $R/2 = 0.3, 0.5, 0.70711, 0.85, 1.00$, and 1.25 units were developed by adding one, two, or three layers of elements, or by doubling part of the existing mesh toward negative z-values. In each case, the coordinate origin was shifted $-R/2$ along z so that plus-minus symmetry about $z = 0$ implicitly generated the second sphere. The two-dimensional cylindrical coordinate meshes were useful for the parallel applied field (along z), but a series of three-dimensional meshes, analogous to those shown in two dimensions, had to be developed for perpendicular (y) field calculations. The three-dimensional results for z-fields were in good agreement with those from the simpler two-dimensional calculations.

Each of the spheres was assigned the linearly varying dielectric function of Equation 11, with the parameters found for the argon atom. Following Oxtoby

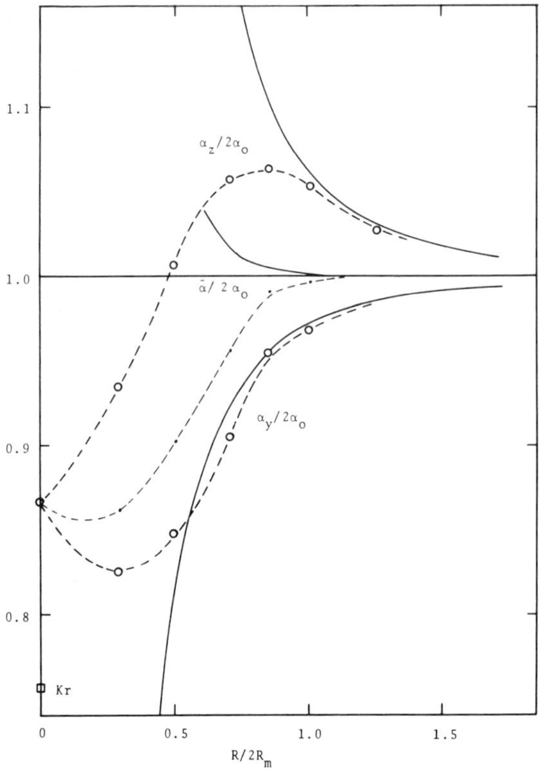

FIGURE 4. Polarizability components for overlapping argon atoms. ○ = calculated points (see text); ● = calculated mean values; —— = theory for point polarizability interaction; □ = data point for krypton atom.

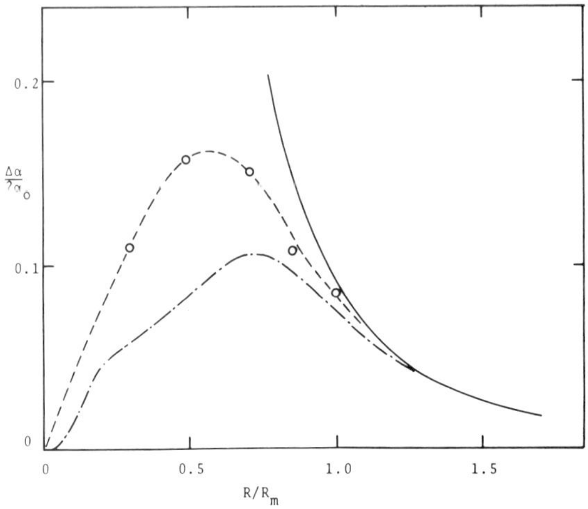

FIGURE 5. Polarizability anisotropies, $\Delta\alpha = \alpha_z - \alpha_y$ for overlapping argon atoms. ○ = calculated points (see text); —— = theory for point polarizability interaction; —·—, Oxtoby and Gelbart.[5]

Gelbart,[5] additivity of polarizability density was assumed in the overlap region. The results of the calculations are shown in FIGURES 4 and 5. All the results are in the form of ratios to the α_0 results obtained for the closely related one-sphere calculation. By coincidence, two argon atoms merge into one krypton atom at $R = 0$. The calculated krypton point on FIGURE 4 is for overlap additivity; i.e., the argon χ_0 of Equation 11 was doubled, but the r_0 was not changed. The experimental krypton polarizability[13] is also shown on FIGURE 4 for comparison.

It will be of interest to repeat these calculations with a polarizability density deduced from the quantum-mechanically derived $P(r, \theta)/E_0$ of Oxtoby and Gelbart.[5] From a broader point of view, calculations of the type described here are a necessary preliminary for a practical or empirical description (within the electrostatic framework) of the polarizability properties and interactions of diatomic molecules, triatomics, and small polyatomic molecules of biological interest.

Reaction Fields of Dipoles in Cavities

The charge density ρ of a point charge q at x_0, y_0, z_0 is approximated by a gaussian function having the same standard deviation σ along each axis:

$$\rho_q(x, y, z) = \frac{q}{(2\pi)^{3/2}\sigma^3} e^{-1/2\sigma^2[(x-x_0)^2+(y-y_0)^2+(z-z_0)^2]} \tag{15}$$

For a charge at the origin, the exponent is simply $\frac{1}{2}(r/\sigma)^2$. If equal and opposite charges $\pm q$ are placed at $\pm d$ along the z-axis, and if they are moved together and increased so that $\mu_z = 2dq$ remains constant, then in the limit $d = 0$,

$$\rho_{\mu_z}(x, y, z) = \frac{\mu_z z}{(2\pi)^{3/2}\sigma^5} e^{-1/2(r/\sigma)^2} \tag{16}$$

This result is generalized to arbitrary orientation by substituting $\mu_\alpha r_\alpha$ for $\mu_z z$, and ρ_μ is referred to as a "gaussian dipole" distribution. It is necessary to use diffuse charge distributions so that the integrals of Equation 8b can be evaluated by standard methods. The integration should include elements within 4σ for adequate precision.

The first calculations were tried with unit gaussian dipole (z-oriented) or unit gaussian charge at the origin of the mesh of FIGURE 1, using $\sigma = 0.25$, and setting $\epsilon = 1$ everywhere. The potential should then be the same as for the point charge or point dipole in vacuum for $r > 4\sigma$. For the dipole calculation, $\phi = 0$ along the $z = 0$ boundary, which is a line of plus-minus symmetry, and the nodes along the symmetry axis are free, as in the preceding examples. For the charge at the origin, however, the surface integral of Equation 4 vanishes along both axis boundaries, and the nodes along both are free.

To achieve 0.1% accuracy in ϕ at unit distance from the origin, it was necessary to use cubic elements out to 3.5 units and quadratic elements from 3.5 to 10 units (the outer boundary). It was also necessary to put the correct limiting potential on the outer boundary (instead of $\phi = 0$). It may be noted that these requirements are considerably more stringent than the requirements for 0.1% accuracy with dielectric objects in a uniform field in the preceding examples. Finally, it was found that precision was lost whenever σ was significantly less than the element size. This effect occurred because the curvature of the potential was too great to be represented by a single cubic function over a larger element. The mesh shown in FIGURE 6 was then designed to allow use of $\sigma = 0.15$ and to allow the charge or dipole to be shifted along the symmetry axis from $z = 0$ to 0.5.

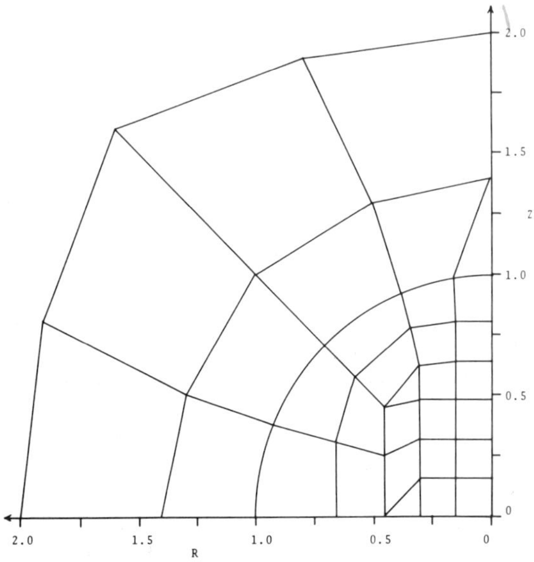

FIGURE 6. Element geometry, as in FIGURE 1, but for dipole reaction field calculations. For calculations lacking plus-minus or reflection symmetry about $z = 0$, the mesh was doubled to negative z.

The reaction field calculations were carried out for cavities in the shape of spheres and ellipsoids of revolution. For each calculation, the potential of the dipole in vacuum was first calculated and stored. Then the calculation was repeated with $\epsilon = 4$ (chosen as a typical value) outside the cavity. The vacuum potential was then subtracted at each node, leaving the potential of the reaction field. The contours of all three potentials were usually plotted by the computer. σ was 0.15 units for all the following cases.

Three calculations were done for unit dipoles along the axis of a spherical cavity. The axial reaction fields are shown in FIGURE 7. The contour plots, which are not reproduced here, showed that for Case 1 (gaussian dipole at the origin), the reaction field was uniform over the sphere. For Case 2 (dipole of discrete charges at ±0.5 units), it was higher along the axis, but lower away from the axis.

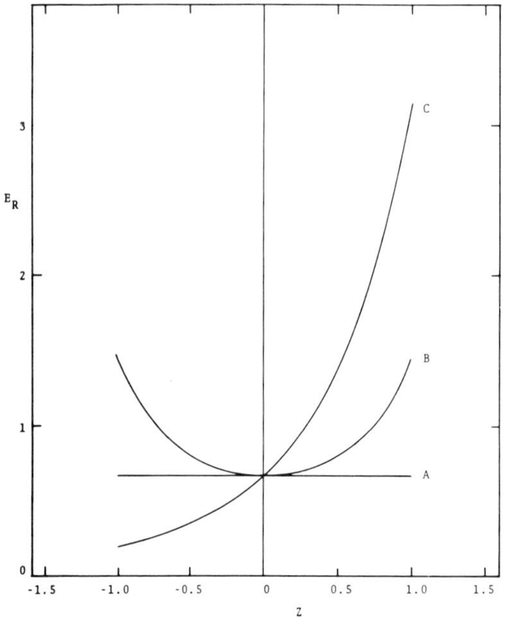

FIGURE 7. Reaction fields along the axis of a spherical cavity in a dielectric of $\epsilon = 4$ for 1) a unit gaussian dipole ($\sigma = 0.15$) at the center; 2) a unit dipole consisting of discrete charges at $z = \pm 0.5$ units; and 3) the unit gaussian dipole shifted to $z = 0.5$ units.

For Case 3 (gaussian dipole shifted to $z = 0.5$), it was also lower away from the axis. FIGURE 7 shows that all three cases have the same reaction field at the center of the cavity. All of these qualitative results for the spherical cavity are known from analytic considerations.[18,19]

A series of calculations was done for unit gaussian dipoles at the center of spheroidal cavities of various axial ratios, as shown in FIGURE 8. It is apparently a new observation that the reaction field changes sign near the tips of a prolate spheroidal cavity when a/b is increased beyond approximately 1.3 for the inner and outer dielectric constants used here (1 and 4). When the reaction fields were averaged over the cavity, the results agreed with the uniform reaction field of a uniformly distributed dipole of the same total magnitude (e.g., a uniformly polarized dielectric ellipsoid), as shown in FIGURE 9. This result was known analyti-

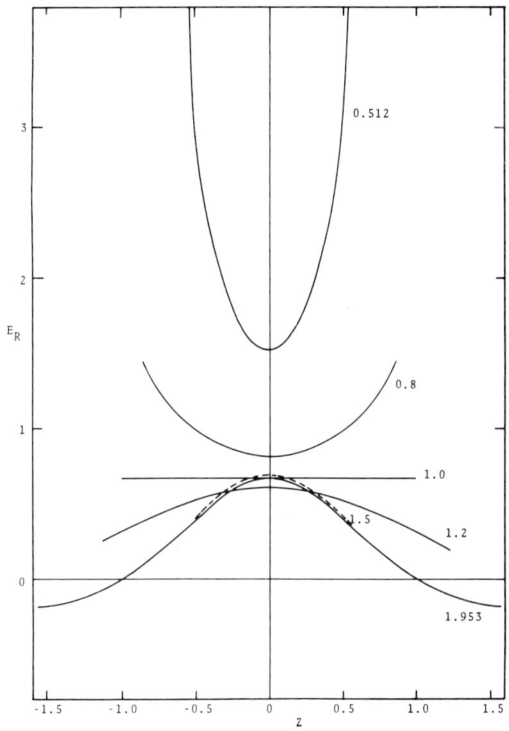

FIGURE 8. Reaction fields along the axis of spheroidal cavities of indicated axial ratio a/b (and $ab^2 = 1$) in a dielectric of $\epsilon = 4$ for unit gaussian dipoles at the center. The curve for $a/b = 1.5$ extends to $z = \pm 1.44$ and coincides with the $a/b = 1.953$ curve except where shown as a dashed line.

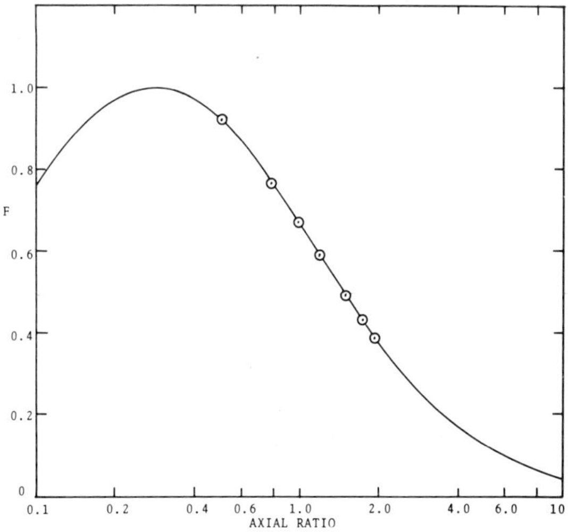

FIGURE 9. Ellipsoidal reaction field factors, $f = E_R/M$. ——, for a uniformly distributed dipole; ○, averaged over the spheroid for a point dipole at the center. The points correspond to the axial ratios shown on FIGURE 8.

cally.[20] For oblate spheroidal cavities, the reaction field along the axis is larger than the average, while for prolate cavities, it is larger only near $z = 0$.

Although most of the qualitative results obtained in this example were known analytically, the demonstrated agreement serves to establish confidence in the finite element approach, which should work just as well for more irregular shapes or for more realistic representations of the environment. The strong variation of the reaction field with position in the cavity suggests that interaction energies should be quite sensitive to the details of the calculations.

Discussion

The four illustrative calculations described in the preceding section demonstrate that the finite element method provides a powerful technique for solving molecular electrostatic problems involving polarizability effects. In each case, analytic results were not available, or were hard to comprehend. The main obstacle to the use of the method is the heavy programming effort required to achieve a practical capability. Subsequent demands on the computer are likely to remain reasonably moderate. In the exploratory stage of a project of this type, the problem is not so much to find a system for which results can be calculated, as it is to decide which calculations give the greatest insight and direction at each step. Following this stage, the ability to evaluate expectations for systems such as specific proteins, or simpler biological molecules and polymers, should be of value.

The problem of justifying the dielectric approach at the atomic level requires more attention than it has received so far. The existence of results such as the Hellmann-Feynman theorem* encourages the view that it may be possible to establish similar results for polarizability interactions between separated parts of the electron cloud (as discussed in an earlier section of this paper on electrostatic theory). Even if the formalism can be justified within certain limits, the form of the polarizability distribution will still be a problem for quantum mechanics. The electrostatic approach might be useful in constructing a more empirical model for combining atoms into bonds, and bonds into molecules.

If a polarizability density $\chi(r)$ can be defined for an atom, then the following qualitative argument suggests that it cannot be isotropic, as has been assumed so far.[5] If $\chi(r)$ is isotropic, then a uniform applied field would shift each spherical shell of the electron cloud (relative to the nucleus) without distortion. However, since there is no net force on a charge within a uniform spherical shell of charge, the shifted charge shell would continue to exert no net force on the nucleus. Only those charge shells close to the nucleus that were shifted past the nucleus would exert a force on it. This conclusion can be avoided if $\chi(r)$ is allowed to be anisotropic, with different values in the radial and angular directions. Then the uniform applied field would shift and distort each charge shell, and all parts of the electron cloud would play a role in retaining the nucleus within the atom in an applied field.

*The Hellmann-Feynman theorem is a quantum mechanical demonstration (within the Born-Oppenheimer approximation) that the forces on the nuclei of a molecule due to the electron cloud may be evaluated from classical electrostatics. (See Ref. 14, pp. 932–937.)

ACKNOWLEDGMENT

The author is indebted to Professor Robert R. Birge for helpful discussions of reaction field problems.

REFERENCES

1. BÖTTCHER, C. J. F. 1952. Theory of Electric Polarization. Elsevier Publishing Company. Amsterdam, The Netherlands.
2. BOTTCHER, C. J. F. 1973. Theory of Electric Polarization. Volume I. Dielectrics in Static Fields. 2nd edit., completely revised by O. C. VAN BELLE, P. BORDEWIJK, & A. RIP. Elsevier Scientific Publishing Company. Amsterdam, The Netherlands.
3. APPLEQUIST, J. 1977. Accts. Chem. Research **10:** 79–85.
4. PITZER, K. S. 1959. Adv. Chem. Phys. **2:** 59–83.
5. OXTOBY, D. W. & W. M. GELBART. 1975. Mol. Phys. **29:** 1569–1576.
6. OXTOBY, D. W. & W. M. GELBART. 1975. Mol. Phys. **30:** 535–547.
7. ZIENKIEWICZ, O. C. 1971. The Finite Element Method in Engineering Science. 2nd edit. McGraw-Hill Publishing Co. Ltd. London, England.
8. KOLATA, G. B. 1974. Science **184:** 887–889.
9. WHITEMAN, J. R. 1975. A Bibliography for Finite Elements. Academic Press Inc. Ltd. London, England.
10. STRANG, G. & G. J. FIX. 1973. An Analysis of the Finite Element Method. Prentice-Hall, Inc. Englewood Cliffs, N.J.
11. INT. J. NUMERICAL METH. ENG. (Many useful articles on the finite element method have appeared in this journal since its inception in 1969.)
12. GEORGE, J. A. 1971. Computer Implementation of the Finite Element Method. Ph. D. Thesis. Stanford University. Stanford, Calif.
13. ORCUTT, R. H. & R. H. COLE. 1967. J. Chem. Phys. **46:** 697–702.
14. HIRSCHFELDER, J. O., C. F. CURTISS & R. B. BIRD. 1954. Molecular Theory of Gases and Liquids. John Wiley & Sons, Inc. New York, N.Y.
15. SCHOLTE, Th. G. 1949. Physica **15:** 437–449.
16. BUCKINGHAM, A. D. & D. A. DUNMUR. 1968. Trans Faraday Soc. **64:** 1776–1783.
17. GELBART, W. M. 1974. Adv. Chem. Phys. **26:** 1–106.
18. DEKKER, A. J. 1946. Physica **12:** 209–216.
19. FROOD, D. G. & A. J. DEKKER. 1952. J. Chem. Phys. **20:** 1030–1033.
20. ABBOTT, J. A. & H. C. BOLTON. 1952. Trans. Faraday Soc. **48:** 422–430.

FIELD FLUCTUATION IN IONIC SOLUTIONS AND ITS BIOLOGICAL SIGNIFICANCE

Fumio Oosawa

Department of Biophysical Engineering, Osaka University
Ōsaka, Japan
Institute of Molecular Biology, Nagoya University
Nagoya, Japan

INTRODUCTION

The response of the membrane to the electric field is involved in the elementary process of excitation of living cells. The electric field across and along the membrane fluctuates thermally because of the Brownian motion of small ions inside and outside the cell. The membrane may respond to this fluctuating field. Actually, most of the sensory cells show spontaneous firing without specific input. The fluctuating field may also contribute to the production of randomness in the swimming behavior of microorganisms. Bacteria and ciliate occasionally change their swimming directions. It is likely that the spontaneous change of direction is induced by depolarization of the membrane potential as a result of thermal fluctuation. Direct measurements by microelectrodes inserted into the *Paramecium* cell showed large fluctuation of the membrane potential.

The average amplitude and the relaxation time of the fluctuating electric field in homogeneous and heterogeneous ionic solutions have been calculated, and the effect of the fluctuating field on the state of molecules in the solution has been analyzed.[1-3] In this paper the result of these calculations is briefly summarized, and then the coupling between the field fluctuation and the state fluctuation is discussed as a possible cause of the amplification of fluctuation to a macroscopic level. In addition, the result of the observation of fluctuation in swimming behaviors of bacteria and paramecia is presented to suggest the correlation to the fluctuation of membrane potential.

FIELD FLUCTUATION

In a homogeneous ionic solution, the Brownian motion of cations and anions produces deviation from neutrality, fluctuating locally with time. If the fluctuating electric field is expressed as a Fourier series of various modes **k**;

$$\mathbf{E}(r) = \sum_{\mathbf{k}} \mathbf{E}_k e^{i\mathbf{kr}} \tag{1}$$

the mean square amplitude of each component is given by

$$\langle \mathbf{E}_k^2 \rangle = (4\pi/\epsilon) k_B T / (1 + (\mathbf{k}^2/\kappa^2))\, V \tag{2}$$

where ϵ is the dielectric constant of the solvent, k_B is the Boltzmann constant, T is the absolute temperature, V is the volume of the solution, and

$$\kappa^2 = (4\pi e_0^2/\epsilon k_B T)(\Sigma z_i^2 n_i/V) \quad (3)$$

where κ is the Debye-Huckel parameter, n_i/V is the number concentration of the i-th ion and z_i is its valence. Then, the spatial correlation of fluctuating field is calculated to be

$$<\mathbf{E}(0)\mathbf{E}(r)> = \sum_{\mathbf{k}} <E_k^2> e^{i\mathbf{kr}} = (k_B T/\epsilon)\kappa^2 e^{-\kappa r}/r. \quad (4)$$

The relaxation time τ_{E_k} of fluctuating field of mode k is given by

$$2/\tau_{E_k} = (D_+ + D_-)k^2 + ((D_+ + zD_-)/(1 + z))\kappa^2 \pm (((D_+ + D_-)k^2 + ((D_+ + zD_-)/(1 + z))\kappa^2)^2 - 4D_+ D_- k^2(k^2 + \kappa^2))^{1/2} \quad (5)$$

for a combination of monovalent cations and z-valent anions, where D_+ and D_- are diffusion constants of cations and anions, respectively. If $D_+ = D_-$ and $z = 1$, the relaxation time of the field fluctuation is given by

$$1/\tau_{E_k} = D(k^2 + \kappa^2). \quad (6)$$

Then, the space and time correlation of the field is

$$<E(0,0)E(r,t)> = (1/2\pi)(k_B T/\epsilon)\kappa^2 \int_{Dt}^{\infty} \exp\,(-(\kappa^2 x + r^2/4x))\,dx/x^{3/2} \quad (7)$$

The mean square field at a point averaged over a finite time δt is given by

$$<\bar{E}^{\delta t 2}> = (k_B T/\epsilon)\kappa(\tau_{E_{k=0}}/\delta t)\kappa^2, \quad (8)$$

which is rewritten as

$$<\bar{E}^{\delta t 2}> = ((k_B T/e_0)^2\kappa^2)((e_0^2/\epsilon k_B T)\kappa)(\tau_{E_{k=0}}/\delta t) \quad (9)$$

where the first term corresponds to the square of the field of $k_B T/e_0$ per length $1/\kappa$ and the second term is of the order of unity, increasing slowly with increasing ion concentration. The third term is the ratio of the relaxation time to the time interval over which the field was averaged. In an ionic solution of ionic strength 0.1 at room temperature, the root mean square field is about (25 mV/10 A) $(\tau_{E_{K=0}}/\delta t)^{1/2}$. With increasing ion concentration, the amplitude of the fluctuating field increases but the relaxation time decreases.

Similar calculation gives the mean square of the potential difference between two points at distance r as

$$<(\psi(r) - \psi(0))^2> = (k_B T/e_0)^2((2e_0^2/\epsilon k_B T)\kappa)(1 - (1 - e^{-\kappa r})/\kappa r)). \quad (10)$$

If fixed charges or dielectrics are put in an ionic solution, fluctuations of various modes have correlation. Then, the spatial correlation of the fluctuating field is given by

$$<\mathbf{E}(\mathbf{r}_1)\mathbf{E}(\mathbf{r}_2)> \; = \Sigma\Sigma <\mathbf{E}_{k_i} \cdot \mathbf{E}_{k_j}> \exp\left(-(\mathbf{k}_i\mathbf{r}_1 - \mathbf{k}_j\mathbf{r}_2)\right) \quad \mathbf{(11)}$$

$$<\mathbf{E}_{k_i}\mathbf{E}_{k_j}> \; = (b_{ij}^{-1})$$

$$b_{ij} = (\epsilon/4\pi k_B T) \int (1 + k_i k_j/\kappa(r)^2) \exp\left(i(k_i - k_j)r\right) dr. \quad \mathbf{(12)}$$

The parameter κ is a function of the spatial coordinate r because of the heterogeneity of fixed charge density or dielectric constant. b_{ij}^{-1} is the ij component of the inverse matrix of b_{ij}. If the heterogeneity is small and localized, the approximate solution of (**12**) is given by

$$<E_{k_i} \cdot E_{k_j}> \; = -\, b_{ij}/b_{ii}b_{jj} \qquad (i \neq j). \quad \mathbf{(13)}$$

For example, the mean square of the fluctuating potential difference between the inside and outside of a small particle suspended in an ionic solution is approximately given by

$$<(\psi(a) - \psi(0))^2> \; = (k_B T/e_0)^2 (2e_0^2/\epsilon k_B T)$$
$$(1 + (1/2)((\kappa_a - \kappa_O)/\kappa_a)(1 - e^{-\kappa}O^a)) \quad \mathbf{(14)}$$

where a is the radius of the particle and κ_O and κ_a are the values of parameter κ outside and inside the particle. The potential difference is smaller around a particle repelling small ions.

State Fluctuation

Let us consider a macromolecule in an ionic solution, the state of which depends on the electric field. The molecule can assume one of the two states, A and B, the free energy of which has a difference

$$f_A - f_B = -\,\mu\mathbf{E}. \quad \mathbf{(15)}$$

The rate constants of the transition from A to B and from B to A are assumed to be given by

$$k_A = k_A^0 \exp\left(-(1/2)\mu\mathbf{E}/k_B T\right)$$
$$k_B = k_B^0 \exp\left(+(1/2)\mu\mathbf{E}/k_B T\right). \quad \mathbf{(16)}$$

If the fluctuation of the field is isotropic and Gaussian, the average of k's is given by

$$<k_A> \; = k_A^0 \exp\left(+\,(1/6)\gamma^2 <E^2>\right)$$
$$<k_B> \; = k_B^0 \exp\left(+\,(1/6)\gamma^2 <E^2>\right) \quad \mathbf{(17)}$$

where $\gamma = (1/2)\mu/k_B T$. Therefore, the average of k's must be larger than k^0's when the field fluctuation is taken into account, even if the average field $<\mathbf{E}>$ is zero. Since the rate constants are defined after averaging over a time longer than the microscopic collision time, $<\mathbf{E}^2>$ in (**17**) must be the mean square given by (**8**).

If we define the probability p_A that a specified molecule is in A,

$$dp_A/dt = -k_A p_A + k_B(1 - p_A). \tag{18}$$

Because of the fluctuation of k's, the probability p_A fluctuates. If k's are sufficiently large, p_A can follow the fluctuating field. If they are small, p_A is determined by the average of k's. That is, the behavior of the molecule depends on the relaxation times of field fluctuation and state fluctuation.

Equation 18 can be rewritten as

$$dp_A/dt = -<k_A>p_A + <k_B>(1 - p_A) + F(t) \tag{19}$$

$$F(t) = -\delta k_A p_A + \delta k_B(1 - p_A).$$

The last term $F(t)$ is the fluctuating force due to the deviation of k's from the average. If we apply (**16**) and (**17**), in the first order of fluctuation, the deviation of p_A from its average $<p_A>$, δp_A, follows the equation

$$d\delta p_A/dt = -(<k_A> + <k_B>)\delta p_A$$
$$+ (<k_A> + <k_B>)<p_A>(1 - <p_A>)2 \sinh(\gamma E). \tag{20}$$

The force $F(t)$ has a time and space correlation. Therefore, if two molecules are at a short distance r in the solution, correlation appears between the fluctuations of these molecules. The probability that they are in A simultaneously becomes larger than the product of probabilities of independent molecules. Calculation gives

$$<p_A(0) \cdot p_A(r)> - <p_A>^2 = <p_A>^2(1 - <p_A>)^2$$
$$4(1/3)\gamma^2<E(0)E(r)> \tau_{E_{k=0}}/\tau_{s_0} \tag{21}$$

where

$$1/\tau_{s_0} = <k_A> + <k_B>. \tag{22}$$

As expected, the excess probability is determined by the sensitivity of the molecules to the field γ and the magnitude of correlating field fluctuation $<E(0)E(r)>$, and the ratio of the relaxation time of field fluctuation $\tau_{E_{k=0}}$ to that of state fluctuation τ_{s_0}.

Thus, the field fluctuation produces a space- and time-correlation between the states of molecules even when they have no direct energetic interaction. Such a spontaneous correlative transition may be significant in the case of membrane channels that consist of multiple-channel molecules sensitive to the field.

In the above calculation, the reaction of the molecule to the field was omitted. If this reaction results in positive coupling between field and state fluctuations, those fluctuations must be amplified and long relaxation times must appear.

Coupling Between State Fluctuation and Field Fluctuation

Positive coupling between these fluctuations is expected, for example, in the case of interaction of channel molecules in a membrane with the electric field across the membrane. Let us consider a membrane that has parallel channels permeable to specific ions. The channel molecules in one kind of channels (channel a) is assumed to be sensitive to the potential difference; it takes one of the two states A and B; the rate constants of transitions between them are expressed in the form of (**16**). The channel is permeable when the molecule is in A and impermeable when it is in B, so that the permeability is proportional to the probability p_A. The chemical potential gradient of ions that can pass through this channel is assumed to be zero, on the average. The channel molecules in other kinds of channels (channel b) are insensitive to the electric potential difference and the chemical potential gradient of those ions which pass through these channels is assumed to give inside negative electric potential on the average.

In this model, let us suppose that the potential difference shifts from negative toward zero because of the Brownian motion of small ions in both sides of the membrane. If the molecules in channel a follow this fluctuation, their state shifts toward A according to (**16**). Then the channel becomes more permeable and the potential shifts more toward zero. Positive coupling thus appears. The coupling becomes stronger with increasing sensitivity of channel molecules to the electric potential difference, and it becomes weaker with shortening of the relaxation time of the potential fluctuation. In order to make a rough estimate of the effect of coupling, let us consider the fluctuation of the potential difference across the membrane, with mode k along the membrane, where both sides of the membrane have the same conductance. The decay of the potential fluctuation across and along the membrane is approximately described by

$$(d/dt + Dk^2)(C_m + C_1 k^2)\delta V = -\rho_1 k^2 \delta V + \delta I_m \quad (\mathbf{23})$$

where δV is the deviation of the potential difference across the membrane from its average $V_m(<0)$ and C_m is the capacitance of the membrane; C_1 and ρ_1 are capacitance and conductance of the solution. The fluctuating current across the membrane δI_m is produced by fluctuation of the potential difference and that of the state of channel molecules. In the same order of approximation as the Hodgkin-Huxley formulation of the potential propagation, the above equation is simplified as

$$d\delta V/dt = -(\rho_m/C_m)\delta V - (\rho_1/C_m)k^2\delta V - (\rho_x/C_m)V_m\delta p_A/\langle p_A\rangle \quad (\mathbf{24})$$

where ρ_m is the average of the total membrane conductance and ρ_x is the average of the conductance of channel a sensitive to the potential difference.

To the deviation of the probability of channel molecules in A from its average δp_A, Equation 20 can be applied; in the present case, γE in (**20**) is replaced by $\gamma'\delta V$, and $\langle k_A\rangle$ and $\langle k_B\rangle$ are rate constants at the average potential V_m. From (**24**) and (**20**) it is found that the relaxation time of the state fluctuation is made longer by coupling with the potential fluctuation, as given approximately by

$$1/\tau_s = 1/\tau_{s0} - (1/\tau_{s0})(\rho_x/(\rho_m + \rho_1 k^2))(2\gamma' \mid V_m \mid). \quad (\mathbf{25})$$

Deviation of the potential from the average produced by the Brownian motion of small ions is canceled again by their motion. A large value of $(\rho_m + \rho_1 k^2)$ means fast cancellation. Usually, this quantity is so large that channel molecules do not follow the potential fluctuation; then the effect of coupling is negligible. However, in the case of small cells in a low salt concentration, where the above quantity is small, the coupling may become appreciable.

A large value of $\gamma' \mid V_m \mid$ is required for strong coupling. If nonlinear dependence of the rate constants on the potential difference given in (**16**) is taken into account, it is expected that strong coupling suddenly occurs when the basic potential fluctuation exceeds a certain critical value. If the coupling is too strong, the fluctuation may diverge.

It must be remarked also that in the above model, the positive coupling required nonzero potential difference across the membrane. Amplification of fluctuation was accompanied by consumption of stored energy.

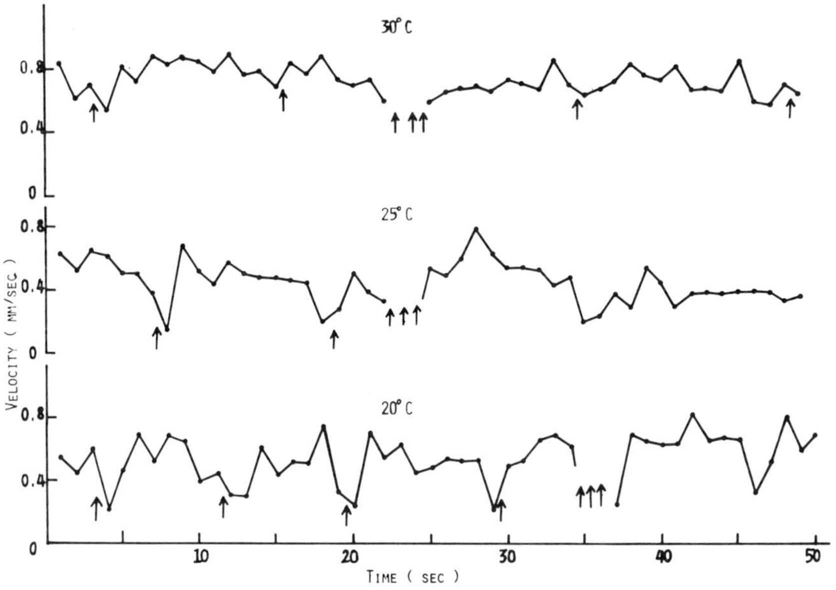

FIGURE 1. Swimming velocity and direction change of a single cell of *Paramecium* measured along the track recorded on a videotape, at three different temperatures. Arrows indicate occurrence of direction change.[5]

Fluctuation in Motility of Microorganisms

Swimming microorganisms such as bacteria and ciliate protozoa show discontinuous direction change, the frequency of which depends on the environmental condition. During straight swimming their velocity is not kept constant but shows large fluctuation.

FIGURES 1 and 2 are the records of velocity and direction change of swimming bacteria and paramecia at different temperatures.[4,5]

In the case of *Paramecium*, the membrane potential can be measured by microelectrodes inserted into a cell fixed on a slide glass. Upon application of mechanical or electrical stimulation, it was found by Naitoh and Eckert that spikelike depolarization is associated with transient reversal of beating direction of cilia, and spikelike hyperpolarization is associated with transient increase of beating frequency of cilia.[6,7] It may therefore be reasonable to suppose that both direction change and velocity fluctuation of free cells are related to the spontaneous fluctuation of membrane potential. The measured membrane potential shows spontaneous fluctuation of very large magnitude with a wide range of correlation time; it contains sharp depolarization and hyperpolarization and also fluctuation of very long periods. FIGURE 3 is a record of fluctuating potential. The time interval of sharp depolarization has an exponential distribution corresponding to that

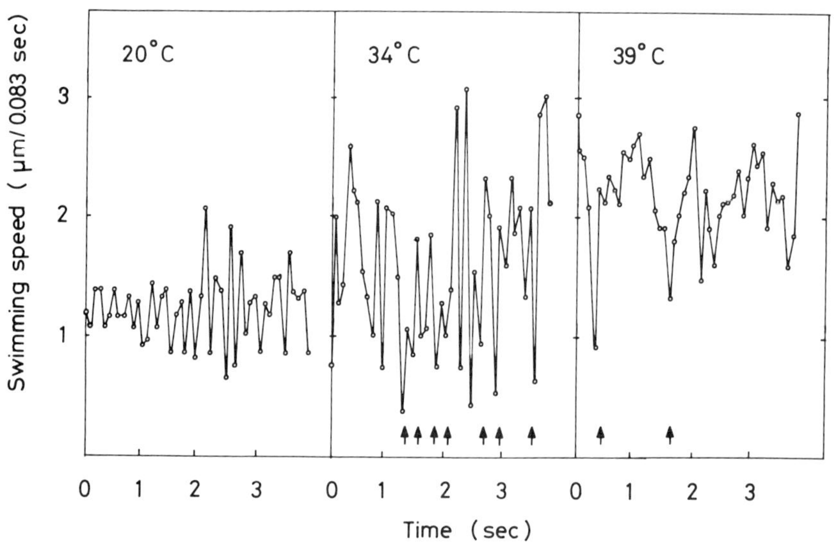

FIGURE 2. Swimming velocity and direction change of a single cell of *E. coli* measured by cinematography, at three different temperatures. Arrows indicate occurrence of direction change. (By permission of the publishers of J. Bacteriol.[4])

of the direction change of free cells.[8] The spectral analysis of the fluctuation of potential and current in a stationary state was made by Moolenaar *et al.*, who suggested participation of potassium ion flux in generation of potential fluctuation.[9] It is likely that both potassium channels and calcium channels, which have properties favorable to positive coupling with potential fluctuation, participate.[10] Details will be reported elsewhere.

Bacteria swim by rotating flagella. If a bacterial cell is tethered to a slide glass by a flagellum using antibody to flagellum, the flagellum cannot rotate, but the cell body does rotate.[11] FIGURE 4 is a record of fluctuating velocity and direction of rotation of a tethered cell obtained by Kobayasi *et al.*[12] The duration time of rotation in a constant direction shows an exponential distribution, suggesting

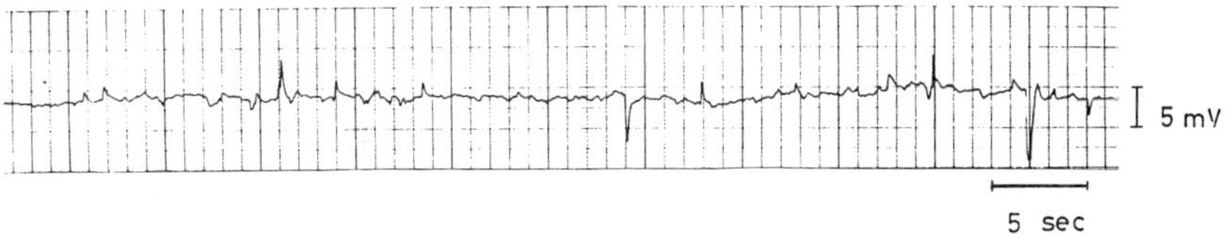

FIGURE 3. A record of membrane potential of a *Paramecium* cell in a culture medium (hay infusion). (By permission of the publishers of Ref. 10.)

random occurrence of switching of the direction of rotation, which is related to the direction change of free cells.

The bacterial cell is too small for direct measurement of membrane potential by microelectrodes. There is, however, some evidence of participation of the membrane potential in the regulation of the direction-change of free cells. It was found by Hosoi and Oosawa[13] that the frequency of direction change is very much increased by replacement of extracellular anions from small to large ones; for example, from chloride ion (Cl^-) to methane-sulfonate ion (MS^-)[13]. If anions are replaced in a reverse way, the direction change is depressed for a while. The replacement of anions from smaller (permeable) to larger (less permeable) ones results in depolarization of membrane potential, and the reverse replacement results in repolarization. Therefore, if the membrane potential has fluctuation and the direction change occurs when the fluctuating potential exceeds a critical value, the frequency of direction change is expected to increase with the replacement of anions from permeable to less permeable ones and decrease with the reverse replacement.

Experimental data suggest also that the membrane potential, or actually the electrochemical potential gradient of proton across the membrane, plays an important role in the generation of force for the rotation of flagella. It was found by Shioi *et al.* that the effect of valinomycin with various concentrations of potassium ion at different pH's on the motility of bacteria can be understood under the assumption that the electrochemical potential gradient of proton above a certain critical value is required for motility.[14] Therefore, the velocity fluctuation of the rotation of flagella is also connected with the fluctuation of membrane potential.

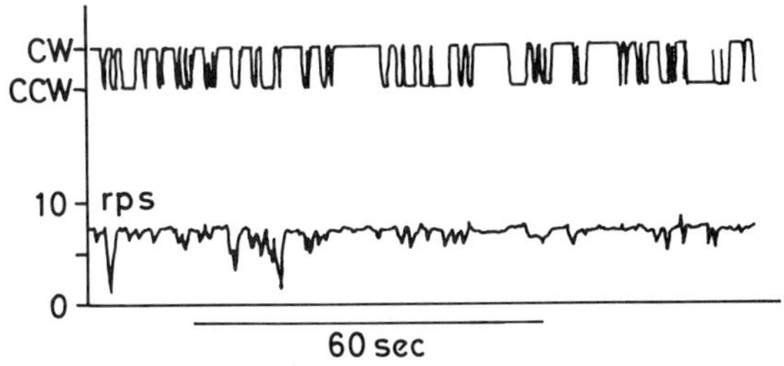

FIGURE 4. The velocity and the direction of rotation of a tethered cell of *E. coli*. (By permission of the publishers of Rev. Sci. Instrum.[12])

Recently, fluorescent dyes have been found to indicate the membrane potential when bound to the membrane. It is desirable to apply these dyes to a free or tethered bacterial cell and measure the fluctuation of fluorescence intensity in order to determine if the potential fluctuation has correlation to velocity fluctuation and direction change of rotation of flagella.

References

1. Oosawa, F. 1973. Field fluctuation in ionic solutions and its biological significance. J. Theor. Biol. **39:** 373–382.
2. Oosawa, F. 1975. The effect of field fluctuation on a macromolecular system. J. Theor. Biol. **52:** 175–186.
3. Oosawa, F. 1976. Field Fluctuation in Ionic Solutions and Membranes, Charged Gels and Membranes. Vol. II. 193–204, ed. by E. Selegny, Ed.: 193–204. D. Reidel Publ. Co. Dordrecht-Holland.
4. Maeda, K., Y. Imae, J. Shioi & F. Oosawa. 1976. Effect of temperature on motility and chemotaxis of *Escherichia coli.* J. Bacteriol. **127:** 1039–1046.
5. Shiomi, T. & Y. Nakaoka. 1977. Doctoral Thesis. Dept. of Biophysical Eng., Osaka Univ. Osaka, Japan.
6. Naitoh, Y. & R. Eckert. 1974. *In* Cilia and Flagella. 305–352, O. Sleigh, Ed. Academic Press. New York, N.Y.
7. Eckert, R. 1972. Bioelectric control of ciliary activity. Science **176:** 473–481.
8. Saji, M. Unpublished data.
9. Moolenaar, W., J. de Goede & A. Verveen. 1976. Membrane noise in *Paramecium.* Science **260:** 344–346.
10. Majima, T. 1976. (Abstr.) Ann. Mtg. Biophys. Soc. Japan. (In Japanese.)
11. Silverman, M. & M. Simon. 1974. Flagellar rotation and the mechanism of bacterial motility. Nature **294:** 73–74.
12. Kobayasi, S., K. Maeda & Y. Imae. 1977. Apparatus for Detecting Rate and Direction of Rotation of Tethered Bacterial cells. Rev. Sci. Instrum. **48:** 407–410.
13. Hosoi, S. & F. Oosawa. 1977. Effect of replacement of extracellular anions on swimming pattern of *Salmonella typhimurium.* Submitted to J. Bacteriol.
14. Shioi, J., Y. Imae & F. Oosawa. 1977. Protonmotive force and motility of *Bacillus subtilis.* Submitted to J. Bacteriol.

REAL-TIME MEASUREMENT OF DIELECTRIC RELAXATION OF BIOMOLECULES: KINETICS OF A PROTEIN-LIGAND BINDING REACTION*

Walter Scheider

Biophysics Research Division, Institute of Science and Technology
University of Michigan
Ann Arbor, Michigan 48109

INTRODUCTION

Some chemical reactions involving biomolecules are known to be accompanied by measurable changes in electric properties. The binding of long-chain fatty acids and similar ligands to human serum albumin (HSA) is one such reaction; a change in the dipole vector of the protein has been documented.[1]

We have used this property to develop a new probe to study the kinetics of these binding reactions.

The probe consists of a measurement of the dielectric properties of a reacting mixture, made by recording the response in real time to a single electric pulse. This technique has the advantage of providing several parameters from each measurement, which, in the experiments we are reporting here, provided us with simultaneous access to two separable measures of reaction progress. This resulted in identification of two distinct steps in the binding reaction.

The observations are made on rapidly mixed solutions of HSA and ligand, with use of stop-flow methods. Each observation requires the recording of data for an interval of 20–50 μsec, and is repeated at intervals from 1 msec to 20 sec after mixing.

SINGLE-PULSE DIELECTRIC RELAXATION MEASUREMENT

Ever since Debye's classical explanation[2] of the mechanism by which polar molecules orient in an electric field and the subsequent discoveries in the 1940s[3,4] that many protein molecules in solution behave in an electric field very nearly like polar molecules of an ideal gas, it has been only a matter of time and technology before the orientation of biologically important molecules could be observed as it happens—in real time.

By a Fourier transformation one can go from a description of a system response in the frequency domain to a description in the time domain, and, as Dr. Wada has pointed out (Husimi and Wada[5]) in the context of his very excellent method of applying a burst of pseudorandom noise, any transient signal that contains as Fourier components the relevant parts of the frequency spectrum, will, in principle, yield the same information as the traditional frequency scan.

Of course the business of finely tuning a test circuit one frequency at a time provides an enormous, useful redundancy in the desired information, allowing one

*Supported by the Institute of General Medical Sciences, NIH grant GM22309.

to amplify the wanted signal greatly while eliminating almost all nonsystematic error.

One reduces this desirable redundancy when one reduces the number of times one repeats the essential features of the test wave shape; one loses in signal-to-noise ratio, and consequently in accuracy, as one shortens the measurement time. Dr. Wada's method, which he will be discussing later, strikes an excellent balance between speed and accuracy.

The method we have developed consists of applying a single square pulse of the order of 20–50 μsec duration, essentially limiting the measurement period to the response time of the sample. In so doing we sacrifice further the ability to filter out noise, but, as it turns out, the information yield is still quite satisfactory for a number of purposes.

We expect the usefulness of the single-pulse method to be primarily in three categories of experiments: 1) those in which it is difficult to maintain the physical environment constant for steady state measurements; 2) those where intentional changes in the sample are induced, and repeated single-pulse measurements are made to observe the resultant sample changes; and 3) those where the pulse itself may cause sample "damage" and one has essentially only one shot at making a measurement. Our present use of the method is in a system of category 2).

Design, Theory, Data Processing

The pulse method involves first of all a choice of pulse shape. The square pulse, or Heaviside function, has a spectral density inversely proportional to the fre-

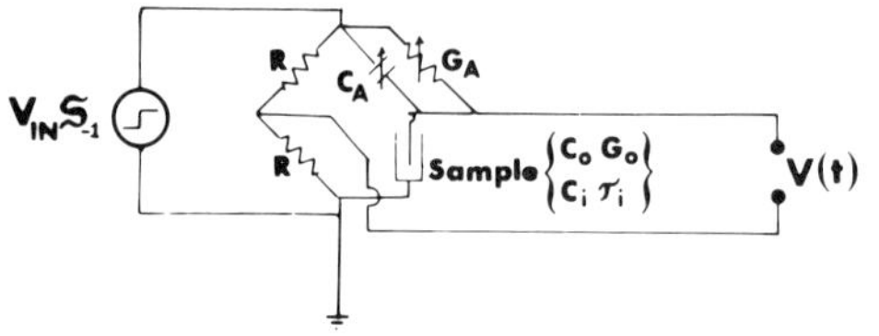

FIGURE 1. Schematic of bridge arrangement for single-pulse response measurement of dielectric relaxation of a sample placed between electrodes (see text for definition of symbols).

quency, and thus is uniform on a log frequency scale. This corresponds most closely to the customary and most efficient sampling density used in frequency scans for obtaining Debye dispersion curves, and so it becomes the pulse shape of choice.

One seeks next to maximize the proportion of response signal that reveals desired information, and one does this by balancing out what might be called the base-line capacitance and conductance of the sample in a bridge arrangement.

For purposes of analysis, the sample is specified in terms of its conductance, G_0; its frequency-independent capacitance, C_0; and dispersion components "i" of increment C_i and relaxation time, τ_i. One aims to balance out C_0 and G_0 by setting the balance arm capacitance and conductance to estimates of these values, C_A and G_A. Fixed and equal low resistances R make up the other two bridge arms, as in FIGURE 1. In this arrangement, the input signal is a step function $\mathfrak{S}_{-1}$ (using the notation of Ref. 6) of amplitude V_{IN}.

The circuit of FIGURE 1 may be solved in the frequency domain. The response $V(t)$ in the time domain to a signal $V_{\text{IN}}\mathfrak{S}_{-1}$ is then obtained by multiplying the frequency response by the spectral density $(j\omega)^{-1}$, where j is the unit imaginary and ω the angular frequency, and obtaining the Fourier transform of the resultant quantity. For the case of a single relaxation component (C_1, τ_1), one obtains

$$\frac{V(t)}{V_{\text{IN}}} = \frac{1}{2}\frac{(G_A - G_0)}{(G_A + G_0)} + \frac{1}{\tau_1(C_A + C_0)(\beta_1 - \beta_2)}\sum_{i=1}^{2} (-1)^i e^{-\beta_i t}(1 - \beta_i\tau_1)\left(C_A - \frac{G_A}{\beta_i}\right) \quad (1)$$

where β_1 and β_2 are the roots of the equation

$$\tau_1\beta^2 - B\beta + \tau_0^{-1} = 0$$

and

$$\tau_0 = \frac{C_A + C_0}{G_A + G_0}$$

$$B = 1 + \frac{C_1}{C_A + C_0} + \frac{\tau_1}{\tau_0}$$

τ_0 can be thought of as a relaxation time of the fixed quantities alone.

For two components of increment C_1 and C_2 and time constants τ_1 and τ_2, one obtains

$$\frac{V(t)}{V_{\text{IN}}} = \frac{1}{W(C_A + C_0)}\prod_{i=1}^{3}\gamma_i^{-1}\sum_{i=1}^{3}\gamma_i\left\{\frac{-G_A}{\beta_i} + \left[\frac{G_A}{\beta_i} - (C_A + G_A T) + \beta_i(WG_A + C_A T) - \beta_i^2 C_A W\right]e^{-\beta_i t}\right\} - \frac{1}{2}$$

where β_i are the roots of

$$W\beta^3 - Q\beta^2 + R\beta - \tau_0^{-1} = 0$$

and

$$\gamma_1 = \beta_2 - \beta_3; \quad \gamma_2 = \beta_3 - \beta_1; \quad \gamma_3 = \beta_1 - \beta_2; \quad (2)$$

$$T = \tau_1 + \tau_2$$

$$W = \tau_1\tau_2$$

$$R = 1 + T/\tau_0 + (C_1 + C_2)/(C_A + C_0)$$

$$Q = T + W/\tau_0 + (C_1\tau_2 + C_2\tau_1)/(C_A + C_0)$$

and τ_0 is defined as in (**1**).

Some simple response characteristics can be described. The value of $V(0)$, at time $t = 0$, is always determined by the balance of C_A and C_0, and if these

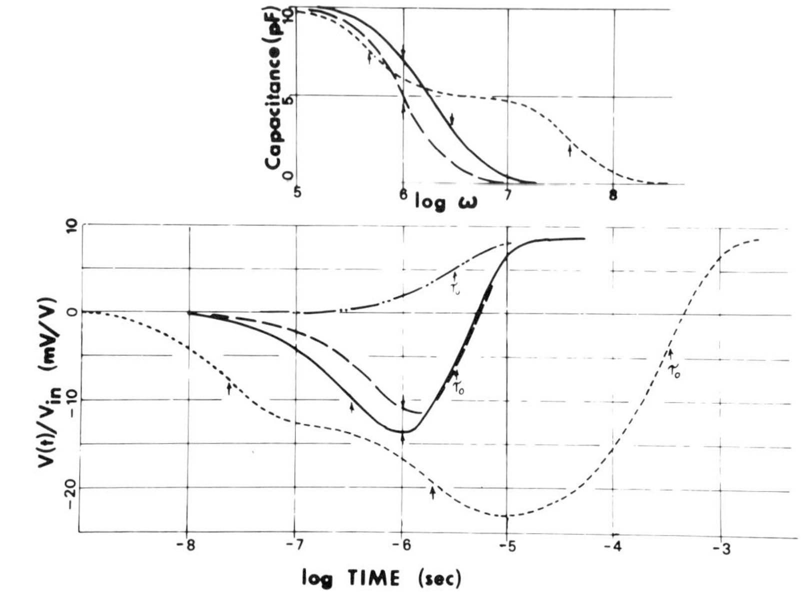

FIGURE 2. Calculated response curves in the time domain and (above) corresponding Debye dispersion curves in the frequency domain for the same parameters. Time domain data calculated from Equation 2: $C_A = C_0 = 100$ pF for all cases. — -- — = no relaxation components, $\tau_0 = 3.4$ μsec; ------- = widely separated dispersion regions, $C_1 = C_2 = 5$ pF, $\tau_1 = .025$ μsec, $\tau_2 = 2$ μsec, $\tau_0 = 340$ μsec; —— = simulated data for globular proteins, $C_1 = C_2 = 5$ pF, $\tau_1 = .333$ μsec, $\tau_2 = 1$ μsec, $\tau_0 = 3.4$ μsec; and — — = single relaxation time, $C_1 = 10$ pF, $\tau_1 = 1$ μsec, $\tau_0 = 3.4$ μsec. Arrows indicate relaxation times or relaxation frequencies.

are equal, the response always begins at the origin (FIGURE 2). At large times, $V(t \rightarrow \infty)$ approaches the asymptotic values $1/2\ (G_A - G_0)/(G_A + G_0)$. If there is no dispersion, it approaches this asymptote exponentially with the time constant τ_0 (dash – dot curve).

If there is one dispersion region, and $\tau_1 \ll \tau_0$, one observes first the polarization due to molecular orientation, the effect of which is to increase the sample capacitance with time. This is followed by the asymptotic approach to $V(\infty)$.

FIGURE 2 contains three simulated-response curves in both time and frequency domains for systems with dispersion. The dotted line represents the case in which τ_1, τ_2 (arrows) and τ_0 are widely separated; this response function is a sum of three exponential decay terms (**2**), and thus the separation of the relaxation domains is qualitatively different from the corresponding separation obtained in the frequency domain, where the response is a sum of Lorentzian terms. The other two response curves shown correspond more nearly to the dispersion properties of globular proteins, where τ_1 and τ_2 span at most one decade, and where the solution conductance at best gives a value of τ_0 somewhat larger than the molecular relaxation times.

FIGURE 3(a) shows experimental time domain response data from a sample of HSA. Such data can be fitted by a one- or two-relaxation time representation directly, using Equations 1 and 2, respectively.

If one is interested in fitting such data with a more complicated model, involving more than two relaxation times, or a continuous distribution of relaxation times as in Cole-Cole[7] equation,

$$\epsilon(\omega) = \epsilon_s + \frac{\epsilon_1}{1 + (j\omega\tau)^\alpha} \tag{3}$$

the lack of linear additivity of relaxation components in the time-domain response functions poses great computational difficulties. In Equation 3, ϵ is the dielectric constant, ϵ_s the solvent dielectric constant, ϵ_1 the dielectric increment, and α is a number between 0 and 1.

Fitting to such more complicated models can be done by computing a Fourier Transform[8] of the time-domain data, and obtaining from that, by straight forward algebraic manipulation, a conventional Debye representation of dielectric increment in the frequency domain, of which the real part is shown in FIGURE 3(b). The transformed data can then be fitted in the frequency domain where relaxation components add linearly. The two principal drawbacks to this method are 1) the ability to assign appropriate weighting factors to the various portions of data is largely lost, and 2) the transformation almost always requires extensive interpolation and extrapolation of the actual data, which is difficult to do without biasing the information.

The electronics used in making these measurements were custom-designed

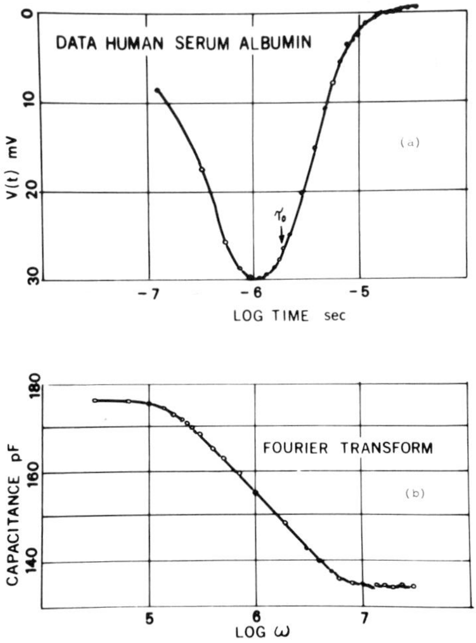

FIGURE 3. (a): Data taken on a sample of Human Serum Albumin, 41.2 g/l in water, Run 26:3, in a cell with cell constant 11.52 cm., at 0°C; (b): Debye dispersion curve from same data calculated by means of a Fourier Transformation.

to permit programming of pulse width, repetition rate, and data-sampling times. The bridge output is amplified and sampled at intervals of 200 nsec, converted to digital form with 8-bit accuracy, and stored in a memory for later retrieval and processing.

Observing Ion Immobilization and Binding in Real-Time

Our initial interest in the ability to make these extremely fast measurements of dielectric relaxation was in its application to kinetic studies of the reactions between human serum albumin and small molecules.

The steady state dielectric properties of serum albumin solutions have been well characterized,[4,9–11] and in particular it is known that the dielectric increment is markedly altered by the binding of fatty acids.[1]

The equilibrium constants of such binding have been studied by chemical methods[12,13] and, for the first several moles of fatty acid bound, are of the order of $10^7 M^{-1}$. This extremely tight binding accounts for the difficulty in completely defatting this protein.[14,15] These and other observations[16–18] have supported the view that two or three of the primary sites for fatty acid and certain other organic ligands are in a portion of the albumin surface which is interior to the molecule in its normal, or *N*, form. Studies of the *N-F* transition, induced by lowering the pH to approximately 4, are consistent with the hypothesis that this interior surface becomes more accessible to the solvent in the *F*-form, accounting for the relative ease with which *in vitro* defatting occurs at lower pH values.

In our current research we use the electric pulse as a probe of the changes occurring in the albumin molecule as free fatty acid is suddenly made available for binding. This condition is created in a stop-flow apparatus, which is conventional in design,[19] except that after the albumin solution and the solution of fatty acid come together, the mixture is observed in an electrode chamber instead of in the customary optical cell.

Our mixing cell with electrode chamber (FIGURE 4) has a mixing time of <2 msec., mixing volume = 40 μl, electrode chamber volume = 14 μl, and a cell constant = 14.8 cm.

Our HSA was obtained as a reworked fraction V [20] from the Blood Fractionation Laboratory of the Michigan Department of Public Health, Lansing, Michigan. It contained a maximum of 3% nonalbumin proteins, and less than 5% of dimer, as characterized both by ultracentrifugation and by gel filtration.

The protein is first defatted by the method of Ref. 21, using XAD-2 resin, and a speeded-up technique for bringing the protein to defatting pH and returning it to pH 5 immediately on exit from the resin column. The defatted protein is found to be labile, and to denature if left at low pH longer than several hours.

Numerous technical problems are associated with the introduction of long-chain fatty acids into the protein solution, related to the fact that these ligands are micelle-formers in water. Consequently also the assignment of such quantities as the free ligand concentration is not straightforward.

We have obtained results from stop-flow mixtures of albumin solutions with 1) dispersions of oleic acid in 10% Ethanol-water mixtures, 2) solutions of sodium oleate, and 3) solutions of the more soluble ligand, *N*-acetyl tryptophan. Similarities in the binding of fatty acids and of *N*-acetyl tryptophan were reported by Dintzis.[22]

In each experiment, one obtains a time-domain response curve such as is shown in FIGURE 3(a) immediately upon stopping of the mixture of HSA and ligand solution in the electrode chamber; one then repeats the application of a pulse and recording of a response curve in a digital memory, at various intervals during the course of the chemical reaction; i.e., the binding of the ligand.

From the series of response curves we routinely calculate values for the parameters G_0, C_0, C_1, τ_1, C_2, τ_2, and we can also obtain by Fourier Transformation a Cole-Cole parameter α in the τ_1 component of increment. However, C_0, τ_1, and τ_2 are assumed not to be parameters of the reaction, and C_1 and C_2 are not even under the best circumstances[1] determinable separately with great certainty.

As reaction-sensitive parameters, we thus obtain the total increment C_{tot} = $C_1 + C_2$, and G_0; these, in fact, provide us with the ability to monitor simultaneously two different events involved in the reaction.

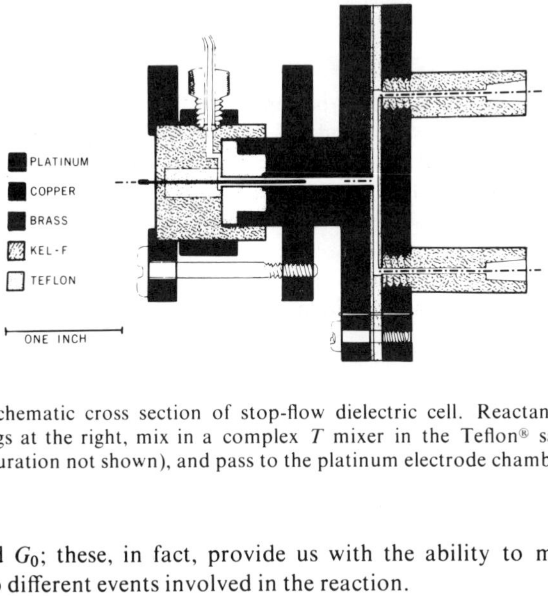

FIGURE 4. Schematic cross section of stop-flow dielectric cell. Reactants enter at the two Luer fittings at the right, mix in a complex *T* mixer in the Teflon® sandwich (details of mixer configuration not shown), and pass to the platinum electrode chamber.

G_0, the solution conductance, changes as an ionic species becomes relatively immobilized by binding to a large molecule.

C_{tot}, the total increment of capacitance due to molecule orientation, gives an indication of the extent to which the reaction has proceeded to its final equilibrium state.[1]

In FIGURE 5 are shown the values of C_{tot} as a function of time after stop-flow mixing, calculated from a sequence of pulse-response curves obtained during the reaction of HSA with a solution of equal molarity of *N*-acetyl-DL-tryptophan. Also shown are results from a control mixture of HSA and water. There are several possible explanations of the data from 1 sec to 20 sec, which will be elucidated in further experiments. In this report we have analyzed the data up to 1 sec

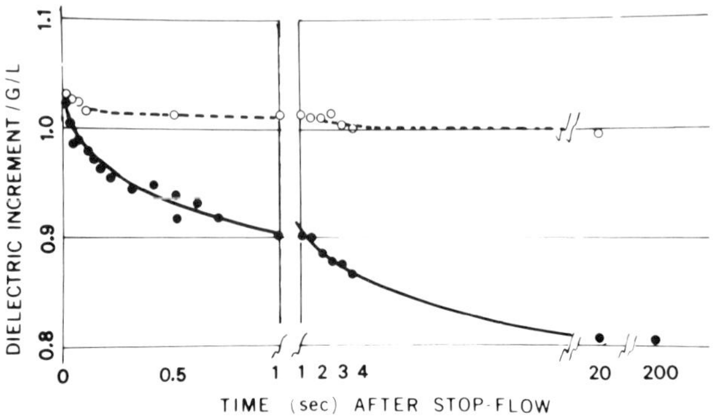

FIGURE 5. Dielectric Increment as a function of time after mixing in a stop-flow cell (FIGURE 4). (●) = Solution of HSA (LHA76QE1), 66.1 g/l in water, mixed with solution 1mM *N*-Acetyl-DL-Tryptophan in water, at 0°C, G_0 = 6.2 μmho/cm; (○) = Control, same HSA solution mixed with water, G_0 = 5.1 μmho/cm.

in terms of a reaction assumed to be separable from the longer-time event. As such, using the initial (high) C_{tot} and the 1 sec (low) value of C_{tot} as parameters reflecting an unreacted and a reacted state for the HSA molecule, we obtain in FIGURE 6 a graphical representation of the extent of reaction as a function of time, indicating a good fit to first-order kinetics with a rate constant of $(3 \pm 1/2)\ \text{sec}^{-1}$.

This result, unexpected to be sure, becomes understandable in the light of the other information from the same sequence of pulse responses. The value of G_0, not presented graphically because it would be a straight horizontal line, is virtually

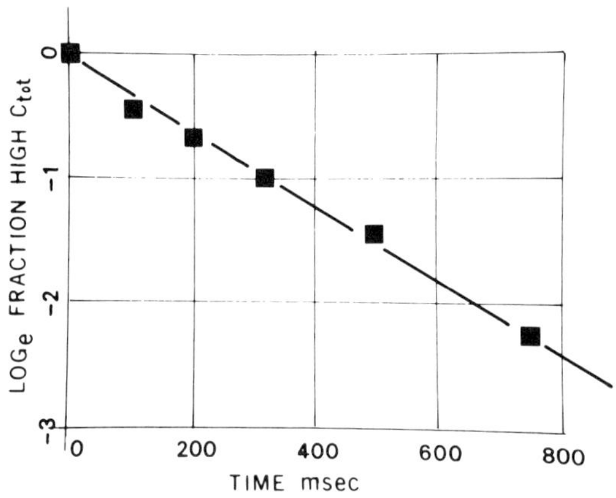

FIGURE 6. Data of FIGURE 5 from $t = 0$ to $t = 1$ sec, displayed as a fraction of protein remaining in the high dielectric increment state. For this calculation, the assumption was made that the decline in dielectric increment after 1 sec is a separable decay.

invariant with time after the first data point, having a value only slightly different from that for the control experiment, and far lower than the total conductance of the unreacted mixture that can be estimated from calculations based on the conductance of free ligand. This is a relatively complicated calculation, involving the degree of ionization of the ligand and the extent of pH buffering by the protein, but the conclusion is clear, that by the time of the first measurement virtually all free ligand has been immobilized, suggesting that it has been bound by the protein, but not yet in the configuration that it reaches at equilibrium.

The inference from the conductivity data is that the time constant of the initial binding step is less than 1 msec, giving a second-order rate constant $k > 10^6 M^{-1} sec^{-1}$. This suggests a diffusion-limited first step. With ligand and protein already in place, it is to be expected that the subsequent step be of first-order kinetics, as indeed the data suggest.

These experiments have also been done with oleic acid as ligand, and the same general kinetics are evident.

Conclusion

The pulse-response technique described has demonstrated that the binding of *N*-acetyl tryptophan and oleic acid by HSA takes place in at least two steps, the first occurring quite rapidly, producing a complex that removes the ligand from the free state but does not involve the characteristic change in dipole moment associated with the final bound state. Presumably, the ligand has found its albumin molecule and attached itself to one of the more accessible low-affinity sites.

Following such an attachment is a second, much slower step, with first-order kinetics, during which the ligand and the protein rearrange their configuration, perhaps involving an opening and reclosing of the *N* configuration, after which the ligand has been bound at the more interior, high-affinity site.

These conclusions are very much consistent with all that equilibrium studies have revealed about the binding of fatty acids to HSA. They are indeed a confirmation of the view of two distinct types of binding sites.

It should be noted that the experiments on the albumin-fatty acid system done thus far are quite limited, and still allow for various interpretations.

Many obvious variations on the experiments described can now be done, and may help develop a better understanding of the functioning of the primary organic binding site or sites on HSA.

The pulse-response method has proved its feasability and its usefulness as a probe of chemical processes. It is our hope that it will also prove versatile enough to access other systems to dielectric measurement, systems too fragile or too changeable to hold still for steady-state measurements.

Acknowledgments

This work would not have been possible without the generous cooperation of Dr. John Mercer at the Blood Fractionation Laboratory, Michigan Department of Public Health, Lansing, Mich. who supplied us with high quality HSA.

References

1. Scheider, W., H. M. Dintzis & J. L. Oncley. 1976. Changes in the electric dipole vector of human serum albumin due to complexing with fatty acids. Biophys. J. **16:** 417.
2. Debye, P. 1929. Polar Molecules. Chem. Cat. Co. New York, N.Y. Reprinted by Dover Publishing Co. New York, N.Y.
3. Wyman, J., Jr. 1936. Polarization and dielectric constant of liquids. J. Am. Chem. Soc. **58:** 1482.
4. Oncley, J. L. 1942. The investigation of proteins by dielectric measurements. Chem. Rev. **30:** 433.
5. Husimi, Y. & A. Wada. 1976. Time-domain measurement of dielectric dispersion as a response to pseudorandom noise. Rev. Sci. Instrum. **47:** 213.
6. Campbell, G. A. & R. M. Foster. 1948. Fourier Integrals. D. Van Nostrand Co. Princeton, N.J.
7. Cole, K. S. & R. H. Cole. 1941. Dispersion and absorption in dielectrics. J. Chem. Phys. **9:** 341.
8. Cooley, J. W. & J. W. Tukey. 1965. An algorithm for the machine calculation of complex Fourier series. Math. Comp. **19:** 297.
9. Moser, P., P. G. Squire & C. T. O'Konski. 1966. Electric polarization in proteins—dielectric dispersion and Kerr effect studies of Isoionic bovine serum albumin. J. Phys. Chem. **70:** 744.
10. Squire, P. G., P. Moser & C. T. O'Konski. 1968. The hydrodynamic properties of bovine serum albumin monomer and dimer. Biochemistry **7:** 4261.
11. Soetewey, F., M. Rosseneu-Motreff, R. Lamote & H. Peeters. 1972. Size and shape determination of native and defatted bovine serum albumin monomers. J. Biochem. **71:** 705.
12. Goodman, D. S. 1958. The interaction of human serum albumin with long-chain fatty acid anions. J. Am. Chem. Soc. **80:** 3892.
13. Spector, A. A., K. John & J. E. Fletcher. 1969. Binding of long-chain fatty acids to bovine serum albumin. J. Lipid Res. **10:** 56.
14. Goodman, D. S. 1957. Preparation of human serum albumin free of long-chain fatty acids. Science **125:** 1296.
15. Chen, R. F. 1967. Removal of fatty acids from serum albumin by charcoal treatment. J. Biol. Chem. **242:** 173.
16. Odell, G. B. 1973. Influence of binding on the toxicity of bilirubin, Ann. N.Y. Acad. Sci. **226:** 225.
17. King, T. P. 1973. Location of the primary organic ligand binding site of bovine plasma albumin. Ann. N.Y. Acad. Sci. **226:** 94.
18. Brown, J. R. 1974. Structure of serum albumin: disulfide bridges. Fed. Proc. **33:** 1389.
19. Chance, B. 1976. Rapid flow methods. Tech. Chem. (N.Y.) **6** (Part 2): 5.
20. Cohn, E. J., J. L. Oncley, L. E. Strong, W. L. Hughes, Jr. & S. H. Armstrong, Jr. 1944. The characterization of the protein fractions of human plasma. J. Clin. Inv. **23:** 417.
21. Scheider, W. & J. K. Fuller. 1970. An effective method for defatting albumin using resin columns. Biochim. Biophys. Acta **221:** 376.
22. Dintzis, H. M. 1952. Dielectric properties of human mercaptalbumin. Ph.D. Thesis. Harvard Univ. Cambridge, Mass.

DISCUSSION

M. Mandel, *Moderator*

Department of Physical Chemistry, Gorlaeus Laboratories
State University of Leiden
Leiden, The Netherlands

UNIDENTIFIED SPEAKER: Dr. Oosawa, I was fascinated with the experiments that have to do with the paramecia and bacteria. I have seen them moving in the straight line. I would have expected that the bacteria would be buffeted around quite a bit, but they were not. I wonder what your feeling is about that. There is a lot of kinetic energy in the solution.

DR. OOSAWA: I think that their movement is produced by their own tissue of flagella over spiral shape and the energy dissipated inside the cell to rotate as if there were a motor at the base of the flagella. Nobody knows why the motor can rotate. Some think that the rotation is induced by a flow of charges from outside into the cell. The movement of bacteria is different from that of simple particle movement consuming the energy inside the cell.

QUESTION: I think what you are saying is that the energy in the flagella and the cilia are much greater than kT?

DR. OOSAWA: Yes. Much greater. But not very large for rotation. The number of ATP used for rotation is not too large, compared to the protein molecules making up flagella. But the energy is much larger than kT.

QUESTION: Dr. Scheider, this question pertains to the response in terms of comparing the whole big molecule such as a protein molecule with a simple dipole, say, HCl, I have tried to get some sort of connection between them. But I had difficulty in doing that, because if I look at a protein molecule, I find it difficult to see the dielectric response in terms of distinction to a simple dipole rotating in the field as it changes. I wonder if you can offer a little enlightment on that?

DR. SCHEIDER: At time equal zero, or very shortly after that, let us say just short enough so that water has had time to undergo the relaxation. At time zero, you have a balance of the bridge as if the large molecules did not have any of the energy storage, the capacity. Large molecules have not started yet to orient themselves in the field, and therefore they have no visible capacity. This is like the high-frequency measurement. If you had large molecules that have long relaxation times, they start orienting in the field and you will then see the energy storage of the rotation as a new capacitance, or as an increase in the capacitance, of the sample. So as the capacitance increases, the voltage across it will decrease, and that is why you see the dip. It represents increasing capacitance as the large molecules do their relaxation.

DR. GREEN: *City College, CUNY, New York, N.Y.:* Dr. Oosawa, we got some evidence in an experiment on noise measurements with ion exchange membrane that we interpreted as a lowering of the activation energy for diffusion in the high electric fields. Now, since you observed that rate constants are increased in high fluctuating fields, which we believe we had in our experiments, I wonder whether your calculations would also apply to the diffusion case that we believe we have observed?

DR. OOSAWA: Yes, I think so.

DR. GREEN: Would it be observed as lowering of activation energy?

DR. OOSAWA: Yes, it may not be equivalent to the lowering, but the rate-constant increase is apparently equivalent to the lowering of the activation.

DR. TAKASHIMA: Dr. Scheider, you are using square pulses having a discrete number of harmonics. Do you have a discrete number of experimental points, rather then a continuous curve? If you use the white-noise input, which has continuous frequency components, you would get a continuous dispersion curve instead of a discrete number of points.

DR. SCHEIDER: We have a continuous response; it is just how often can we measure it. As long as we measure it often enough to get a curve that gives as much information as we need, then that is not a limitation and one has as many points as one can use.

DR. T. ROSENBERG *(College of Physicians and Surgeons, Columbia University, New York, N.Y.)*: Dr. Scheider, is there information in the shape of a single pulse with respect to events that do occur in a microsecond range? Could you look at events that occur in the microsecond range as a homogeneous system in that time interval?

DR. SCHEIDER: It depends on how many microseconds you are talking about. You can repeat this experiment, but you have to have a quiescent period in between in order not to bias your next result. If you are talking about processes that happen at a scale of time of the relaxation of the molecule itself, then you are talking about steady-state fluctuations of dipole moment. Then one does not use the same technique, but one can use the same measurements. You can make calculations of what the effects of such fluctuations or changes would be on the measurement.

DR. ROSENBERG: That is what I was referring to. You had a system in which the chemical reaction was occurring in microseconds.

DR. SCHEIDER: What would happen is that you would superimpose chemical reactions on the orientation. As the capacitance changes, due to polarization, and if this were changing continuously after relaxation, you would simply be stretching out the relaxation time. You can make calculations of this, because it is just another relaxation process in essence.

DR. C. O'KONSKI: Dr. Scheider, you mentioned that the binding reactions of *N*-acetyl triptophane can be decomposed in two steps. I wonder if you can say just a little more about the observations in the time domain, which perhaps involve repeated pulses, and their analysis of the changes that were occurring in the conductivity and capacitance, which led you to decompose the process in the two steps?

DR. SCHEIDER: If you mix albumin solution with *N*-acetyl triptophane, the conductivity at time zero is very high because of hydrogen ions of triptophane. As time goes on, the albumin is capable of buffering all of the hydrogen ions, or most of them. It may bind the unionized *N*-acetyl triptophane, which may drive the ionized triptophane to recombine the hydrogen ion. We do not know exactly what happens, but we can enumerate the possibilities. If the ligands were not immediately tied to immobilize, there would be a very large addition to the final conductivity.

PART II. THEORETICAL AND TECHNICAL ASPECTS OF DIELECTRIC RELAXATION

DIELECTRIC THEORY AND PROPERTIES OF DNA IN SOLUTION

Robert H. Cole

Department of Chemistry
Brown University
Providence, Rhode Island 02912

INTRODUCTION

In this paper, we discuss three problems in dielectrics that are relevant to the use of dielectric measurements as a tool for the better understanding of the behavior of biopolymers in solution. In each case, the work is unfinished and will be published in detail elsewhere, hence the paper is in the nature of a progress report presented in the hope that the incomplete results to date will be of interest.

The first problem concerns the effect of electrostatic dipole interaction forces on the static dielectric constant (permittivity) of polar liquids, especially in relation to predictions based on the Onsager and Kirkwood equations. The second concerns what conclusions can be drawn from the differences in Kerr effect response (time-dependent birefringence) of a polar liquid following application or removal of a strong electric field. The third part is a discussion of experimental problems in measuring low-frequency dielectric relaxation of charged biopolymers in solution, with a description of an improved four-terminal method and some results for DNA solutions obtained by M. S. Tung and R. J. Molinari.

DIPOLE INTERACTIONS

Onsager's equation[1] for the dielectric constant ϵ of a polar liquid can be written

$$(\epsilon - \epsilon_\infty)\left(\frac{2\epsilon + \epsilon_\infty}{3\epsilon}\right) = \left(\frac{\epsilon_\infty + 2}{3}\right)^2 \frac{4\pi n\mu^2}{3kT} \tag{1}$$

where μ is the permanent dipole moment of the polar molecule, ϵ_∞ is related to the polarizability α by $(\epsilon_\infty - 1)(\epsilon_\infty + 2) = 4\pi n\alpha/3$, n is the number density of molecules, and kT has the usual meaning. It has long been recognized that this result is, so to speak, the best that can be done with use of macroscopic continuum arguments and is a useful approximate formula that works surprisingly well. Kirkwood[2] obtained an equally celebrated, formally more general, expression, one form of which can be written

$$\frac{\epsilon - 1}{\epsilon + 2} - \frac{\epsilon_\infty - 1}{\epsilon_\infty + 2} = \langle \mu^2 + \boldsymbol{\mu} \cdot \Sigma_i(\boldsymbol{\mu}_i + e\mathbf{r}_i)\rangle \frac{4\pi n\mu^2}{9kT} \tag{2}$$

In this equation, the effect of dipole interactions appears in the average correlation of a permanent dipole $\boldsymbol{\mu}$ with the sum of permanent and induced moments

$\mu_i + e\mathbf{r}_i$ of other molecules in a finite spherical sample. If this correlation is approximated by a macroscopic continuum, Equation 2 reduces to Onsager's result.

It has often been tactily assumed that purely dipolar forces between polar molecules lead to no average correlation effects other than those imposed by macroscopic boundary conditions at the surface of the sample, and hence that Onsager's Equation takes account of the short-range effects of such forces. However, theoretical treatments of two different models by M. Lax[3] and by Wertheim,[4] taking explicit account of dipole forces, do not support this conclusion. These results seem to be unknown to most dielectricians, partly because the predicted dependences of dielectric constant on dipole moment, density, and temperature seem to have been given only in part or not at all. Our purpose here is to call attention to these theories and present numerical calculations of some of their predictions with an indication of their implications.

The Spherical Model

In the Toupin and Lax[5] rigid cubic lattice model of polarizable permanent dipoles with moments $\mathbf{m}_i = \mathbf{u}_i + e\mathbf{r}_i$, the potential energy U of the system is given by

$$U = -\sum_i \mathbf{m}_i \mathbf{E}_0 - \sum_i \sum_{j \neq 2} \mathbf{m}_i \mathbf{G}_{ij} \mathbf{m}_j + \frac{1}{2\alpha} \sum_i (e\mathbf{r}_i)^2 \tag{3}$$

where $\mathbf{E}_0$ is the uniform field of external charges. The basic statistical mechanical problem is then to evaluate the partition function Z and from it the polarization $P = kT\partial \ell n\, Z/\partial E_0$. The former is given by

$$Z = \prod_i \int_{-\mu}^{\mu} d\mu_i \prod_j \int_{-\infty}^{\infty} dr_i \exp\,(-U/\mathrm{kT})$$

in which the phase-space integrations are over the restricted permanent dipole orientations, as symbolized by the limits $-\mu$ and μ, and induced dipole harmonic oscillator displacements r_i. Integration over the latter only presents no difficulty because, as Van Vleck[6] pointed out many years ago, one can reduce U to a sum of squares by changes of variable and then carry out the integrations over the new variables because their ranges are from $-\infty$ to $+\infty$. By this procedure, Van Vleck provided the first dynamic proof of the validity of the Clausius-Mossotti formula for this model of induced polarization.

Toupin and Lax accomplished the first part of a similar treatment with permanent dipoles included and approximated the 3n fold integrations over restricted orientations of n dipoles by the limits $-\infty$ to $+\infty$ and the "spherical" constraint $\sum_i \mu_i^2 = n\mu^2$, which, when represented by an integral delta function, permitted evaluation of Z and P. The normal coordinate transformation

$$\mu_i = \sum_k Y_k \exp\,(i\mathbf{k}\mathbf{R}_i)$$

introduces "dipole wave sums" λ_k given by

$$\boldsymbol{\lambda}_k = \sum_i \sum_j \mathbf{G}_{ij} \exp\left[i\mathbf{k}(\mathbf{R}_i - \mathbf{R}_j)\right] \tag{4}$$

where $\mathbf{R}_i$ is the lattice vector position of dipole $\boldsymbol{\mu}_i$.

Evaluation of the polarization results in the dielectric constant formula

$$\sum_k \left[\frac{3}{\epsilon - 1} + 1 - \frac{3}{4\pi}\boldsymbol{\lambda}_k\right] = \frac{4\pi}{3}\frac{n\mu^2}{3\mathrm{kT}} \tag{5}$$

Here, for simplicity, only the permanent dipole terms have been included, but the complete formula for $\alpha \neq 0$ presents no difficulty.

A remarkable feature of this result is that it reduces to Onsager's formula if the $\boldsymbol{\lambda}_k$ are approximated by their values $-8\pi/3$ (longitudinal) and $+4\pi/3$ twice (transverse) in the limit $k \rightarrow 0$. (As expected, setting all $\boldsymbol{\lambda}_k = 0$ and thereby neglecting dipole coupling altogether gives the Clausius-Mossotti formula.) To determine the effect of dispersion of the $\boldsymbol{\lambda}_k$ for finite k, Toupin and Lax considered explicitly only a simple cubic lattice by complicated analytical evaluations of the $\boldsymbol{\lambda}_k$ and sums involving the $\boldsymbol{\lambda}_k$.

We have used Cohen and Keffer's[7] numerical calculations of $\boldsymbol{\lambda}_k$ to evaluate the k sums in Equation 4 for the more realistic body-centered lattice and for a dipole continuum outside an "Onsager cavity" of radius a satisfying $4\pi a^3 n/3 = 1$ with the correct domain in k space. These are given graphically in *Predictions of the Models* after a brief description of Wertheim's treatment and result for a liquid model.

Wertheim's Mean Spherical Model Theory

Wertheim[4] considered the problem of evaluating dipole correlation effects as expressed in Equation 2 for a model of a dipolar fluid. In terms of the pair distribution function $g(r_{12})$, the contribution to be evaluated is proportional to the integral

$$\int d\Omega_1 \int dV_{12}\, \mu_1\, \mu_2\, e\, g(r_{12})$$

where the integrations are over orientations and vector separation r_{12} of dipoles μ_1 and μ_2, induced dipoles and interactions being neglected. The problem solved exactly is a generalization of the "mean spherical model" for hard-sphere fluids, which takes account of the dipole interaction term in Equation 3 for U. In this model, the direct correlation function $C(r_{12})$ is related to $g(r_{12})$ by a generalized Ornstein-Zernike equation

$$g(r_{12}) - 1 = C(r_{12}) + n \int dr_{13}\, C(r_{13})[g(r_{12}) - 1] \tag{6}$$

and the approximation of the model is taking $C(r_{12}) = -U(r_{12})/\mathrm{kT}$. Wertheim, by an elegant analysis, reduced the solution to that of the Percus-Yevick equation for hard spheres. The result for ϵ is given by the parametric equations

$$\frac{\epsilon - 1}{\epsilon + 2} = \frac{q(K\eta) - q(-K\eta)}{q(2K\eta) + 2q(-k\eta)} \tag{7}$$

$$\frac{4\pi}{3}\frac{n\mu^2}{3kT} = \frac{1}{3}[q(2K\eta) - q(-K\eta)]$$

where the second equation determines K in terms of $q(\eta) = (1 + 2\eta)^2/1 - \eta)^4$, obtained from solution of the Percus-Yevick equation with $\eta = \pi R_0{}^3 n/6$, R_0 being the hard-sphere radius. The dielectric constant is then obtained as a function of the reduced dipole temperature $T/T_c = 9kT/4\pi n\mu^2$ by assigning values of $K\eta$ in the range 0 to 1/2.

Predictions of the Models

In FIGURE 1, calculated values of ϵ from the results of the various theoretical models are plotted as a function of the reduced temperature T/T_c. It is seen that the dipole interaction models give values much closer to the Onsager result than the Lorentz field (Clansius-Mossotti) values. With the exception of the simple cubic lattice behavior at low temperatures (with a phase transition indicated by the arrow), the predicted values of ϵ are increasingly larger than from Onsager's Equation at lower temperatures. The similarity of the results for the Wertheim hard-sphere fluid and the "continuum" dipole lattice in the spherical model is also striking, and the body-center lattice values are also quite similar. (Incomplete calculations for the face-centered lattice indicate that the results will not be greatly different from those for the body-centered lattice).

The behavior of the various results at low temperatures is better seen from the plots of $1/(\epsilon - 1)$ against T/T_c in FIGURE 2. The continuum, body-, and face-

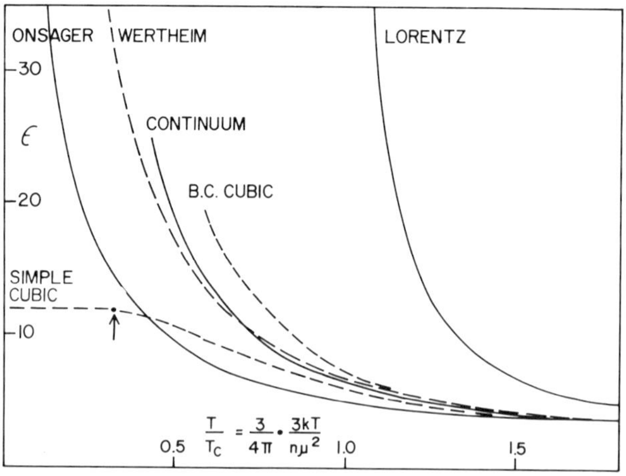

FIGURE 1. Dielectric constant ϵ versus reduced temperature T/T_c from theories of rigid dipole interaction effects discussed in text.

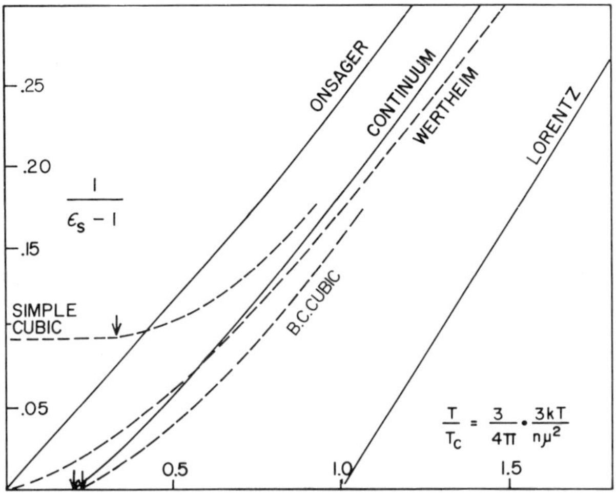

FIGURE 2. Reciprocal of $\epsilon - 1$ versus reduced temperature T/T_c from theories of rigid dipole interaction effects discussed in the text.

centered cubic lattice sums all indicate an infinite permittivity and phase transitions at $T/T_c \simeq 0.2$ to 0.25, i.e., at about one-fifth the Curie temperature for the Lorentz field "$4\pi/3$ catastrophe." These predictions and questions of the significance to be attached to them need further study. The hard-sphere fluid values show no transition; $\epsilon - 1$ approaches infinity only as $T \rightarrow 0$, and is more like the Onsager result in this respect. (But a little-known result of Pirenne[8] should be noted: below a critical temperature of order 0.1 T_c, the electrostatic equations of Onsager's model do admit a solution with finite spontaneous polarization in zero field). The simple cubic lattice behavior is quite different and associated with anomalous dispersion of the $\lambda(\mathbf{k})$ values and related lattice sums; this has been discussed by Fulton.[9] As the more closely packed structures are presumably more realistic for most systems, these peculiarities are unlikely to be of great interest.

For simplicity, we have given results only for rigid dipoles ($\alpha = 0$). Real molecules are, of course, polarizable with $(\epsilon_\infty - 1)/(\epsilon_\infty + 2)$ typically of order 0.25 to 0.5. The Toupin and Lax results for $\epsilon - 1$ with polarizability included is a bit tedious to work out but involves no real difficulty once the $\lambda(\mathbf{k})$ eigenvalues are known; calculations indicate that the principal effect is to increase the temperatures for a given ϵ by a factor $[(\epsilon_\infty + 2)/3]^2$ without major changes in shape of the curves in FIGURES 1 and 2. Necessary modifications to generalize Wertheim's analysis to include polarizability will require further study.

The principal conclusion from the approximate calculations of dipole interaction effects is seen to be that the interactions result in *larger* dielectric constants than predicted by Onsager's Equation or by Kirkwood's Equation if net correlations of a dipole with its neighbors are *assumed* to vanish. The Kirkwood g factor expressing the difference is correspondingly significantly greater than 1 for quite weak dipole forces, but detailed discussion of the implications should await more complete evaluations of the polarizability contributions. In any case, it should be clear that Onsager's Equation, or modifications of it from similar electrostatic continuum models, should be used with caution in attempts at quantitative calculations of ϵ from dipole moments or vice versa.

Transient Kerr Effect

Studies of anisotropy in electric polarization produced by a strong electric field can give useful information about molecular parameters and intermolecular forces, and if the field is time-dependent about the kinetics of molecular reorientation. One measure of the anisotropy is the difference in refractive indices parallel and at right angles to the orienting field $E(t)$. This can be determined by sending linearly polarized light through a sample in the x direction at right angles to $E(t)$ as shown in Figure 3, and evaluating the difference Δn of refractive indices n_z and n_y from the birefringence (elliptical polarization) it produces. For axially symmetric molecules with a difference $\Delta\alpha = \alpha_p - \alpha_s$ of polarizabilities α_p parallel and α_s perpendicular to the molecular symmetry axis at an angle Θ with the orienting field $E(t)$ along z, the refractive index difference Δn is necessarily an even function of $E(t)$, i.e., independent of the sign of $E(t)$.

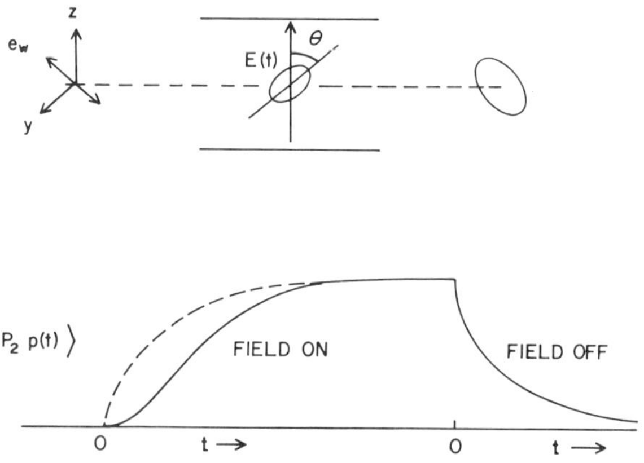

Figure 3. *Top:* Schematic arrangement of Kerr effect measurements. *Bottom:* Transient Kerr effect following application and removal of the polarizing field. The dashed curve is the mirror image of the decay function following removal of the field.

For not too intense fields, $\Delta n = KE^2$ and the Kerr "constant" K is related to $\Delta\alpha$ and degree of orientation induced by the field by an expression of the form

$$K = A\,\Delta\alpha < P_2(\cos\Theta)\rho(t) >$$

In this expression, numerical factors, number density of molecules, and instrumental constants have been represented by the factor A. (See Beevers *et al.*,[10] for example, for the complete expression.) The quantity of interest here is the average

$\langle P_2(\cos\Theta)\rho(t)\rangle$, where $P_2(\cos\Theta) = (3\cos^2\Theta - 1)/2$ is the second Legendre polynomial, resulting from the projection of the optical field components onto molecular axes and the projection of the moments they induce back onto laboratory axes, and $\rho(t)$, is the time-dependent distribution function.

In the equilibrium case for constant field E, the equilibrium distribution function ρ_E is given for polar molecules with dipole moment μ by

$$\rho_E = N \exp\left(-\beta\left[H_0 - \mu P_1(\cos\Theta)E - \frac{1}{3}\Delta\alpha P_2(\cos\Theta)E^2\right]\right) \quad \textbf{(8)}$$

where $\beta = 1/kT$, N is the normalization constant (partition function), H_0 is the Hamiltonian for $E = 0$, and for simplicity, intermolecular dipole forces have been neglected. The well-known solution[10] for the quadratic field dependence is

$$\langle P_2(\cos\Theta)\rho_E\rangle$$

$$= \left(\frac{1}{3}\beta\Delta\alpha \langle P_2^2 \exp(-\beta H_0)\rangle + \frac{1}{2}\beta^2 \langle P_2 P_1^2 \exp(-\beta H_0)\rangle\right) E^2$$

$$= \frac{1}{15}\beta(\Delta\alpha + \beta\mu^2)E^2, \quad \textbf{(9)}$$

showing the dependence of the effect on both $\Delta\alpha$ and μ^2.

For time-dependent orienting fields $E(t)$, the appropriate time-dependent distribution function $\rho(t)$ must be determined if the relation of the time-dependent birefringence to molecular reorientations is to be used to learn about the latter. Benoit[11] solved this problem for the model of forced rotational diffusion when a constant field E is either applied or removed at time $t = 0$. His results are

(a) Field removed at $t = 0$,

$$\langle P_2\rho(t)\rangle = \frac{1}{15}\beta(\Delta\alpha + \beta\mu^2)E^2 \exp(-6Dt) \quad \textbf{(10)}$$

(b) Field applied at $t = 0$

$$\langle P_2\rho(t)\rangle = \frac{1}{15}\beta(\Delta\alpha + \beta\mu^2)E^2[1 - \exp(-6Dt)]$$

$$- \frac{1}{10}\beta^2\mu^2E^2[\exp(-2Dt) - \exp(-6Dt)]$$

where D is the rotational diffusion coefficient.

The remarkable feature of Benoit's results that has made the transient Kerr effect of such interest is that the response and decay curves lack the symmetry of linear polarization effects. As shown in FIGURE 3, the rising response is slower than the decay, and for the diffusion model the decay is associated entirely with the exponential loss of correlation of P_2 (cos Θ) with rate constant $6D$. The response, on the other hand, is not just the mirror image of the decay of P_2 correlation. The extra effect results from a difference of the P_2 and P_1 correlations, with the latter decaying more slowly (rate constant $2D$), and clearly arises from the second-order effect of dipole orientation energy $-\mu\cos\Theta E$ and corresponding torque, as it is proportional to μ^2.

The rotational diffusion model can thus be tested for a given system by determining whether the polarization does decay exponentially, and if it does, the relative importance of permanent and induced dipole polarizations can be determined if the difference between response and decay of polarization is consistent with the model. The questions we consider here concern what can be inferred more generally, without a specific model, from the observed transients. Thus if the decay function is not exponential, as it often is not, is it still true that this function is determined only by $P_2(\cos\Theta)$ time correlation, and is it possible to infer relative magnitudes of permanent and induced dipole contributions with useful information about $P_1(\cos\Theta)$ time correlation from the difference of response and decay functions? This sort of problem should be amenable to more general nonequilibrium theory, but presents difficulties not present in linear response theory because interactions of the molecules with the applied field and with the lattice or heat bath become coupled in second- and higher-order approximations that are necessary to take account of permanent dipole torques and energies. An analysis of this kind has not yet been fully carried out, but enough progress has been made to permit some useful conclusions, and a brief outline of the present state can be given.

Although either classical or quantum treatments should be possible, the latter has advantages, even for the classical limit, of greater ease of manipulation. The distribution function is then the density matrix $\rho(t)$ satisfying $d\rho(t)/dt = -(i/\hbar)[H(t), \rho(t)]$, where the bracket is the commutator and $H(t)$ the time-dependent Hamiltonian. This can be taken to be

$$H(t) = H_0 + H_c(t) - \mu P_1(\cos\Theta)E(t) - \frac{1}{3}\Delta\alpha P_2(\cos\Theta)E^2(t) \qquad \textbf{(11)}$$

where it is important to recognize that the Hamiltonian for zero field is not just the time-independent H_0 appearing in the Boltzmann factor $\exp(-\beta H_0)$, but also includes coupling to the "lattice" or heat bath denoted by $H_c(t)$. This is assumed, as in spin-lattice NMR theory, for example,[12] to have zero-time average and only short-time correlation such that the time average $\overline{H_c(t)H_c(t+\tau)} \rightarrow$ constant for time $t \gg \tau$ of interest. General perturbation solutions for $\rho(t)$ through terms in $E_2(t)$ present considerable complication, but there is more hope for the simpler cases that a constant field E is applied or removed at $t = 0$.

If the field E is removed at $t = 0$, then $\rho(t)$ relaxes to the final equilibrium ρ_0 as

$$\frac{d}{dt}(\rho - \rho_0) = -(i/\text{h})[H_0 + H_c(t), \rho - \rho_0] - (i/\text{h})[H_c(t), \rho_0]$$

with the initial condition $\rho = \rho_E$ given above. Because the field-dependent terms in $H(t)$ are not involved in the rate equation, the solution is readily obtained and expressed in terms of the "natural motions" of $P_1(t)$ and $P_2(t)$ satisfying

$$\frac{dP}{dt} = +(i/\hbar)[H_0 + H_c(t), P].$$

The result for the relaxation function of interest is

$$<P_2\rho(t)> = \frac{1}{3}\beta\Delta\alpha <P_2P_2(t)\exp(-\beta H_0)> E^2$$

$$+ \frac{1}{2}\beta^2\mu^2 <P_2P_1^{\,2}(t)\exp(-\beta H_0)> E^2$$

$$= \frac{1}{15}\beta(\Delta\alpha + \beta\mu^2)\phi_2(t)E^2$$

where $\phi_2(t)$ is the normalized time-dependent correlation function for $P_2(\cos\Theta)$. The result is then just the generalization of the diffusion model solution (a) that one might expect from replacing $\exp(-6Dt)$ by $\phi_2(t)$.

If the field E is applied at $t = 0$, the relaxation of $\rho(t)$ to its final equilibrium value ρ_E is given by

$$\frac{d}{dt}(\rho - \rho_E) = -(i/\mathrm{h})[H_0 + H_c(t) - \mu P_1 E - \frac{1}{3}\Delta\alpha P_2 E^2, \rho - \rho_E]$$

$$-(i/h)[H_c(t), \rho_E]$$

with $\rho = \rho_0$ at $t = 0$. The motion is now perturbed by the field, complicating the second-order perturbation methods considerably. We have so far succeeded in obtaining a partial solution only of the form

$$<P_2\rho(t)> = \frac{1}{15}\beta(\Delta\alpha + \beta\mu^2)(1 - \phi_2(t)E^2$$

$$- \frac{1}{15}\beta\mu^2\left[\int_0^t dt'\phi_2(t')\phi_1(t - t') + F(t)\right]E^2$$

where the function $F(t)$ in the second term has so far resisted reduction to any simple form in terms of $\phi_2(t)$ and $\phi_1(t)$ for the time-dependent correlation of $P_1(\cos\Theta) = \cos\Theta$. Comparison with the diffusion solution (b) above shows that the first term is the generalized relaxation of $P_2(t)$, whereas the second involving both $P_2(t)$ and $P_1(t)$ arises from permanent dipole polarization only.

At this stage of development, the response-theory treatment thus shows that the decay function depends only on $P_2(t)$ relaxation, but the response function depends on both the mirror image of this relaxation plus terms depending on response of permanent dipole polarization. One can thus answer the question of whether permanent dipole polarization is present by comparing the response and decay curves, but the answer to the question of whether analysis of the difference between the two curves can give useful information about both kinds of relaxation must await further analysis of response-function integrals.

Low-Frequency Relaxation of Biopolymers in Solution

Although dielectric relaxation processes at audio- and subaudiofrequencies have been known or inferred for many biopolymers of high molecular weight in aqueous solution and may arise from the major contributions to orientational polarization, relatively little is known about these processes in many systems because of experimental difficulties.

There are four related kinds of difficulty. The first is that the dispersion and absorption currents for a given applied alternating field decrease with frequency of the field, requiring high-current sensitivity, low-frequency measurements. This problem is not too serious by itself, but becomes much more so if there is appreciable ionic conductance of the solution, since ion currents for a given field are independent of frequency and hence increasingly dominant dielectric currents by as much as a factor of a million at frequencies of a few hertz. The third complication is from space charge or electrode polarization if the metal electrodes are partly or wholly blocking for ion-electron charge transfer, as they usually are; this increasingly acts to reduce the polarizing field in the body of the solution at low frequencies. Finally, biopolymer solutions often fail to reach a completely steady state, with a consequent drift of real and apparent conductance with time.

Four-Terminal Null Methods

These various difficulties can be overcome, to a large extent, and the range of useful measurements extended to much lower frequencies by making use of AC four terminal methods. In these, currents are driven in the solution between two current electrodes across which an alternating field is applied, as shown in FIGURE 4, while the potentials at two points in the solution with negligible electrode space charge are picked up by probe electrodes, which ideally do not draw current

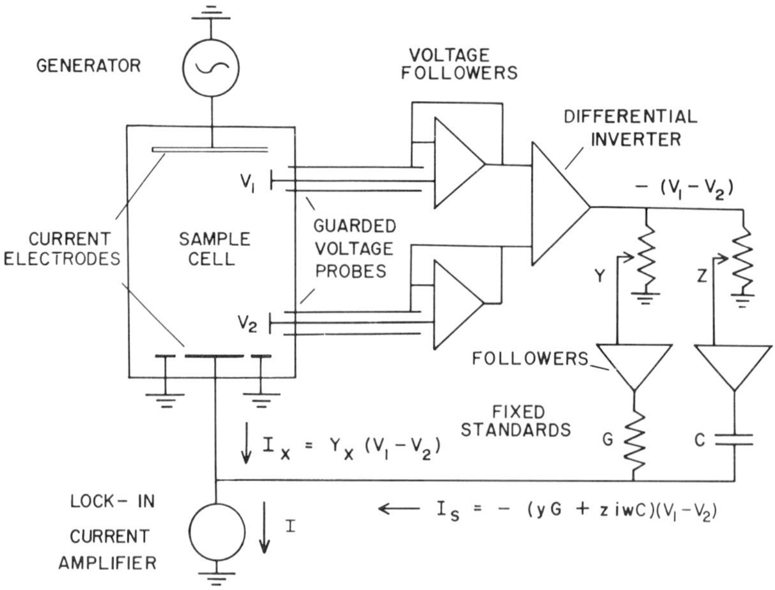

FIGURE 4. Simplified schematic diagram of cell and electronic instrumentation for four-terminal null measurements of dielectric properties of conducting solutions.

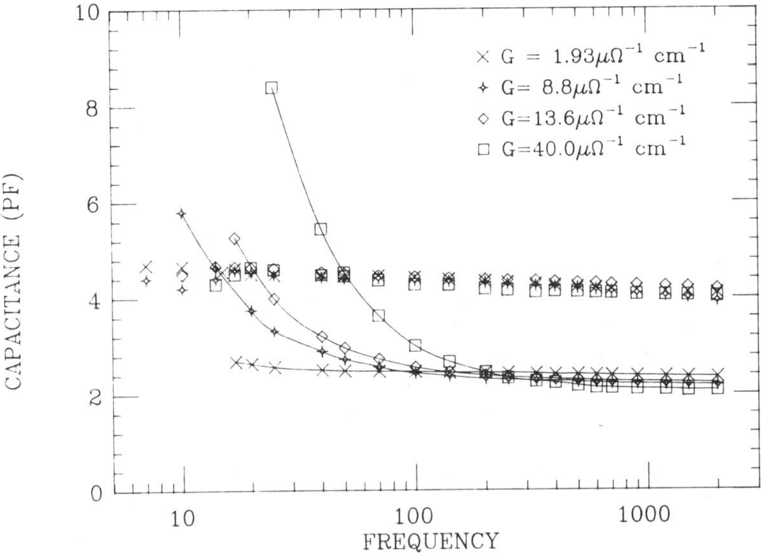

FIGURE 5. Uncorrected calibration data. Measured capacitances of acqueous KCl solutions of four specific conductances as a function of frequency. The frequency-independent values are from four-terminal null measurements; the curves from three-terminal bridge measurements show increases at low frequency as a result of electrode polarization.

or otherwise distort the current flow. For such conditions, the ratio of the current to the probe potential difference is directly proportional to the sample admittance $y = i\omega\epsilon_t^*$, where ϵ_t^* is the "total permittivity" usually and somewhat arbitrarily written as the sum of ohmic conduction and dielectric currents: $\epsilon_t^* = \epsilon^* + \sigma/i\omega E$, with E a conversion factor for consistent units.

The four-terminal method can reduce electrode polarization errors enormously, leaving the problem of measuring small currents of interest when much larger ionic currents are superposed. Various electronic null methods to accomplish this have been proposed, with sensitivity and resolution increasing as more sophisticated solid-state devices become available. A four-terminal "bridge" for the purpose, originally developed by Berberian and Cole[13] and since refined considerably by R. J. Molinari, is shown schematically in FIGURE 4. Ideal voltage probes are approximated by small platinum electrodes in the solution connected to voltage followers with very small FET input currents (10^{-13} amp) by coaxial lines with driven shields to virtually eliminate probe to ground stray capacitances. Variable fractions x and y of the inverted difference $V_2 - V_1$ of probe potentials are applied to a standard capacitance C and conductance G producing current I_s, which by adjustment of x and y balances the solution current I_x. For this condition, $I = I_x + I_s = 0$ as determined by the null detector and the sample admittance $Y = I/(V_1 - V_2)$ between the probes is given by $Y = x(i\omega C) + yG$.

For this scheme to work successfully down to a few Hertz or less, care must be taken to determine and compensate or correct for extremely small phase shifts. The degree to which this has been accomplished is shown by results plotted in

FIGURE 5 for water and KCl solutions of several specific conductances, as indicated. The measured capacitances by the four-terminal method are virtually independent of frequency down to 5 Hz or less, whereas the values by three-terminal measurements (probe input circuits connected to generator and detector) show the familiar rapid rise at low frequencies and increasing with solution conductance.

The necessary sensitivity and phase discrimination in determining the null condition is realized satisfactorily by a PAR-124A lock-in amplifier, and the only serious limitation of the method is difficulty in balancing drifting ionic conductances of biopolymer solutions and determining the small components attributable to dielectric absorption. In this respect, the ingenious but electronically rather complicated scheme of Hayakawa *et al.*[14] for automatic balance of conductance at a fixed reference frequency and determination of conductance differences at other frequencies has significant advantages.

The four-terminal method just described has been used in a series of studies of aqueous DNA solutions by M. S. Tung and R. J. Molinari. We outline here some of the results to date, which are being published in more detail elsewhere.[15] In this work, three samples of calf thymus DNA as the sodium salt were used, with stated or inferred average molecular weights of 1.4, 2, and 9×10^6 and unknown but doubtless broad distributions. The principal interests in these studies were the effect of added salt and temperature variation on the limiting low-frequency, steady-state permittivity for comparison with the theoretical treatment of counterion fluctuation polarization by McTague and Gibbs.[16]

The observed dielectric behavior for a 3.8×10^{-4} solution of average molecular weight 2×10^6 at 25° is shown in FIGURE 6, together with the effect of added NaCl (4×10^{-5}N). It is seen that the relaxation is within experimental error described by the circular arc function $\epsilon^* = \epsilon_0 + (\epsilon_0 - \epsilon_\infty)/[1 + (i\omega\tau)^\beta]$[17] with β of order 0.75 and central relaxation frequency 10 Hz. These values are both lowered somewhat by added NaCl, but the principal effect is the decrease of extrapolated static permittivity ϵ_0 by about 40 percent. The effect of divalent positive ions is even more marked, a 1.5×10^{-4} normality of Mg^{++} ion reducing the increment in ϵ_0 over that of water virtually to zero.

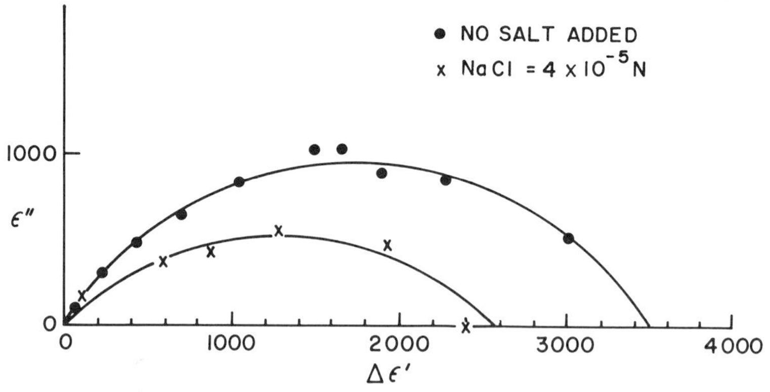

FIGURE 6. Complex plane loci of dielectric constants of salt-free DNA solution and DNA solution with added NaCl.

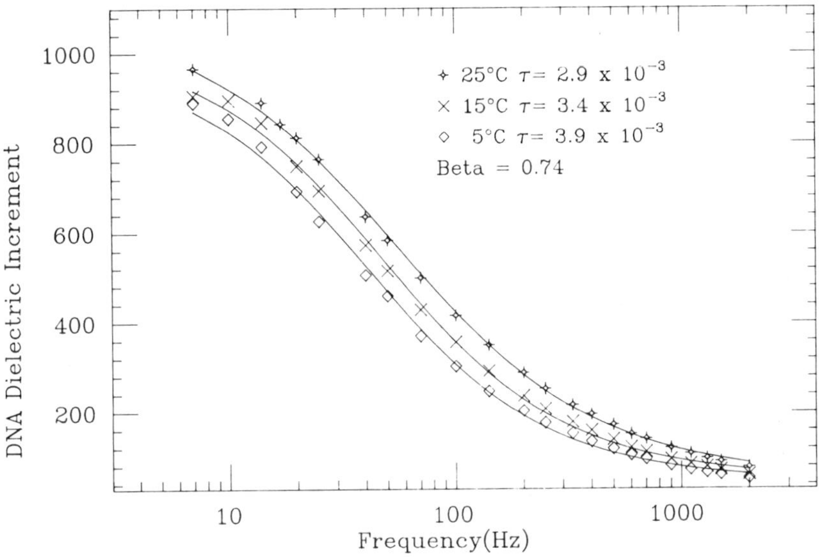

FIGURE 7. Dispersion curves of ϵ' versus frequency for DNA solutions (molecular weight 9×10^6, concentration 6.2×10^{-4} M) at 5°, 15°, and 25°. The solid curves are circular arc functions with $\beta = 0.74$.

The effect of temperature on the dispersion of ϵ' for solutions of concentration 6.2×10^{-4}M and average molecular weight 9×10^6 is shown in the plots of ϵ' versus frequency at 25°, 15°, and 5° in FIGURE 7. These have been computer least squares fitted by the solid curves for circular arc functions, with $\beta = 0.74$ having no significant temperature-dependence in this limited range. The static permittivities from the fitted curves are plotted against temperature in FIGURE 8 and show a significant *positive* temperature coefficient, in contrast to the negative coefficient ($1/T$ temperature-dependence) characteristic of permanent dipole polarization.

That these large low-frequency polarizations are characteristic of the helical form of DNA is shown by the fact, established by several other studies as well as these, that partial denaturation by heating to temperatures above 60° reduces their magnitude markedly, heating to 95° for 15 minutes sufficing to reduce the dielectric increments observed at room temperature virtually to zero.

The evidence just cited and other measurements are qualitatively consistent with explanations of the low-frequency polarization in terms of field-induced displacements of counterions attached to a fraction of available negative charge sites on the DNA helix. This fraction (one minus the charge fraction) is expected to be somewhat greater than one-half for the salt-free DNA samples at the concentrations used studied. The McTague-Gibbs Theory predicts, as one would expect intuitively, that the static polarization is a maximum for half the sites occupied, decreasing to zero for none or all filled. One then expects that added positive counterions will decrease the polarization as observed because more sites are occupied, whereas an increase in temperature will reduce the number and increase the polarization, again as observed.

The magnitude of polarization predicted by the McTague-Gibbs Theory is also in at least qualitative agreement with the observed values. For an estimated mean square end-to-end length of 3400 Å for DNA of molecular weight 2×10^6, assuming a contour length of 10,000 Å and persistence length of 600 Å, the theory predicts $\Delta\epsilon/c = 3.8 \times 10^6$, which is comparable with the observed value of 9.8×10^6 at a concentration of 0.93×10^{-3} molar.

The measurements of Tung and Molinari also indicate an increase in the specific dielectric increment $\Delta\epsilon/c$ as the concentration is reduced. This is consistent with a decrease in fraction of occupied sites on dilution and increase with polarization per molecule, but is somewhat ambiguous because the concentrations are rather high, with appreciable concentration-dependent solute interactions likely as a result and because slow denaturation after dilution was also observed.

The frequency dependence of the low-frequency relaxation process differs con-

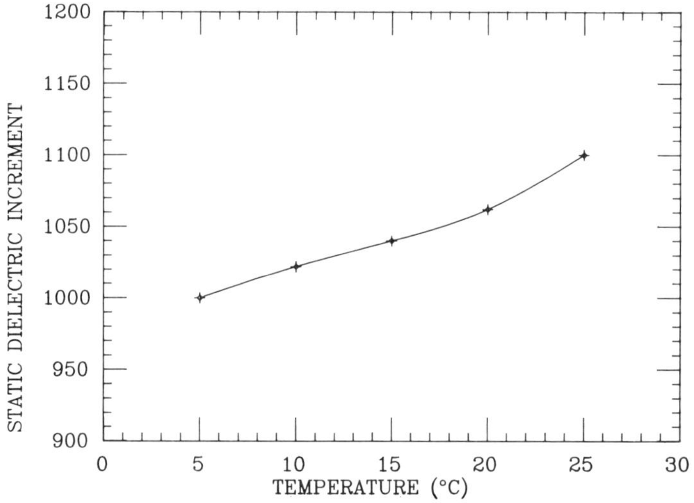

FIGURE 8. Static dielectric constants from the data plotted in FIGURE 7 as a function of temperature.

siderably from simple Debye behavior for all the DNA solutions so far studied by Tung, Molinari, and other workers. It does not seem possible to conclude with any assurance the extent to which the rather broad frequency range of relaxation is the result of heterogeneity in size, concentration-dependent solute interactions, cooperative counterion motions, or other effects, and more extensive studies will be required to throw light on such questions. The average or central relaxation frequencies in the subaudiorange are at least of the order predicted by kinetic theories of counterion fluctuations,[18-20,22] and one can secure agreement of their predictions with observed values by assigning plausible values to such parameters as counterion jump rates or mobilities appearing in the theories, but the question of whether real significance can be attached to such values or the models is unanswered as yet.

The results summarized above agree with measurements of other workers at higher frequencies and those of Sakamoto *et al.*[21] at room temperature to the extent comparison is possible for samples of different origins, molecular weights, and weight distributions. They are consistent with predictions from models of counterion fluctuation polarization, which appear to be more satisfactory from several points of view than alternative explanations that have been put forward. The principal difficulty in reaching definitive conclusions is that DNA samples of high molecular weights and broad distribution of molecular weights have been studied at rather high concentrations, with no clear indications of effects of polydispersity and solute interactions. Extensive measurements are needed on better and more sharply characterized samples over wide ranges of concentrations, added salt, temperature, and so on, if more quantitative conclusions are to be reached.

References

1. Onsager, L. 1936. J. Am. Chem. Soc. **58:** 1486.
2. Kirkwood, J. G. 1939. J. Chem. Phys. **7:** 911.
3. Lax, M. 1952. J. Chem. Phys. **20:** 1351.
4. Wertheim, M. S. 1971. J. Chem. Phys. **55:** 4291.
5. Toupin, R. A. & M. Lax. 1957. J. Chem. Phys. **27:** 458.
6. Van Vleck, J. H. 1937. J. Chem. Phys. **5:** 556.
7. Cohen, M. H. & F. Keffer. 1954. Scientific Paper 60-94469-2P2, Westinghouse Research Laboratories.
8. Pierenne, J. 1949. Helv. Phys. Acta **22:** 479.
9. Fulton, R. L. 1975. J. Chem. Phys. **62:** 3676.
10. Beevers, M., J. Crossley, D. C. Garrington, & G. Williams. 1976. J. Chem. Soc. Faraday II **72:** 1482.
11. Benoit, H. 1951. Ann. Physique **6:** 561; 1952. J. Chim. Phys. **49:** 517.
12. Slichter, C. P. 1963. Principles of Magnetic Resonance, Chapter 5. Harper and Row. New York, N.Y.
13. Berberian, J. G. & R. H. Cole. 1969. Rev. Sci. Inst. **40:** 811.
14. Hayakawa, R., H. Kanda, M. Sakamoto, & Y. Wada. 1975. Jap. J. App. Phys. **14:** 2039.
15. Tang, M. S., R. J. Molinari, R. H. Cole & J. H. Gibbs. Submitted to Biopolymers.
16. McTague, J. P. & H. H. Gibbs. 1966. J. Chem. Phys. **44:** 4295.
17. Cole, K. S. & R. H. Cole. 1941. J. Chem. Phys. **9:** 341.
18. Oosawa, F. 1970. Biopolymers **9:** 677.
19. Schwartz, G. 1962. J. Phys. Chem. **66:** 2646.
20. Mandel, M. 1961. Mol. Phys. **4:** 489.
21. Sakamoto, M., H. Kanda, R. Hayakawa & Y. Wada. 1976. Biopolymers **15:** 879.
22. van der Touw, F. & M. Mandel. 1974. Biophysical Chemistry **2:** 218.

DIELECTRIC PROPERTIES OF CHARGED LINEAR MACROMOLECULES WITH PARTICULAR REFERENCE TO DNA*

M. Mandel

Department of Physical Chemistry, Gorlaeus Laboratories
State University of Leiden
Leiden, The Netherlands

Since the classic work of Lorenz and Debye, it is well known that information about molecular structure and molecular behavior may be obtained through the study of the dielectric properties of pure systems or solutions. It is not surprising, therefore, that a large number of different classes of substances have been investigated by this physical-chemical method, the techniques of which have been highly developed in the last thirty years to cover a larger frequency range with always-increasing accuracy. During this period it has also become possible to study the dielectric properties of conducting aqueous solutions, particularly aqueous solutions of charged macromolecules, of both synthetic and natural origin.

The dielectric behavior of synthetic charged macromolecules (polyelectrolytes) in aqueous solutions differs considerably from the behavior observed with solutions of ordinary electrolytes as well as with solutions of uncharged macromolecules in both the nature and the amplitude of the effects observed. This specificity corresponds to the specificity observed for many physical-chemical properties of polyelectrolyte solutions that can be understood, at least qualitatively, if it is realized that in such systems a nonadditive combination occurs between macromolecular and electrolyte behavior, the former characterized by the importance of molecular dimensions in conjunction with conformational statistics, the latter by the importance of strong long-range electrostatic interactions. Most strikingly, these specific properties are exhibited in dilute solutions of highly charged macromolecules in the absence or at low concentrations of additional, low molecular, salt. Because of intramolecular charge repulsions, the average conformation of polyelectrolytes will deviate considerably from the ideal coillike one, and, in fact, successful theories explaining the thermodynamic properties of polyelectrolyte solutions often start from a rodlike model for the polyion.[1-4] As a consequence of the strong intermolecular interactions between the charged macromolecule and the counterions in solution, the phenomenon of *association* is observed. This implies that on the average, a certain fraction of the counterions stays in the close neighborhood of the macromolecular chain, being unable to move independently through the solutions and thus reducing the effective charge on the polyion. (The occurrence of associated counterions has been predicted theoretically and has also been demonstrated experimentally by independent experimental methods.[1-5]) Both phenomena are important in understanding the dielectric properties of charged, synthetic polyelectrolytes, as we shall see in what follows.

*These studies have been carried out under the auspices of the Netherlands Foundation for Chemical Research (SON) and with financial aid from the Netherlands Organization for the Advancement of Pure Research (ZWO).

For natural, charged macromolecules in aqueous (salt) solutions the situation may be much more complicated, particularly if many different charged groups occur along the chain and very specific interactions are responsible for highly specific average conformations, such as in globular proteins. The double-stranded DNA molecule at neutral pH, however, has been shown to exhibit physical chemical properties in aqueous salt solutions that often can be satisfactorily understood in terms of polyelectrolyte behavior.[6] It is the purpose of this paper to demonstrate that the dielectric properties of DNA also can be explained in the same manner as those of synthetic, charged linear macromolecules.

The investigation of the dielectric properties of DNA probably goes back to the

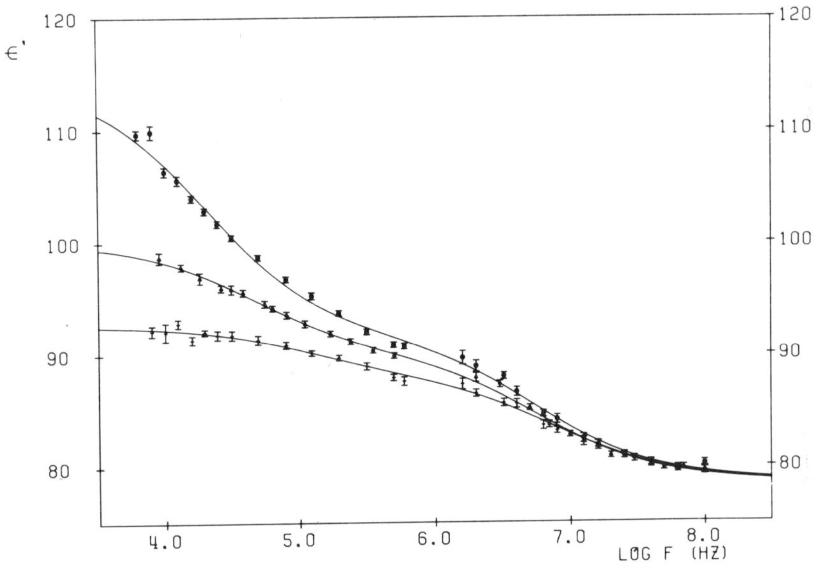

FIGURE 1. Molecular weight-dependence of the electric permittivity ε versus the frequency F, for poly(acrylic acid), degree of dissociation $\alpha = 0.4$, $C_p = 5.2 \times 10^{-3}$ monomol l^{-1}. Degree of polymerization $DP = 640$ (+), 1530 (▲) and 7500 (○). (By permission of the publishers of Biophys. Chem.[21])

work of Jugner and coworkers in 1940–1950, and since then a reasonable number of papers have been devoted to this topic (some papers are given by References 7–16), but with contradicting conclusions. In most cases high molecular weight DNA($M > 10^6$) has been used and the dielectric properties studied either below or above 1 MHz. In our laboratory, we recently investigated Na-DNA of $M = 3 \times 10^5$ (obtained by sonication of calf thymus Na-DNA) in a frequency range stretching from roughly 10 kHz to 100 MHz. (For the experimental methods used see References 15–17.) The results obtained showed a complete analogy to what is observed with synthetic linear polyelectrolytes of comparable molecular weights in aqueous solution studied in the same frequency range.

Let us first summarize the experimental results obtained so far with synthetic linear polyelectrolytes of sufficiently high-chain charge densities in salt-free solutions with simple monovalent counterions at room temperature.[18] At frequencies below 100 MHz, such systems show an appreciable concentration and frequency-dependent increment of the electric permittivity with respect to pure water, even at low concentrations (10^{-3} monomol l^{-1} or smaller), and generally two dispersion regions. Both the amplitude of the low-frequency dispersion region and its critical frequency $f_{c,1}$ (and thus also the mean relaxation time $\tau_1 = 1/2\pi f_{c,1}$) depend on the *molecular weight*, the amplitude increasing and the critical frequency decreasing with increasing degree of polymerization (FIGURE 1). The high-frequency

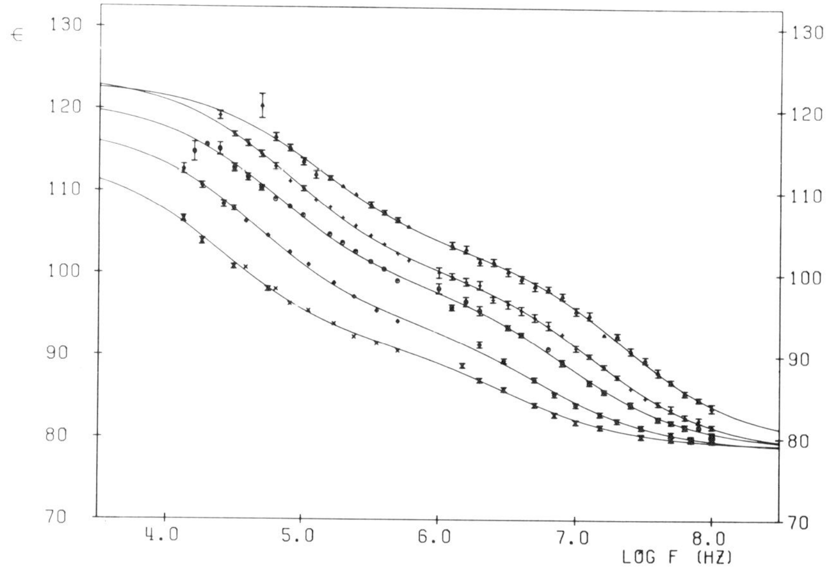

FIGURE 2. Concentration dependence of the electric permittivity versus the frequency for Na-polystyrenesulphonate, $DP = 500$; $C_p = 1.5 \times 10^{-3}(x)$, 3×10^{-3} (◇), 6×10^{-3}(○), 10^{-2} (+) and 2×10^{-2} (▲) monomol l^{-1}. (By permission of the publishers of Biophys. Chem.[21])

dispersion curve, at the end of which (near 100 MHz) the permittivity has reached the value of the pure solvent, does not depend appreciably on this parameter (as has been pointed out by Sachs *et al.*[19]). The electric permittivity depends considerably on the macromolecular concentration C_p (FIGURE 2). The specific increments $(\Delta\epsilon_s/C_p)$ and $(\Delta\epsilon_2/C_p)$ are furthermore not constant down to concentrations below 10^{-3} monomol l^{-1}, probably indicating strong intermolecular interactions; here $\Delta\epsilon_s = \epsilon_s - \epsilon_s^0$ is the static dielectric increment of the solution with respect to the static electric permittivity ϵ_s^0 of the solvent, and $\Delta\epsilon_2$ stands for the maximum amplitude of the high-frequency dispersion curve. (It should be noted that so far we have been unable to measure electric permittivities

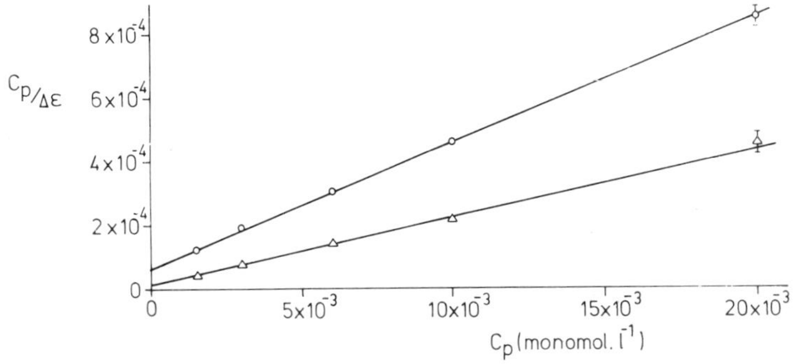

FIGURE 3. Concentration-dependence of the specific increments for Na-polystyrenesulphonate, $DP = 500$; $C_p/\Delta\epsilon_s$ (Δ) and $C_p/\Delta\epsilon_2$ (0). (By permission of the publishers of Biophys. Chem.[21])

of conducting solutions below 1 kHz and in our laboratory, the value for the static electric permittivity ϵ_s sometimes has to be obtained by an extrapolation procedure, using analytical expressions for the dispersion curve such as the Cole-Cole Equation,[20] which takes into account a certain distribution of relaxation times as observed experimentally. Such a procedure neglects the possible occurrence of a third dispersion region at very low frequencies, for which no experimental evidence has been found.) It has been demonstrated[21] that the concentration dependence of the specific increments can be represented by the following empirical equation,

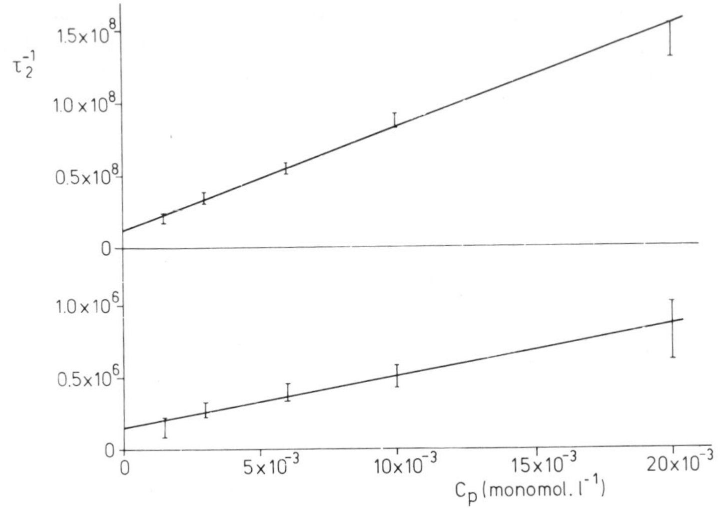

FIGURE 4. Concentration-dependence of the mean relaxation times τ_1 and τ_2 for Na-polystyrenesulphonate, $DP = 500$. (By permission of the publishers of Biophys. Chem.[21])

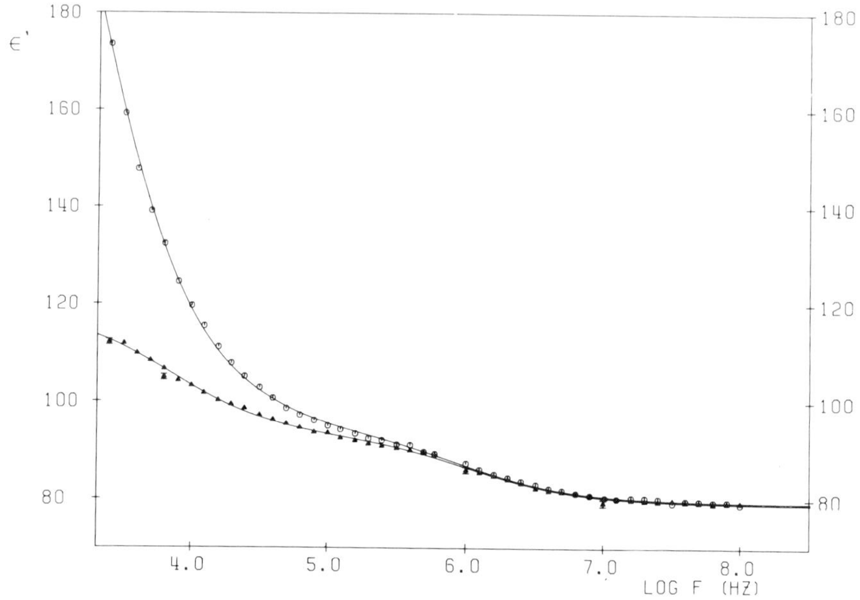

FIGURE 5. Molecular weight-dependence of the electric permittivity versus the frequency for calf thymus Na-DNA in aqueous solution, $C_e = 1$ eq.m^{-3}; native sample (0) and sonicated sample (Δ).

$$(\Delta\epsilon_i/C_p) = (\Delta\epsilon_i/C_p)_0/1 + B_i C_p \quad \textbf{(1)}$$

where $\Delta\epsilon_i$ stands for either $\Delta\epsilon_s$ or $\Delta\epsilon_2$, and $(\Delta\epsilon_i/C_p)_0$ represents the specific increment at infinite dilution and B_i a (constant) interaction parameter. Therefore, *linear* extrapolations can be performed by plotting $C_p/\Delta\epsilon_i$ against C_p (FIGURE 3) The same seems to be true for the mean relaxation times τ_1 and τ_2 of the low-frequency and high-frequency dispersion curves, respectively (FIGURE 4).

$$\tau_i = (\tau_i)_0/1 + B_i' C_p \quad \textbf{(2)}$$

These extrapolation procedures make it possible to obtain dielectric parameters referring to an infinite dilution situation in which the interactions between polyions, in principle, may be neglected.

The same features have been observed in our laboratory with solutions of calf thymus Na-DNA in the same frequency range of 10 kHz–100 MHz. Here again, a concentration-dependent dispersion curve has been obtained exhibiting the two distinct low- and high-frequency regions observed with synthetic polyelectrolytes. Most striking is the molecular weight dependence as found for natural and sonicated Na-DNA in aqueous solutions, respectively; there is a considerable overlap in the high-frequency end of the dispersion curve for both samples, but in the lower frequency region the electric permittivities definitely diverge, emphasizing the molecular weight dependence of both the amplitude and the critical frequency of the low-frequency dispersion curve for DNA (FIGURE 5). This is in agreement with previous findings of Takashima[9] for DNA in aqueous NaCl solutions in the frequency range 50 Hz to 200 kHz. It prompted us to concentrate our efforts on sonicated calf thymus DNA of relatively low molecular weight in view of the lower frequency limit of our measuring equipment. Full details of these measurements will be published elsewhere.[22] Results quoted below all refer to sonicated Na-DNA, with $M_w = (3.0 \pm 0.2) \times 10^5$ as found from light-scattering determinations, in aqueous 3×10^{-4} M NaCl solutions. (Salt has been added to reduce denaturation of the DNA at low C_p, although the dielectric behavior of DNA in pure water has been found not to differ markedly from that at the low salt concentration.) The concentration dependence of the dispersion curves is quite analogous to what is found with synthetic linear polyelectrolytes (FIGURE 6), and the extrapolation of the specific increments and the relaxation times to infinite dilution can be performed according to Equations 1 and 2, respectively (FIGURES 7 and 8).

Having thus established the similarity between the dielectric behavior of sonicated calf thymus Na-DNA and simple polyelectrolytes in the frequency range studied, let us now turn to the interpretation of these results. For synthetic linear charged macromolecules, it has become increasingly obvious that $\Delta\epsilon_s$ is not primarily determined by the (average) permanent dipole moment of the macromolecular chain. The increments observed would lead to extremely high values of the dipole moment, which cannot be accounted for by the structure of these polyelectrolytes. Also, $\Delta\epsilon_s$ becomes vanishingly small if the charge on the macromolecule tends to zero, as can easily be checked with weak polyelectrolytes of

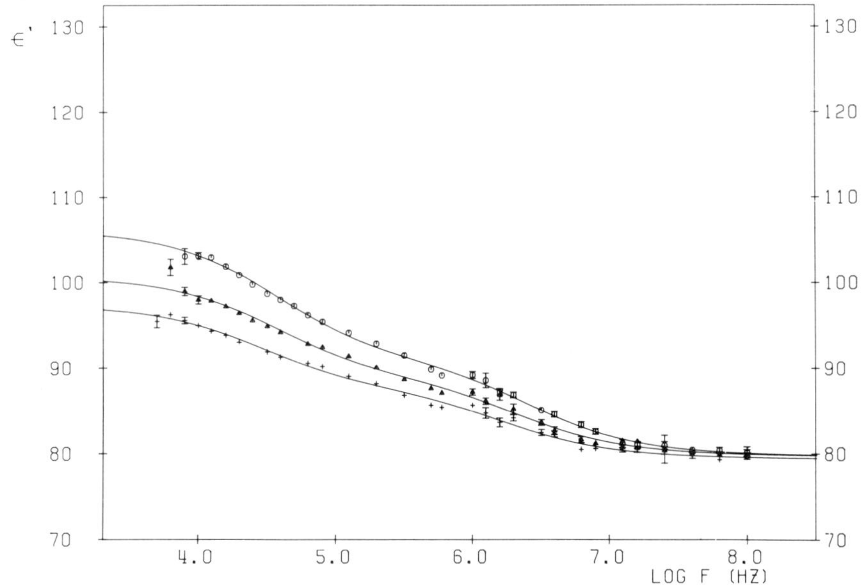

FIGURE 6. Concentration-dependence of the electric permittivity versus the frequency for sonicated calf thymus Na-DNA ($M_w = 3 \times 10^5$) in 3×10^{-4} M NaCl; C_e = 1.5 (0), 0.9 (Δ), and 0.6 (+) eq.m^{-3}

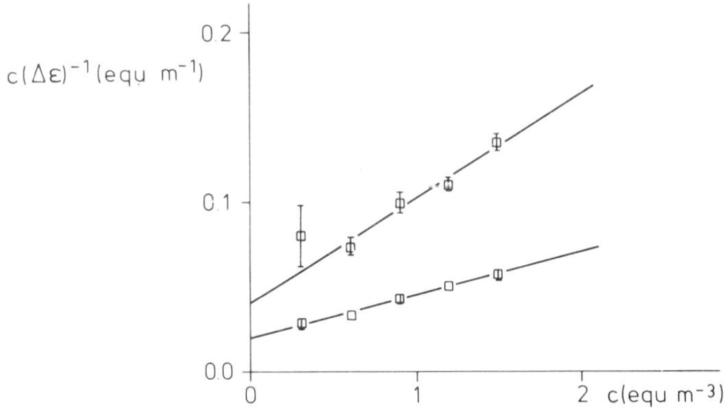

FIGURE 7. Concentration-dependence of the specific increments for sonicated calf thymus DNA ($M_w = 3 \times 10^5$) in 3×10^{-4} M NaCl. Upper curve for $C_e/\Delta\epsilon_2$ and lower curve for $C_e/\Delta\epsilon_s$.

which the charge can be controlled by addition of either strong low-molecular base or acid. (Additionally, for DNA its structure as a double helix warrants the absence of a total, permanent dipole moment.) Therefore, the dielectric effects observed are rather attributed to large, *induced dipole moments* originating in the distribution of strongly associated counterions around the polyions. It is assumed that these associated counterions may move more or less freely along the stretched

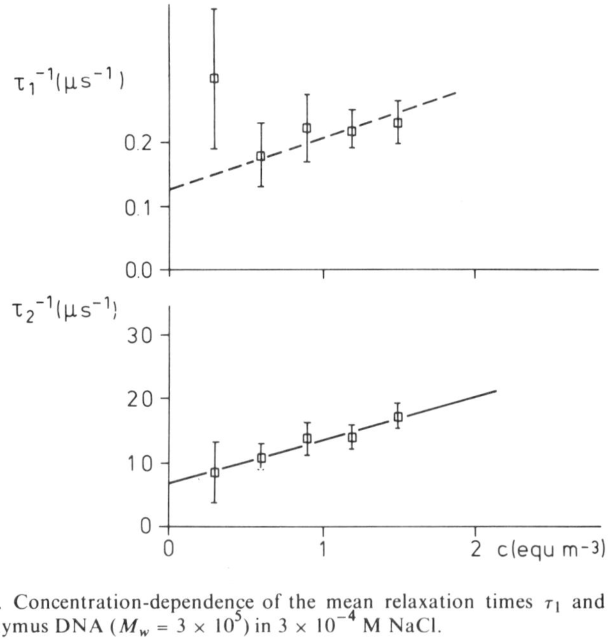

FIGURE 8. Concentration-dependence of the mean relaxation times τ_1 and τ_2 for sonicated calf thymus DNA ($M_w = 3 \times 10^5$) in 3×10^{-4} M NaCl.

macromolecular chain as occurring in dilute solutions at very low salt concentration or in the absence of added salt. The equilibrium distribution of these counterions along the chain will be slightly disturbed by the application of an external electric field and will give rise to an induced dipole moment roughly proportional to the square of the average extension of the macromolecule. Such field effects on the ion atmosphere also occur in ordinary electrolyte solutions, but owing to the small dimensions of the simple ions and the spherical symmetry of the atmosphere, the resulting increments are very small (this is the effect predicted by Debye and Falkenhagen[23]); recently it was shown in our laboratory that for chlorides of monovalent cations in water, the static increments calculated theoretically by Debye and Falkenhagen are verified indeed up to concentrations of 2.5×10^{-2} M[24]).

Several theoretical calculations,[25–28] using a rigid rod with uniform distribution of *fixed charges* as a model for the linear polyion and differing essentially only in the way interactions between the associated counterions are taken into account, have shown that the observed experimental static increments can qualitatively, at least, be interpreted in this way. For a truly rigid and elongated particle, the Alfalfa mosaic virus, the agreement with the most simple theory[26] is even semi-quantitative,[29] in spite of all the assumptions in the theory. For that particle, only one single dispersion region is observed in the complete frequency range 10 kHz–100 MHz, and $\Delta\epsilon_s/C_p$ is found to be constant in the concentration interval, where for synthetic polyelectrolytes, it is not. However useful, the rigid-rod model was found to be inadequate for understanding the details of the dielectric behavior of linear polyelectrolytes, particularly insofar as the occurrence of the high-frequency, molecular weight-independent dispersion region is concerned.

Recently, with van der Touw,[30] we used a more complicated although still rough model, which has turned out to be very helpful in understanding all the dielectric results observed with synthetic polyelectrolytes of high-charge density. This model takes into account the fact that through some flexibility of the macromolecular chain the average conformation may deviate from a rigid, rodlike one. Furthermore, it incorporates the possibility that the motion of the associated counterions along the chain may be characterized by two different diffusion times. Short-time fluctuations may occur over distances of only a limited number of monomeric units, because local potential barriers opposing this ion motion may arise because of the flexibility of the chain. Fluctuations over the complete macromolecular domain refer to a much longer time scale. Only sufficiently diluted polyelectrolyte solutions are considered, so that its properties can be derived additively from those of an electroneutral subvolume containing only one polyion. In such a subvolume the distribution of the counterions is described according to the so-called *two-phase model*,[3] in which the strong interactions between a polyion and its counterion leads to the "binding" of an average fraction $\bar{f}$ of all counterions to the macromolecular phase, leaving an average fraction $(1 - \bar{f})$ unaffected by the polyion in the solvent phase.

The polyion is now represented by a sequence of p thin, rodlike subunits of identical length b. The relative orientation of each subunit with respect to its preceding one is assumed, in first approximation, to be fixed and time-independent. The *fixed charges* on each subunit are distributed symmetrically with respect to its center of mass, which coincides with the middle of the rodlike subunit. Associated counterions can move freely along each subunit but can cross over from one subunit to a neighboring one only by overcoming a potential barrier that arises from the "break" in the average conformation at the junction of two subunits. This implies that a fluctuation in the distribution of the associated counterions along the chain relaxes toward the equilibrium distribution in two steps: 1) a process characterized by a relaxation time τ_b leads to an uniform distribution of the associated counterions along each subunit but does not affect the number of counterions on each subunit; and 2) a relatively slower process equilibrates the number of associated counterions between the different subunits with a relaxation time τ_q, leading to a uniform distribution of the counterions along the whole molecule. In the presence of a small external electric field, this uniform distribution is perturbed and also gives rise to an induced dipole moment that is assumed to be responsible for the polarization increment with respect to the polarization of the solvent not affected in this frequency range by the polyions. Neglecting *correlations between the associated counterions* distributed along the macromolecular chain, we have found by a statistical mechanic calculation an expression for the static dielectric increment of the highly diluted solution as well as an expression for the dielectric dispersion function. In the calculation of the latter, two other relaxation mechanisms besides those for the redistribution of associated counterions have been taken into account as responsible for the disappearance of the additional polarization due to the induced dipole moment after the field is switched off. These are the relaxation of the counterion distribution between the "associated" and the "free" state with relaxation time τ_f, and the orientational relaxation of the complete, rigid macromolecule with a relaxation time τ_r. The contribution of the latter arises from the fact that the additional polarization depends on the induced dipole moment's component in the direction of the field and thus depends on the orientation of the macromolecule. For full details of the calculation one should consult the original paper,[30] in which only the main results are summarized.

According to this model the static dielectric increment of a highly diluted polyelectrolyte solution (containing C_m macromolecules per m^3) is given by the following expression.

$$\Delta\epsilon_s = (e^2\bar{f}NC_m\gamma S^2)/3kT\epsilon_0 \qquad (3)$$

Here N stands for the total number of (monovalent) counterions of charge e per polyelectrolyte molecule, ϵ_0 for the absolute electric permittivity of free space, γ for the ratio of the effective electric field acting on a polyion and the average Maxwell field in the solution, and S for the radius of gyration of the polyion.† (Note that the concentration C_m can be expressed into the concentration C_e in equivalents per m^3 by $NC_m = N_{Av}C_e$ or in C_p in monomol/liter by $NC_m =$

†In Reference 30 Equation 3 was written with $R_g{}^2 + (b^2/12)$, replacing S^2. There R_g was a nonconventionally defined radius of gyration of the *subunits* with respect to the center of mass of the macromolecule, $R_g{}^2 = p^{-1}\,\Sigma_{(\alpha)}\,q_\alpha{}^2$, with q_α the distance between the middle of a subunit α and the center of mass. Assuming a uniform distribution of the centers of mass of the monomeric units along the length b of a subunit, it is easy to show that the relation between $R_g{}^2$ and the conventionally defined radius of gyration of the macromolecule is given by $S^2 = R_g{}^2 + (b^2/12)$, provided the distance between the center of mass of neighboring monomeric units is small compared to b.

$10^3 N_{Av}\alpha C_p$ with α the fraction of monomeric units bearing an elementary charge and N_{Av} Avogadro's Constant.) Equation 3 is of the same form as the expression derived for the static dielectric increment of a dilute solution of polyions represented by rigid rods of length L, S^2 being replaced by $L^2/12$ in the latter case. The calculations also show that two or more dispersion regions can be predicted for the dielectric increment, depending on the order of magnitude of the different relaxation times involved. From reasonable estimates for these, and taking into account that the experimentally observed low-frequency dispersion region has a molecular weight-dependent mean relaxation time but that the high-frequency one does not, it could be concluded that the high-frequency dispersion is controlled by the relaxation mechanism of the distribution of associated counterions along the *subunits* and that its amplitude $\Delta\epsilon_2$ should be proportional to the square of the length of a subunit (assumed to be independent of molecular weight for a macromolecule of sufficiently high degree of polymerization).

$$\Delta\epsilon^2 = (e^2\bar{f}NC_m\gamma b^2)/36\text{kT}\,\epsilon_0 \tag{4}$$

$$\tau_2 = \tau_b \tag{5}$$

The low-frequency dispersion region is controlled by the two relaxation mechanisms characterized by τ_r and τ_q, respectively, both relaxation times being molecular weight-dependent in contrast to τ_b and τ_f. In principle, τ_f also could contribute to τ_1 if its order of magnitude is comparable to τ_r, but this possibility is not likely for macromolecules of not too high molecular weight but relatively large charge density.

$$\tau_1^{-1} = \tau_r^{-1} + \tau_q^{-1}. \tag{6}$$

Theoretical *estimates* for τ_b and τ_r are available. For the latter, formulas proposed for the rotational relaxation time of prolate ellipsoids of revolution or cylindrical particles rotating about the short axis may be used, neglecting, however, the flexibility of the chain.

$$\tau_r = (\pi\eta_s L^3)/6\text{kT}(\ln 2p - 0.5). \tag{7}$$

Here η_s stands for the viscosity of the solvent, L for the length of the particle (measured along its major axis for the ellipsoid), $p = L/d$ with d the length of the minor axis or the diameter of the cylinder. An expression for τ_b may be obtained through the solution of the unidimensional diffusion equation of an associated counterion along a subunit bounded by impenetrable barriers at both ends, as counterion have been neglected throughout.

$$\tau_b = b^2/\pi^2 kTu \tag{8}$$

with u defined as the mobility of the associated counterion moving along the rigid linear subunit of length b. For τ_q no theoretical estimate fitting into the framework of the model is yet available.

This theoretical analysis shows that is should be possible to estimate the average dimensions of the polyion at infinite dilution from the specific static dielectric increment. Values for $\bar{f}$ may be taken from thermodynamic measurements; γ may be assumed to be close to unity, as is usually done. It should be emphasized that the values for the average dimensions thus obtained are only estimates in view of the assumptions in the theoretical calculations and of the empirical extrapolation procedure used. Nevertheless, for several synthetic polyelectrolytes reasonable values for the average dimensions at infinite dilution have been obtained which demonstrate that the polyion, although considerably extended, is not fully stretched.[21,31] Values for b can be estimated analogously from (**4**) and the specific increments referring to the amplitude of the high-frequency dispersion curve. In general, large values of b are found, as compared to the contour length L_c, with L_c/b of the order 2 to 3, indicating that the chains are stiff. This follows from the fact that b should represent the average distance over which an associated counterion can move without being opposed by a potential barrier (higher than kT) arising from local deviations of the conformation from a linear one. The larger the value b/L_c, the more rigid should be the chain, with, in the limit of a completely rigid rod $b/L_c \simeq 1$. From these two findings it may be concluded that these polyelectrolytes at infinite dilution come close to a Kratky and Porod "wormlike chain."[32]

The consistency of the interpretation of these dielectric results, particularly insofar as the second high-frequency dispersion region is concerned, follows from the comparison between $(\tau_2)_0$ and the value calculated according to (**5**) and (**8**), using for b the value derived from $(\Delta\epsilon_2/C_p)_0$ and for u the mobility u_0 of the counterion in an infinitely diluted electrolyte solution. In all cases investigated, the agreement was more than reasonable.[21,31] Additional confidence about this analysis is gained from the fact that the values for τ_r calculated according to (**7**) with reasonable estimates for L, deduced from $<S^2>$ as obtained from (**3**) and d, are in general of the same order of magnitude as but somewhat larger than τ_1 in agreement with (**6**). Therefore, the conclusion seems justified that the theory proposed, although still rather crude, provides a consistent picture for the dielectric behavior of charged linear macromolecules of sufficiently high degree of polymerization in infinitely diluted solution, thus emphasizing both the importance of the axial motion of the associated counterions and of the deviations from a rodlike conformation insofar as the dielectric properties are concerned. Although the same probably remains true, at least qualitatively, at higher concentrations, a quantitative approach is not yet available and remains difficult, particularly in the range where partial overlap between the macromolecular regions occurs, since it is not easy to take into account in the theory all the interactions that are likely to occur at these concentrations.

We have applied the same analysis to the dielectric results obtained with our sonicated calf thymus Na-DNA. From the light-scattering measurements in 0.2 M NaCl, a molecular weight of $(3.0 \pm 0.2) \times 10^5$ and a square root of the Z-average mean square radius of gyration $(<S^2>)_Z^{1/2}$ equal to 45 ± 2 nm was obtained. From the former a weight average contour length $\overline{L_c^w}$ of 153 ± 12 nm was calculated (using the value of 1950 dalton nm^{-1} for a perfect double-stranded helix). This value is in good agreement with the value of $\overline{L_c^w}$ = 167 nm deduced from electron microscopy pictures of the DNA molecule, which definitely showed a slight heterodispersity of the sonicated sample ($\overline{L_c^w}/\overline{L_c^n}$ = 1.2, with $\overline{L_c^n}$ the number average contour length), in agreement with recent findings by Godfrey.[33] The dielectric measurements were performed in 3×10^{-4} M NaCl (higher concentrations in salt being prohibited for their large conductivity) at 5 different concentrations $0.3 \leq C_e \leq 1.5$ eq.m^{-3}. The values of the four dielectric parameters extrapolated to infinite dilution are given in TABLE 1 with their standard deviation, as obtained from the least-square procedure. For $(\tau_1)_0$ the result should be considered with proper care in view of the large errors on the individual low-frequency mean relaxation times as deduced from a curve-fitting of the electric permittivity data, particularly at the lower concentration end. The value quoted in the table has been calculated not taking into account the result for $C_e = 0.3$ eq.m^{-3}; therefore, no standard deviation is given.

The value of b (strictly speaking, a weight-average value of b^2 is found from (**4**) and the same also applies to $\langle S^2 \rangle$) has been computed from $(\Delta\epsilon_2/C_e)_0$, using $\gamma = 1$ and $\bar{f} = 0.54$, the latter being deduced from activity measurements of Na^+ with a sodium-specific glass-electrode and assuming the validity of the additivity rule for the activity of the counterions.[1,2] With this value for $b = 62 \pm 7$ nm and with u_0 the high-frequency relaxation time at infinite dilution was estimated according to (**8**) to be $(3.1 \pm 0.7) \times 10^{-7}$ s; this compares favorably with the extrapolated value $(\tau_2)_0 = 1.5 \times 10^{-7}$ s in view of all the simplifying assumptions used in the computations and the theoretical approach. The large value of b

TABLE 1

DIELECTRIC PARAMETERS FOR SONICATED CALF THYMUS NA-DNA IN 3×10^{-4} M NACL, EXTRAPOLATED TO INFINITE DILUTION

$(\Delta\epsilon_s/C_e)_0$ in m^3 eq.$^{-1}$	$(\Delta\epsilon_2/C_e)_0$ in m^3 eq.$^{-1}$	$(\tau_1)_0$ in 10^{-5} s	$(\tau_2)_0$ in 10^{-7} s.
51 ± 9	25 ± 5	(0.8)	1.5 ± 0.7

points to a rather stiff chain for DNA, in agreement with general observations. From $(\Delta\epsilon_s/C_e)_0$ the mean square radius of gyration was estimated. The value thus found, $(\langle S^2 \rangle)^{1/2} = 26 \pm 2$ nm, although not unrealistic, seems to be somewhat small with respect to the value found from light-scattering at much higher ionic strength. The latter, which comes close to the weight average radius of gyration $\overline{S_c^w} = \overline{L_c^w}/12 = 48$ nm for the fully stretched chain, may be too high, since no correction for optical anisotropy has been applied.[34] It corresponds to a persistence length $a = 75 \pm 15$ nm if the DNA molecule at infinite dilution is considered as a wormlike chain, a model found to apply reasonably well to DNA. This value of a was calculated with the Benoit-Doty relation[35] between S^2, L_c and a taking into account the heterodispersity of the sample. It nicely overlaps with the most recent estimate for DNA at the same ionic strength by Godfrey and Eisenberg,[34] $a = 60 \pm 6$ nm, and is a strong indication for a rather stiff chain, in agreement with the large value found for b. The fact that the mean square of gyration as deduced from the analysis of the dielectric results is too low is not surprising, however, in view of all the approximations and simplifying assumptions underlying the estimate of this value. It is rather gratifying that the right order of magnitude is obtained! Furthermore, it should not be forgotten that the dielectric measurements were performed at low DNA and salt concentrations and that some denaturation may have occurred (this possibility is certainly not to be excluded, because results in pure water do not differ markedly from those at 3×10^{-4} M NaCl). Such a denaturation might introduce additional flexibility into the chain—i.e., lower the persistence length of the DNA molecule. Another point to be taken into consideration is the possibility that the molecular geometry determining the scattering of light is not necessarily the same one as "seen" by the counterions during their fluctuations. However, starting from $\langle S^2 \rangle$ as obtained from the static dielectric increment and estimating the mean square end-to-end distance according to the formula for a wormlike chain, $(\langle L^2 \rangle)^{1/2} \simeq 73$ nm is found. With this value and the assumption that for its rotational relaxation the DNA molecule may be treated as a rigid ellipsoid with its longer axis equal to $(\langle L^2 \rangle)^{1/2}$ and $3 \leq d \leq 20$ nm, one may estimate τ_r to lie between 1.5×10^{-5} s and 2.4×10^{-5} s according to (**7**). Such a value is again of the right order of magnitude as compared to the extrapolated value $(\tau_1)_0 = 8 \times 10^{-6}$ s found from the dielectric results, particularly if one takes into account the fact that the latter should be smaller than τ_r according to (**6**).

The conclusion is thus reached that not only does Na-DNA in aqueous solutions exhibit a dielectric behavior completely analogous to what is observed with other polyelectrolytes of sufficiently high degree of polymerization and charge density along the chain, but that the analysis of the dielectric results, at least in the limit of infinite dilution, can be performed consistently according to the theory proposed by van der Touw and Mandel for linear, charged macromolecular chains with some internal flexibility. This implies that this dielectric behavior may also be explained in terms of an induced dipole moment arising from the perturbation of the associated counterion distribution along the macromolecular chain by the electric field, taking into account the deviation of the average conformation from the rigid rodlike one.

Far be it from us to claim that the last word has been said in this difficult matter by the proposed analysis. The theory is still a crude one and certainly needs further improvement. In particular, it is necessary to take into account the fact that both the length of a subunit and the conformation of the macromolecule are fluctuating quantities and thus not time-independent on the time scales considered. This undoubtedly will have some bearings on the dynamics of the dielectric behavior. Furthermore, the potential barriers opposing the motion of the associated counterions along the chain, introduced *ad hoc* in the theory, also need to be treated in a more fundamental way, thus making estimates for τ_q available. Finally, the theory should be extended so as to include interactions both between the counterions along the chain and between polyions. This would probably help one to understand the strong concentration effects observed.

Acknowledgments

The work presented here is the result of the efforts of many of my coworkers, particularly of Dr. F. van der Touw and also Dr. W. M. van Beek and Dr. T. Vreugdenhil.

References

1. Katchalsky, A., A. Alexandrowicz & O. Kedem. 1966. Polyelectrolyte solutions. *In* Chemical Physics of Ionic Solutions. B. E. Conway & R. G. Barradas, Eds. John Wiley & Sons, New York, N.Y.
2. Katchalsky, A. 1971. Pure Appl. Chem. **26:** 327–373.
3. Oosawa, F. 1971. Polyelectrolytes. Marcel Dekker. New York, N.Y.
4. Manning, G. S. 1974. Limiting laws for equilibrium and transport properties of polyelectrolyte solutions. *In* Polyelectrolytes. E. Sélegny, Ed. D. Reidel. Dordrecht, The Netherlands.
5. Nagasawa, M. 1974. Ion binding phenomena of polyelectrolytes. *In* Polyelectrolytes. E. Sélegny, Ed. D. Reidel. Dordrecht, The Netherlands.
6. Bloomfield, V. A., D. M. Crothers & I. Tinoco Jr. 1974. Physical Chemistry of Nucleic Acids. Harper & Row. New York, N.Y.
7. Jugner, G., I. Jugner & I. G. Algen. 1949. Nature **163:** 849–850.
8. Takashima, S. 1963. J. Mol. Biol. **7:** 455–467.
9. Takashima, S. 1966. J. Phys. Chem. **70:** 1372–1380.
10. Takashima, S. 1967. Biopolymers **5:** 899–913.
11. Takashima, S. 1973. Biopolymers **12:** 145–155.
12. Hanss, M. & J. C. Bernengo. 1973. Biopolymers **12:** 2151–2159.
13. Goswami, D. N. & N. N. Das Gupta. 1974. Biopolymers **13:** 1549–1556.
14. Sakamoto, M., H. Kanda, R. Hayakawa & Y. Wada. 1976. Biopolymers **15:** 879–892.
15. van der Touw, F., J. de Goede, W. M. van Beek & M. Mandel. 1975. J. Phys. E. Sci. Instrum. **8:** 840–844.
16. van der Touw, F., G. Selier & M. Mandel. 1975. J. Phys. E. Sci. Instrum. **8:** 844–846.
17. van Beek, W. M., F. van der Touw & M. Mandel. 1976. J. Phys. E. Sci. Instrum. **9:** 385–391.
18. Mandel, M., & F. van der Touw. 1974. Dielectric properties of polyelectrolytes. *In* Polyelectrolytes. E. Sélegny, Ed. D. Reidel. Dordrecht, The Netherlands.
19. Sachs, S. B., A. Raziel, H. Eisenberg & A. Katchalsky. 1969. Trans. Farad. Soc. **65:** 77–90.
20. Cole, K. S. & R. H. Cole. 1941. J. Chem. Phys. **9:** 341–351.
21. van der Touw, F. & M. Mandel. 1974. Biophys. Chem. **2:** 231–241.
22. Vreugdenhil, T., F. van der Touw & M. Mandel. To be published.
23. Falkenhagen, H. 1971. Theorie der Elektrolyte. 2nd edit. Hirzel. Leipzig, Germany.
24. van Beek, W. M. & M. Mandel. To be published.
25. Schwarz, G. 1959. Z. Physik. Chem. N.F. **19:** 286–314.
26. Mandel, M. 1961. Mol. Phys. **4:** 489–496.
27. McTague, J. P. & J. H. Gibbs. 1966. J. Chem. Phys. **44:** 4295–4301.
28. Oosawa, F. 1970. Biopolymers **9:** 677–688.
29. van der Touw, J. W. H. Briede & M. Mandel. 1973. Biopolymers **12:** 111–119.
30. van der Touw, F. & M. Mandel. 1974. Biophys. Chem. **2:** 218–230.
31. Muller, G., F. van der Touw, S. Zwolle & M. Mandel. 1974. Biophys. Chem. **2:** 242–254.
32. Kratky, O. & G. Porod. 1949. Rec. Trav. Chim. **5:** 285–299.
33. Godfrey, J. E. 1976. Biophys. Chem. **5:** 285–299.
34. Godfrey, J. E. & H. Eisenberg. 1976. Biophys. Chem. **5:** 301–318.
35. Benoit, H. & P. Doty. 1953. J. Phys. Chem. **57:** 958–963.

DISCUSSION

Edward H. Grant* and Fumio Oosawa,† *Moderators*

**Physics Department*
Queen Elizabeth College
London W8 7AH England

†Department of Bioengineering
Faculty of Basic Engineering Science
Osaka University
Toyonaka, Osaka, Japan

DR. O'KONSKI: Dr. Mandel, have you considered the extension of the Maxwell-Wagner polarization model, which treats the counterions as occupying a thin layer into this entity of the polyelectrolyte? Of course, this is an approximation, but it does give, roughly, the dielectric behavior.

DR. MANDEL: Yes, but this interpretation would never give you these values of the relaxation times I found here.

DR. S. SANFORD *(Eastman Kodak Co., Rochester, N.Y.)*: Prof. Cole, could you avoid manual adjustment of your admittance bridge by simply comparing the output of the differential amplifier with an in-phase or out-of-phase square wave?

DR. COLE: Yes. I am sure you could, but drift is also a problem, and there is no point in having a discrimination of one part in 10^7 if a great deal of drift is present.

DR. SANFORD: What is the drift rate?

DR. COLE: It is of the order of 0.02% per minute, or a little less.

DR. GIBBS: J. McTague and I predicted some years ago that dielectric increment of polyelectrolytes would go through a maximum at the ionic strength such that the sites are half covered. With no added salt, it is already enough to cover more than half of the sites in DNA molecules. Therefore, with increasing ionic strength, the dielectric increment decreases. Secondly, we observed that the dielectric increment increases with increasing temperatures. If you have already got more counterions than would give you the maximum ionic polarization, then by increasing the temperature you decrease it down closer to the optimum value, and you have the dielectric constant increasing with temperature.

DR. MANDEL: These are done at finite concentrations, and you know that concentration effects can be very strong at low concentrations. With regard to the inverse temperature effect, you should stop first to extrapolate and make measurements at the lowest concentration possible.

DR. HOUSSIER: Professor Mandel, our birefringence measurements indicate that the dielectric measurements on DNA are mainly due to the orientation of segments in the macromolecular chain. I think the next thing to do would be to try to measure the whole frequency dispersion of the Kerr effect and especially go to very high frequencies. Until now only the very low frequencies range has been covered.

DR. MANDEL: Although I agree that the model we used is not unique, there are other arguments in its favor, e.g., from our measurements with a rigid molecule like the alfalfa virus particle, which could not be interpreted according to a Maxwell-Wagner effect. With DNA, the fact that there are two dispersion regions must be associated with the flexibility of the chain.

DR. LAMBERT *(Princeton University, Princeton, N.J.):* How far do the counterions move from the DNA molecule?

DR. MANDEL: In theory, nothing is assumed about the distance from the chain of the counterion. It could be 10Å or 20Å. Moreover, there could be several layers of counterions moving, and this may explain why a distribution of relaxation times is always observed.

DR. IBERAR: In view of the fact that you have a strong underlying association between dielectric properties and viscous effects, I was wondering whether a macroscopic mechanical model involving a "Knudsen's slip" type of mechanism might be appropriate?

DR. STOCKMAYER: There is a great deal of slip in rotatory motion of the small liquid molecules, as is clear from dielectric results. The only thing I say is that if you are going to work on molecular theory, the overwhelming influence of the Brownian motion cannot be ignored. Macroscopic mechanics alone will not suffice.

DR. BENDER *(Fairleigh Dickinson University, Rutherford, N.J.):* Professor Mandel, if you add extra electrolytes to your DNA solution, would you observe any extra effect?

DR. MANDEL: You probably would. However, for experimental reasons, it is very difficult to measure up substantial concentrations of salt. For DNA in solution, we have measured concentrations up to 5×10^{-4} M NaCl and have found no substantial effects due to the addition of the salt.

SUBSECOND MEASUREMENT OF DIELECTRIC DISPERSION

Yuzuru Husimi,* Haruki Nakamura, and Akiyoshi Wada

Department of Physics, Faculty of Science
University of Tokyo
Hongo, Tokyo, 113, Japan

INTRODUCTION

The easy availability of digital computers and the development of digital techniques have made it easy to acquire and process vast amounts of data, and all the measuring methods established in the precomputer age have been reexamined. The first remarkable result of such reexamination was the development of the Fourier transform (FT) infrared spectrometer.[1] It was followed by the FT-NMR,[2] Hadamard transform (HT) infrared spectrometer,[3] HT-NMR,[4] HT-chromatography,[5] and so on. All these methods have the "multiplex advantage" (Fellgett advantage) (advantage due to simultaneous measurements of many frequency components), and the measuring time is reduced by the number of multiplexing for a measurement of data having the same signal-to-noise ratio.

In the dielectric measurement, the first example of methods having the "multiplex advantage" may be the time-domain refractometer (TDR)[6-8] in the microwave region. The extension of this method to the low-frequency region characteristic to biopolymers was limited by the length of a coaxial cable used. Moreover, the primary aim of the TDR was to avoid the difficulties of making microwave oscillators of variable frequencies and microwave bridges, and in the low-frequency region these difficulties do not exist. These facts have left the time-domain method undeveloped for the low-frequency region. (But in the very low frequency region, these difficulties reappear, and time-domain methods have been used.) The first method, which truly exhibits the advantage, is the pseudorandom noise dielectric spectrometer (PND-spectrometer)[9,10] developed recently in our laboratory. The purpose of the present article is to describe the PND method and to present some examples showing its uniqueness.

Hayakawa *et al.*[11,12] pointed out the problems to be solved in measuring the dielectric permittivity of aqueous solutions as (1) the thermal drift of d.c. conductivity, and (2) the electrode polarization. They developed a kind of a frequency-difference method with a four-electrode cell (FSFE method). Using a frequency difference method, the problem (1) is solved, and using a four-electrode cell, the difficulty (2) is solved. The PND-spectrometer is a more radical method of solving the difficulty (1). The distortion originating from the drift may be avoided if responses of all the frequencies of interest are acquired simultaneously. This can be performed by present-day digital technology, taking data as rapidly as possible, and as many as possible, in the time-domain.

*Present address: Department of Environmental Chemistry, Faculty of Engineering, Saitama University, Urawa, 338, Japan.

The fact that the simultaneous measurement of many frequency responses can be performed means also that the dispersion profile can be measured during the time required to take one fixed data point by the frequency-domain method. The reduction of the measuring time by a factor of 10^2 may be possible. Because of this high speed, kinetic measurements, such as tracing the time course of the characteristic frequency change, are possible.

Therefore the PND is a complement of the TDR in the low-frequency region and also a complement of the precise low-frequency bridge method in high-speed measurement. A wide range of applications to static and kinetic studies of biological macromolecules and materials having low characteristic frequencies is expected.

Principles of the Method

Many solution systems of biologically interesting molecules have linear and shift- (on the time axis) invariant properties upon excitation by an electric field of 100 Vm^{-1}. According to the linear response theory, a linear and shift-invariant system is completely described by the impulse response function $h(t)$.[13] Here, "completely" means that the response $g(t)$ to any excitation $f(t)$ can be described in terms of $h(t)$ and $f(t)$ only,

$$g(t) = h(t) * f(t), \tag{1}$$

where the asterisk denotes the convolution of the two functions on either side of them. On Fourier transforming this equation, we get

$$G(\omega) = H(\omega) \cdot F(\omega) \tag{2}$$

where G, H, and F are the Fourier transforms of g, h, and f, respectively. From (2), it is derived that $g(t) = H(\omega_0) \exp(i\omega_0 t)$ is the response to a sinusoidal excitation $f(t) = \exp(i\omega_0 t)$. This fact is the basis of the frequency-domain methods, since the system can be described completely by the inverse Fourier transform of the transfer function $H(\omega)$.

In the method having the "multiplex advantage," we have to consider a case where the superposition of many sinusoidal waves is taken as an excitation $f(t)$, i.e.,

$$f(t) = \sum_k A_k \exp(i(\omega_k t + \phi_k)). \tag{3}$$

(Here and in the following equations the real-part operation is omitted for simplicity.) For simplicity we put A_k's as constant A and take ω_k to be equally spaced,

$$f(t) = A \sum_k \exp(i(\omega_k t + \phi_k)). \tag{4}$$

(These conditions are slightly modified later.) When all $\phi_k = 0$, (**4**) expresses an impulse at time 0. When ϕ_k is randomly distributed between 0 and 2π, (**4**) expresses a white noise. The system response to this $f(t)$, using the linearity property, becomes

$$g(t) = A \sum_k H(\omega_k) \exp(i(\omega_k t + \phi_k)). \tag{5}$$

In this function many frequency responses are simultaneously recorded and superposed on one another as a multiplexed signal. In order to get $H(\omega_k)$ for each ω_k, we have to demultiplex $g(t)$. Using the orthogonality of sinusoidal functions, we get

$$H(\omega_k)\exp(i\phi_k) = A^{-1}\int g(t)\exp(-i\omega_k t)\,dt$$
$$= A^{-1}G(\omega_k). \tag{6}$$

Since the phases ϕ_k of the excitation are known, $H(\omega)$ is obtained from the *FT* of $g(t)$. In other words, (**6**) is no other than (**2**), and

$$H(\omega_k) = A^{-2}G(\omega_k)\cdot F(\omega_k)^*. \tag{7}$$

$G(\omega)\cdot F(\omega)^*$ is called the cross-power spectrum between f and g.

How to choose ϕ_k is dependent upon technical aspects of a given experiment. In the pulse excitation, the excitation energy is concentrated in a short time; the requirement of linearity of the system response is more severe. Moreover, because the most of the information from the system is clustered near the excitation pulse, the dynamic range of the detector and the bit number of the A/D converter must then be larger than in the noise excitation. In other words, for the same peak power of excitation, the noise method gives the signal energy an N times larger than that in the pulse method. Any other choice of ϕ_k gives an intermediate situation between these two methods. For this reason we have adapted the noise excitation to the present system. An example of the noise excitation and the system response to it are shown in FIGURE 1. The real and imaginary parts of the Fourier transforms of these functions are shown in FIGURE 2. The random-phase situation can be seen from this figure.

In conclusion, a linear and shift-invariant system is completely described by the cross-power spectrum between a white noise and the response of the system to it. This method makes it possible to measure all the frequency components simultaneously (the "multiplex advantage"). The Fourier transformation corresponds to the demultiplexing process. Since the primary advantage of this method is the fact that the time-consuming demultiplexing process can be performed after a

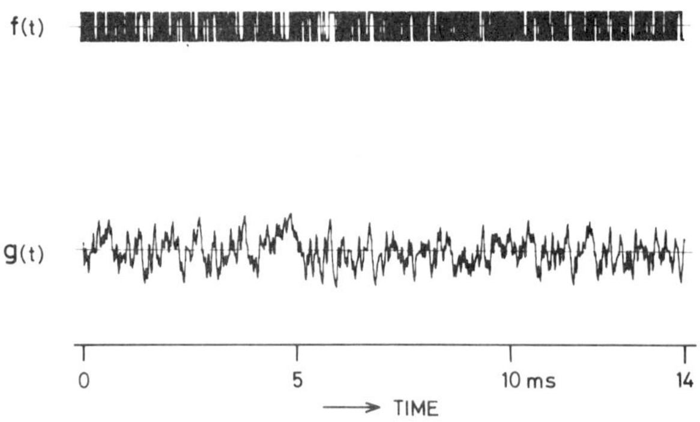

FIGURE 1. Binary pseudorandom noise as an excitation $f(t)$, and a system response to it, $g(t)$. Clock time $\Delta T = 13.72\ \mu s$. The figures correspond to one complete cycle of the BPN.

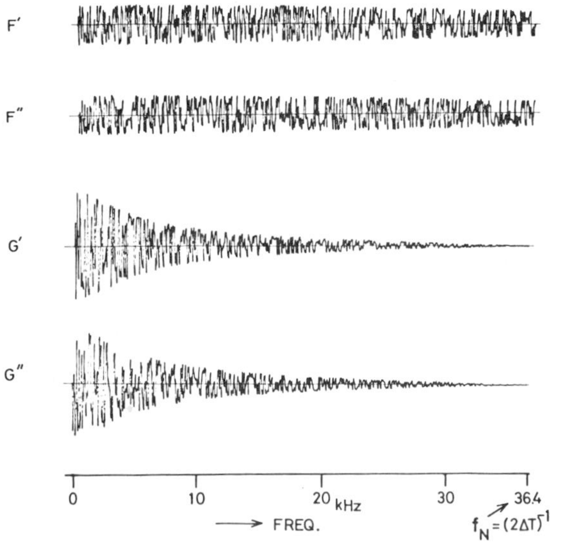

FIGURE 2. a: Fourier transform (analytical) $F(\omega)$ of the excitation $f(t)$ in FIGURE 1 F' and F'' are the real and the imaginary parts of $F(\omega)$. b: Fourier transform (discrete) $G(\omega)$ of the response $g(t)$ in FIGURE 1. G' and G'' are the real and the imaginary parts of $G(\omega)$.

measurement, the excitation time can be shortened in comparison with frequency-domain methods.

To realize this advantage in an experimental apparatus, we have to solve two problems. One is the generation of a periodic noise, because the cross-power spectrum has to be obtained in a finite time for the present purpose of short-time measurements. In other words, the interval of integration of (**6**) must be finite, and the period T of the periodic noise should coincide with this interval. Since the periodic noise used is artificial, it is called a pseudorandom noise. The other problem is the insertion of fast-buffer memories between a bridge and a data processor to record $g(t)$ in the time-domain. Since the recording length must be equal or longer than T, the number of memories is larger than $N = T/\Delta T$, where ΔT is a sampling interval. Fortunately, fast transient-recorders are commercially available as buffer memories.

In TABLE 1, various methods of linear-system analysis are summarized. The last two methods should be considered as noise methods, in a broad sense. After the inverse Fourier transformation of (**7**), $h(t)$ is obtained as the cross-correlation function between f and g. However, the calculation of the cross-correlation function from the stored data in a computer is time-consuming. In fact, the correlation function is usually calculated via fast Fourier transformations, and this procedure is the one carried out in our PND-spectrometer itself. Although the use of hard-wired correlators for this purpose is an alternate way, most commercial correlators are designed so as not to use almost all available information, as they are of the delay time-sweep type or batch type. They are inadequate for the present

TABLE 1

COMPARISON OF VARIOUS METHODS OF LINEAR SYSTEM ANALYSIS

Method	Measuring Time	Excitation Power	Detector Dynamic Range	Phases of Various Freq. Components	Computer	Reference
Freq. sweep	long	low	very narrow	—	not necessary	12
Pulse or Relaxation (ex. TDR)	short	high	wide	same at time zero	necessary	6 7
Pseudorandom noise (PND)	short	low	narrow	randcm	necessary	9 10
Correlation	long short*	low	narrow	random	necessary	
Thermal noise analysis	long short*	zero	narrow	—	not necessary or necessary	14

*In case a multichannel nonbatch type correlator is used.

purpose of short-time measurements. A thermal noise analysis of the system may be another way to measure the permittivity. The power spectrum this obtained, however, cannot describe the system completely, because the phase information is not included.

GENERATION OF PSEUDORANDOM NOISE

There are several methods of generating a periodic noise. One method is to collect a noise from a physical source, e.g., a noise diode. This method is not adequate for the present purpose because of a large fluctuation in the power spectrum for a short-time window. For the same reason, a time series of random numbers generated in computer software is not adequate.

A well-characterized noise of a finite length may be generated by some regular generation rule. Two rules have been examined.

Generation Rule: PN-sequence

The PN-sequence generator used in our apparatus essentially consists of a serial-in parallel-out shift register and an exclusive OR gate. In this study, a 10-bit shift register that has two outputs at bit 3 and bit 10 is employed. The sum of modulo-2 addition of two outputs is put into bit 1 at the next clock time. This output sequence $\{a_i\}$ has a period of maximum length, $2^{10} - 1$. The PN-sequences have well-characterized properties of randomness.[15,16] (1) "1" occurs 512 times, where "0" occurs 511 times; (2) one-half of the runs have length 1, a quarter are of length 2, an eighth are of length 3, and so on; (3) the autocorrelation of the PN-sequence is as follows:

$$\sum_{i=0}^{1022} a_i a_{i+k} = \begin{cases} 512 & k = 0, 1023, 2046, .. \\ 256 & \text{otherwise} \end{cases}$$

When the output levels of the PN-sequence at logic "1" and logic "0" are shifted to $-v$ and $+v$ volts, respectively, by a shaper circuit, the time course of the output $f(t)$ of the circuit is a periodic binary pseudorandom noise (BPN) of period $T = 1023 \cdot \Delta T$, where ΔT is the clock time of the shift register. The details of the shaper circuit are described elsewhere.[9] With use of CMOS-IC analog switches, two voltage sources (mercury batteries, $\pm v = \pm 1.35$V) are switched according to the PN-sequence and a gate signal. The binary pseudorandom noise thus obtained and its inversion ($f(t)$ and $-f(t)$) are continuously transmitted to a sample cell and a reference admittance, respectively. A typical wave form of BPN has been shown in FIGURE 1.

The power spectrum of the BPN is

$$P_f(\omega) = 1024 \cdot (v/1023)^2 \cdot (\sin(\omega\Delta T/2)/(\omega\Delta T/2))^2$$
$$\text{for } \omega = 2\pi k/T, \quad k = 1, 2, 3, \ldots \qquad \textbf{(8)}$$

As a result, the BPN is not a white noise, but is smooth and nearly white up to the

Nyquist frequency, $f_N = \omega_N/2\pi = (2\Delta T)^{-1}$. The power spectrum was measured by an analog wave analyzer and is shown in FIGURE 3. The analytical Fourier transform $F(\omega)$ of the BPN was calculated and stored on a floppy disk. (In FIGURE 2, $F(\omega)$ is shown.)

When the time window for data sampling is selected to be equal to the period T of the binary pseudorandom noise, (7) is modified to be

$$\begin{aligned} H(\omega) &= F(\omega)^* \cdot G(\omega)/P_f(\omega) \\ &= 1/1024 \cdot (1023/v)^2 \cdot (\omega\Delta T/2/\sin(\omega\Delta T/2))^2 \cdot F(\omega)^* \cdot G(\omega), \\ &\text{for } \omega = 2\pi k/T, k = 1, 2, 3, \ldots, 511. \end{aligned} \quad (9)$$

In conclusion, the procedure to obtain the transfer function $H(\omega)$ is to compute the Fourier transform of $g(t)$ and then to perform the complex multiplication of $G(\omega)$ with the stored coefficients.

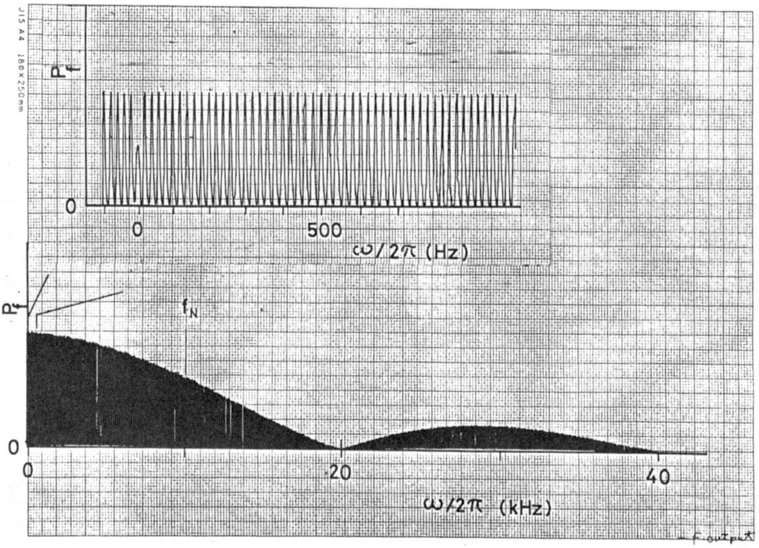

FIGURE 3. Power spectrum of a BPN excitation. In this case, clock time = 50 μs. Output of the shaper circuit was measured by a wave analyzer. The upper figure represents the enlarged portion of the low-frequency region. The region between d.c. and Nyquist frequency f_N is used for measurement.

The shift register has a specific internal state once per period, because there are $2^{10}-1$ internal states. Therefore, a synchronization pulse is easily obtained by detecting a specific internal state, and it is possible to accumulate $g(t)$ because of its periodicity.

An advantage of the PN-sequence method is that it is generated easily. Disadvantages are that (1) it has high-frequency components and in some cases a low pass filter is required to avoid the aliasing in the discrete Fourier transformation;

(2) its period is not $2^n \cdot \Delta T$ where n is an integer, and it is not suited for the fast Fourier transform algorithm (FFT); and (3) its power spectrum is nearly white on a linear frequency scale instead of on a logarithmic frequency scale.

Generation Rule: Fourier Synthesis

In a broad sense, this includes the PN-sequence method. In a narrow sense, this method consists of the generation of $f(t)$ in computer software according to (**3**) and outputting it through a D/A converter. Many kinds of power spectra can therefore be generated. For example, according to the following equation,

$$f(m\Delta T) = \sum_k 1/\sqrt{k}(\cos\theta_k \cdot \cos(mkA) + \sin\theta_k \cdot \sin(mkA)) \qquad (\mathbf{10})$$

where $A = \pi/1536, m = 1, 2, \ldots, 3072$, and θ_k are random values between 0 and 2π, the power spectrum of $f(t)$ is nearly proportional to $1/\omega$. This means that the power spectrum on the logarithmic frequency scale is nearly flat, which is preferable for dielectric dispersion profile measurement. Moreover, it has a sharp cut-off at the maximum frequency, and the aliasing problem is less severe than in the PN-sequence method. It is suited for FFT because of the presence of the 2^n-terms.

Another example is synthesized as follows;

$$f(m\Delta T) = A\sum_k (\cos\theta_{\psi(k)} \cdot \cos(m\psi(k)) + \sin\theta_{\psi(k)} \cdot \sin(m\psi(k))) \qquad (\mathbf{11})$$

where $\psi(k) = 2\pi/1024 \cdot [2^{k/10}]$. In this case data points are nearly equally spaced on the logarithmic frequency scale.

A disadvantage of the Fourier synthesis method is that it requires a number of ROMs and two well-balanced, fast D/A converters. This method is being studied in our laboratory in cooperation with the Ando Electric Co. (Tokyo, Japan). A slightly modified method is also being studied by Hayakawa.[17]

Measuring System

A block diagram of the hardware of the present system is shown in Figure 4. The minicomputer used is a YHP 2100A (Yokogawa-Hewlett-Packard, 16bits, 16Kw, 0.98μs memory-cycle time). The bridge circuit used is described elsewhere.[9] The shaper with an inverter constitutes two arms of the bridge, and the other two arms consist of the sample cell and the reference admittance. The deviation from the balancing condition was amplified by a negative feedback, wide-band, and fast-settling FET operational amplifier. In the summing amplifier configuration, the relationship between $H(\omega)$ and $\epsilon(\omega)$ is simple, although the circuit is not fast enough for the measurements of MHz region.

Provided that (1) the polarization effect is negligible, (2) the OP-amp is ideal, and (3) the bridge is balanced at the Nyquist frequency, the following formulas are given,

$$\Delta\epsilon' \equiv \epsilon'(\omega) - \epsilon'(\infty) = (R_f C_{vac})^{-1} H''(\omega)/\omega$$

$$\Delta\epsilon'' \equiv \epsilon''(\omega) = (R_f C_{vac})^{-1}(H'(\omega) - H'(0))/\omega \qquad (12)$$

where ϵ', ϵ'', H', H'', C_{vac}, and R_f are the real and the imaginary parts of the permittivity, the real and the imaginary parts of the transfer function, the air capacitance of the cell, and the feedback resistance, respectively.

Two kinds of two-electrode cell are used. One consists of platinum concentric cylindrical electrodes fixed in a glass cell. The other was designed by Parry-Jones[24] and consists of stainless-steel parallel electrodes, one of which constitutes the container itself.

The output signal from the summing amplifier was fed into a transient recorder (DL905, Data Laboratories, 8 bits, 1 Kw, 5 MHz), directly or via an active low-pass filter. The transient recorder is externally advanced by the same clock as the

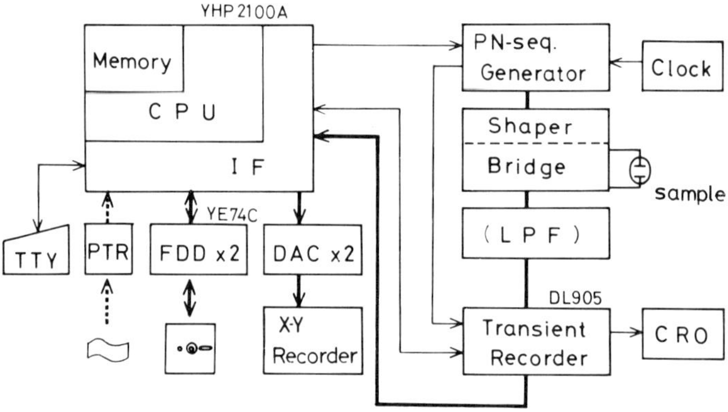

FIGURE 4. Block diagram of the system. IF = interface circuit between CPU and external devices; TTY = teletypewriter; PTR = photo tape reader; FDD = floppy disk drive; DAC = digital-to-analog converter; LPF = low-pass filter (option); CRO = monitoring oscilloscope.

binary pseudorandom noise. When frequency components higher than the sampling rate exist, the low-pass filter is required to eliminate the aliasing and to increase the duty factor. The time required for data accumulation by the minicomputer is 20 ms, and hence the duty factor is $\frac{1}{2}$ up to 25 kHz and $\frac{1}{4}$ up to 100 kHz.

The sampling time at each data point is delayed by τ_d (larger than the settling time of the summing OP-amp) from the BPN clock pulse. This causes a phase-shifting and amplitude depression in the transfer function, which are corrected in the computer software. The balancing of the bridge in the high-frequency limit is performed by observing the wave through a monitoring oscilloscope.

The floppy disk drives are used for an external mass memory (ca. 121 Kw). They store the programs, the tables of the constants for calculation, and the processed data. They play a crucial role in kinetic measurements as a buffer mass memory.

The flow chart of the software is shown in FIGURE 5. The measurement programs are written in the conversational HP-BASIC language, which has been ex-

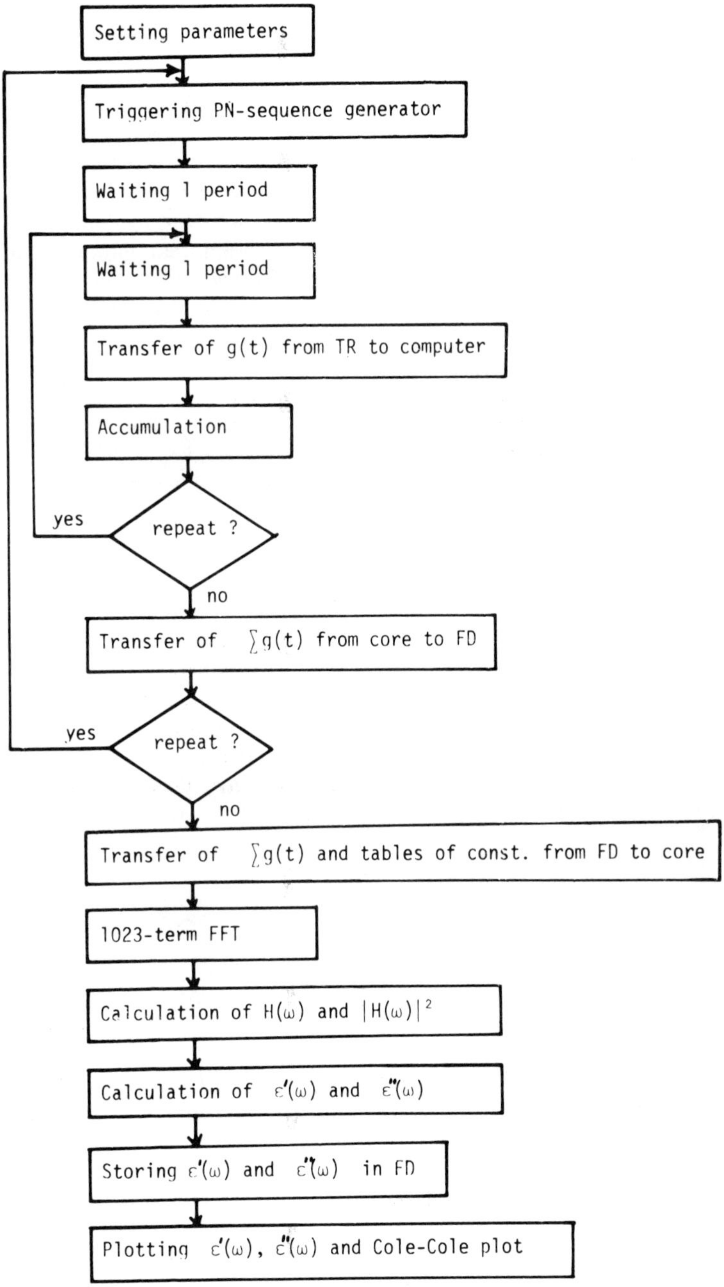

FIGURE 5. Flow chart of the software. TR = transient recorder; core = computer main memory; FD = floppy disk.

tended with the inclusion of several driver subroutines. All calculations are performed in floating point (mantissa:24 bits, exponent:8 bits). The FFT algorithm should be mentioned briefly. Although the conventional fast Fourier transformation algorithm is written to treat 2^n terms (n is an integer), in the PN-sequence method the number of data points is $2^{10}-1$. If one data point is augmented in order to have 2^{10} terms, the "leakage" problem of discrete Fourier transformation will appear; in other words, the randomness property (3) described under GENERATION OF PSEUDORANDOM NOISE will be destroyed. One method of avoiding this difficulty is to augment with $2^{10}+1$ zeros to obtain the noncyclical convolution and to perform the conventional 2^{11} term FFT (two times) and the complex multiplication (three times). However, this method requires a large number of memories and is not suited for a minicomputer.

In this study, by finding a relation $2^{10} - 1 = 3 \times 11 \times 31$, the 1023-term FFT can be completed by performing 31-term, 11-term, and 3-term normal FTs successively. In this algorithm, the size of the occupied memory block is only 1023 + 31, and the number of operation is 1023 × (3 + 11 + 31), which is comparable to the 2048 × (11 + 3) operations of the above-mentioned 2^{11}-term method.

RESULTS AND DISCUSSION

Two examples of the dielectric relaxation measurements on the solutions of synthetic polypeptides are shown. One is a measurement of a solution system in equilibrium states, and the other is that of a transient phenomenon. The former type of measurement used to be done by the traditional bridge method in the frequency domain, but it was very time-consuming. In the latter case, our PND-spectrometer has made it possible to obtain the dielectric dispersion curves at any given time-points during a transient phenomenon.

Example 1: Observation of the Helix-Coil Transition of Poly(ε-Carbobenzoxy-L-Lysine) in m-Cresol

Poly(ε-carbobenzoxy-L-lysine) (PCBL) in *m*-cresol is known to undergo a sharp thermal helix-coil transition of the inverse type around 30°C.[18-20] By means of the PND-spectrometer, the dilute solution of the PCBL sample M-1† ($\overline{M_w} = 47{,}200$) is measured at various temperatures. The solvent is freshly distilled *m*-cresol, and solute concentration is 5.0×10^{-3}g/ml. The helix-coil transition of this sample has been studied by Omura *et al.*,[20] and the value of the cooperative parameter $\sigma^{1/2}$ is either 0.0025 from an optical rotation measurement[18] or 0.0027 from a heat capacity measurement.[19]

The real and imaginary parts of permittivities of the solution at various temperatures are shown in FIGURE 6, a and b, and c shows their Cole-Cole plots. In these figures, permittivities are plotted against the reduced parameter $f \times (\eta_T/\eta_{298}) \times (298/T)$, in order to offset the trivial effect of the viscosity of the solvent η_T(P) and temperature T(K) to the rotary diffusion of the solute molecule.

These results indicate that the rotational frictional coefficients are almost the same in the transition range, decreasing little as the temperature is raised, but a

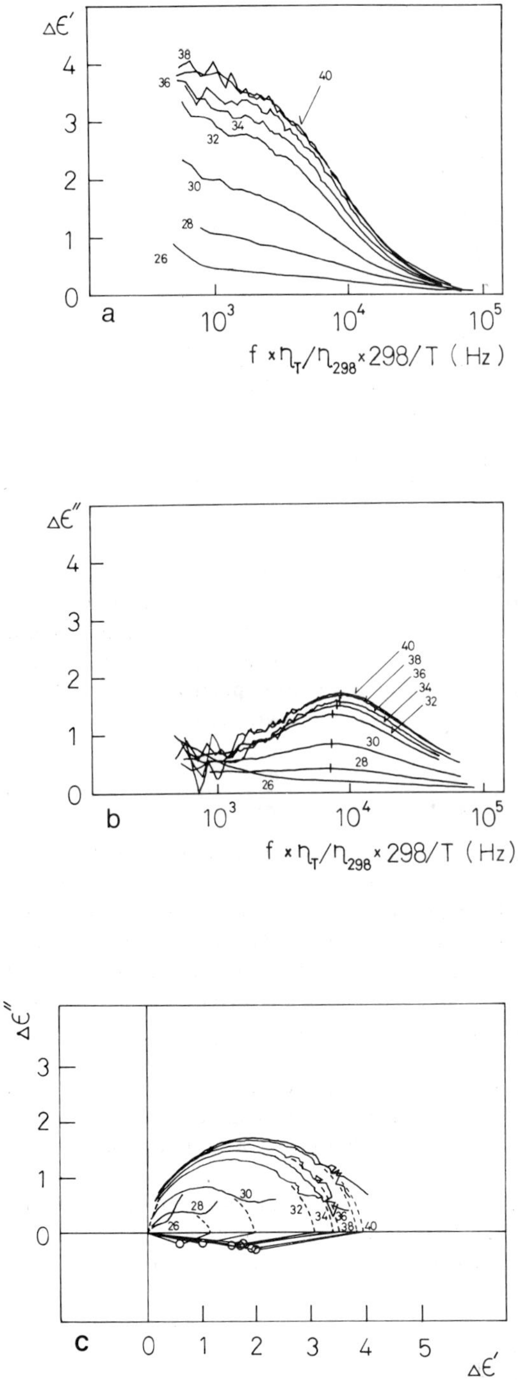

FIGURE 6. Complex permittivities of PCBL ($\overline{M_w}$ = 47,200) in *m*-cresol at various temperatures (conc. is 5.0×10^{-3}g/ml). a = real parts of permittivities; b = imaginary parts of permittivities; c = the Cole-Cole plots. The numbers in the figures represent temperatures (°C), and η_T the viscosity of the solvent (P) at temperature T(K).

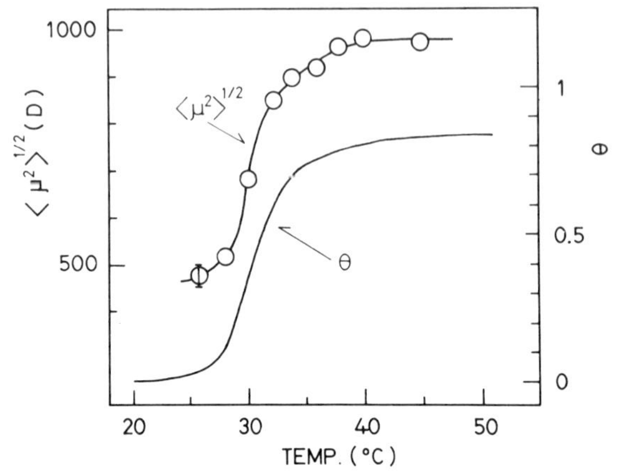

FIGURE 7. Plots of the "vacuum" dipole moment $<\mu^2>^{1/2}$ and the helical content θ against temperature (°C). [I] represents the error bar.

large increase is observed in the static dielectric constants $\Delta\epsilon_s$ $(=\lim_{\omega\to 0}\Delta\epsilon'(\omega))$ determined from the extrapolations of the Cole-Cole plots[20,22] (see FIGURE 6, c).

As seen in a and b of FIGURE 6, the dispersion curve has one inflection point, and the chemical polarization due to the helix-coil transition[19] may be hidden behind the major contribution coming from the orientation polarization. The "vacuum" dipole moment $<\mu^2>^{1/2}$ of the PCBL sample is written as

$$<\mu^2>^{1/2} = \left[\frac{3kT}{4\pi}\cdot\frac{M}{N_{AV}}\cdot\frac{\Delta\epsilon_s}{c}\cdot\frac{q}{(fg)}\right]^{1/2} \tag{13}$$

where M is the molecular weight of the solute, c the solute concentration (g/ml), k the Boltzmann constant, T the absolute temperature (K), and N_{AV} the Avogadro number. The factor $q/(fg)$ has been proposed by Applequist and Mahr[20,22] by modifying Buckingham's Equation.[23] In this case it is difficult to estimate the correct value of (fg), because it is a function of the axial ratio of the rodlike molecule.[20] However, for a large axial ratio and high solvent polarity, (fg) is approximately $\frac{2}{3}$ (independent of the axial ratio of the solute molecule), which we use in the present report.

The temperature dependency of $<\mu^2>^{1/2}$ and the helical content θ estimated from $[m']_{436}$ of an optical rotation measurement[20] are compared in FIGURE 7. This indicates that the molecular dipole moment of PCBL increases in almost the same way as θ.

From this evidence, together with the invariance of the rotational frictional coefficient, we may surmise that the transition is an all-or-none type (σ is small enough) or one helical type, in which the random coils at the end of the helical part obstruct the rotational motion.[20,21]‡

†This sample is a gift of Dr. A. Teramoto.
‡Detailed discussion will be published elsewhere.

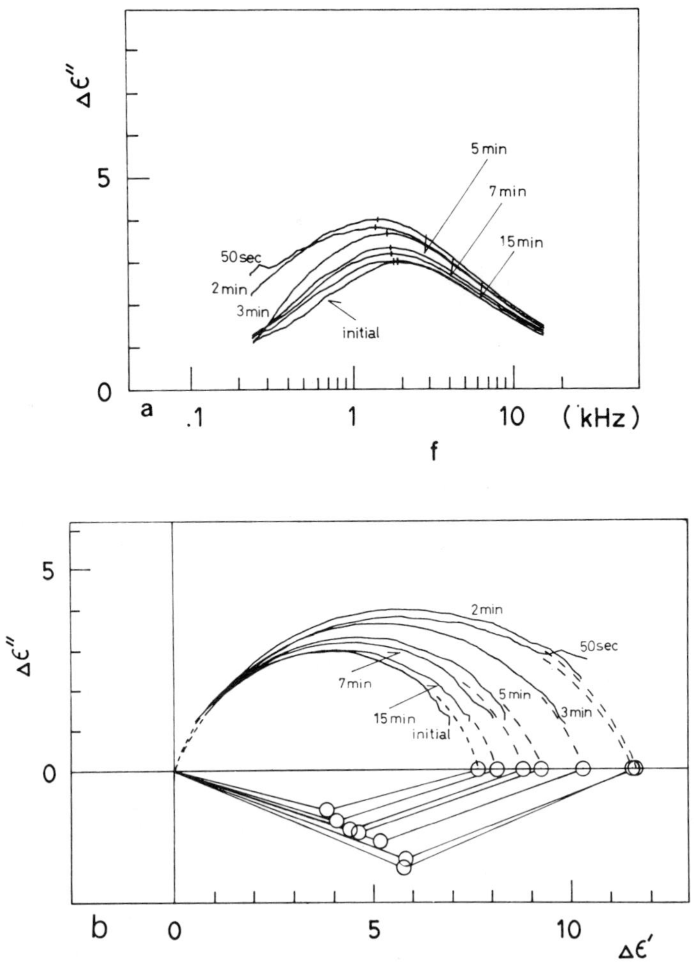

FIGURE 8. Complex permittivities of PBLG ($\overline{M_w}$ = 150,000) in 1,2-dichloroethane at several stages after the application of a large field at 30°C (conc. is 2.0 × 10^{-3}g/ml). "Initial" represents the one before the application of the field, and the numbers in the figures represent the times after the removal of the field. a = imaginary parts of permittivities; b = Cole-Cole plots.

Example 2: Observation of a Time Course of the Deaggregation of Poly (γ-benzyl-L-glutamate)

The application of a large electric field to a dilute solution of helical Poly (γ-benzyl-L-glutamate) (PBLG) gives rise to its aggregation, which produces a large increment in permittivity. When the external field is removed, the permittivity (real part) at a fixed low frequency decreases gradually to the initial zero field value, indicating deaggregation.[24]

In order to analyze this phenomenon by a dielectric measurement, only $\Delta\epsilon_s$ was measured, and neither the complex permittivity spectrum nor the rotational relaxation time could be obtained, except in the case of an extremely slow process.[24] In the present study, however, a series of measurements by the PND-spectrometer along the time axis has been performed, each of which carried out 10 data accumulation within 1 second.

FIGURE 8, a, shows the time course of the changing ϵ'' profiles of the helical PBLG ($\overline{M_w}$ = 150,000) in 1,2-dichloroethane (conc. is 2.0×10^{-3} g/ml) after the application of a high field of 20 kV/cm for 1 minute. FIGURE 8, b, shows their Cole-Cole plots. For the analysis of the time-dependence of deaggregation, $(\Delta\epsilon_s - \Delta\epsilon_{is})/\Delta\epsilon_{is}$ and $(\tau^{1/3} - \tau_i^{1/3})/\tau_i^{1/3}$ at 30°C are plotted against time t(min) after the removal of the field in FIGURE 9, where $\Delta\epsilon_{is}$ and τ_i are the values of $\Delta\epsilon_s$ and τ before the application of the field.

Under a large electric field, a parallel aggregation is most likely to occur, and if so, from (**13**) $\Delta\epsilon_s$ is proportional to $n \cdot \langle\mu_s^2\rangle$, n being the degree of aggregation and $\langle\mu_s^2\rangle$ the mean square dipole moment of a single helical polymer. On the other hand, from Perrin's equation, rotational relaxation time τ is nearly proportional to the cube of the axial length of the rodlike aggregated molecules. Then, if the aggregation is head-to-tail and linear, both $(\Delta\epsilon_s - \Delta\epsilon_{is})/\Delta\epsilon_{is}$ and $(\tau^{1/3} - \tau_i^{1/3})/\tau_i^{1/3}$ indicate the increase in the degree of aggregation normalized by the values before the application of the field.

It can be seen in FIGURE 9 that at the initial stages of the deaggregation, both data are almost on a straight line, respectively. The least-square method applied to the first five data points gives the rate constants and the degrees of aggregation just after the removal of the field as follows:

$$(\Delta\epsilon_s(t) - \Delta\epsilon_{is})/\Delta\epsilon_{is} = 0.68 \cdot \exp(-t/4.7) \quad \textbf{(14)}$$

$$(\tau(t)^{1/3} - \tau_i^{1/3})/\tau_i^{1/3} = 0.14 \cdot \exp(-t/4.2) \quad \textbf{(15)}$$

where t is expressed in minutes, and the solid lines in FIGURE 9 show these relations.

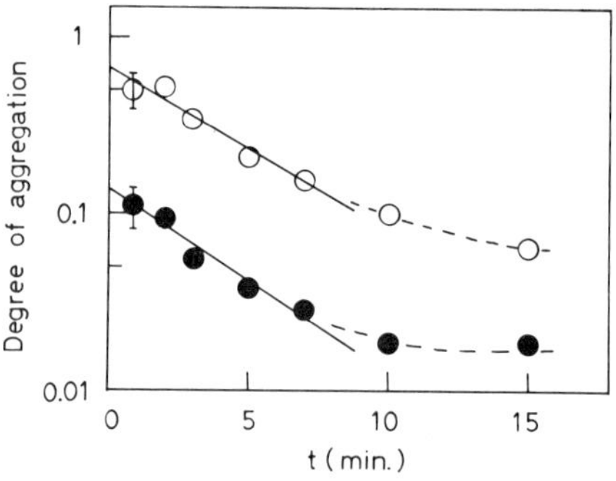

FIGURE 9. The time-dependence of the degree of aggregation. ○ represents $(\Delta\epsilon_s - \Delta\epsilon_{is})$ and ● = $(\tau^{1/3} - \tau_i^{1/3})/\tau_i^{1/3}$, where $\Delta\epsilon_{is}$ and τ_i are the values before the application of the field. The solid lines show (**14**) and (**15**). I represents the error bar.

It is interesting to find that although (**14**) and (**15**) have nearly the same rate constants, the degree of aggregation just after the removal of the field (at $t = 0$) shows a greater difference. This result may suggest the mechanism of aggregation due to the high field and also the mechanism of deaggregation,[25] as follows.

From the values in (**14**) and (**15**), there seems to be not only a head-to-tail-type of aggregation, which has a great effect on the relaxation time, but also a side-by-side type one, which contributes little to the relaxation time. Even at the initial state before applying the high field, these two types of aggregation seem to occur in this solvent system. ($\Delta\epsilon_{is}$ and $\tau_i^{1/3}$ are 2.1 and 1.4 times, respectively, as large as the expected values, whereas the values of $\Delta\epsilon_{is}$ and $\tau_i^{1/3}$ of the same sample in *m*-cresol has been measured, indicating no aggregation.[10]) The almost equal rate constants in $\Delta\epsilon_s$ and $\tau^{1/3}$ suggest that the initial stage of deaggregation process is due to the deaggregation of the head-to-tail type.

ACKNOWLEDGMENTS

We wish to thank Dr. G. Parry Jones for his interest and for kindly lending the special sample cell for a large electric field. We also thank Dr. A. Teramoto for his gift of the PCBL sample.

REFERENCES

1. VANASSE, G. A. & H. SAKAI. 1967. Prog. Opt. **6:** 259–330.
2. ERNST, R. R. & W. A. ANDERSON. 1966. Rev. Sci. Instrum. **37:** 93–102.
3. DECKER, J. A. 1971. Appl. Opt. **10:** 510–514.
4. KAISER, R. 1974. J. Mag. Reson. **15:** 44–63.
5. HUSIMI, Y. To be published.
6. FELLNER-FELDEGG, H. 1969. J. Phys. Chem. **73:** 616–623.
7. NICHOLSON, A. M. 1968. IEEE Trans. Instrum. Meas. **IM-17:** 395–402.
8. VAN GEMERT, M. J. C. 1973. Phillips Res. Rep. **28:** 530–572.
9. HUSIMI, Y. & A. WADA. 1976. Rev. Sci. Instrum. **47:**213–219.
10. NAKAMURA, H., Y. HUSIMI, G. PARRY JONES & A. WADA. To be published.
11. HAYAKAWA, R., H. KANDA & Y. WADA. 1974. Rep. Prog. Poly. Phys. Jap. **17:** 673–676.
12. HAYAKAWA, R., H. KANDA, M. SAKAMOTO & Y. WADA. 1975. Jap. J. Appl. Phys. **14:** 2039–2052
13. CHAMPENEY, D. C. 1973. The Fourier Transforms and their Physical Applications. Academic Press. New York, N.Y.
14. MICHERON, F. & L. GODEFROY. 1972. Rev. Sci. Instrum. **43:** 1460–1465.
15. KORN, G. A. 1966. Random-process, Simulation and Measurements. Chap. 7. Mc-Graw-Hill. New York, N.Y.
16. GOLOMB, S. M. 1967. Shift Register Sequences. Chapt. 4. Holden-Day. San Francisco, Calif.
17. HAYAKAWA, R. 1976. J. Fac. Engng. Univ. Tokyo **A14:** 42.
18. MATSUMOTO, M., T. NORISUYE, A. TERAMOTO & H. FUJITA. 1973. Biopolymers **12:** 1515–1532.
19. NAKAMOTO, K., H. SUGA, S. SEKI, A. TERAMOTO, T. NORISUYE & H. FUJITA. 1973. Macromolecules **7:** 784–789.

20. OMURA, I., A. TERAMOTO & H. FUJITA. 1975. Macromolecules **8:** 284–290.
21. WADA, A., T. TANAKA & H. KIHARA. 1972. Biopolymers **11:** 587–605.
22. APPLEQUIST, J. & T. G. MAHR. 1966. J. Amer. Chem. Soc. **88:** 5419–5429.
23. BUCKINGHAM, A. D. 1953. Aust. J. Chem. **6:**93–103.
24. GREGSON, M., G. PARRY JONES & M. DAVIES. 1970. Chem. Phys. Lett. **6:** 215–216.
25. PARRY JONES, G., H. NAKAMURA, Y. HUSIMI & A. WADA. To be published.

DIELECTRIC DISPERSION OF POLYELECTROLYTES DUE TO ION FLUCTUATION

Akira Minakata

Department of Physics
Hamamatsu University School of Medicine
Hamamatsu, Japan

Introduction

Polyelectrolytes are macromolecules having many ionizable groups. In aqueous solution a polyelectrolyte molecule dissociates into a polyion and a number of counterions. Major portions of the counterions are condensed to the polyion forming an ionic atmosphere. These bound ions are relatively freely mobile in a domain around the polyion, under an external electric field, but their motion is substantially limited within this domain because of the strong attraction of the polyion. This limited movement of bound ions induces large polarizability of the polyion and is responsible for the large dielectric constant of the polyelectrolyte solution.[1,2]

The above description is supported by the following two facts. First, the contribution of electrophoretic movement of the polyion to the dielectric increment is negligible, or too small to be observed, because it provides an opposite dispersion with increasing frequency, i.e., an increase in dielectric constant with frequency.[3] This opposite dispersion has never been observed. Second, the contribution of the orientation of the permanent dipole seems to be quite small, since the permanent dipole is easily canceled by ionic charge, and also because, in most cases, a polyelectrolyte molecule has no permanent dipole along its main chain from a structural point of view.

As mentioned before, more than half the counterions are bound to the polyion under ordinary conditions, even at very low concentrations of the polyion. The dielectric properties of the polyelectrolyte solutions have been considered to be controlled by the degree of ion binding or its specificity. The specificity of ion binding is determined by many factors and is easily affected by changing the degree of ionization or by adding simple electrolytes. The dielectric increment is, however, characterized simply as follows. It is enhanced by loosely bound ions but reduced by strongly bound ions.[4-6] This qualitative statement suggests that the dielectric increment is related to the looseness of binding or to the fluctuation of bound ions, rather than to the degree of binding.

On the basis of the above-mentioned characteristics of the dielectric increment, we can derive the polarizability $\alpha(\omega)$ of a polyion under an alternating electric field with frequency ω by using the fluctuation-dissipation theorem[7] as

$$\alpha(\omega) = (3kT)^{-1}\left[<\mu^2(0)> - i\omega\int_0^\infty \exp(-i\omega t)<\mu(0)\mu(t)>dt\right], \quad (1)$$

where $<\mu^2(0)>$ and $<\mu(0)\mu(t)>$ represent the mean square and the mean time-dependent correlation of the dipole moment along the rodlike polyion in the absence of external field. In the above expression, the polyion is assumed to have no permanent dipole moment, i.e., $<\mu>_{E=0} = 0$, where E is the external field intensity. This assumption indicates that the mean square fluctuation of the dipole moment is simply given by $<\mu^2(0)>$, which is generally nonvanishing even if $<\mu>_{E=0} = 0$.

According to this idea, Oosawa has calculated the frequency-dependent dielectric properties of a rodlike molecule having a counterion cloud along the rod by the Fourier transform of the fluctuation of the cloud and the diffusion equation of the fluctuation.[8] We have also calculated the static dielectric increment of the polyion expressed by a linear lattice with a mixture of mono- and divalent counterions by a matrix method.[5] Later we derived the frequency-dependent properties by using the linear lattice model, and clarified the role of the counterion-counterion interaction energy in the expression of the dielectric properties.[2]

In the present article, the above results were extended so that the effect of the outside medium was taken into account. The two-phase model[1,9] was applied to the case of flexible polyelectrolytes; that is, the whole solution is divided into the polymer domain of an elongated ellipsoid of revolution and the outer medium. Motion of the condensed ions in the polymer domain was classified to the parallel and perpendicular processes to the polymer chain. The perpendicular diffusion process reduces the relaxation time of some modes of the ion fluctuation, which provides an additional dispersion in the high-frequency region.

Fluctuation of Counterions

Let us consider a one-dimensional linear lattice having n lattice points. These lattice points have an interval b corresponding to the distance between neighboring ionic groups of a polyion. The counterions are distributed in these lattice points. The number of counterions at point j is denoted as n_j, which is, at present, limited to 0 or 1.

The number n_j is assumed to be controlled by interaction energy ϵ_j of the counterion at point j, as shown in Figure 1. Here, the interaction energy is considered to be the sum of the binding energy ϵ^0 and the counterion-counterion interaction energy $\phi(|j - k|)$ between the points j and k as

$$\epsilon_j = \epsilon^0 + \sum_{k \neq j} \phi(|j - k|) n_k. \tag{2}$$

To calculate the mean degree of ion binding and its fluctuation, it is convenient to define a quantity Ξ_j, the partition function of a single lattice point either occupied or not occupied by a counterion

$$\Xi_j = 1 + \exp[(\mu - \epsilon_j)/kT] \tag{3}$$

where μ is the chemical potential of the counterion.

If ϵ_j is fixed, the average value of n_j is

$$<n_j>_{\epsilon_j} = \left(\frac{kT}{\Xi_j}\right)\left(\frac{\partial \Xi_j}{\partial \mu}\right)$$
$$= \{1 + \exp[(\epsilon_j - \mu)/kT]\}^{-1}. \quad (4)$$

On the other hand, according to the Weiss approximation, the average value of n_j over all lattice points in equilibrium, n_0, is assumed to be expressed by the average interaction energy e^0 as

$$n_0 = <n_j>_{e^0}$$
$$= \{1 + \exp[(e^0 - \mu)/kT]\}^{-1}. \quad (5)$$

where

$$e^0 = \epsilon^0 + \sum_{k \neq j} \phi(|j - k|)n_0. \quad (6)$$

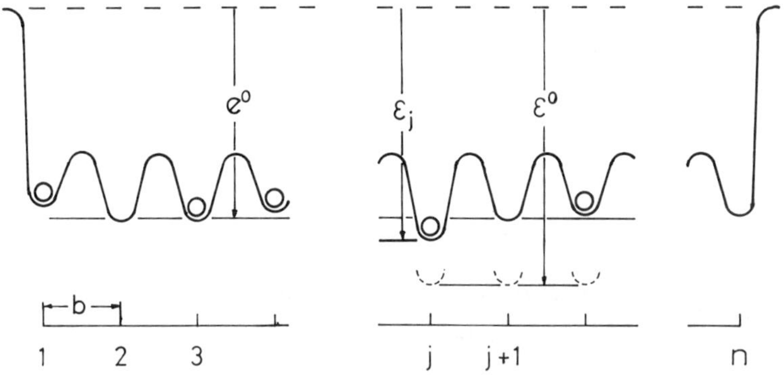

FIGURE 1. Schematic representation of the potential profile of a rodlike polyelectrolyte molecule. The interaction energy ϵ_j at point j consists of the binding energy ϵ^0 and the counterion-counterion interaction energy which is determined by ion distribution in the neighboring lattice points. The average interaction energy e^0 is the ensemble average of ϵ_j.

The fluctuation of the bound ions in other lattice points δn_k $(= n_k - n_0)$ induces the fluctuation of the interaction energy ϵ_j as

$$\delta\epsilon_j = \sum_{k \neq j} \phi(|j - k|)\delta n_k. \quad (7)$$

By use of the above relation, $<n_j>_{\epsilon_j}$ is given by

$$<n_j>_{\epsilon_j} = \frac{n_0}{n_0 + (1 - n_0)\exp(\delta\epsilon_j/kT)}. \quad (8)$$

The number of bound ions n_j is controlled by the fluctuation of the interaction energy in this manner.

The Fourier transform of the fluctuation of the number of bound ions at point j (δn_j) is expressed as

$$\delta n_q = \left(\frac{1}{n}\right) \sum_j \delta n_j \exp(iqj), \tag{9}$$

where q stands for the wave mode number and given by

$$q = 2\pi m/n \, (m = 0, \pm 1, \pm 2, \ldots, n/2), \tag{10}$$

and conversely

$$\delta n_j = \sum_q \delta n_q \exp(-iqj). \tag{11}$$

The periodic boundary condition of periodicity n is assumed in the above expressions.

The mean square fluctuation of mode q is expressed as[10]

$$\begin{aligned} < |\delta n_q|^2 > &= \left(\frac{1}{n^2}\right) \sum_j \sum_k <\delta n_j \delta n_k> \exp[iq(j-k)] \\ &= (1/n)[<\delta n_j^2> + \psi(q)]. \end{aligned} \tag{12}$$

where

$$\psi(q) = \sum_{j \neq k} <\delta n_j \delta n_k> \exp[iq(j-k)]. \tag{13}$$

The terms $<\delta n_j^2>$ and $\psi(q)$ are calculated[2] for sufficiently large n and given by

$$<\delta n_j^2> = n_0(1 - n_0) \tag{14}$$

and

$$\psi(q) = -nn_0^2(1 - n_0)\phi(q)/(kT)[1 + nn_0(1 - n_0)\phi(q)/(kT)]^{-1} \tag{15}$$

where

$$\phi(q) = (1/n) \sum_{j \neq k} \phi(|j - k|) \exp[iq(j-k)] \tag{16}$$

is the Fourier transform of the counterion-counterion interaction energy. If the Debye-Hückel potential is assumed for $\phi(|j - k|)$ with parameter κ and the lattice is sufficiently long ($n \gg 1$), then

$$\phi(q) = -(e^2/\epsilon_0 1) \ln[1 - 2(\cos q) \exp(-\kappa b) + \exp(-2\kappa b)]. \tag{17}$$

Substitution of (**14**) and (**15**) into (**12**) leads to

$$< |\delta n_q|^2 > = [n_0(1 - n_0)/n][1 + nn_0(1 - n_0)\phi(q)/kT]^{-1}. \tag{18}$$

The dipole moment produced by the fluctuation δn_j is

$$\mu = eb \sum_j [j - (n+1)/2] \delta n_j. \tag{19}$$

If no correlation between different modes, i.e. random-phase approximation,[10] is assumed, the mean square of the dipole moment is obtained from (**11**) as

$$<\mu^2> = (ebn)^2 \sum_q <|\delta n_q|^2> (1 - \cos q)^{-1} \quad (\mathbf{20})$$

where q takes only positive values because of the symmetry of $\phi(q)$ and $\cos q$. The term for $q = 0$ does not contribute to the dipole moment. Substituting (**18**) and (**20**) into (**1**), we have the polarizability along the rodlike molecule in a static field, as

$$\alpha = (3\mathrm{kT})^{-1}(eb)^2 \sum_q \frac{nn_0(1 - n_0)}{[1 + nn_0(1 - n_0)\phi(q)/kT](1 - \cos q)} \quad (\mathbf{21})$$

The static dielectric increment is expressed as

$$\begin{aligned}\Delta\epsilon &= (4\pi/3)NBe^2\ell^2 n_0(1 - n_0)/nkT) \\ &\times \sum_q \{[1 + nn_0(1 - n_0)\phi(q)/kT](1 - \cos q)^{-1}\end{aligned} \quad (\mathbf{22})$$

where N is the number concentration of the polyion, B the internal field correction factor, and $\ell = bn$ the length of the rod.

Frequency-Dependence of the Polarizability

Longitudinal motion of the bound ions results in the relaxation of the polarizability. This process is characterized by the average flow of the number concentration of bound ions at point j. A simple scheme is given in Figure 2, where an ion is jumped between adjacent lattice points with rate constants $W_{j-1,j}$ and $W_{j,j-1}$.

As mentioned before, the average number n_j is controlled by the potential depth at each point defined by the interaction energy. In addition, the rate constant is assumed to be controlled by the interaction energy, since it is considered to be inversely proportional to exponential of the potential barrier between adjacent points. That is, the value $W_{j-1,j}$ is proportional to $\exp[-(w_0 - \epsilon_{j-1})kT]$, where w_0 is the energy at the top of the barrier. Two possible examples are considered, as shown in Figure 2: one is the case in which the value of w_0 is constant, and the other, the case where the energy is given by $w_0 = [w_0' + (\epsilon_j + \epsilon_{j-1})/2]$, with constant w_0'. In the first case, the rate constant $W_{j-1,j}$ is written as,

$$W_{j-1,j} = W \exp(\delta\epsilon_{j-1}/kT) \quad (\mathbf{23})$$

where W is the value under the average interaction energy. In the second case, we have

$$W_{j-1,j} = W \exp[(\delta\epsilon_{j-1} - \delta\epsilon_j)/2\mathrm{kT}]. \quad (\mathbf{23'})$$

By use of the rate constants defined above, the time-dependence of the mean value of $n_j(t)$ is

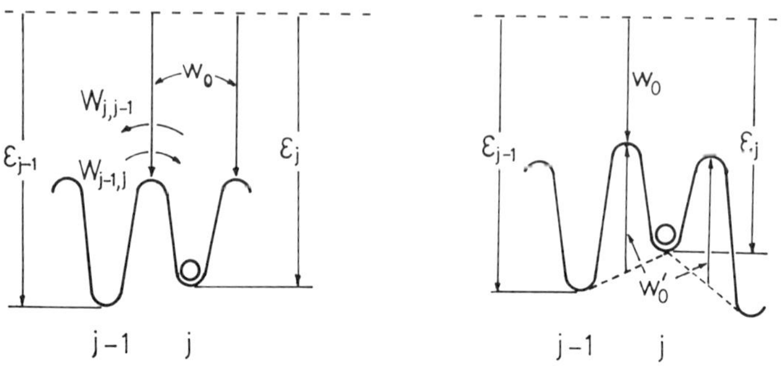

FIGURE 2. Two possible jumping processes of a bound counterion described by rate constants $W_{j-1,j}$ and $W_{j,j-1}$; *left*, the maximum energy of the barrier w_0 is assumed to be constant; *right*, the average height of the barrier w_0' is constant, i.e. $w_0 = w_0' + (\epsilon_{j-1} + \epsilon_j)/2$.

$$(d/dt) <n_j(t)>_t = <W_{j-1,j}n_{j-1}(1 - n_j)>_t - <W_{j,j-1}n_j(1 - n_{j-1})>_t + <W_{j+1,j}n_{j+1}(1 - n_j)>_t - <W_{j,j+1}n_j(1 - n_{j+1})>_t \quad (24)$$

where $<X>_t$ is the short-time average of the quantity X. The time interval is assumed to be sufficiently longer than that of the elementary process of the transition but sufficiently smaller than the observed relaxation time.

Substituting (**23**) or (**23'**) into (**24**) and taking the first-order terms of $<\delta n>_t$ and $<\delta\epsilon>_t$ by expansion of the exponential terms, we have the same results for both cases, i.e., (**23**) and (**23'**), as

$$(d/dt) <\delta n_j(t)>_t = W(<\delta n_{j-1}>_t + <\delta n_{j+1}>_t - 2<\delta n_j>_t) + [Wn_0(1 - n_0)/kT](<\delta\epsilon_{j-1}>_t + <\delta\epsilon_{j+1}>_t - 2<\delta\epsilon_j>_t). \quad (25)$$

Using (**7**), (**9**), and (**16**), we have

$$(d/dt) <\delta n_q>_t = -2W(1 - \cos q)[1 + nn_0(1 - n_0)\phi(q)/kT] <\delta n_q>_t. \quad (26)$$

The relaxation time of mode q is

$$\tau_q^{-1} = 2W(1 - \cos q)[1 + nn_0(1 - n_0)\phi(q)/kT]. \quad (27)$$

The complex polarizability of mode q is defined as

$$\alpha_q(\omega) = \alpha_q/(1 + i\omega\tau_q). \quad (28)$$

By use of this definition and (**20**), the real part of the polarizability $\alpha(\omega)$ is given by

$$\alpha(\omega) = e^2\ell^2 n_0(1 - n_0)/(3nkT) \times \sum_q \{[1 + nn_0(1 - n_0)\phi(q)/kT](1 - \cos q)(1 + \omega^2\tau_q^2)\}^{-1}, \quad (29)$$

which gives dielectric increment due to the ion fluctuation as

$$\Delta\epsilon(\omega) = 4\pi NB\alpha(\omega). \quad (30)$$

The dielectric loss increment $\Delta\epsilon''$ is derived by the imaginary part of (**28**) as

$$\Delta\epsilon'' = (8\pi/3)NBWe^2\ell^2 n_0(1 - n_0)/(nkT) \sum_q \frac{\omega\tau_q^2}{1 + \omega^2\tau_q^2}. \quad \textbf{(31)}$$

From this equation, total conductivity increment $\sigma_\infty - \sigma_0$ is easily obtained

$$\sigma_\infty - \sigma_0 = \frac{NB}{3\zeta} e^2 n^2 n_0(1 - n_0) \quad \textbf{(32)}$$

where ζ is the friction constant of a counterion defined by kT/b^2W.

FIGURE 3 shows the calculated dielectric increment against log $(\omega\tau_1)$, logarithm of the reduced frequency, where τ_1 is the lowest relaxation time. Here the following parameters are used; $n = 10^3$, $\ell = 4 \times 10^{-5}$, $n_0 = 0.6$, and $\epsilon_0 = 80$. In this case, the coulombic interaction is considered as the counterion-counterion interaction, which is given by (**17**) in the limit of $\kappa b \rightarrow 0$. The dominant contribution is made by the lowest mode in the low frequency limit (52.1%). The second, third, and fourth modes contribute 14.6, 7.0, and 4.2%, respectively. Remaining higher frequency modes contribute only 22.1%, but broaden the dispersion pattern on the high-frequency side. This is illustrated in FIGURES 3 and 4. In FIGURE 4, the relative contribution of each Fourier mode to the static dielectric increment is plotted for two cases; one is the case with coulombic interaction between counterions, and the other, without interaction between counter-

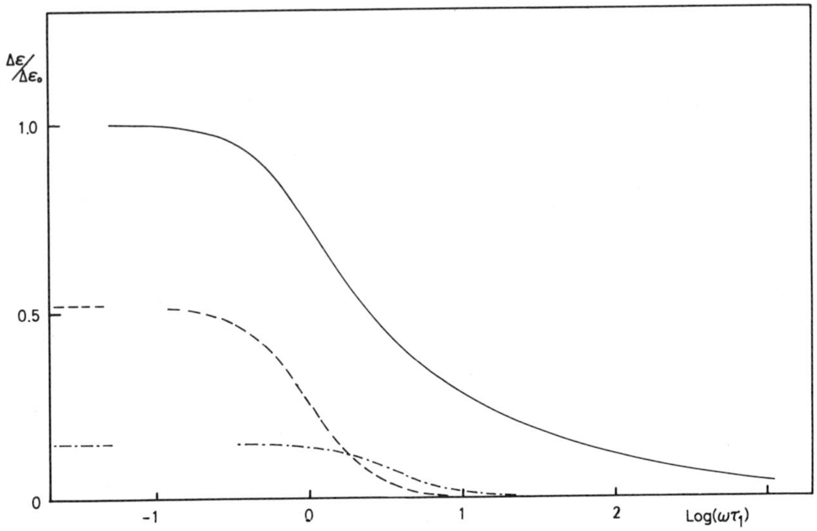

FIGURE 3. Calculated dielectric increment of polyelectrolytes against $\log(\omega\tau_1)$, logarithm of the reduced frequency, where τ_1 is the lowest relaxation time. The following parameters are used; $n = 10^3$, $\ell = 4 \times 10^{-5}$, $n_0 = 0.6$ and $\epsilon_0 = 80$. The counterion-counterion interaction is assumed to be coulombic, i.e. $\kappa b = 0$. —— = whole dispersion profile; – – – = contribution of the lowest mode of the fluctuation; and -·—·- = that of the second mode.

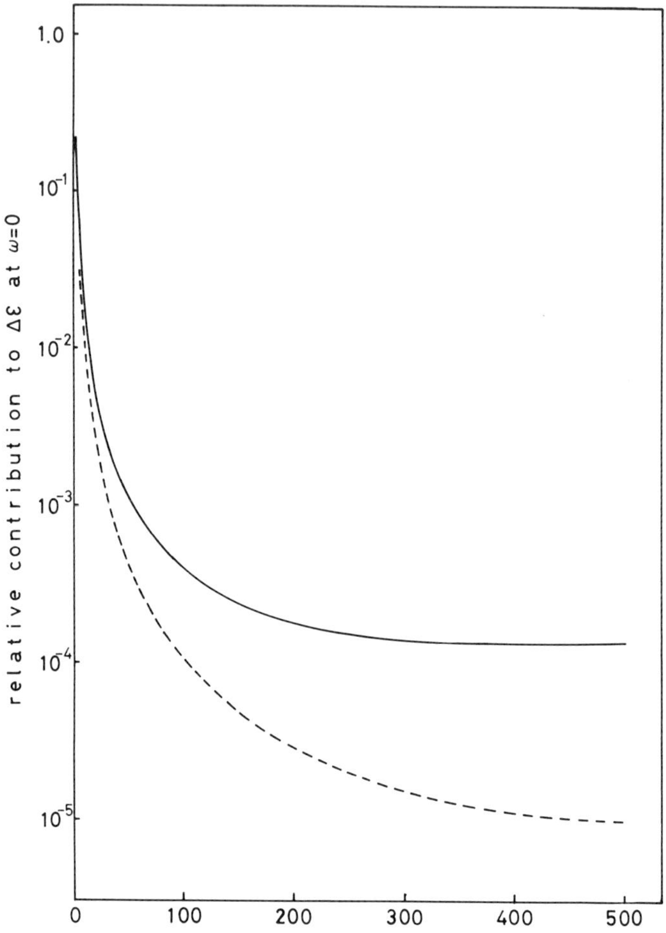

FIGURE 4. Relative contribution of each mode to the dielectric increment in the low-frequency limit; log (relative contribution) vs. mode number, for coulombic interaction between counterions, $\kappa b = 0$, ——, and for no counterion-counterion interaction, $\kappa b \to \infty$, – – –.

ions. The latter case shows more rapid decay of the contribution with mode number than the former.

For high-frequency region, however, intermediate Fourier modes with shorter wave lengths contribute mainly to the dielectric increment. FIGURE 5 shows that the number of the mode that makes the maximum contribution shifts toward higher mode numbers with increasing frequency. In FIGURE 5, each mode is assumed to have its own amplitude and relaxation time, given by (**18**) and (**27**). If the amplitude or relaxation time of some modes are changed by the changes in

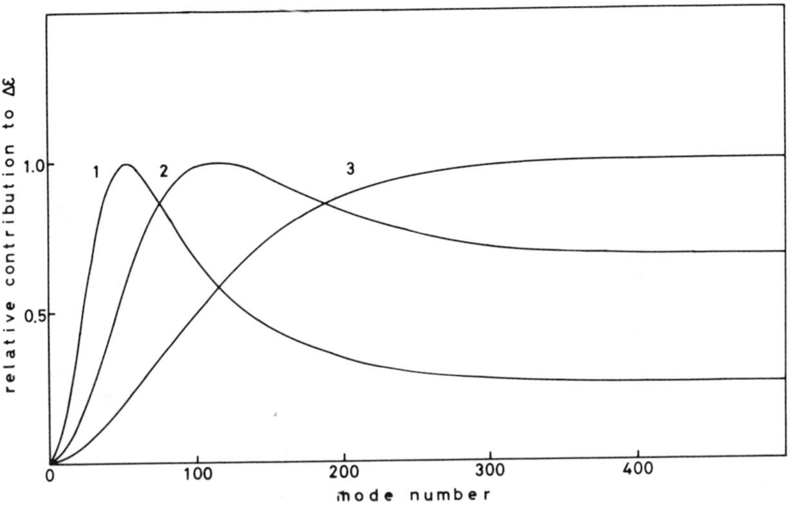

FIGURE 5. Relative contribution of each mode to $\Delta\epsilon$ for high frequency region. Details are the same as in FIGURE 4; 1, $\omega\tau_1 = 10^3$, 2, $\omega\tau_1 = 3 \times 10^3$, and 3, $\omega\tau_1 = 10^4$.

the external conditions, this may alter the dispersion profile, as will be discussed later.

The relative contribution of each mode to the conductivity is also shown in FIGURE 6. In this case, each mode makes the same contribution to the conductivity increment at high-frequency limit. FIGURE 6 indicates that at an intermediate frequency, only small numbers of modes with longer wavelengths contribute to the conductivity.

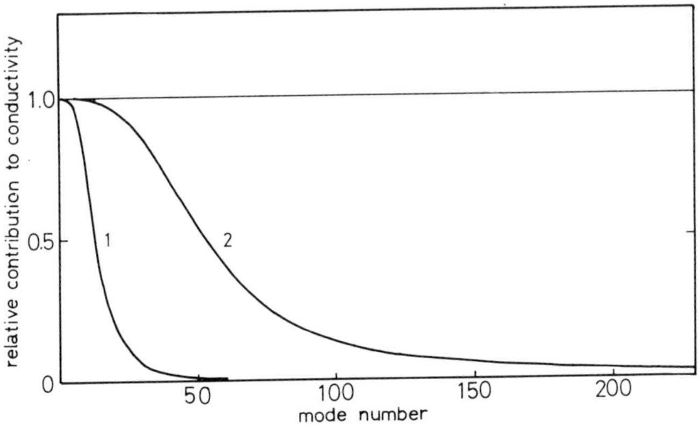

FIGURE 6. Relative contribution of each mode to the conductivity; 1, $\omega\tau_1 = 10^2$, and 2, $\omega\tau_1 = 10^3$

Effect of the Outer Medium

The effect of the outer medium is not considered in the above derivation of the polarizability of a polyion, except for an introduction of the parameter B, which represents the internal field correction. In a conductive medium, however, not only the longitudinal diffusion but transversal diffusion may take place with respect to the rod, which may affect the mode of the fluctuation. The mean square fluctuation of mode q, $<|\delta n_q|^2>$, which represents the deviation of the counterion concentration along the rod from the average value, is thought to be reduced by the increase of the ion flow. The counterion-counterion interaction, however, is also reduced because of the decrease in $\phi(q)$ of low mode numbers. It is difficult to determine which effect is stronger.

Besides the above consideration, it is certain that the main mechanism of the polarization is due to the limited motion of the bound counterions, as mentioned previously. This means that the polarization of the polyion is not changed so much by the presence of the outer medium. Experimentally, however, a decrease in dielectric increment was reported for DNA solution by addition of salts.[11] It therefore seems appropriate that the amplitude of each mode of the fluctuation is thought to be depressed equally by the presence of the conductive medium.

On the other hand, the relaxation process is altered by the presence of the medium. This is explained by introduction of the perpendicular diffusion process to the rod. This process reduces the relaxation time of the fluctuation, since it provides another way of movement of the condensed ions. The perpendicular diffusion is treated theoretically by adding the following two terms to the right hand side of (**25**):

$$<B_j(1 - n_j)>_t - <A_j n_j>_t \qquad \textbf{(33)}$$

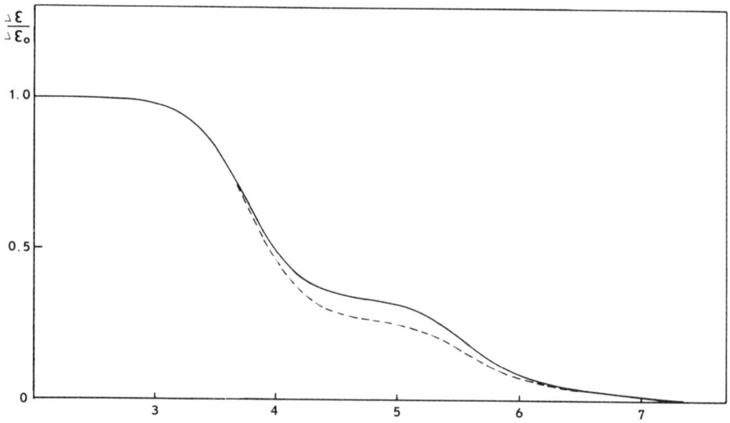

FIGURE 7. Effects of the outside medium on dielectric dispersion. $W = 3 \times 10^7$; ——, $c/W = 10^{-4}$ for 1st and 2nd modes; $c/W = 10^{-2}$ for higher modes; -----, $c/W = 10^{-4}$ for 1st, 2nd, and 3rd modes; $c/W = 10^{-2}$ for higher modes. Other parameters are the same as those in FIGURE 3.

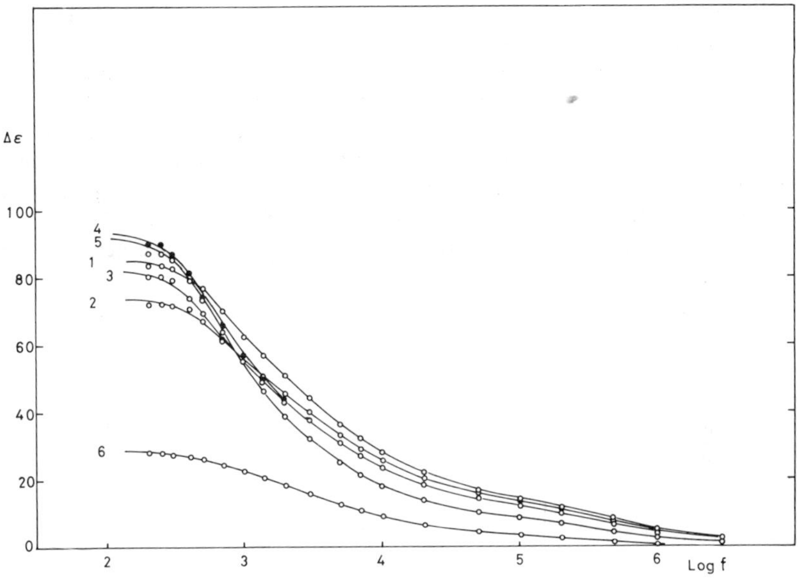

FIGURE 8. Dielectric dispersion of poly(maleic acid-co-methyl vinyl ether) neutralized with mixtures of Na^+ and Ca^{++} ions; $C_p = 0.5$ mN, degree of neutralization, $\beta = 1$. Counterion ratios (Na^+:Ca^{++}) are: 1, 10:0, 2, 9:1, 3, 8:2, 4, 7:3, 5, 5:5, 6, 2:8.

The rate constant B_j is assumed to be constant, since an ion in the outer medium can move under averaged interaction energy over some wide domain, whereas A_j depends on the interaction energy ϵ_j in the same way as the case of $W_{j-1,j}$, so that

$$B_j = c$$

$$A_j = c' \exp(\delta\epsilon_j/kT). \quad \textbf{(34)}$$

The principle of detailed balance for this process leads to the relation between the two constants c and c' as

$$c(1 - n_0) = c'n_0. \quad \textbf{(35)}$$

After substitution of (**34**) and (**35**) into (**33**), (**26**) becomes

$$(d/dt)\langle\delta n_q\rangle_t = -[2W(1 - \cos q) + c/n_0] \times [1 + nn_0(1 - n_0)\phi(q)/kT]\langle\delta n_q\rangle_t. \quad \textbf{(36)}$$

The relaxation time of mode q is

$$\tau_q^{-1} = [2W(1 - \cos q) + c/n_0][1 + nn_0(1 - n_0)\phi(q)/kT]. \quad \textbf{(37)}$$

The addition of the term c/n_0 means the decrease of the relaxation time. However, the perpendicular diffusion does not immediately result in the decrease of the dipole moment in the direction along the polyion, because the mean square fluctuation $<|\delta n_q|^2>$ is given by (**18**).

In the case of the rodlike molecule, the perpendicular diffusion is mostly prohibited because of the large barrier of the interaction energy ϵ_j to the outside medium. On the other hand, a flexible polyelectrolyte molecule takes a fairly stretched but randomly coiled conformation forming a polymer domain. On the basis of the two-phase model of the polyelectrolyte solution,[1,9] this region is characterized by the potential trough due to the charge of the polyion, where ions are condensed but freely mobile within this region. In this case, the perpendicular dif-

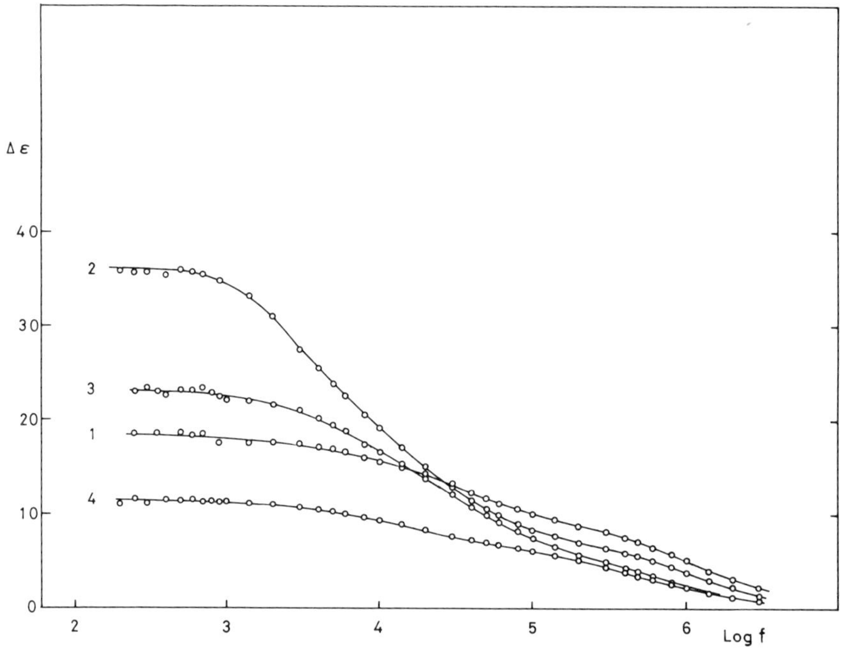

FIGURE 9. Dielectric dispersion of poly(maleic acid-co-ethyl vinyl ether) neutralized with mixtures of Na^+ and Mg^{++} ions; C_p = 0.8 mN, β = 0.5. Counterion ratios (Na^+:Mg^{++}) are: 1, 10:0, 2, 8:2, 3, 5:5, 4, 0:10.

fusion takes place more easily. It may contribute to the formation of the relatively higher modes of the fluctuation by the ion movement in a short-cut way. If we assume that the region where the perpendicular diffusion process is effective is about the same as the width of the ellipsoidal polymer domain, we can obtain much larger values of the constant c for the modes, the wavelength of which is comparable or smaller than the width. This results in the decrease in τ_q of these modes. As a result, the high-frequency side of the dispersion curve is shifted toward higher frequency, leading to the separation of the dispersion curve into two. The dispersion in the high-frequency region is thus characterized by the perpendicular diffusion process.

Numerical calculation is performed to demonstrate the separation by assuming the axial ratio of the polymer domain as 3 and 4, and by assuming the value of the parameter $W = 3 \times 10^{7}$ and $c/W = 10^{-4}$ for the lower modes and $c/W = 10^{-2}$ for the higher modes, as plotted in FIGURE 7. Such a dispersion profile is usually observed experimentally, as seen in FIGURE 8 for poly(maleic acid-co-methyl vinyl ether) neutralized with mixtures of Na^+ and Ca^{++} ions, and in FIGURE 9 for poly(maleic acid-co-ethyl vinyl ether) neutralized with mixtures of Na^+ and Mg^{++} ions.

DISCUSSION

As previously stated, the dielectric dispersion process for the polyelectrolyte solutions is divided into two parts. The dielectric increment in the low-frequency limit is now considered to be due to the ionic polarization resulting from the fluctuation of bound ions in the longitudinal direction.[1,2,4–6,8,12,13] The total increment is roughly proportional to the square of the length of the polyion, and is increased by the addition of some kinds of divalent ions, as shown in FIGURES 8 and 9. These characters are well explained by several models.[1,2,5,8,12,13]

On the other hand, the dispersion in the high-frequency region (HFD) has been believed to be due to the Maxwell-Wagner effect.[4,14,15] Several qualitative characters of HFD have been explained by this mechanism. For example, the decrease in the relaxation time with increasing concentration was attributed to the increase in the conductivity of both domains.[14]

Van der Touw and Mandel,[13] however, proposed that this dispersion was due to the relaxation of the fluctuation in the subunits of flexible polymers. The size of the subunit was characterized by either the dielectric increment of HFD or the relaxation time of HFD in the limit of low polymer concentration. At finite concentration, these two quantities gave us largely different values for the size of the subunit. The main difficulty is, however, the detection of such a subunit structure, which has never been reported by other experiments. The Rouse-Zimm model,[16,17] which is widely applied to viscoelastic properties of polymers, is not shown to predict such structures.

The present model explains several features of HFD, as follows. A sharp decrease in the relaxation time with increasing polymer concentration is due to the shrinkage of the polyions resulting in the extension of the effective region of the perpendicular diffusion, as well as increase in the potential depth of the polymer domain. Insensitivity to the presence of multivalent ions is due to the nature of the perpendicular diffusion; that is, this process is understood, without introduction of the discrete lattice. On the other hand, anomalous increase in the dielectric increment by the presence of divalent ions is explained only in terms of the lattice model.[5,12] The idea that an ion moves within the polymer domain freely but largely restricted to the outer medium is the same as the concept of the interfacial polarization. This is why introduction of the perpendicular diffusion process satisfies several features of the interfacial polarization mechanism such as the Maxwell-Wagner effect.

Presence of the outer medium results in another process: longitudinal motion of ions at the end of the polyion. This process is thought to be similar to the perpendicular process, but must be coupled with the longitudinal diffusion along the rod. At present, it seems difficult to formulate theoretically this process from the microscopic point of view.

References

1. Oosawa, F. 1971. Polyelectrolytes. Marcel Dekker, Inc. New York, N.Y.
2. Warashina, A. & A. Minakata. 1973. J. Chem. Phys. **58:** 4743–4749.
3. Schwan, H. P., G. Schwarz, J. Maczuk & H. Pauly. 1962. J. Phys. Chem. **66:** 2626–2635.
4. Minakata, A. & N. Imai. 1972. Biopolymers **11:** 329–346.
5. Minakata, A., N. Imai & F. Oosawa. 1972. Biopolymers **11:** 347–359.
6. Minakata, A. 1972. Biopolymers **11:** 1567–1582.
7. Kubo, R. 1957. J. Phys. Soc. Japan **12:** 570–586.
8. Oosawa, F. 1970. Biopolymers **9:** 677–688.
9. Oosawa, F. 1957. J. Polymer Sci. **23:** 421–430.
10. Brout, R. H. 1965. Phase Transitions. Benjamin. New York, N.Y.
11. Sakamoto, M., H. Kanda, R. Hayakawa & Y. Wada. 1976. Biopolymers **15:** 879–892.
12. Mandel, M. 1961. Mol. Phys. **4:** 489–496.
13. Van Der Touw, F. & M. Mandel. 1974. Biophys. Chem. **2:** 218–230.
14. Dintzis, H. M., J. L. Oncley & R. M. Fuoss. 1954. Proc. Natl. Acad. Sci. USA **40:** 62–70.
15. O'Konski, C. T. 1960. J. Phys. Chem. **64:** 605–619.
16. Rouse, P. E. 1953. J. Chem. Phys. **21:** 1272–1280.
17. Zimm, B. H. 1956. J. Chem. Phys. **24:** 269–278.

DISCUSSION

Robert H. Cole, *Moderator*

Department of Chemistry
Brown University
Providence, Rhode Island 02912

DR. MANDEL: I would like to comment about what Dr. Minakata said, and ask him some questions. First of all, I am glad to see that, just as I did, he needed to introduce flexibility of the chain in order to understand the dispersion behavior of his polyelectrolytes.

One can argue, of course, about the model one uses. We were certainly aware that the model we used was highly artificial. On the other hand, it is striking that consistent results are obtained. I think this shows that the model, although it seems artificial as it is described on paper, has some bearing on the physics of what is happening.

I do not think we can involve the Rouse-Zimm Equation, because we are not treating the motion of the chain, although it may also have some very interesting properties. We are looking, instead, at the motion of the counterions along the chain, and I wonder whether introducing subunits is any more artificial than introducing longitudinal and perpendicular diffusion models. In either case, you are artificially decomposing the overall motion of the counterions into two parts. That may be roughly equivalent to decomposing the motion of ions of the whole molecules into two parts. I do not know which is the better approach. Probably one should treat, in the most general way, the overall motion that takes into account deviation from linear behavior.

Also, if I understand correctly, you introduced the perpendicular motion of the counterions but still kept the rigid rod model. That, of course, is a very strong approximation for the flexible macromolecule. These are my comments; now I have a question.

When you take into account the interaction between the counterions, you introduce electrostatic forces only. Don't you think that that approach strongly exaggerates the interaction, since the influence of the opposite charge on the chain is not taken into account? On the other hand, is it not wise when you treat interactions between the counterions along the chain, also to consider interactions between them and the polyions?

DR. MINAKATA: I will try to respond to the question. Polyions are flexible in the dynamical sense, and the chain configurations fluctuate, but in this case the chain configurations should be averaged as compared with the motion of counterions. Also, the existence of a definite subunit structure seems to be doubtful. This is why I think the description of the chain configuration is of secondary importance. In my analysis I assumed that the longitudinal diffusion process is a special one, which takes place by a successive counterion jumping mechanism to adjacent counterion sites along the polyion. The other process is different, however, in that the jumping is directly from the counterion sites to the outside medium. This describes the perpendicular diffusion. So the longitudinal and perpendicular jumping mechanisms are not assumed to induce their respective polarizations in the same manner. Interaction between an ion and fixed charges on the polyion is included in my analysis as a constant term.

DR. R. MOLINARI (*Brown University, Providence, R.I.*): Dr. Minakata, and, perhaps, Dr. Mandel, the specific dielectric increment in these polyelectrolytes depends very strongly on concentration. As an example, Professor Mandel proposed in his talk an empirical extrapolation of concentration. The problem is that, as concentration changes, other things, such as charge fraction, mean square end-to-end lengths, and polymer-salt interactions, also change. All these phenomena affect the dielectric increment. Therefore, my question is, what exactly do we assume or imply in extrapolating (as in Dr. Mandel's paper) to zero concentration? More generally, how do we account for all of these simultaneously changing quantities by any one single extrapolation designed to eliminate concentration effect? The goal obviously is to find a convenient dielectric-related quantity for comparison of the dielectric properties of different samples or sizes of polyelectrolytes independent of concentration effects.

DR. MINAKATA: Concentration dependence is a very complicated problem. This is one case, and I can show you another case for polyelectrolyte solutions. That is, this is a very difficult problem in the case of viscosity. The same is true in the case of dielectric behavior: the concentration dependence is remarkable not only in the total increment, but also in the increment in the high-frequency region. Even if the measured properties have no direct theoretical significance at finite concentrations, we must nevertheless define molecular parameters at (perhaps rather high) finite concentrations.

DR. MOLINARI: Restated, perhaps my question should be "Is there any theoretical basis to separate out from the concentration dependence of the dielectric increment the component parts of, say, charge-fraction dependence, polymer-polymer or polymer-salt interactions, and so on, all of which change simultaneously with concentration?"

DR. MANDEL: I don't think any theory up to now takes into account the interaction effect between the polyions. If you want these effects, you should make a theory in which these come in explicitly. I don't think I know of any properties of polyelectrolytes where the interactions are really put into theory.

On the other hand, that these concentration effects are very important is certainly shown in the effect of divalent counterions. If you dilute the solution sufficiently, you can see reversals of the effects which are considered normal behavior. In that case, certainly, the interaction between the polyelectrolytes plays a very important role in determining even the direction of the concentration dependence.

DR. MINAKATA: When polyelectrolyte solutions are measured at very high dilution, a single dispersion, instead of the familar two-step process, tends to fit the data best. The extrapolation of the observed dispersion curve to infinite frequency supports the assertion that there is only a single broad dispersion. This means that even the number of recognizable dispersions changes as a function of concentrations, providing further complication to any theoretical analysis of the concentration dependence of the specific dielectric increment.

AGGREGATION AND CONFORMATIONAL CHANGES IN SOLUTION STUDIED BY DIELECTRIC ABSORPTION OF POLYPEPTIDES AND SUBSTITUTED POLYSACCHARIDES

Eve Marchal

Centre National de la Recherche Scientifique
Centre de Recherches sur les Macromolécules
6, rue Boussingault
67000 Strasbourg, France

INTRODUCTION

In 1964, Tobolsky[1] pointed out that isolated polymer chains of extended single-strand helices that are unstabilized by electrical charges are a relatively rare phenomenon. Definitely established cases seem to be caused by unusual solvent effects. On a thermodynamic basis, Tobolsky suspected that in dilute solution, many macromolecules that are capable of crystallization arrange themselves (fold up) into "lattices" or associates of single molecules. The objective of this work was to study, by dielectric absorption measurements, polymer solutions in "usual" solvents. In this way, association of rodlike polar macromolecules, mainly uncharged polypeptides that are in helical configuration in solution and that have been—or can be—characterized in their isolated state, can be conveniently investigated. We have limited our investigations to concentrations lower than those for which liquid crystalline phases occur. Some factors leading to increasing aggregation are described first. This will be followed by the study of a disaggregation agent which, in a further step, induces the conformational change of the isolated helix to a coil. Finally, we shall mention that other types of conformational transitions may occur in biological macromolecules such as substituted polysaccharides.

METHODS

All the isolated polymers under study here bear high permanent dipole moments along the direction of the end-to-end distance and thus show low-frequency molecular weight-dependent relaxations in dilute solution due to orientation of the rigid particles. Basically, we compare the dipole moment μ and critical frequency of relaxation f_c of the aggregate to those found for the unaggregated macromolecules. For these last, μ and f_c were found to obey laws of the type

$$f_c = kM^{-\beta}$$

$$\mu^2/M = k'M^{\alpha}$$

where α and β are constant in the limited range of molecular weights M examined. The experimentally accessible values by dielectric absorption are f_c (the frequency at which the real part of the dielectric constant of the solution minus that of the solvent has decreased to half its value) and $\Delta\epsilon$ (the change of dielectric constant due to absorption). Calling w the weight fraction of the solute, $\Delta\epsilon/w$ is proportional to μ^2/M; one thus has

$$\Delta\epsilon/w = K f_c^{-\alpha/\beta}.$$

This is reflected on a log $\Delta\epsilon/w$ versus log f_c plot by a straight line. Although neither its slope nor its ordinate are characteristic of the rigidity of a polymer, it can be used on a qualitative basis for the study of aggregate structure in a given polymer-solvent system. The dipole moment and critical frequency of a head-to-tail aggregate having the same rigidity as the isolated species will be such that these values can be represented by a point on the line. Besides this type (A), other basic patterns of aggregation are

linear head-to-head (B);
side-by-side with parallel orientation (C);
side-by-side with antiparallel orientation (D);
random (E).

Type B aggregation is highly improbable. For type C, the point would be above the line, with f_c not appreciably changed. For type D, $\Delta\epsilon/w$ would be 0 for an even number of aggregates. For an odd number, $\Delta\epsilon/w$ is unchanged, and f_c hardly changed. For type E, $\Delta\epsilon/w$ is unchanged, f_c would be lower. For mixtures of different aggregation types, the position of the point will give information on the predominant behavior.

The dielectric measurements were carried out in the frequency range 20 Hz–2 MHz by the General Radio Type 1616 Capacitance Bridge and two impedance bridges described elsewhere.[2] Cylindrical stainless-steel cells of 11, 89, and 196 pf air capacitance were used.

Aggregation

Effect of Side Chains and Solvent on Aggregation of Polypeptides

The determination of the dipole moment of the macromolecules from $\Delta\epsilon/w$ measured in solvents whose dielectric constant varies in a broad range is not straightforward, since it depends on the relation between the applied field, the directing field, and the internal field. This relation is still under discussion.[3] When solvent is adsorbed on the polymer, an apparent dipole moment can be calculated that includes the dipole moment of the solvent layer and the dipole moment of the polymer. Quantitative comparisons can be obtained only in solvents of low dielectric constant, or by using different polymers in one solvent.

We first compare two polypeptides, poly(D,L-phenylalanine) (PDLPA) and poly (β benzyl-L-aspartate) in chloroform;[4] the asymetric carbon atom bears a nonpolar radical in the former; it has a polar ester group in the second. FIGURES 1 and 2* show that in both cases the aggregates formed in chloroform at low concentration ($10^{-4} < w < 4.10^{-4}$) fall above the line found when chloroform + 0.5%

*Throughout this paper, η is the solvent viscosity at temperature T.

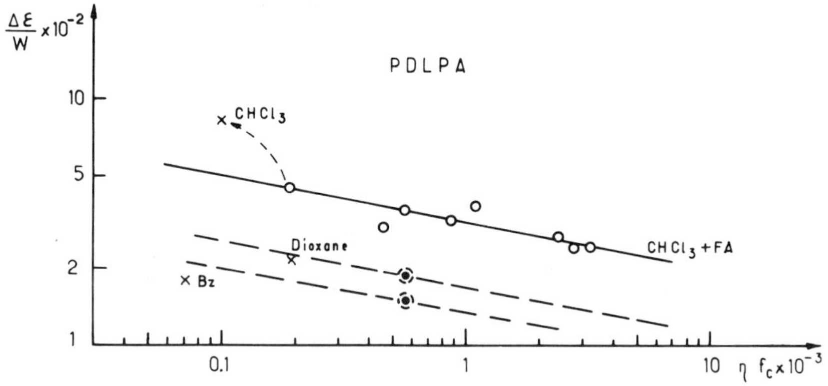

FIGURE 1. Log ($\Delta\epsilon/w$) vs log (ηf_c) for PDLPA in chloroform +0.5% FA(○), benzene and dioxane (●). Aggregated species in these solvents (x).

formamide (FA), where no aggregation occurs,[50] is used. Since f_c has decreased simultaneously, elongated bundles of side-by-side helices are formed.

We now compare PDLPA in benzene and dioxane. FIGURE 1 shows that the points lie on the line characteristic of α-helices, calculated from the dielectric measurements in chloroform +0,5 % FA. The aggregates are formed by a small average number n (1.7 in dioxane, 2.7 in benzene) of head-to-tail molecules. Thus, at low concentration, solvent seems to have a stronger influence on aggregate structure than the polarity of the side chain, as far as the ester group is concerned.

On the other hand, a pronounced effect of the side chain is observed when comparing poly (γ phenacyl-L-glutamate) (PPLG) and poly (γ benzyl-L-gluta-

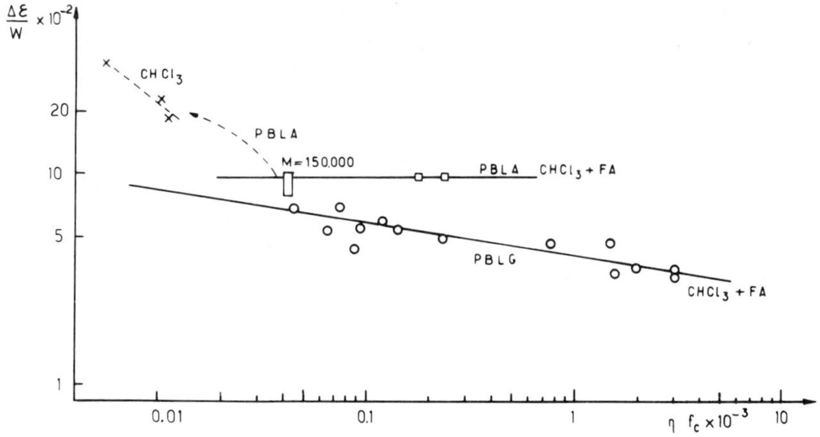

FIGURE 2. Log ($\Delta\epsilon/w$) vs log (ηf_c) for isolated PBLA helices (top curve). Aggregated PBLA in chloroform (x).

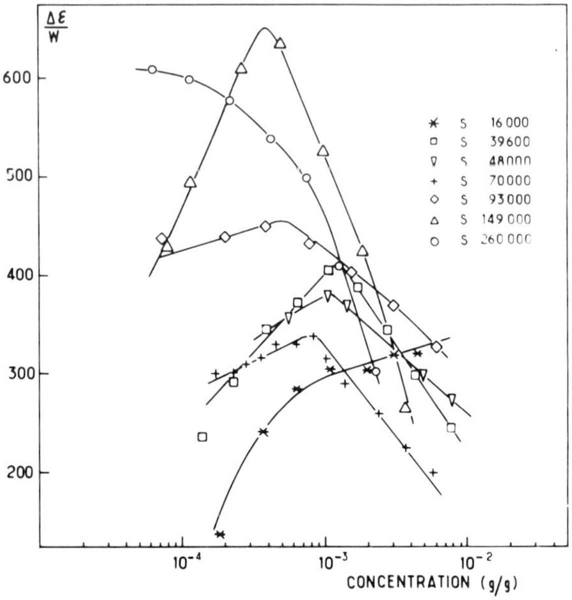

FIGURE 3. Variation of $\Delta\epsilon/w$ with concentration for various samples of PBLG in dioxane.

mate)[5] (PBLG). Strong side chain-side chain interactions occur in PPLG. In pyridine where PBLG is unaggregated, the dielectric behavior of PPLG in dilute solution depends on molecular weight. For $M = 26{,}000$, a side-by-side parallel-type aggregation is found, whereas for $M = 60{,}000$, the simultaneous presence of both isolated folded helices and aggregates of these accounts for the observations. Unfolding and disaggregation are achieved above 38°C, and the same dielectric behavior as PBLG is observed.

For PBLG, Watanabe,[6] Powers and Peticolas,[7] and Tsuji[8] by electrooptical investigations, and Wada,[9] Kihara,[10] and Block *et al.*[11] by dielectric measurements have studied aggregation in various solvents and solvent mixtures. Although, as will be shown in the following section, aggregation depends on concentration and *M* in certain cases, in a sufficiently high concentration range in dioxane and benzene (or a sufficiently high *M* sample), aggregation is found to be antiparallel side by side, whereas for the same samples and concentration ranges PBLG tends to form head-to-tail linear aggregates in ethylene dichloride and chloroform.

Thus we may conclude that at constant polypeptide concentration, in the absence of specific strong side-chain interactions, the solvent has a dominant effect on the type of aggregation.

Influence of Concentration and Molecular Weight

PBLG

In our laboratory, Gupta[12] examined the change in aggregate structure of PBLG in dioxane with concentration and molecular weight. FIGURE 3 shows the variation of $\Delta\epsilon/w$ vs. concentration for various samples. In this figure, the numbers after the letter S stand for M. Except for the two extreme molecular weights, a maximum is observed; in the same concentration ranges, critical frequencies decrease continuously, except for sample S 260,000, where no change is observed. Using again the double logarithmic plot discussed above, FIGURE 4 shows that when the concentration is lower than that (w_c or critical concentration) at which $\Delta\epsilon/w$ is maximum, aggregation proceeds in a head-to-tail fashion, since the points are situated near the line. Above w_c, antiparallel structures are observed of increasing overall size as concentration increases. Takashima,[13] by examining the dielectric behavior of high molecular weight PBLG in shear gradients, seems to have found that a small transverse dipole moment exists in the PBLG aggregates, and he estimated the transverse dimension of these aggregates to be 150 Å in dioxane. Barisas[48] made similar measurements in a mixture of polychlorinated biphenyl acid and tetrachloroethane and did not interpret his measurements this way.

These results were confirmed by light-scattering measurements:[14] 1) the mass/unit length of the particles are invariant below w_c and conforms to the value of the unaggregated helices; 2) below w_c the molecular weight obtained by light-scattering measurements is equal to that (hereafter called Magg) which can be determined from the dielectric measurements by using the log f_c vs. log M and log $\Delta\epsilon/w$ vs. log M plots for unaggregated PBLG.[50] The ratio Magg$/M$ allows the determination of the average degree of aggregation n; 3) the radius of gyration increases with increasing w for $w < w_c$; 4) above the critical concentration a sharp change in the zero angle scattering intensity indicates the change in structure of the solution and is in good agreement with the assumption of an increasing order.

FIGURE 3 also shows that the critical concentration increases as M is raised. The molecular weights, as well as the average degrees of aggregation at w_c, are given in TABLE 1. It appears that the head-to-tail aggregation proceeds as long as $M(n - 1) \leq 220{,}000$. Thus, if the molecular weight of the unaggregated PBLG

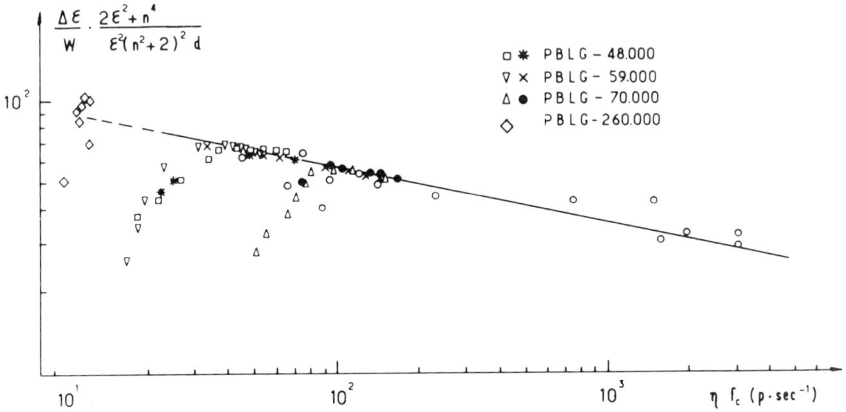

FIGURE 4. Effect of concentration on $K\mu^{-2}/M$ (ordinate) and $\eta_0 f_c$ (abscissa) for various PBLG samples in dioxane. Concentration increases as $\eta_0 f_c$ decreases, except for ◇, where concentration increases as $K\mu^{-2}/M$ decreases. ○ = unaggregated PBLG.

TABLE 1

VALUES OF MAXIMUM MOLECULAR WEIGHTS OF HEAD-TO-TAIL AGGREGATES OF PBLG IN DIOXANE

Sample	Critical Concentration (g/g)	Max. Mol. Wt. of Head-to-Tail Aggregates	Average Degree of Aggregation
S-39600	1.2×10^{-3}	215000	5.4
S-48000	1.1×10^{-3}	210000	4.4
S-70000	8.5×10^{-4}	170000	2.4
S-93000	5.5×10^{-4}	225000	2.4
S-149000	3.8×10^{-4}	300000	2.0

exceeds 220,000, it will aggregate only in the antiparallel fashion. For low molecular weight samples, this antiparallel aggregation will occur at very high concentrations. In order to study the mechanism of head-to-tail aggregation occurring if $w < w_c$ we apply the following scheme, in which particles of any size (aggregates of any degree of aggregation) may aggregate with each other:

$$\begin{aligned} M_{\mathrm{I}} + M_{\mathrm{I}} &\rightleftharpoons M_{\mathrm{II}} \\ M_{\mathrm{II}} + M_{\mathrm{I}} &\rightleftharpoons M_{\mathrm{III}} \\ M_{\mathrm{II}} + M_{\mathrm{II}} &\rightleftharpoons M_{\mathrm{IV}} \\ &\vdots \\ M_i + M_j &\rightleftharpoons M_n \end{aligned} \tag{1}$$

$n = i + j$. Here the subscripts denote the degree of aggregation such that M_{I} represents an unaggregated molecule (unimer), M_{II} an aggregate of two unimers, and so on. Assuming identical equilibrium constant K for each of these reactions, one can write:

$$\frac{c}{M} = [M_{\mathrm{I}}] + 2K[M_{\mathrm{I}}]^2 + 3K[M_{\mathrm{I}}]^3 + \ldots + nK^{n-1}[M_{\mathrm{I}}]^n \tag{2}$$

If $\bar{n}$ be the average degree of aggregation, we have:

$$\frac{c}{\bar{n}M} = [M_{\mathrm{I}}] + K[M_{\mathrm{I}}]^2 + K^2[M_{\mathrm{I}}]^3 + \ldots + K^{n-1}[M_{\mathrm{I}}]^n \tag{3}$$

where c is the concentration, M the number average molecular weight of the unaggregated molecules, and the quantities in the bracket are the molar concentrations. For $K[M_{\mathrm{I}}] < 1$ and for sufficiently large values of n, (**2**) and (**3**) can be summed to give the following:

$$\frac{c}{M} = \frac{[M_{\mathrm{I}}]}{(1 - K[M_{\mathrm{I}}])^2} \tag{4}$$

and

$$\frac{c}{\bar{n}M} = \frac{[M_{\mathrm{I}}]}{1 - K[M_{\mathrm{I}}]}. \tag{5}$$

These two equations, (**4**) and (**5**), yield:

$$K = \frac{\bar{n}(\bar{n} - 1)M}{c}. \qquad (\mathbf{6})$$

The experimental results on sample S-16000 in the investigated concentration range 0.19 g/l $< w <$ 4.4 g/l ($4 < \bar{n} < 14$) gave a constant value of the equilibrium constant, $K = 7.6 \times 10^5$. Average n was obtained from the ratio of the aggregate molecular weight determined according to the procedure described above and the sample molecular weight. We assume that $\bar{n}$ thus determined and M_{agg} determined by light-scattering (weight average) are not far from their number average values. For the other higher molecular weight samples, the degrees of aggregation were very low and the above expression cannot be used. The free energy of aggregate formation is $\Delta G = -RT\ln K = -7$ Kcal/mol for PBLG in dioxane. Furthermore, temperature induces disaggregation.[3] From the temperature variation of n in dioxane the enthalpy of aggregation ΔH was found to be ~ -3 Kcal/mol.

Weill[15] has shown that it is possible to account for the fact that above a certain length head-to-tail aggregates are energetically favored up to a rather large size, above which side-by-side antiparallel aggregates appear on a simple electrostatic basis, taking into account the difference in the distance between the helices.

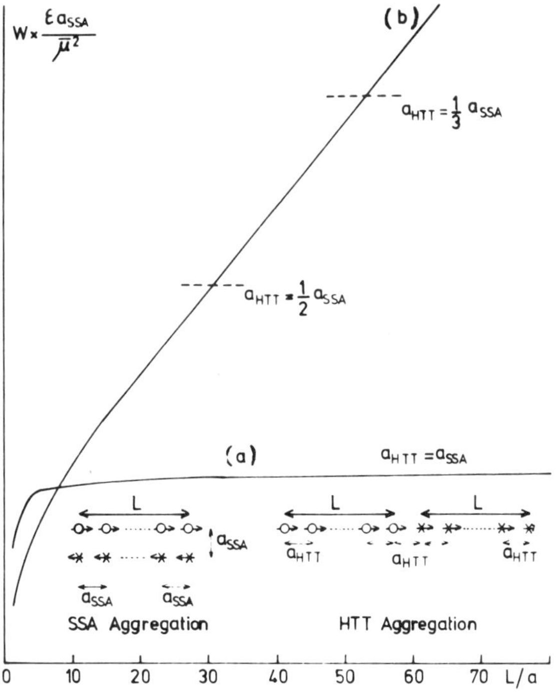

FIGURE 5. Interaction energy versus the number of point dipoles in helices calculated for (a) head to tail aggregation; (b) side-by-side aggregation. Bottom: schematic model used.

He calculated the interaction energy W of head-to-tail (HTT) as well as side-by-side antiparallel arrangements (SSA) of point dipoles regularly spaced along a helix of length L at distances equal to the helix separation a. This model is shown in FIGURE 5. Assuming a dipole moment $\bar{\mu}$ per unit length the variation of interaction energy in terms of $(W/\mu^2)\epsilon a$ as a function of L/a is also shown in this figure for HTT (curve a) and for SSA arrangements (curve b). The intersection of these curves, which defines the helix length above which SSA is preferred, occurs at $L_c/a = 9$. Taking $\bar{\mu} \sim 3$ Debye/1.5 Å, $\Delta H = 3$ Kcal/mol $\sim$ 5RT and $\epsilon = \epsilon_{\text{dioxane}} = 2.2$ one gets $a \sim 9$ Å. Assuming now $a_{\text{SSA}} = 2\, a_{\text{HTT}}$ corresponding to a helix diameter of 18 Å (a reasonable value compared to the crystallographic pack-

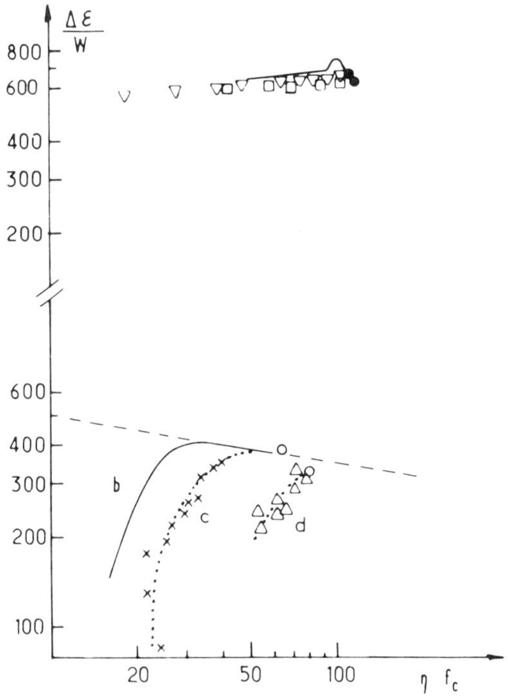

FIGURE 6. Log ($\Delta\epsilon/w$) vs. log ηf_c

Bottom: PBLG
---- = α helix;
b = without PS;
○ = without PS;
cx = PS (M = 80,000);
$d\Delta$ = PS ($M = 4 \times 10^6$).

Top: PHIC
—— = without PS;
○ = without PS;
▽ = PS (M = 80,000);
□ = PS ($M = 4 \times 10^6$).

Increasing PS concentrations are shown from right to left.

ing distance 14.5 Å), $L_c/a \sim 30$ (see dotted line in FIGURE 5) and $L_c = 540$ Å corresponding to $M \sim 80{,}000$. This value is smaller than experimentally found. However, the assumption that point dipoles of $2 \times 9 = 18$ Debyes are regularly spaced 9 Å apart is crude. Furthermore, the value $\bar{\mu} = 3$D may be wrong, and/or the helices may be more highly separated in the SSA arrangement. Weill also points out that disalignment of the helices very quickly decreases the attractive energy, and this may explain the very restricted flexibility of the aggregates. A decrease between the gap between helix ends, as would occur if they start to unfold or if foreign molecules accumulate, would decrease the aggregation enthalpy and therefore the limiting length. This could, in fact, be the situation when small amounts of dichloroacetic acid (DCA) are added to the solution (see DISAGGREGA-

TION AND HELIX COIL TRANSITION OF POLYPEPTIDES). It would be interesting to verify this by studying the thermal disaggregation and concentration effect in dioxane in the presence of small amounts of DCA.

It seems interesting to mention at this stage that Reibel *et al.*[16] showed that in a three-block copolymer consisting of two helical PBLG blocks separated by a central coiled block of poly(ethyleneoxide), the PBLG helices are antiparallel head-to-head, with slight overlapping in dilute pyridine solutions. Thus the constraint brought by the central coiled block prevents the expected head-to-head or antiparallel side-by-side conformations. It would be interesting to study the aggregation of such copolymers.

For homo-PBLG, another way to monitor aggregation is by adding polystyrene (PS).[17] Even though the PBLG concentration is constant and lower than w_c, increasing PS induces increasing aggregation of the antiparallel type. This is shown on the lower half of FIGURE 6, where downward curvatures are obtained for increasing amounts of PS, in a qualitatively similar manner as the full line in FIGURE 6, which is obtained for increasing concentrations of PBLG above w_c. It is thus possible to induce antiparallel aggregation of high critical frequencies, since the PS entanglements seem to prevent the initial head-to-tail growth of the PBLG aggregates in dioxane. This aggregation by a foreign nonpolar component may have some implications in liquid crystalline-phase occurrence and regulation in living substances.[18]

The effects described above do not alone regulate association. Brown and Jones[19] and Tsuji[8] have observed that high electric fields promote aggregation. In dichloroethane, field effect persists after the removal of the field. Electric and magnetic fields are known to induce phase changes in liquid crystals of polypeptides.[20] High electric fields may occur in living substances due to, for example, charge accumulation at membrane interfaces. Temperature also plays an important role, as we have already stated. Adachi *et al.*[21] have observed a discontinuity in the ϵ versus T curve for PBLG in dichloromethane at low temperature in dilute solution, which they attributed to a phase transition.

*Poly(*n *hexylisocyanate)*

The solution properties of polyalkylisocyanates are very similar to those observed for polypeptides. An apparent helical conformation and high-chain stiffness have been found to describe both poly *n*-butyl and *n*-hexylisocyanates. Electric birefringence[22,23] and dielectric relaxation[24–26] measurements have shown that they posess high dipole moments. We have observed,[27] on the basis of dielectric relaxation and electric field birefringence, that the trends of the results for poly(*n* hexylisocyanates) (PHIC) in toluene and for PBLG in dioxane are the same when concentration and molecular weight vary. The overall magnitude of the changes are, however, smaller for PHIC. It seems noteworthy to mention here that the maximum length and maximum dipole moment supported by a linear head-to-tail aggregate are of the same order for the two systems.

The behavior of a dilute PHIC solution when polystyrene is present[17] is shown on the upper curve in FIGURE 7. It resembles that of the PBLG-PS-dioxane sys-

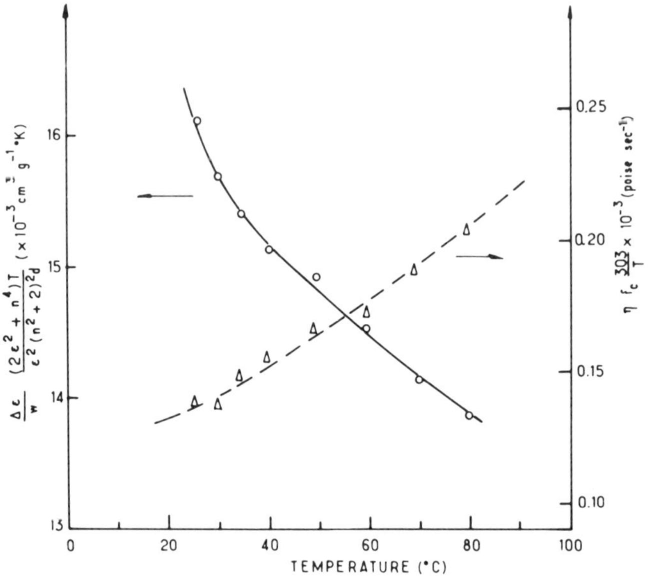

FIGURE 7. Change in dielectric increment and critical frequency with temperature for PBLG in dioxane.

tem: antiparallel aggregation increases as PS is progressively added. Thus the above features of aggregation seems to be general for polar particles of high asymetry.

DISAGGREGATION AND HELIX-COIL TRANSITION OF POLYPEPTIDES

Poly(benzyl-L-glutamate)[3]

We mentioned earlier that disaggregation of PBLG in dioxane can be produced by heating. This is illustrated in FIGURE 7, where the change in dielectric constant and critical frequency are reported for $w = 5.4 \times 10^{-4}$. The factor given with $\Delta\epsilon/w$ on the ordinate axis in this figure is the proportionality factor appearing in the relation between $\mu^2/M(4\pi N_A/9kT)$ and $\Delta\epsilon/w$ when the Barriol and Weissbecker[28] formulation is used. N_A is Avogadro's number, ϵ and n the solvent dielectric constant and refractive index, respectively, and d the solution density. The factor given with f_c includes η, since $\eta f_c/T$ is constant for a constant particle shape. These experimental results are shown in FIGURE 8. They fall on the straight line determined for α helices, meaning that the process involves breaking of the head-to-tail hinges.

Destruction of the aggregates also occurs with the addition of dichloroacetic acid (DCA) to dioxane solutions. FIGURES 8 and 9 show that it is complete when 30% DCA are added, since $\Delta\epsilon/w$ and f_c have then reached their respective values for the isolated polypeptide. No dipole moment change which would be due to protonation is observed. In the range 30–80% DCA, the high $\Delta\epsilon/w$ and the constant f_c values reflect the preferential DCA absorption onto the helices. The sharp decrease of $\Delta\epsilon/w$ and increase of f_c above 80% DCA are due to the helix-coil transition which is in agreement with optical rotatory dispersion (ORD) measurements, indicating that the Moffitt parameter b_0 suddenly changes in the same concentration range. The thermodynamic parameters characteristic of the thermal transition were determined from ORD measurements in a DCA 97.5%-dioxane mixture by examination of PBLG samples of different molecular weights.[29]

The characteristic thermodynamic parameters can also be determined on a single sample from ORD and dipole moment measurements during the thermal transition.[30] This has been done for PBLG in the chloroform-DCA system; it was shown that the $\Delta\epsilon/w$ versus T curves are shifted with respect to the degree of helicity curves. The magnitude of the shift depends on the rigidity of the polypeptide in its helical conformation.

In various solvent mixtures, as the thermal coil → helix transition proceeds, we have observed that the solution conductivity suddenly increases. This is an unexpected trend; it even occurs in the ADC-heptane mixture, where light scattering measurements[31] indicate that DCA is preferentially adsorbed (though to a much smaller degree) onto the helix.

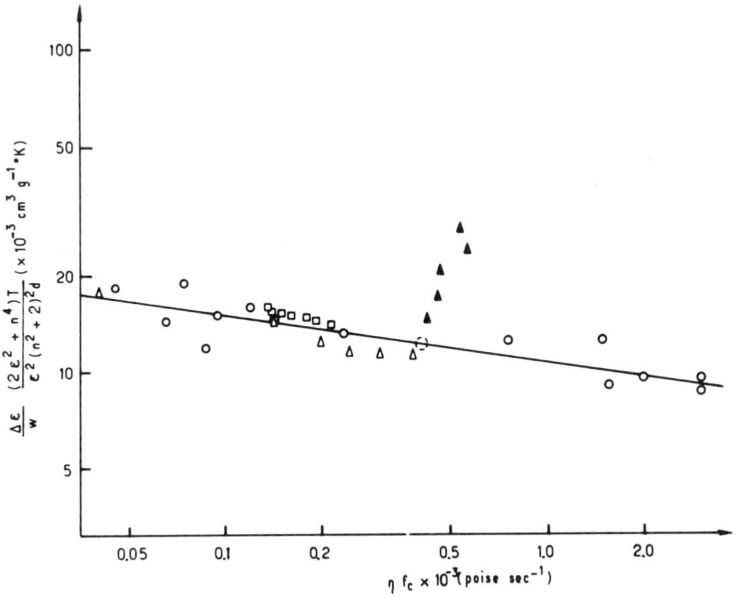

FIGURE 8. Plot of $(\Delta\epsilon/w)\,(2\epsilon^2 + n^4)T/\epsilon^2(n^2 + 2)^2d$ vs ηf_c on double logarithmic scale for PBLG. O = unaggregated α-helix (Marchal & Marchal[50]); Δ = O-25 wt% DCA in dioxane; ▲ 30–70 wt % DCA in dioxane; □ thermal disaggregation in dioxane. Dotted circle corresponds to the unaggregate sample, M = 59,000.

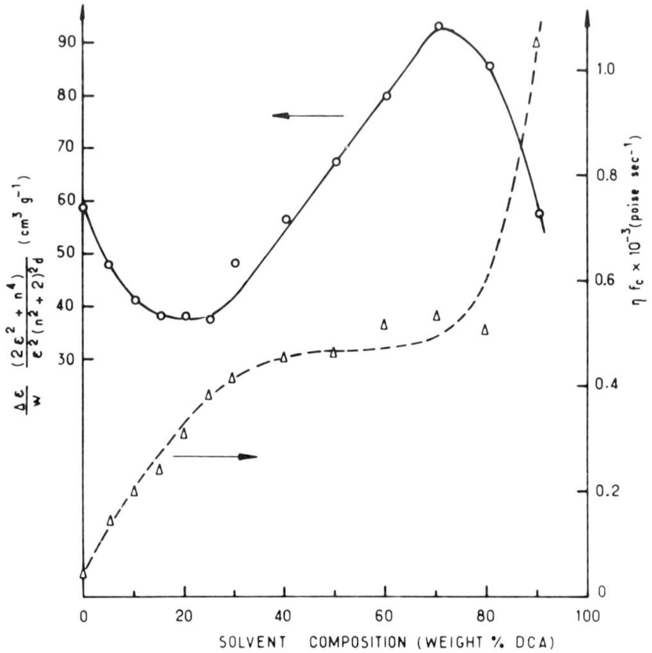

FIGURE 9. Variation of $(\Delta\epsilon/w)[(2\epsilon^2 + n^4)/\epsilon^2(n^2 + 2)^2d]$ and ηf_c for PBLG in dioxane-DCA mixture at 30°C with solvent composition.

Thus, the change in dipole moment with binary solvent content (and the change in ηf_c, provided M is far enough from that at which $(\eta f_c)_{\text{helix}} = (f_c)_{\text{coil}}$ which occurs for PBLG if $M \sim 70{,}000$) can be used to compare helix stability of polypeptides. This is especially useful when other measurements cannot be applied, as ORD for optically inactive polypeptides. Examples for PBDLG are given elsewhere.[32]

Poly(D,L-phenylalanine and poly(L-benzylaspartate)[4]

In FIGURE 10 the variation of $\Delta\epsilon/w$ versus DCA content in chloroform-PDLPA solutions is shown. At the early stages of DCA addition, $\Delta\epsilon/w$ decreases and ηf_c increases as expected for the destruction of highly polar and large aggregates. The scatter of the experimental points when aggregation is present denotes the difficulty in reproducing a particular state of aggregation owing to its sensitivity with regard to the purity of the solvent. This comment holds for all the polypeptides we have examined. The aggregate destruction is complete for 0.6% DCA content. The $\Delta\epsilon/w$ value determined in the chloroform-FA mixture where no aggregation takes place is shown on the ordinate axis by the heavy line. Between 1.1 and 5% DCA, a decrease of dipole moment and an increase of f_c are observed, because of the helix → coil transition. The same trend is observed for

PBLA: the sudden decrease of $\Delta\epsilon/w$ when the DCA concentration in chloroform is 8% reflects the helix-coil transition.

The thermal transition of PBLA in *m* cresol has been investigated by dielectric measurements by Wada *et al.*[33] and Marchal.[34] The two sets of measurements differ by the occurrence, in the first set, of a second high-frequency absorption caused by chemical relaxation.

Dielectric measurements of the helix-coil transition of poly(ϵ carbobenzoxy-L-lysine) in *m* cresol are reported by Omura *et al.*[35] These authors based their theoretical calculation of dipole moment variations on the theory of Nagai.[45]

Conformational Changes in Substituted Polysaccharides

Cellulose derivatives also belong to the class of polymers whose repeat unit has a dipole component parallel to the chain contour and thus will show low-frequency molecular weight-dependent dielectric absorption. This was reported to be so for cellulose acetate and ethylcellulose.[36,49] Here, we examine the dielectric behavior of cellulose[37] and amylose[38] tricarbanilates in dioxane and ethylacetate.

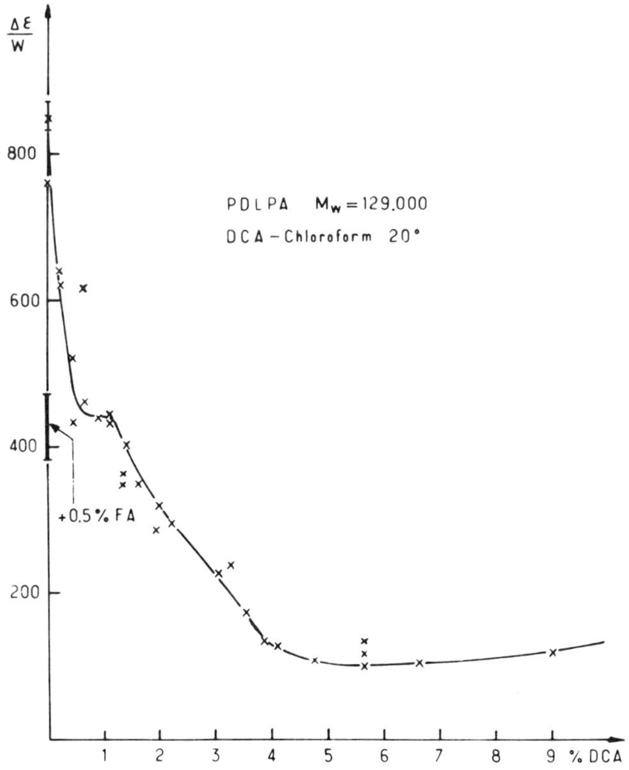

FIGURE 10. Variation of $\Delta\epsilon/w$ for PDLPA vs chloroform-DCA content.

These two isomers differ only in the way the anhydroglucose rings are linked together. In cellulose, the rings of the glucose units are joined in their equatorial planes (β-glucosidic bond), whereas in amylose the subsequent rings are attached axially. The cellulose tricarbanilate (CC) structure has some similarity with the zig-zag chain of alkanes, whereas amylose tricarbanilate (AC) has a tendency for helix information.

It seems interesting to determine the stability of these conformations with regard to temperature. Previously, Peticolas[39] gave transmission and electronic

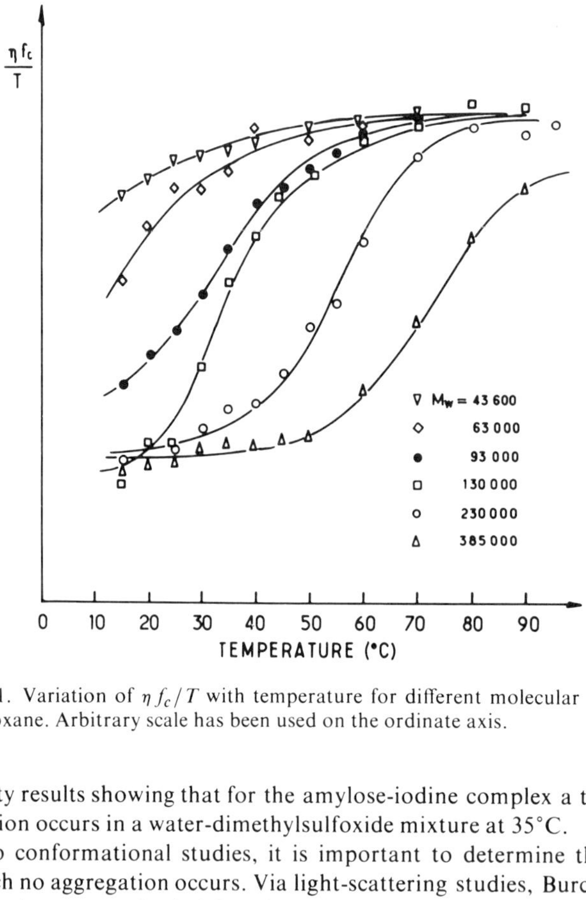

FIGURE 11. Variation of $\eta f_c/T$ with temperature for different molecular weight samples of CC in dioxane. Arbitrary scale has been used on the ordinate axis.

conductivity results showing that for the amylose-iodine complex a thermal helix-coil transition occurs in a water-dimethylsulfoxide mixture at 35°C.

Prior to conformational studies, it is important to determine the conditions under which no aggregation occurs. Via light-scattering studies, Burchard[40] found that in organic solvents, both CC and AC behave like wormlike coils whose stiffness is characterized by persistence length values lower than those obtained for PBLG or PHIC, yet much higher than for other coiled synthetic polymers. Thus, on the basis outlined above, aggregation might also easily occur in these fairly rigid macromolecules.

This is the case for both CC[37] and AC.[38] In ethyl acetate and in the concentration range examined $2 \times 10^{-3} < W < 2 \times 10^{-2}$, $\Delta\epsilon/w$ continuously decreases

TABLE 2

VALUES OF $\eta f_c/T$ FOR CC IN ETHYL ACETATE

Temperature °C	$\eta f_c/T$ (poise sec^{-1} °K^{-1})	
	M = 130000	M = 230000
0	0.469	0.132
20	0.528	0.183
40	0.599	0.275
64	1.284	—
65	—	0.545

with increasing concentration, whereas f_c is constant up to $w \sim 7 \times 10^{-3}$. This may reflect antiparallel side-by-side aggregation even at the lowest concentrations used. In dioxane, $\Delta\epsilon/w$ and f_c are constant up to $w \sim 7.10^{-3}$ for CC and AC (at least if degree of polymerization < 1500), then both these values decrease. The conformational studies were made in both solvents, but those in dioxane, in the low concentration range just defined, are more meaningful, since no polymer interaction occurs.

Both in dioxane and ethylacetate the quantity $\eta f_c/T$ was found to vary with temperature. This is shown for several CC samples in dioxane in FIGURE 11, where the ordinate has arbitrary scales for the various curves. The $\eta f_c/T$ values at various temperatures are given in TABLES 2 and 3. The changes of $T\Delta\epsilon/w$ and of $\eta f_c/T$ with temperature are given for an AC sample, degree of polymerization (DP) 560, in FIGURE 12. These changes are reversible. The increase of $\eta f_c/T$ with increasing temperature indicates that the molecules have larger overall dimensions at lower temperatures. Furthermore, the S-shaped character of the curve could imply some cooperativity in this conformational change, although it is not a sufficient condition. It may be due to the breaking of intramolecular side chain-side chain hydrogen bonds. The range of temperature in which this change occurs, i.e. 40°C, is rather broad, compared, for example, with that of the helix-coil transition of polypeptides. Furthermore, the temperature range of the conformational change was found to depend on molecular weight. The higher the molecular weight, the higher was the temperature of such a transition. In addition, for the same molecular weight sample it was higher in ethyl acetate than in dioxane for CC.

TABLE 3

VALUES OF $\eta f_c/T$ FOR CC IN DIOXANE

Temperature °C	$\eta f_c/T$ (poise sec^{-1} °K^{-1})					
	M = 43600	M = 63000	M = 93000	M = 130000	M = 230000	M = 385000
15	2.690	1.385	0.769	0.370	0.149	0.058
25	2.825	1.466	0.868	0.386	0.154	0.062
40	2.875	1.667	1.092	0.460	0.169	0.067
60	2.940	1.666	1.193	0.493	0.226	0.086
80	—	—	1.224	0.506	0.268	0.141

For an AC sample, DP 760, the variations of $\eta f_c/T$ and of $T(\Delta\epsilon/w)$ with T in ethyl acetate are broader than in dioxane and good plateau regions are not reached in the range $15°C < T < 90°C$.

For CC in dioxane, the range of temperature in which $\Delta\epsilon/w$ changes (a threefold decrease is observed for $M = 130.000$) is broader than that in which $\eta f_c/T$ changes, and the two corresponding curves are somewhat shifted with respect to each other, showing larger effects of conformation changes on dipole moments at the early stage of flexibility change of the rigid particles. For AC such a shift is

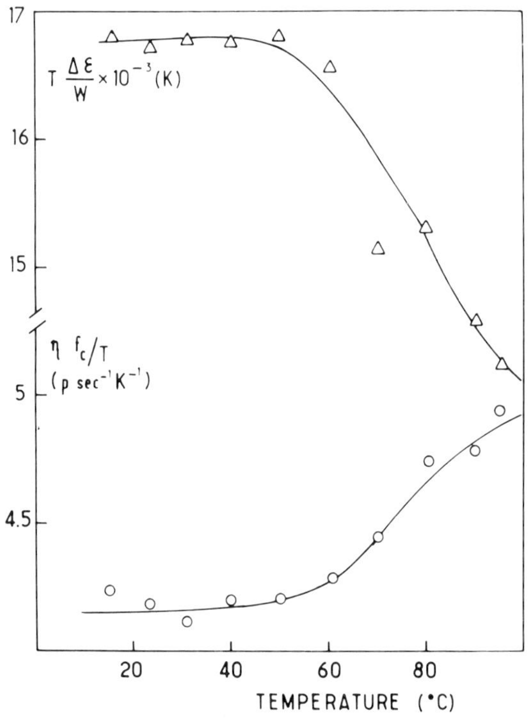

FIGURE 12. Variation of $T\,(\Delta\epsilon/w)$ and $\eta_0 f_c/T$ with temperature for AC (DP-560) in dioxane.

not observed. The reason for the difference between the behavior of CC and that of AC should be due to different chain stiffness in the two polysaccharides.

The striking feature of the $\eta f_c/T$ curves is the strong chain-length dependence of the temperature range where a conformational change from a comparatively stiff chain to a more flexible one occurs. As already mentioned, the appearance of two plateau regions and the S-shaped transition curves are strong indications for well-defined conformational states of the chains similar to those of the helical and coil states of polypeptides and polynucleic acids. Hence our experiments confirm structural peculiarities of CC and AC, as were suggested by Burchard as a result of light-scattering measurements in various solvents.[41] A closer comparison of our results with those obtained from polypeptides and oligo adenylic acid (oligo A)[42] reveals that the chain-length dependence of the transition temperature observed with CC and AC is also typical of strong linear nearest neighbor interaction. In fact, the theory of strongly cooperative processes of order-disorder along chains of finite lengths predicts a more or less pronounced chain-length dependence of the transition temperature.[43–45] Its magnitude is caused partly by the difference in the free energy content of the chain elements at the end of the chain to that of an internal chain element in the H-bonded state. This different free energy content is caused, on the one hand, by the fact that the chain units at the end can be fixed in space by only half of the H bonds; on the other hand, it has its origin in the bulky additional carbamate residues in the C_1 and C_4 positions, respectively, of the glucose rings at the ends.

The different chemical structure of CC and AC with respect to polypeptides should also affect the characteristic cooperativity parameters, which again affect the chain-length dependence. For polypeptides there is a drastic change in persistence length as the thermal transition proceeds (of the order of 100-fold), whereas the change is small for CC (about a factor of 2). In other words, the nucleation parameter σ of the helix-coil transition theory would be considerably larger for the present polymers than for polypeptides.

The chain-length dependence of the transition region is unusually large compared with those known for polybenzylglutamate and oligo A. While for polypeptides and polynucleic acids its effect remains restricted to relatively low DP's, it is still pronounced at high DP's for CC. FIGURE 11 demonstrates that at 25°C the CC molecular weight must exceed 200000, i.e., DP 380, before the chains are stabilized in their stiffened state. We note that below DP 380 at 25°C, however, the chains exist in states somewhere between the two limits of the plateau regions. We may say that in a plateau region the molecules have the same structures.

Taking only plateau f_c values, for AC in dioxane, one gets:

$$f_c = 3.7 \times 10^{13} M^{-1.77} \qquad \text{(low temperature);}$$

$$f_c = 9.7 \times 10^{11} M^{-1.36} \qquad \text{(high temperature).}$$

The lower value of the exponent in the second relation with respect to the first shows that the AC molecules are more flexible at high temperatures. For CC we do not have sufficient plateau values for determining a precise relation between f_c and M.

The tendency of increasing flexibility with increasing temperature for CC is established by the determination of the persistence dipole moment μ_a from a relation similar to the one proposed by Benoit and Doty[46] for the radius of gyration of wormlike coils:

$$\langle \mu^2 \rangle = 2\mu_a^2 \left[\frac{N\mu_0}{\mu_a} - 1 + \exp(-N\mu_0/\mu_a) \right]$$

μ_0 is the monomer dipole component along the chain, N the degree of polymerization, and μ the molecular dipole moment. A twofold decrease in μ_a occurs for

CC, M = 130.000, in dioxane when temperature increases from 15 to 60°C. A decrease in persistence length with increasing temperature has also been observed by low-angle neutron-scattering measurements.[47]

At room temperature, the number of monomeric units per persistence dipole moment, i.e. the ratio μ_a/μ_0, is 38 for AC and 21 for CC. This result shows that AC is more rigid, as was also concluded from light-scattering measurements by Burchard.[40] The dipole moment per monomeric unit in the direction of the chain is μ_0 = 0.9 Debye for CC, whereas it is 1.8 Debye for AC. This is further proof that the conformations of the two polysaccharides are not identical.

Another absorption above 1 Mc is likely to occur for CC and AC, since $(\Delta\epsilon/w)_{HF}$, the high-frequency limiting value of the dielectric constant corresponding to the low-frequency absorption, is found to be higher than that of the pure solvent notwithstanding atomic and electronic polarization. For AC in dioxane $(\Delta\epsilon/w)_{HF}$ increases from 1.0 to 4.0 with increasing temperature; the change in $(\Delta\epsilon/w)_{H.F.}$ occurs at lower temperatures than that in $(\Delta\epsilon/w)_{L.F.}$ and in ηf_c.

This absorption must be due to dipole moment components perpendicular to the chain. At low temperature, due to the high-chain stiffness, they are strongly correlated and for a sufficiently long molecule will cancel, leading to a zero average perpendicular dipole moment. As the temperature increases, the decreasing correlation of the chain segments, as shown from the first absorption, will allow segmental motion to set in. At the same time, the decreasing correlation of the perpendicular dipole moments will increase the dielectric increment relative to that motion. Thus, increasing flexibility of the AC chain with temperature strongly supports the observed behavior, which may very well be a consequence of hydrogen-bond breaking. The temperature shift of $(\Delta\epsilon/w)_{HF}$ seems to show that the second absorption is very sensitive to the first stages in the hydrogen bond-breaking mechanism when the overall molecular conformation is still frozen in. For CC in dioxane, the change of $(\Delta\epsilon/w)_{HF}$ with T is found to be greater than for AC.

Conclusion

At first sight, dielectric dispersion studies of dilute macromolecular solutions in the low-frequency range could seem rather formal. We hope we have been able to show that they can bring new and important information on the physical chemistry of these systems. By this simple technique it is possible to obtain useful information on structural changes arising during conformational transitions, and aggregation that occurs in many cases at finite concentration. These studies on model substances can shed light on similar phenomena which are found in more complicated systems as biological macromolecules *in vivo*.

References

1. Tobolsky, A. V. 1964. J. Phys. Chem. **68:** 2267–2272.
2. Marchal, E. 1964. Doctoral Dissertation. University of Strasbourg. Strasbourg, France.
3. Gupta, A. K., C. Dufour & E. Marchal. 1974. Biopolymers **13:** 1293–1308.
4. Marchal, E. 1974. Biopolymers **13:** 1309–1316.
5. Heitz, F., E. Marchal & G. Spach. 1975. Macromolecules **8:** 145–151.

6. WATANABE, H. 1965. Nippon Kagaku Zasshi **86:** 179–185.
7. POWERS, J. C. & W. L. PETICOLAS. 1967. *In* Ordered Fluids and Liquid Crystals (Adv. in Chem. 63), Amer. Chem. Soc. Washington, D.C.: 217–231; 1970. Biopolymers **9:** 195–203.
8. TSUJI, K. 1975. Ph.D. Thesis. Chapt. III. University of Tokyo. Tokyo, Japan.
9. WADA, A. 1960. J. Polym. Sci. **45:** 145–153.
10. KIHARA, H. 1975. Polym. J. (Japan) **7:** 406–408.
11. BLOCK, H., E. F. HAYES & A. M. NORTH. 1970. Trans. Faraday Soc. 569, 66, Part 5: 1095–1105.
12. GUPTA, A. K. 1976. Biopolymers **15:** 1543–1554.
13. TAKASHIMA, S. 1970. J. Phys. Chem. **74:** 4446–4452.
14. GUPTA, A. K., C. STRAZIELLE, E. MARCHAL & H. BENOIT. 1977. Biopolymers **16:** 1159–1165.
15. WEILL, G. & J. J. ANDRÉ. Submitted to Biopolymers.
16. REIBEL, L., G. SPACH & C. DUFOUR. 1973. Biopolymers **12:** 2391–2408.
17. GUPTA, A. K., H. BENOIT & E. MARCHAL. To be published.
18. STEWART, G. T. 1967. *In* Ordered Fluids and Liquid Crystals (Adv. in Chem **63**): 141–156. Amer. Chem. Soc. Washington, D.C.
19. BROWN, B. L. & G. P. JONES. 1975. J. Polym. Sci. Pol. Phys. Ed. **13:** 599–605.
20. IIZUKA, E. 1975. *In* Adv. Polymer Science. **20:** 79–107 and references therein. Springer Verlag, New York, N.Y.
21. ADACHI, K., Y. HIROSE & Y. ISHIDA. 1975. J. Polym. Sci. Polym. Phys. Ed. **13:** 1491–1499.
22. TSVETKOV, V. N., I. N. SHTENNIKOVA, E. I. RIUMTSEV, Y. P. GETMANCHUK. 1970. Eur. Polym. J. **7:** 767–774.
23. JENNINGS, B. R. & B. L. BROWN. 1971. Eur. Polym. J. **7:** 805–826.
24. LOCHHEAD, R. Y. & A. M. NORTH. 1972. J. Chem. Soc. Faraday Trans. 2, **68:** 1089.
25. TSVETKOV, V. N., E. I. RIUMTSEV, F. M. ALIEV & I. N. SHTENNIKOVA. 1974. Eur. Polym. J. **10:** 55–60.
26. BUR, A. J. & D. E. ROBERTS. 1969. J. Chem. Phys. **51:** 406–420.
27. COLES, H. J., A. K. GUPTA & E. MARCHAL. 1977. Macromolecules **10:** 182–187.
28. BARRIOL, J. & A. WEISSBECKER. 1964. C.R. Acad. Sci. **259:** 2831–2833.
29. GUPTA, A. K., C. DUFOUR, E. MARCHAL & H. BENOIT. 1975. Biopolymers **14:** 641–648.
30. DUFOUR, C. & E. MARCHAL. 1972. Biopolymers **11:** 1021–1030.
31. STRAZIELLE, C., C. DUFOUR & E. MARCHAL. 1970. Eur. Polym. J. **6:** 1133–1145.
32. MARCHAL, E., C. DUFOUR & G. SPACH. 1971. J. Chim. Phys. **68:** 831–834.
33. WADA, A., T. TANAKA & H. KIHARA. 1972. Biopolymers **11:** 587–605.
34. MARCHAL, E. 1971. Chem. Phys. Lett. **12:** 9–11.
35. OMURA, T., A. TERAMATO & H. FUJITA. 1975. Macromolecules **8:** 284–290.
36. SCHERER, P. C., D. W. LEVI & M. C. HAWKINS. 1957. J. Polym. Sci. **24:** 19–26; 1959. **37:** 369–374.
37. GUPTA, A. K., E. MARCHAL & W. BURCHARD. 1975. Macromolecules **8:** 843–849.
38. GUPTA, A. K., E. MARCHAL & W. BURCHARD. To be published. GUPTA, A. K. 1976. Doctoral Dissertation. Thesis. University of Strasbourg. Strasbourg, France.
39. PETICOLAS, W. 1963. Nature **197:** 898–899.
40. BURCHARD, W. 1971. Br. Polym. J. **3:** 214–225.
41. BURCHARD, W. 1964. Z. Phys. Chem. **42:** 293–313.
42. APPLEQUIST, J. & V. DAMLE. 1965. J. Am. Chem. Soc. **87:** 1450–1458.
43. ZIMM, B. H. & J. K. BRAGG. 1959. J. Chem. Phys. **31:** 526–535.
44. LIFSON, S. & A. ROIG. 1961. J. Chem. Phys. **34:** 1963–1974.
45. NAGAI, K. 1961. J. Chem. Phys. **34:** 887–904.
46. BENOIT, H., & P. Doty. 1953. J. Phys. Chem. **57:** 858–963.
47. GUPTA, A. K., J. P. COTTON, E. MARCHAL, W. BURCHARD & H. BENOIT. 1976. Polymer **17:** 363–366.
48. BARISAS, B. G. 1974, Macromolecules **7:** 930–933.
49. BLOCK, H., W. D. IONS, G. POWELL, R. P. SINGH & S. M. WALKER. Private communication.
50. MARCHAL, E. & J. MARCHAL. 1967. J. Chim. Phys. **64:** 1607–1615.

DIELECTRIC PROPERTIES OF NORMAL AND ABNORMAL LIPOPROTEINS IN AQUEOUS SOLUTION

C. G. Essex, E. H. Grant, R. J. Sheppard,
G. P. South, M. S. Symonds

Physics Department
Queen Elizabeth College
London W8 7AH, United Kingdom

G. L. Mills

Courtauld Institute of Biochemistry
Middlesex Hospital Medical School
London W1P 5PR, United Kingdom

Joan Slack

MRC Clinical Genetics Unit
Institute of Child Health
London WC1N 1EH, United Kingdom

Dielectric dispersion techniques have been in use for nearly forty years[1] as a means of characterizing the molecular properties of proteins[2-4] and smaller biological molecules[5,6] in aqueous solution. More recently dielectric measurements have been made on larger biological molecules such as DNA,[7] and this has been paralleled by investigations on polypeptides and artificial biopolymers.[8,9] By contrast, very little dielectric work appears to have been carried out on natural lipoproteins (apart from a small pilot study[10]), although there have been some recent studies on phospholipid vesicles and similar systems.[11,12]

In this paper we report dielectric measurements made on aqueous solutions of human and bovine serum low-density lipoproteins (LDL). Permittivity and conductivity measurements were made over the frequency range 0.15–1000 MHz and as a function of temperature, concentration, ionic conductivity, and solution pH. Particular emphasis was placed on determinations carried out at 800 MHz, it having been shown in previous work[10] that this is in the optimum frequency region for locating any differences in hydration that may exist between the various forms of lipoprotein. The dielectric decrement at 800 MHz was measured for samples of LDL isolated from the sera of patients with familial hyperbetalipoproteinaemia and compared with that determined for LDL obtained from the controls.

Analysis of the results shows the presence of three separate dispersion regions in the range 0.15–1000 MHz, which, in order of ascending frequency, may be ascribed to counterion relaxation, Maxwell-Wagner effects arising from the heterogeneous nature of the lipoprotein particle, and bound water. These three dispersions may be respectively termed the α, β, and δ regions, and they exist in addition to the γ dispersion which is due to the relaxation of free water molecules and which occurs at frequencies in the region of 1–100 GHz. Of particular interest in the present work is the β-dispersion, where it is shown that the magnitude and relaxation frequency are compatible with the structure of the lipoprotein molecule comprising a conducting core surrounded by an insulating shell. Thus the dielectric data are consistent with a lipid bilayer model but are inconsistent with a lipid core model.

The results at 800 MHz suggest an association between permittivity and clinical condition. The most likely explanation of the observed differences in permittivity is that the quantity of bound water (water of hydration) associated with the lipoproteins arising from the patients with hyperbetalipoproteinaemia is larger than that present in the controls.

Materials and Methods

LDL is usually defined as that group of lipoproteins having a density range 1.006–1.063 and a flotation rate (S_f) 0–20 Svedberg units. (Flotation rate is equivalent to velocity/acceleration and 1 Svedberg unit equals $10^{-13} s$). In human plasma LDL is the most abundant lipoprotein and is normally present in a concentration of about 350 mg/100 ml of plasma.

The isolation of the lipoproteins was carried out as described by Grant *et al.*,[10] with the following modification. After the crude lipoprotein concentrate obtained by precipitation had been dialysed against 0.15 M NaCl solution for 48 hours, it was added to one fifth of its volume of NaCl solution of density 1.079 *g*/ml. This raised the density of the solvent phase to 1.019 *g*/ml. Six ml of this solution were added to each of an appropriate number of centrifuge tubes, and 3 ml of NaCl solution of density, 1.019 *g*/ml layered on top. The prepared tubes were then centrifuged for 16 hours at 12°C in the 30.2 Spinco ultracentrifuge rotor at 30 000 rev/min (79420 *g*), after which the supernatant lipoproteins were aspirated off in 1 ml and discarded. The next three ml were also removed separately. Four ml of NaCl solution of density 1.0955 *g*/ml were added to each tube, and thoroughly mixed. After centrifugation under the same conditions as before, the supernatant lipoprotein layers were aspirated off in a combined volume of about 6 ml and dialyzed for 48 hours against a solution containing 0.17×10^{-4} M KH_2PO_4 and 3.17×10^{-4} M Na_2HPO_4. Before determining the permittivity, the samples were filtered to prevent any sedimentation occurring during the measurement procedure. Since the filtration could slightly alter the concentration, a small amount of the final solution (~ 1 ml) was set aside for concentration determination. To eliminate decomposition, all the measurements were made within two days of preparation, a trial run having shown that consistent values of permittivity are obtained over this period of time.

The relative permittivity ϵ' and conductivity (σ) were determined for the LDL solutions by three experimental techniques. In the frequency range 0.15–9 MHz, a Wayne Kerr B201 admittance bridge was used with a cell design and calibration procedure described previously.[13,14] At frequencies between 1 and 100 MHz, measurements were made with a Boonton 33A admittance bridge[15] and in the high frequency range (300–1000 MHz), an automated coaxial line system was employed.[16,17] A few confirmatory measurements were made in other laboratories[18,19] at frequencies below and above the range covered by our apparatus.

Results and Analysis

Typical dispersion curves in permittivity (ϵ') and conductivity (σ) are shown in Figures 1 and 2, the full data being given by Essex.[20] The conductivity plotted in Figure 2 is the incremental value over and above the contribution due to ionic conductivity, which is assumed to be frequency-independent. The quantity S which is plotted as the ordinate in Figure 2 is equal to the measured conductivity (σ) divided by the permittivity of free space (ϵ_0). Use of S rather than σ facilitates the calibration of the bridge.[13] At the weaker concentrations the incremental conductivity is a small change in a large quantity, and for this reason the analysis was carried out with the permittivity data only.

It is assumed that the permittivity curves can be described mathematically by

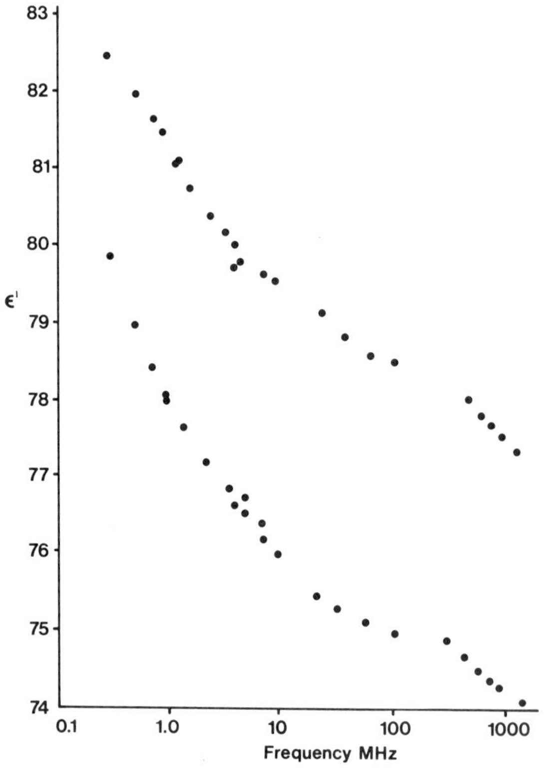

Figure 1. Variation with frequency of the permittivity of human LDL solution. (Concentration = 48.17 mg/ml; pH = 6.8; upper curve, 10°C; lower curve, 20°C.)

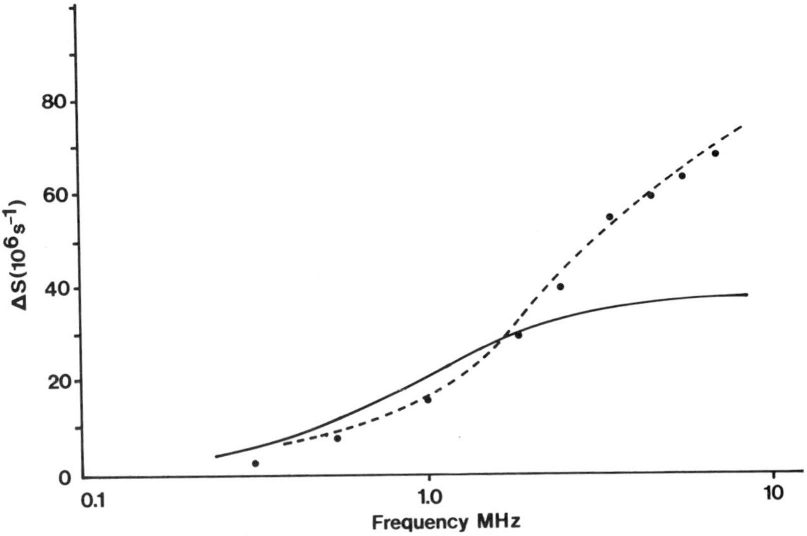

FIGURE 2. Variation with frequency of incremental conductivity of human LDL solution. ----- = theoretical curve for two Debye dispersions; ——— = theoretical curve for one Debye dispersion.

$$\epsilon_f' = \sum_{i=1}^{n} \frac{\Delta_i}{1 + \omega^2 \tau_i^2} + \epsilon_\infty \qquad (1)$$

where ϵ_f' is the permittivity of the solution at a frequency f, Δ_i is the magnitude of the ith dispersion, ω is the angular frequency ($\omega = 2\pi f$), τ_i is the relaxation time of the ith dispersion and ϵ_∞ has its usual significance. The experimental curves were analyzed with the aid of a computer, and with a least mean squares minimization technique.[21]

To correct for the effect of the γ dispersion, which is caused by the relaxation of free water molecules having a relaxation frequency of about 20 GHz at 25°C,[22] the following procedure was adopted. With use of Time Domain methods,[23] measurements were made[19] at frequencies up to 10 GHz on a few samples to confirm the form of the γ dispersion. As expected, the results indicated that the contribution of the γ dispersion to changes in the permittivity with frequency is negligible below 400 MHz. At higher frequencies small corrections were made to the lipoprotein data with the Debye equations which are known to give an adequate description of the dielectric data of the γ dispersion. After correction (0.1% at 400 MHz and 0.5% at 800 MHz) the values of permittivity of the lipoprotein solutions were fitted to Equation 1.

The data between 0.15 and 1000 MHz are found to be characterized by three dispersions with relaxation frequencies of 0.5 MHz, 4–6 MHz and 200–500 MHz, which, following previous classification,[24,25] may be termed the α, β and δ dispersions, respectively. The computer analysis predicts a leveling of the permittivity at frequencies below 0.15 MHz, and this was confirmed by measurements made[18] on a few samples with other apparatus.

In using Equation 1, various values of n were tried, and since, in general, an increase in the number of parameters will lead to a reduction in the error of fit, the values obtained were tested for statistical significance. No parameter was introduced that did not yield a significant improvement in the fit at the 95% level. With the computer technique used, it may be possible to fit too few parameters but difficult, if not impossible, to fit too many. If an extra component is introduced the effect of which on the curve is not detectable, it will invariably result in a correlation coefficient near unity; i.e., the extra parameter will be found to be highly correlated with another parameter. Although three Debye dispersions have been assumed in the frequency range 0.15–1000 MHz, the possibility cannot be excluded that small additional dispersions may also be present or that one or more of the observed dispersion regions may have a distribution of relaxation times. With the experimental accuracy available in the present work, the introduction of any extra parameters does not result in any significant improvement; i.e., additional components cannot be detected in the presence of the experimental error. Thus the three-component fit suggested is found to be the most appropriate analysis of the data.

The results taken at 800 MHz are shown in FIGURES 7 and 8 and will be discussed separately at the end of the paper.

MOLECULAR INTERPRETATION OF RESULTS

As mentioned earlier in this paper, the dielectric measurements were carried out over a wide range of values of various physical parameters: temperature, concentration, and solution pH and ionic conductivity. The purpose of this was to establish the variation of the dielectric increment (Δ_i) and the relaxation frequency [$f_i = (2\pi\tau_i)^{-1}$] with the physical parameters listed above. By this means it is possible to eliminate particular molecular processes from being responsible for the observed dispersions while at the same time narrowing down the number of acceptable possibilities. From the observed dielectric behavior we conclude that the best explanation for the origin of the α-dispersion is to attribute it to counterion relaxation of the type proposed by Schwarz.[26] Full details of the observations and calculations that lead to this conclusion will appear in a subsequent publication,[27] but the essential features of the argument can be summarized as follows.

The linear dependence of the dielectric increment (Δ_α) upon concentration (up to a certain critical concentration) is shown in FIGURE 3. Observations also show Δ_α to be independent of pH and concentration and only slightly dependent on temperature. With regard to the relaxation frequency (f_α) the experiments show that this is independent of concentration, conductivity, and pH. The relationship between f_α and the inverse of absolute temperature is exponential, and a logarithmic plot gave an activation enthalpy of 4 kJ/mole. All these observations, coupled with a measured relaxation frequency of 0.5 MHz are compatible with a counterion mechanism but are incompatible with any of the other processes that give rise to dielectric relaxation:[28] Debye rotation, proton fluctuation, Maxwell-Wagner effects, surface conductivity, ion atmosphere, and structured water. Moreover, the assumption of a counterion relaxation mechanism enables the charge on the lipoprotein molecule (e_0) to be calculated from the dielectric data. Assuming a molecular radius of 11 nm the value of e_0 turns out to be 2×10^{-18} C, which agrees well with the value obtained from electrophoretic studies.

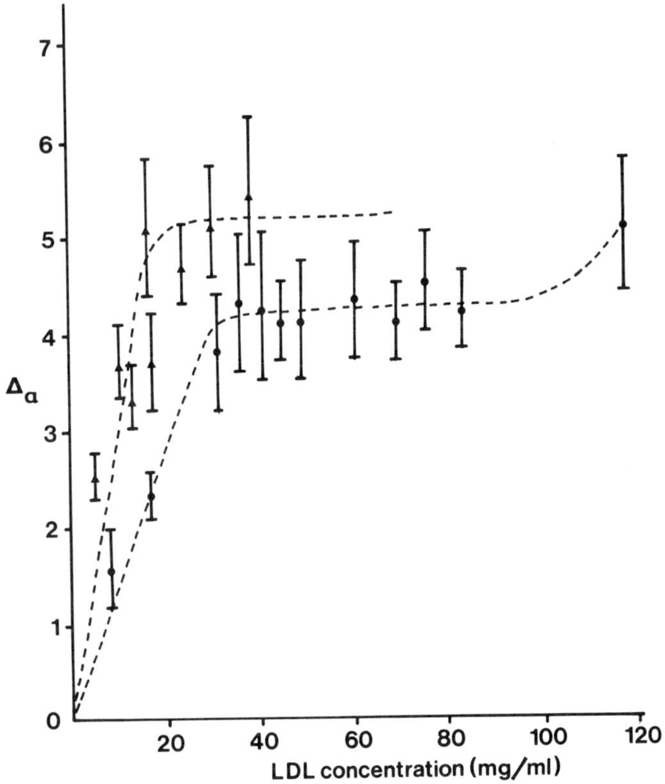

FIGURE 3. Variation with concentration of the dielectric increment of the α-dispersion (T = 20°C; pH = 6.8). ▲ = bovine LDL. ● = human LDL.

The abrupt change in shape of the increment-concentration curve at such a low concentration of 30 mgm/ml is a very unusual kind of phenomenon and is presumably due to the formation of micellular structure.[29]

The dielectric behavior of the β-dispersion is very similar to that observed for the α-dispersion, except that the increment (Δ_β) is smaller and the relaxation frequency is about ten times as high. The variation of Δ_β with concentration is approximately linear (FIGURE 4), but there is little dependence on conductivity,

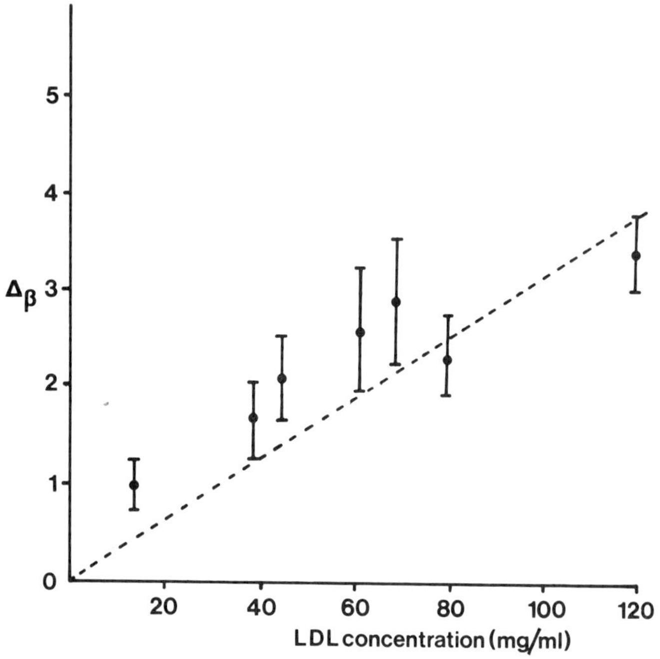

FIGURE 4. Variation with concentration of the dielectric increment of the β-dispersion (T = 20°C; pH = 6.8). (Measurements for human LDL.) ------- = theoretical curve predicted by Pauly-Schwan Equations.

pH or temperature. The relaxation frequency (f_β) is independent of concentration, pH and conductivity and its variation with temperature predicts an activation enthalpy of about 4 kJ/mole. The best explanation (full mathematical details will appear later[27]) for the origin of the β-dispersion is to attribute it to a Maxwell[30]-Wagner[31] type of mechanism. The LDL molecule is a highly complicated entity, but it is known to have a spherical shape and it may be assumed that the protein and lipid constituents tend to concentrate in different specific regions. For the purpose of interpreting the β-dispersion, it is necessary to employ a model for the LDL molecule that arranges the lipid and protein regions so that the electrical properties of the overall structure would predict the existence of a dielectric dispersion region of appropriate dielectric increment and relaxation frequency of about 5 MHz. To satisfy these conditions the requirements are that the LDL molecule should have a conducting core surrounded by an insulated shell. Of the various models proposed for the structure of a LDL molecule, one that complies with the above requirements is a lipid bilayer model such as the one proposed by Mateu *et al.*,[32] in which the lipoprotein molecule is considered to consist of a protein core 4.6 nm in radius surrounded by a lipid shell 3.5 nm in width. Outside the shell are further protein subunits that are sufficiently spaced so as not to make a continuous shell. For the purpose of discussing electrical behavior the model can be represented by FIGURE 5, where ϵ_w' and σ_w are the permittivity and conductivity of the water continuum, and ϵ_i', σ_i, and ϵ_s', σ_s refer to the protein interior and lipid shell, respectively. Assuming realistic values for these six parameters and using the above values of R and d, we have calculated the dielectric increment (Δ_β) and the relaxation frequency (f_β) by using the Pauly-Schwan equations.[33] The agreement between theory and experiment was found to be very satisfactory, and FIGURE 4 shows the predicted dependence of increment on concentration. The capacitance/unit area (C_m) of the bilayer was calculated from the dielectric results, and the value obtained is 4.4 ± 1.1 $\mu F/cm^2$. This is in good agreement with the figure of 3.3–4.2 $\mu F/cm^2$ obtained for phospholipid vesicles by Redwood *et al.*[12] Typical values for other biological membranes are 0.5–5 $\mu F/cm^2$.[34]

The amplitude of the δ dispersion is small but significant, and, following previous proposals,[4] may be interpreted as being due to the relaxation of bound water of an amount in the range of about 0.02 to 0.14 g/g lipoprotein. The properties of the δ-dispersion will be more fully discussed subsequently,[27] and the question of lipoprotein hydration as deduced from measurements at a single frequency will be discussed now.

PERMITTIVITY MEASUREMENTS AT 800 MHz

The relationship between relativity permittivity and frequency for an aqueous solution of biological macromolecules is shown in FIGURE 6. At frequencies in excess of 10 MHz the permittivity of the solution is seen to be less than that of pure water, and for a solution of unit concentration this fall in permittivity below

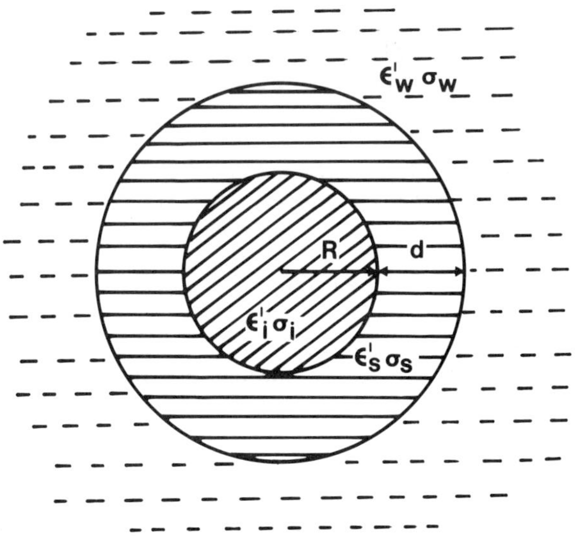

FIGURE 5. Simple three-layer model of a lipoprotein molecule in water. R = radius of core; d = thickness of shell. ϵ_i', ϵ_s', ϵ_w' = permittivity of core, shell, and solvent, respectively; σ_i, σ_s, σ_w – conductivity of core, shell, and solvent, respectively.

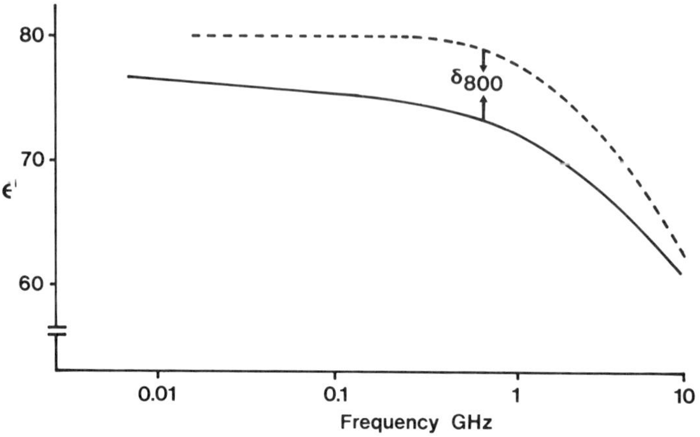

FIGURE 6. Permittivity of a biological solution of unit concentration, showing decrement of 800 MHz (δ_{800}). ------- = permittivity of pure water.

the pure-water value is termed the dielectric decrement (δ). At frequencies around 800 MHz, it has been customary to assume that the solute molecules contribute to the overall permittivity in respect of their atomic and electronic polarization only. However, it was established by Buchanan *et al.*[35] in 1952, and by others more recently,[4,36] that for protein solutions the value of δ_{800} is too large to be accounted for by the effect of the solute molecules alone, and that the reason for this is that the water molecules tightly bound to the protein (water of hydration) do not contribute fully to the polarization. Thus by measuring the dielectric decrement at frequencies around 500–1000 MHz and adopting a suitable dielectric mixture formula, estimates of the water of hydration for various proteins have been published.[4,10,35,36]

The weakness with this approach is that the value of hydration obtained from the dielectric data is critically dependent on the choice of mixture formula and is very sensitive to the value assumed for the partial specific volume of the solute. The shape of the solute molecule also figures in the calculation, and this may not be known accurately. For these reasons it has been considered preferable to calculate the hydration from the amplitude of the δ-dispersion,[4,37] rather than from the dielectric decrement. In the present work, however, we are concerned with comparing the values of hydration for various types of lipoprotein of similar molecular shape and partial specific volume; therefore the disadvantages normally associated with the method disappear and the advantages of a measurement at one frequency only become immediately apparent.

The relative permittivity at 800 MHz was measured for aqueous solutions of lipoproteins isolated from the sera of sixteen normal subjects and fifteen patients with type II familial hyperbetalipoproteinaemia (FH), a condition characterized by high concentrations of LDL and associated with an increased risk of ischaemic heart disease.[38] Of the fifteen patients considered in this particular study, seven were homozygotes and eight heterozygotes, although in general the proportion of homozygotes is very much smaller than that of the heterozygotes. Patients in the latter group are prone to develop ischaemic heart disease at an early age, the expectation of coronary death for heterozygous men being about 50% before the age of 55.[39] The much rarer homozygotes are likely to die in their late teens or early twenties.

The results obtained for the controls are shown in FIGURE 7, where the expected linearity between relative permittivity and solute concentration is seen. The dielectric decrements for the normals and for both categories of patient are shown in FIGURE 8.

A statistical analysis of the decrements was carried out in two stages. First, an analysis of variance was performed on the data and the value of the F ratio (variance between the samples/variance within the samples) gave $F = 8.82$. Tabulated

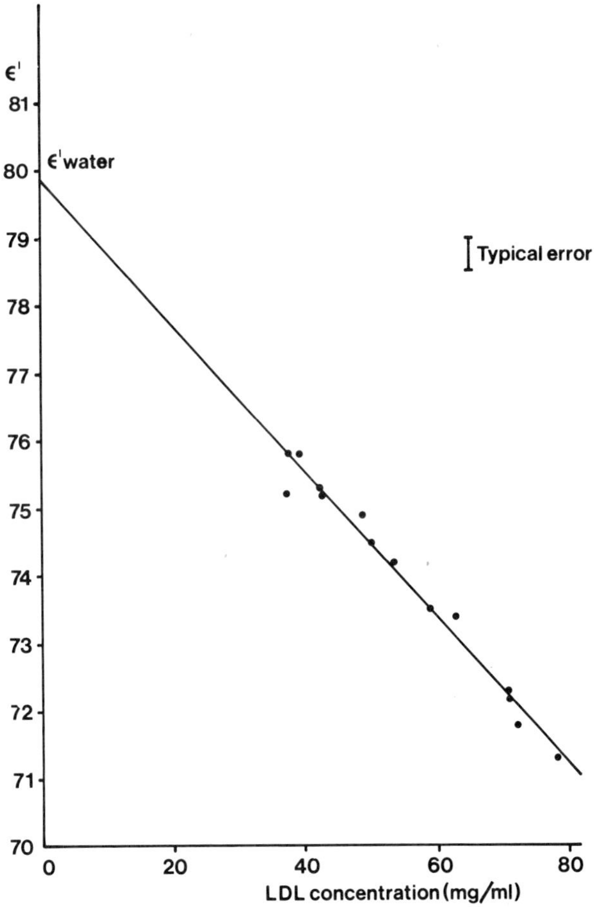

FIGURE 7. Variation with concentration of permittivity of normal human LDL solution at 800 MHz.

values[40] of F with (2,28) degrees of freedom at the 99% level give $F = 5.45$. 95% confidence intervals then evaluated, using a t-distribution, showed the following mean values and upper and lower 95% confidence limits for δ_{800}: Normal, $105.9 < 107.4 < 109.0$; heterozygote, $109.8 < 112.2 < 114.7$; homozygote, $110.8 < 115.6 < 120.4$. Hence there is strong evidence that δ_{800} depends upon the clinical condition of the patient and, in particular, that abnormals can be distinguished from nor-

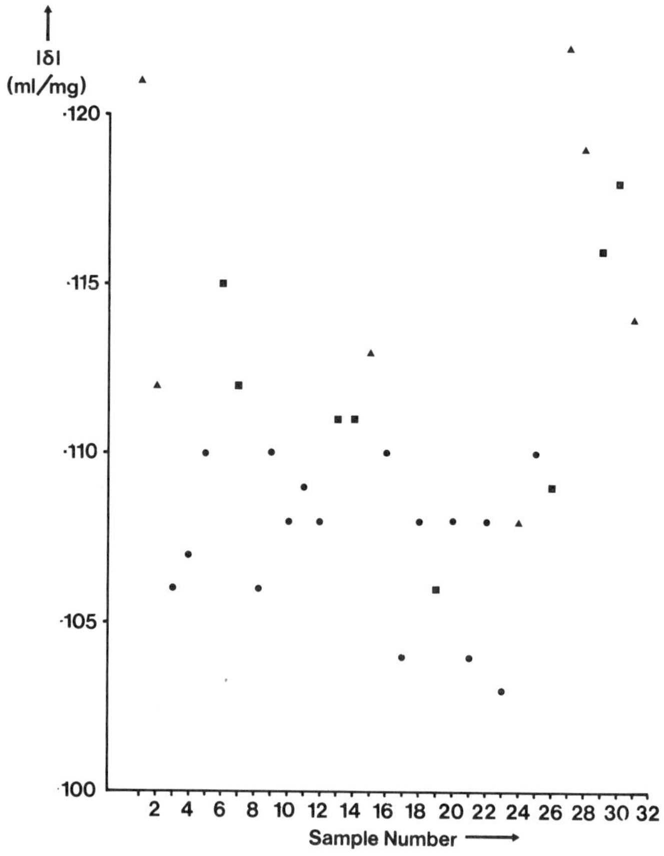

FIGURE 8. Values of δ_{800} for aqueous solutions of LDL. ● = normals; ■ = heterozygotes; ▲ = homozygotes.

mals. Following the analysis of variance, a subsequent t-test carried out between the means of the normals and heterozygotes gave a value of $t = 4.03$, as compared with a value of 2.84 for 22 degrees at the 99% confidence level obtained from statistical tables. The statistically significant differences in δ_{800} may be interpreted at a molecular level as follows.

For a suspension of macroscopic spheroids of relative permittivity ϵ_p in a continuum of relative permittivity ϵ_w the relative permittivity of the mixture is given[41] by ϵ, where

$$\frac{\epsilon - \epsilon_w}{\epsilon + x\epsilon_w} = p\,\frac{\epsilon_p - \epsilon_w}{\epsilon_p + x\epsilon_w}. \qquad (2)$$

In this equation p is the volume concentration of the spheroids and equals two if the particle is a sphere, but is greater than two for either a prolate or an oblate spheroid. In the present context ϵ_p refers to a hydrated particle and the numerator of the left-hand side of Equation 2 is the product of the decrement and the solute concentration expressed as a fraction by weight. In general, ϵ_p is some assumed function of the relative permittivity (ϵ_h) of the bound water (water of hydration) and the relative permittivity (ϵ_L) of the solute core, the precise nature of the function depending upon the assumed model for the hydrated particle. In the present work, however, we are concerned with measuring *differences* in hydration (absolute values being relatively unimportant), and so the approximation $\epsilon_p \cong \epsilon_h \cong \epsilon_L \simeq 5$ at a frequency of 800 MHz can be made. Using the fact that all solutions are dilute ($p < 0.08$), Equation 2 can be reduced[10,37] to the following convenient approximation

$$\bar{v} + w = \frac{k}{K}\,\delta_{800} \qquad (3)$$

where w is the water of hydration expressed as a weight fraction, $K = (1 + x)/x$ and k is a constant of value 13.5 for the concentrations being considered in the present work. The partial specific volume (psv) of the solute particle is represented by $\bar{v}$ and takes a value of 0.97 for LDL.

Using Equation 3, the values of hydration can be calculated for the three categories of LDL investigated. The shape of the LDL molecules as judged from electron micrographs appeared to deviate from spherical to a negligible extent, in which case $K = 1.5$ and the values of hydration are 0.04 ± 0.02 (Normal), 0.09 ± 0.02 (heterozygote) and 0.12 ± 0.04 (homozygote). The errors are the 95% confidence intervals, which, as seen from Equation 3, are directly proportional to the 95% confidence intervals obtained for the values of δ_{800}. Since δ_{800} is dependent on $\bar{v}$ and K as well as w, the question arises whether the dielectric results could be explained by a change either in shape or in partial specific volume, between the three forms of LDL. The first proposition can be rejected by noting that ultracentrifuge experiments show[38] that the abnormal lipoprotein has an increased S_f (flotation rate) over that of the normal, which would require a change of shape in the opposite direction to that needed to explain the dielectric data. Therefore the observed differences in δ_{800} must mean that the quantity ($\bar{v} + w$) varies between the three categories of LDL studied, and since appreciable variations in $\bar{v}$ appear unlikely,[43] the best interpretation of the dielectric measurements is that the lipoprotein hydration depends on whether the LDL originates from normal serum or whether it is produced by the gene responsible for hyperbetalipoproteinaemia. This further supports the hypothesis proposed previously in a pilot study[10] carried out on the dielectric properties of normal and abnormal LDL.

ACKNOWLEDGMENTS

Thanks are due to the Wellcome Trust for supporting CGE, GPS and MSS during the investigations and for financing the purchase of apparatus. Acknowledgment is also due to the Central Research Fund of London University for an equipment grant.

We are indebted to Dr. A. Suggett, Unilever Ltd, Colworth House, Sharnbrook, U.K., and to Mr. R. Lamote, of the Simon Stevin Instituut, Brugge, Belgium, for making the TDS measurements and low-frequency bridge determinations, respectively.

REFERENCES

1. ONCLEY, J. L. 1943. *In* Proteins, Amino-Acids and Peptides. E. J. Cohn and J. T. Edsall. Eds.: Chapt. 22. Reinhold Publishing Corporation. New York, N.Y.
2. PENNOCK, B. E. & H. P. SCHWAN. 1969. J. Phys. Chem. (Ithaca) **73:** 2600–2610.
3. MOSER, P., P. Q. SQUIRE & C. T. O'KONSKI. 1966. J. Phys. Chem. (Ithaca) **70:** 744–755.
4. GRANT, E. H., BARBARA G. R. MITTON, G. P. SOUTH & R. J. SHEPPARD. 1974. Biochem. J. **139:** 375–380.
5. AARON, MARGARET W. & E. H. GRANT. 1967. Trans. Faraday Soc. **63:** 2177–2180.
6. SHEPHERD, J. C. W. & E. H. GRANT. 1968. Proc. R. Soc. Lond. **A307:** 335–357.
7. MANDEL, M. This volume.
8. STOCKMAYER, W. H. This volume.
9. MARCHAL, E. This volume.
10. GRANT, E. H., R. J. SHEPPARD, G. L. MILLS & JOAN SLACK. 1972. Lancet i:1159–1161.
11. SCHWAN, H. P., S. TAKASHIMA, V. K. MIYAMOTO & W. STOECKENIUS. 1970. Biophys. J. **10:** 1102–1119.
12. REDWOOD, W. R., S. TAKASHIMA, H. P. SCHWAN & T. E. THOMPSON. 1972. Biochim. Biophys. Acta **255:** 557–566.
13. SOUTH, G. P. 1970. Ph.D. Thesis. University of London. London, England.
14. GRANT, E. H., G. P. SOUTH, S. TAKASHIMA & H. ICHIMURA. 1971. Biochem. J. **122:** 691–699.
15. ESSEX, C. G., G. P. SOUTH, R. J. SHEPPARD & E. H. GRANT. 1975. J. Phys. E. **8:** 385–389.
16. SHEPPARD, R. J. 1972. J. Phys. D. **5:** 1576–87.
17. SHEPPARD, R. J. & E. H. GRANT. 1972. J. Phys. E. **5:** 1208–12.
18. LAMOTE, R. Private communication. (See ACKNOWLEDGMENTS.)
19. SUGGETT, A. Private communication. (See ACKNOWLEDGMENTS.)
20. ESSEX, C. G. 1976. Ph.D. Thesis. University of London. London, England.
21. SHEPPARD, R. J. 1973. J. Phys. D. **6:** 790–794.
22. SCHWAN, H. P., R. J. SHEPPARD & E. H. GRANT. 1976. J. Chem. Phys. **64:** 2257–2258.
23. SUGGETT, A. & A. H. CLARK. 1976. J. Solution Chem. **5:** 1–15.
24. SCHWAN, H. P. 1957. Advances in Medical and Biological Physics **5:** 147–209.
25. SCHWAN, H. P. 1974. *In* Biological Effects and Health Hazards of Microwave Radiation. P. Czerski, Ed.: 152–59. Polish Medical Publishers. Warsaw, Poland.
26. SCHWARZ, G. 1962. J. Phys. Chem. (Ithaca) **66:** 2634–42.
27. ESSEX, C. G., E. H. GRANT, G. L. MILLS, R. J. SHEPPARD, J. SLACK & G. P. SOUTH. To be published.
28. SOUTH, G. P. & E. H. GRANT. 1972. Proc. R. Soc. Lond. **A328:** 371–87.
29. BEARD, R. B., T. F. MCMASTER & S. TAKASHIMA. 1974. J. Colloid. Interface Science **48:** 92–99.

30. MAXWELL, J. C. 1892. A Treatise on Electricity and Magnetism. Oxford Univ. Press, Oxford, England.
31. WAGNER, K. 1913. Ann. Phys. **40:** 817–29.
32. MATEU, L, A. TARDIEU, V. LUZZATI, L. AGGERBECK & A. M. SCANU. 1972. J. Mol. Biol. **70:** 105–116.
33. PAULY, H. & H. P. SCHWAN. 1959. Z. Naturforsch. **146:** 125–38; **6:** 621–633.
34. COLE, K. S. 1968. Membranes, Ions and Impulses. University of California Press. Berkeley, Calif.
35. BUCHANAN, T. J., G. H. HAGGIS, J. B. HASTED & B. G. ROBINSON. 1952. Proc. R. Soc. Lond. **A213:** 379–91.
36. GRANT, E. H. 1957. Phys. Med. Biol. **2:** 17–28.
37. GRANT, E. H., S. E. KEEFE & S. TAKASHIMA. 1968. J. Phys. Chem. (Ithaca) **72:** 4373–80.
38. SLACK, J. & G. L. MILLS. 1970. Clinica Chim. Acta **29:** 15–25.
39. SLACK, J. 1969. Lancet ii: 1380–82.
40. LINDLEY, D. W. & J. C. P. MILLER. 1964. Cambridge Elementary Statistical tables. Cambridge Univ. Press. Cambridge, England.
41. FRICKE, H. 1924. Phys. Rev. **24:** 575–88.
42. TORO-GOYCO, E. 1958. Ph.D. Thesis. Harvard University. Cambridge, Mass.
43. FISHER, W. R., M. G. HAMMOND & G. L. WARMKE. 1972. Biochemistry **11:** 519–25.

DISCUSSION

Gerhard Schwarz* and Chester T. O'Konski,† *Moderators*

**Department of Biophysical Chemistry*
Biocenter of the University of Basel
Basel, Switzerland

†Department of Chemistry
University of California, Berkeley
Berkeley, California 94720

DR. MEVES: Dr. O'Konski, have you any explanation what happens if acetyl choline reacts with the acetyl choline receptors and why this reaction should leave an increase of the ionic permeability of the postsynaptic membrane? There are, perhaps, changes in the orientation of the receptor molecule, or does one of its various chemical groups bend?

DR. O'KONSKI: Dr. Devor has made the preliminary experiments, looking for evidence that some conformational change which might underlie the initiation of the nerve impulse in the postsynaptic neuron. He saw no change in the relaxation time, but there seemed to be a change in the Kerr constant for the acetyl choline receptor on binding of a small quantity of acetyl choline.

DR. SCHERAGA: Dr. Marchal, I would like to ask two questions, one about conformation and one about aggregation. Using conformational energy calculations, Heslink and I showed that a regular repeating DL-polymer with a variety of different types could assume an alpha helical conformation, and I wonder what the experimental evidence is that you have: an alpha helix in poly-DL-phenylalanine, or poly-DL-benzyl glutamate. Another question. Is it possible to obtain an accurate enough measure of the degree of polymerization from the dielectric method, and do they agree with the one from independent methods?

DR. MARCHAL: I shall answer the second question first. They do agree. We have measured below the critical concentration only. Where we have the head-to -tail structure it is very easy to determine the molecular weight by two independent methods, the one by critical frequency and the other by light-scattering. We found the same molecular weight by these methods. As to the first question, Spach and Heintzes have some evidence from x-ray measurements. They suspected that under certain conditions, the DL-3 helix can occur. We checked that in solution by heating the alpha helix transforms into the DL-3 helix, because in that helix the CO groups are antiparallel and this leads to a zero dipole moment which actually does occur, so this was only to confirm what they had observed in the solid states.

DR. FOSTER *(University of Pennsylvania, Philadelphia, Pa.)*: Dr. Grant, you found from the measurements at high frequencies that the apparent hydration number of the abnormal low-density lipoproteins was higher than that of the normal lipoproteins. Have you made any attempt to correlate your dielectric findings with NMR measurements in the solutions?

DR. GRANT: No, we have not. The point is that the water of hydration you measure from NMR studies is not the same as the hydration water you get from dielectric studies.

DR. STOCKMAYER: Dr. Grant, or perhaps, Dr. Schwarz, Why is the temperature coefficient of the counterion mobility inferred in the alpha relaxation process so low? The counterion mobility "u" has a very low temperature coefficient, corresponding to an apparent activation energy of only 4 kilo joulds per mole. That surprises me, because I should have thought that the frictional force is specifying the magnitude, and the temperature coefficient should be comparable to that of the viscosity of water.

DR. SCHWARZ: I agree with you, that this is a rather small value. I would also expect a little more than that.

DR. TAKASHIMA: Dr. O'Konski, the receptor protein is dissolved in the aqueous media containing the detergent, and I presume that the receptor protein has the cluster of the detergent molecules surrounding it. Do you know how many detergent molecules are associated with the receptor protein? Does it cause any errors in the value of the dipole moment?

DR. O'KONSKI: The binding of the Triton X-100 has not been studied quantitatively, and that is one of reservations of the interpretation not only for the dipole moment, but also for the hydrodynamic properties. May I make a comment about the question of the temperature coefficient of the counterion mobility? I believe that there are some approximations, and one of the important questions of any dielectric theory is the question of the internal field. If one looks at the limiting case in the Maxwell-Wagner formulation where the internal conductivity or the surface conductivity is very high, then the local field goes toward zero. If the temperature coefficient is low, as may be because one is in that region, I am not sure what expression you use for the electric field on the ion in your theory. Dr. Schwarz, was it not the mean electric field in the medium, rather than the local field?

DR. SCHWARZ: This was a local field calculated on some electrostatic assumption.

DR. O'KONSKI: Did you put in the conductivity in evaluating the local field?

DR. SCHWARZ: In this theory, calculation of the electric field acting on the surface of the spherical particles was done taking into account the complex dielectric properties of the system; i.e., the electrolyte and also the bulk material of particles. So there is a local field involved in the theory, but of course you could use some other assumptions on this.

DR. O'KONSKI: My suggestion is that perhaps I should say that recently I have made a comparison of the relaxation time expression from your theory and that from the modified Maxwell-Wagner theory, and I notice that there is a difference in form involving the conductivity of the surrounding medium. I suspect it is due to a local field effect in which the conductivity was not entirely included in the expression you gave.

DR. SCHWARZ: What conductivity do you mean? The conductivity of the ions in solution, or some kind of surface conductivity?

DR. O'KONSKI: The conductivity of both the bulk medium and the particle. Whether one represents it as a volume conductivity or a surface conductivity, there should be some appropriate function of conductivities appearing in the expressions of the relaxation time. It does appear in the surface conductivity model, but it does not appear in your theory.

DR. SCHWARZ: In my model the mobility of the ions on the surface is included, but not a special surface conductivity. In that case, for which this theory has been developed, there is no such special surface conductivity, because it is included in the mobility of the ions.

DR. O'KONSKI: In the surface-conductivity model, the mobility of the ions is included in calculating the surface conductivity so that the ion mobility is present there.

DR. SCHWARZ: Some kind of surface conductivity comes out of it, but it is a complex surface conductivity.

DR. LAMPERT (*Princeton University, Princeton, N.J.*): Dr. Grant, the final model that you come up with is a rather formal model, with the conducting core and the nonconducting, presumably lipid, layer surrounding it. Did you check what the packing density of the lipids would be in the available volume? Is it comparable, for example, to normal membrane packing density, or is it much rarer in density?

DR. GRANT: There is a model that suggested that the lipoprotein had a lipid protein core with a lipid bilayer. Some x-ray work also suggested this. The answer to your question is that the volume composition of the lipoproteins in the table does line up with that of the model. There are, however, workers in the x-ray field who think that the lipid is in the center and protein outside. This model is not consistent with the dielectric evidence. There is some NMR evidence that suggests that the conducting part should be on the inside and the lipids outside.

DYNAMICS OF SUPERHELICAL DNA AND ITS COMPLEXES WITH ETHIDIUM BROMIDE FROM ELECTRO-OPTIC RELAXATION MEASUREMENTS*

Arthur E. Pritchard and Chester T. O'Konski

Department of Chemistry
University of California, Berkeley
Berkeley, California 94720

INTRODUCTION

The interaction between ethidium bromide and DNA has previously been studied in relation to the conformation properties of covalently closed circular DNA. It is thought that ethidium and other dyes bind to DNA by intercalation[1-3] and thereby produce changes in the superhelicity of the DNA. Large effects of ethidium binding on homogeneous superhelical PM 2 DNA were already measured in sedimentation velocity[4-6] and viscometric[7] experiments. This enhanced our interest in this system for studies using the electro-optic technique. Previous electric birefringence relaxation experiments on (heterogeneous) calf thymus DNA showed no detectable effects.[8,9] Some earlier preliminary measurements made here by N. F. and D. Ellerton in 1967 showed marginal decreases of the electro-optic relaxation times of calf thymus DNA upon binding of proflavine, but heterogeneity of the samples and experimental uncertainties precluded drawing conclusions. In Houssier's published study,[8] which involved ethidium binding to calf thymus DNA, the mean relaxation time did not vary significantly with binding, although the intrinsic viscosity increased appreciably. In this report we present a study of the electric birefringence relaxation of superhelical PM2 DNA, including the effects of binding of ethidium. This research was carried out with a computerized dedicated on-line electro-optic relaxation spectrometer. Large effects of enzymatic "nicking" and of ethidium binding were measured.

EXPERIMENTAL

Chemicals. Ethidium bromide (Calbiochem) concentrations were calculated from the absorbance (in water) at 480 nm, with use of a molar absorptivity of 5.6×10^3 cm^2/mol.[8]

DNA's. The PM2 DNA was graciously donated by R. Depew, T. S. Hsieh, and L. Liu of J. Wang's Laboratory. The growth of the phage and purification of DNA are described elsewhere.[10,11] PM2 DNA containing approximately one single-chain scission per molecule ("nicked" DNA) was also given. It was prepared by endonucleolytic treatment of the superhelical DNA with pancreatic DNase I.[9] The DNA was stored in 2M NaCl, 0.1 M EDTA and dialyzed against

*Supported by a Grant-in-Aid from the National Institutes of Health, CA 12,540-04 and -05.

the desired medium just prior to the electric birefringence measurements. DNA concentrations were determined spectrophotometrically, using a molar absorptivity at 260 nm of 6.60×10^3 cm^2/mol of phosphate.[12,13]

Electric Birefringence Measurements

The principles and methods of the pulsed electro-optic relaxation technique have been described.[14-19] Briefly, one uses an electronic pulse generator that applies short electrical pulses to the macromolecules in a solution through which polarized light is passed. The transient signals caused by macromolecular orientation effects are picked up with a photomultiplier and recorded electronically.[18] An on-line computer (PDP 11/10 with an AED 2500 flexible disk system) is used to analyze the signals by least squares fitting procedures. The apparatus and the mathematical methods are described elsewhere.[18,19] The birefringence decay signals, $S(t)$, were modeled in this work as a sum of three exponential terms,

$$S(t) = \Sigma s_i^0 \exp(-t/\tau_i) \qquad (1)$$

where s_i^0 is the amplitude associated with each discrete relaxation time τ_i at the time of the removal of the pulse ($t = 0$).

The DNA concentration was close to 15 μg/ml in all of the experiments. Exploratory experiments led to the following "standard" applied electric field "step-up, step-down" pulse: 7 kV/cm; 200 μsec duration; rise and decay times were 3–4 μsec exponentials. The temperature was kept at 25° during measurements.

Because of conductivity effects, electro-optic studies require low salt concentrations. Most studies of the physical-chemical properties of superhelical DNA have been made in about 0.1 M salt concentrations, where the somewhat flexible DNA macromolecule is more contracted, on the average. We chose low concentration Tris buffers at pH 8 and added some Mg^{++}, which binds to DNA, diminishes its polyelectrolyte character, and tends to stabilize the native structure. We also varied Mg^{++} to see its effects on the E-O, relaxation.

To determine whether the superhelical DNA was irreversibly altered by application of the pulses, fluorescence assays[20] were performed on samples before and after several pulses were applied. This technique measures the amount of covalently closed circular DNA. The fluoresence of ethidium bound to double-stranded DNA was determined before and after heat denaturation. The difference in fluorescence intensities is due to the fact that only covalently closed circular DNA will renature upon cooling, and that the fluorescence of ethidium is enhanced when it is complexed with native DNA. It was found that 95% of the DNA was in the covalently closed form if fewer than six "standard" pulses were applied. Most signals were recorded with only one to three pulses on fresh DNA samples. The fluorescence measurements also showed that the DNA stayed in the native form in the low ionic strength buffers used in the electro-optic experiments. After electric field pulses, we found that a small fraction of the DNA was irreversibly denatured, although most of the molecules were covalently closed and therefore could spontaneously renature.

Fluorescence Measurement. Fluorescence intensities were measured with an

Aminco-Kiers spectrofluorimeter, using excitation at 525 nm and emission at 580 nm, at native PM2 DNA concentrations of 15 μg/ml. The fluorescence intensity was corrected for the absorption of excitation light as described by LePecq and Paoletti.[21]

RESULTS

The birefringence relaxation times of covalently closed circular PM2 DNA and of circular PM2 DNA with a few single-strand scissions ("nicked" PM2) are

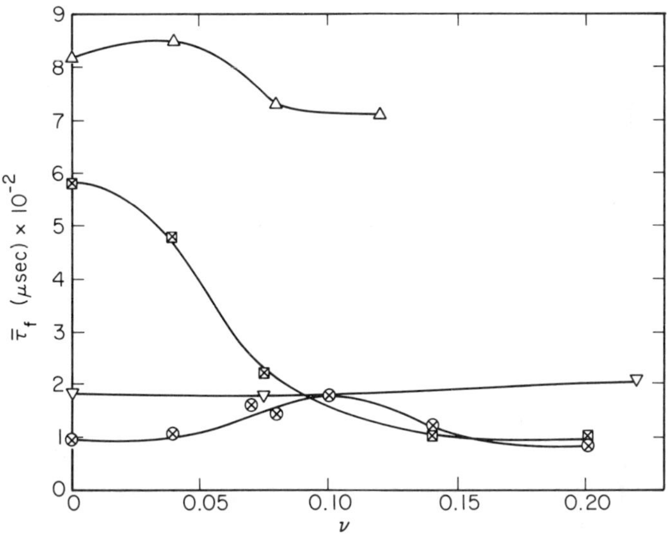

FIGURE 1. Fitted average birefringence relaxation time, $\bar{\tau}_f$, as a function of bound ethidium ion/DNA-phosphate ratio, ν. Δ = nicked PM2 DNA in 0.5 mM Tris; ⊠ = superhelical PM2 DNA in 0.5 mM Tris; ▽ = nicked PM2 DNA in 0.5 mM Tris, 0.1 mM Mg^{++}; ⊗ = superhelical PM2 DNA in 0.5 mM Tris, 0.1 mM Mg^{++}

shown in FIGURE 1 as a function of added dye, ethidium bromide. Three discrete relaxation terms as given in Equation 1 were used to calculate the fitted average relaxation time $\bar{\tau}_f$, which is defined by the equation[18]

$$\bar{\tau}_f = \frac{\sum_{i=1}^{N} s_i^0 \tau_i}{\sum_{i=1}^{N} s_i^0} \qquad (2)$$

where N is the number of exponential fitting terms; three were always computed in this research.

Ethidium was added to DNA buffered with 0.5 mM Tris, pH 8.0, solutions that either contained 0.1 mM Mg^{++} or had no Mg^{++} present. The amount of bound dye was calculated from the DNA and total dye concentrations, by the Scatchard equation[22]

$$\frac{\nu}{C_f} = K(n - \nu) \tag{3}$$

where ν is the average number of bound dye molecules per DNA phosphate, n is the maximum number, K is the intrinsic association constant of a site, and C_f the free dye concentration.[23] Values of $K = 6.0 \times 10^6$ for the solutions containing 0.5 mM Tris, 0.1 mM Mg^{++}, and $K = 6.6 \times 10^6$ for the solutions containing no Mg^{++}, were extrapolated from the data of Le Pecq and Paoletti,[21] who determined the ionic dependence of the binding constant. These values are approximate, but since the binding constant is so large, they are adequate to calculate a value of ν (around $\nu = 0.10$) that is accurate to within ± 0.01. A magnitude around 6×10^6 is substantiated by Waring,[24] who got values up to 2.5×10^6 using a buffer of higher ionic strength (40 mM Tris). For the binding of ethidium to covalently closed circular DNA, an exponential factor that reflects the superhelical free energy is required in Equation 3,[10] but since the binding constant K is so large in the solutions used here, the effect of this factor on the calculation was found to be negligible within experimental error.

FIGURE 1 shows that the effect of added ethidium on the DNA relaxation times is strongly dependent on the presence or absence of Mg^{++} in the buffer solution. With no Mg^{++} present, the nicked DNA relaxation time is always greater than that of the supercoiled DNA and decreases only slightly with increasing dye concentration. The addition of ethidium bromide to superhelical DNA causes a large monotonic decrease in the average relaxation time. When the Mg^{++} concentration is 0.1 mM, the curves for both the nicked and covalently closed DNA are shifted to lower relaxation time values, but at $\nu = 0$, the nicked DNA still relaxes more slowly than superhelical DNA. In this medium the average relaxation time of superhelical DNA increases to a maximum and then decreases with increasing dye concentration, whereas the nicked DNA relaxation time is relatively independent of added ethidium.

The influence of Mg^{++} on the fitted average relaxation times for nicked and superhelical PM2 DNA was determined at different Mg^{++} concentrations, as shown in FIGURE 2. To plot these curves, the number of bound Mg^{++} per DNA phosphate, r, was calculated from the total DNA and Mg^{++} concentrations by the Scatchard equation

$$\frac{r}{C} = 6 \times 10^3(0.5 - r) \tag{4}$$

where C is the free Mg^{++} concentration. The numerical values for the binding constant and maximum number of binding sites per DNA phosphate in this calculation were taken from Sander and Ts'o.[25] (Again, we do not expect the binding constant to be exact, since its value depends on the ionic strength. It is useful, however, to use this approximation so that the stoichiometry of the effects of binding to Eth^+ and Mg^{++} can be compared. The conclusions drawn in this work do not depend on a precise value for the amount of *bound* Mg^{++}.) For both the nicked and superhelical PM2 DNA samples, the addition of Mg^{++} causes a decrease in the relaxation times. The slopes of these curves gradually decrease with increasing *r*.

FIGURES 3–5 give the discrete relaxation rate spectra for several of the experiments shown in FIGURES 1 and 2. The distribution was obtained by fitting the birefringence relaxation curve to a sum of three exponential terms.

Broad ranges of relaxation times were found for both nicked and superhelical PM2 DNA. When three-term fits were made, the shortest relaxation times are 8–50 μsec and the longest about 3 mseconds. FIGURES 3–5 show that decreases in

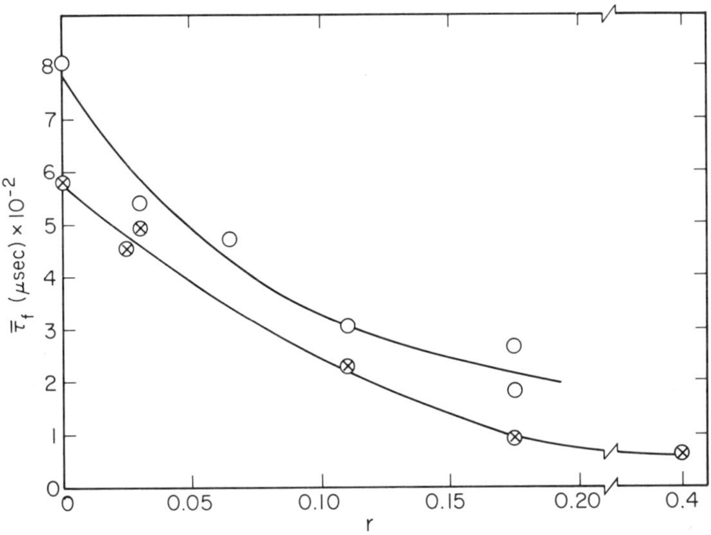

FIGURE 2. Fitted average birefringence relaxation time, $\bar{\tau}_f$, as a function of bound Mg^{++} ion/DNA-phosphate ratio, *r*. ○ = nicked PM2 DNA in 0.5 mM Tris plus added Mg^{++}; ⊗ = superhelical PM2 DNA in 0.5 mM Tris plus added Mg^{++}.

the fitted average relaxation time resulting from the addition of Mg^{++} or ethidium bromide (shown in FIGURES 1 and 2) are due primarily to decreases in the longest relaxation time.

DISCUSSION

Birefringence Relaxation in the Presence of Mg^{++}

The curves for the variation of the fitted average relaxation time, $\bar{\tau}_f$, of nicked and superhelical PM2 DNA with added ethidium bromide (FIGURE 1), are similar to titration curves previously obtained by measurements of sedimentation coefficients.[5,6] The sedimentation results have been interpreted as a decrease in superhelical density caused by intercalation by ethidium bromide, which causes an unwinding of the duplex turns, and a concommitant change in the number of superhelical turns. As the concentration of dye increases, the number of negative superhelical turns, and consequently the sedimentation coefficient, decrease until the minimum is reached, corresponding to "relaxed" DNA or circular molecules with no superhelical turns. The nicked DNA and relaxed superhelical DNA have approximately equal sedimentation coefficients and equal $\bar{\tau}_f$ values at this point. Further addition of ethidium causes the formation of *positive* superhelical turns and an increase in the sedimentation coefficient.

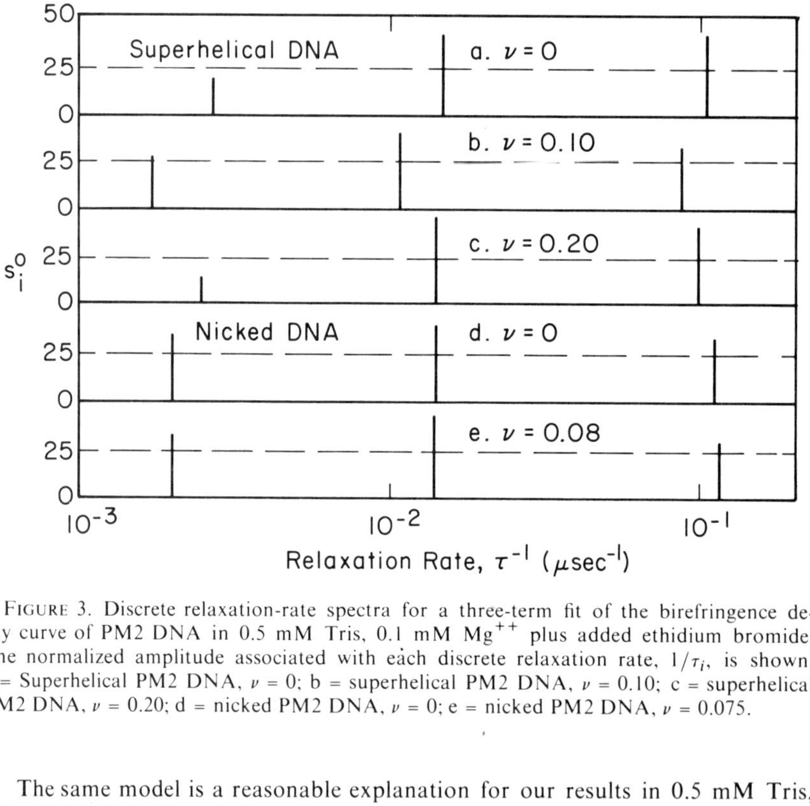

FIGURE 3. Discrete relaxation-rate spectra for a three-term fit of the birefringence decay curve of PM2 DNA in 0.5 mM Tris, 0.1 mM Mg^{++} plus added ethidium bromide. The normalized amplitude associated with each discrete relaxation rate, $1/\tau_i$, is shown. a = Superhelical PM2 DNA, ν = 0; b = superhelical PM2 DNA, ν = 0.10; c = superhelical PM2 DNA, ν = 0.20; d = nicked PM2 DNA, ν = 0; e = nicked PM2 DNA, ν = 0.075.

The same model is a reasonable explanation for our results in 0.5 mM Tris, 0.1 mM Mg^{++} buffer solutions, because of the qualitative similarities of the relaxation time and sedimentation coefficient titration curves. The sedimentation velocity is inversely proportional to the hydrodynamic (translational) friction factor, whereas the relaxation time is directly proportional to the rotational friction factor. Both factors go through a maximum at zero superhelical turns. The sedimentation coefficient reaches a minimum at ν = 0.075,[10] and the maximum for the relaxation time (ν = 0.1) is at a slightly higher value (FIGURE 1). The difference between 0.075 and 0.1 is not significant for three reasons. 1) The uncertainty in the

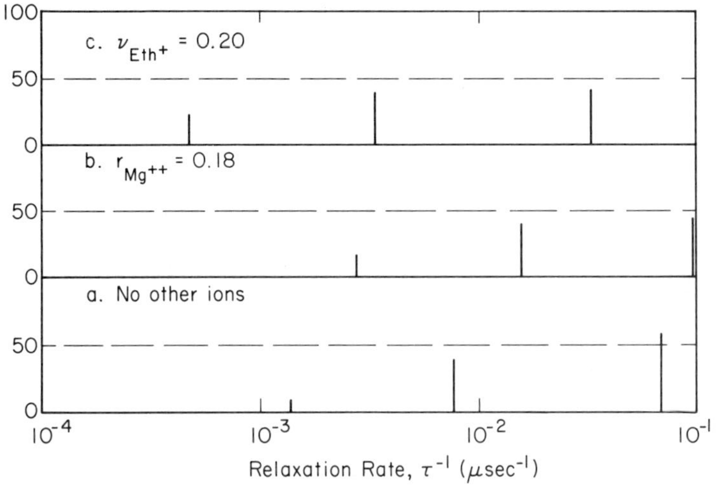

FIGURE 4. Discrete relaxation-rate spectra for superhelical PM2 DNA in 0.5 mM Tris plus added ions: a = no other added ions besides $Tris^+$; b = Mg^{++} added to a level such that r = 0.175; c = ethidium bromide added to a level such that ν = 0.20.

value of $\bar{\tau}_f$ is estimated to be ±10% (based on repeated measurements under the same conditions) and the difference in $\bar{\tau}_f$ at ν = 0.075 and ν = 0.1 is less than 20%. 2) Data points were taken at intervals of $\nu \simeq 0.02$. 3) Rather than directly measuring the amount of bound ethidium, it was calculated with use of an extrapolated value for the binding constant (See EXPERIMENTAL).

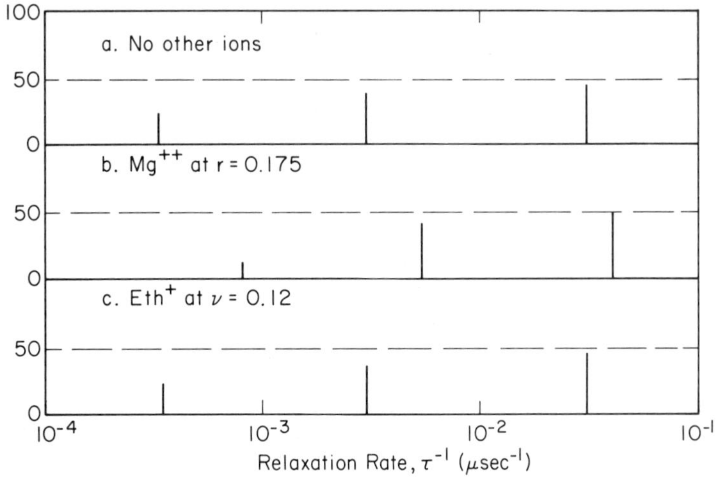

FIGURE 5. Discrete relaxation rate spectra for nicked PM2 DNA in 0.5 mM Tris plus added ions: a = no other added ions besides $Tris^+$; b = Mg^{++} added to r = 0.18; c = ethidium bromide added to ν = 0.12.

Birefringence Relaxation without Mg^{++}

In Tris buffer alone, the initial ($\nu = 0$) fitted average and discrete relaxation times of the nicked PM2 DNA are greater than those of the superhelical DNA, but increasing dye concentration causes a large monotonic decrease in the covalently closed DNA relaxation times, while the nicked DNA relaxation times decreases only slightly. These results are surprising, and an unequivocal explanation is not possible at this time; let us therefore consider two possibilities: (1) a nonintercalative mode of binding to the DNA in 0.5 mM Tris; and (2) an intercalative mode of binding to the closed circular DNA which (at 25° in 0.5 mM Tris with no ethidium added) has undergone the "early" helix-coil transition[26] and therefore contains no superhelical turns. Addition of ethidium in case (2) would cause the formation of positive superhelical turns and a decrease in the relaxation time as observed in 0.5 mM Tris, 0.1 mM Mg^{++}, with dye concentrations above $\nu \simeq 0.10$. These possibilities will be discussed in order.

Nonintercalative Mode of Binding. The possibility of more than one binding process for the association of ethidium bromide and DNA was discussed previously. Waring[24] observed that Scatchard plots of ν/C_f vs. ν(in 0.04M Tris) are not linear, the deviation from linearity occurring mostly at high values of ν. This was explained by postulating the existence of secondary sites to which the dye binds with a low affinity compared with "primary site" binding (intercalation). The secondary binding apparently becomes significant only after a certain fraction of the primary sites has been filled. External binding has been found to contribute only a small percentage of the total binding in high salt, but is reported to be much more important in low salt concentrations.[21]

It is known that Mg^{++} binds externally to the phosphate on the DNA backbone.[25] The effect of a nonintercalative mode of ionic binding on the DNA relaxation times was measured here by the addition of varying amounts of Mg^{++} to the DNA solutions buffered with 0.5 mM Tris, pH 8.0. The results of these experiments, given in FIGURE 2, show that both nicked and superhelical DNA relaxation times decrease as a function of added Mg^{++}. This is probably due to a shielding of the charge repulsions of the phosphate groups and an increased chain flexibility and a decreased average persistence length. It was seen, when $\bar{\tau}_f$ for superhelical DNA was plotted both as a function of bound Mg^{++} and Eth^+ per DNA phosphate on the same graph (not shown) that the Mg^{++}-induced decrease in $\bar{\tau}$ was quantitatively similar to that caused by Eth^+. However, with nicked DNA, bound Mg^{++} decreases $\bar{\tau}_f$ much more than bound Eth^+ does, both on a molar basis and on a number of charges basis (see FIGURES 1 and 2). This suggests different kinds of binding for Mg^{++} and Eth^+ to nicked DNA, and implies an intercalative mode for Eth^+.

Intercalative Mode of Binding. Intercalated ethidium bromide has been reported to have a greater fluorescence intensity than the externally bound dye.[21] Therefore, we measured the fluorescence properties of ethidium bromide bound to superhelical DNA. The results, shown in FIGURE 6, indicate intercalation by Eth^+, since there is an increase in fluorescence increment between $\nu = 0$ and $\nu \simeq 0.20$. The decrease in fluorescence increment with further increase of ethidium concentration has been interpreted as due to either quenching by energy transfer from intercalated dyes to externally bound dyes or to a decrease in the amount of intercalated dye when the external sites are occupied.[21] But in the range encompassing our relaxation time experiments, the fluorescence increment increases, presumably because of intercalation.[21]

We consider that the mode of Eth^+ binding is largely intercalative. Coulombic shielding by Eth^+ will tend to decrease relaxation times, similar to added Mg^{++}. Earlier studies also suggest that dye binding increases the flexibility of linear DNA.[27,28] The situation is intricate, because there are an intrinsic base-stacking

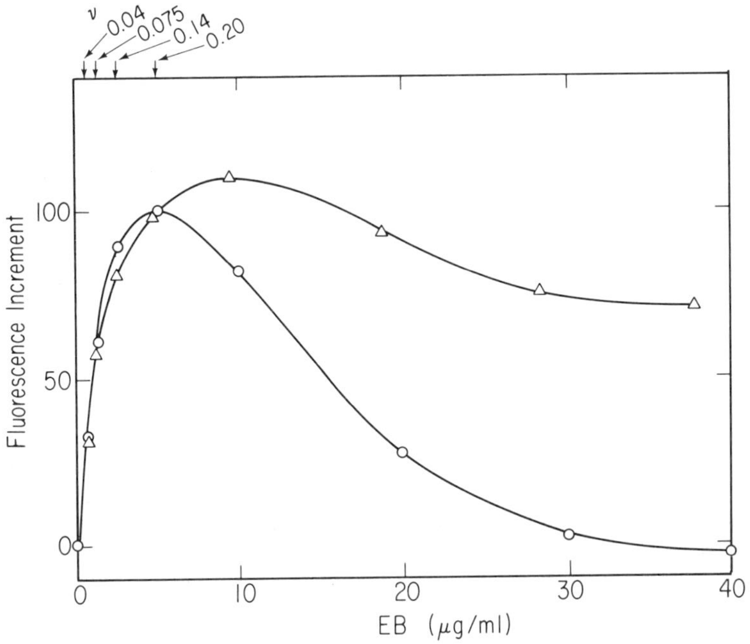

FIGURE 6. Spectrofluorimetric measurements of ethidium bromide binding to native PM2 DNA. The fluorescence increment is the fluorescence intensity for ethidium in the presence of DNA minus the fluorescence intensity of ethidium in the absence of DNA. The maximum value of this increment obtained with PM2 DNA in 0.5 mM Tris was arbitrarily taken to be 100. Also shown is the amount of bound dye per DNA phosphate, ν, which was calculated for the measurements made in 0.5 mM Tris as described in the text. -0- = 0.5 mM Tris, pH 8.0; -△- = 0.5 mM Tris, 0.1 mM Mg^{++}, pH 8.0.

rigidity effect and the coulombic repulsions tending to increase the relaxation times of the DNA, as well as changes in superhelicity that can increase or decrease them.

There is a one-to-one correspondence between the unwinding of secondary (duplex) and tertiary (superhelical) turns in closed circular duplex DNA, as expressed by the relation[26,29]

$$\alpha = \beta + \tau$$

where α, the topological winding number, is fixed upon closure of the last phosphodiester bond and remains invariant as long as the molecule remains covalently closed. The "duplex winding number" β is the number of revolutions made by one strand about the duplex axis; τ, the "superhelix winding number," is the number of revolutions made by the duplex about the superhelix axis.[26] There are a number of ways that β and therefore τ can be altered, including intercalation by ethidium, which unwinds the duplex helix by 26° per bound dye.[30]

Let us recognize that the superhelical form of closed circular DNA can be converted to the relaxed form with no superhelical turns by melting a small fraction of the base pairs in the molecule. Then we may define a melted region winding number γ, and the topological equation becomes

$$\alpha = \beta + \gamma + \tau.$$

Here γ is the number of turns that each strand makes about an axis projected through the melted region. The Gibbs free-energy variation with γ, $dG/d\gamma$, is expected to be far less than $dG/d\beta$ and $dG/d\tau$; thus superhelical turns would be released by premelting.

Since superhelical molecules have a higher free energy than the relaxed species, there may be an "early" helix-coil transition in closed circular DNA that is driven by the positive free energy of superhelix formation.[26] At higher temperatures, the nicked molecules denature close to the temperature expected for the denaturation of the linear molecule. At even higher temperatures, the "late" transition of superhelical DNA occurs. For polyoma DNA in 7.2 M $NaClO_4$, the melting temperatures for the nicked and closed circular forms are 48°C and 73°C, respectively.[26] The thermally induced hyperchromicity associated with the early helix-coil transition, however, is too small to be detected, and therefore the early transition must be measured by alkaline titration in a cesium chloride density gradient.[26] Assuming that temperature and pH increments are linearly related, the onset of the early transition for closed polyoma DNA in 7.2 M $NaClO_4$ is estimated to be about 25°.[26] Since the superhelical density of native PM2 DNA is 66% greater than that of polyoma DNA and since the midpoint for the late helix-coil transition of native PM2 DNA occurs at about 70° in 0.5 mM Tris,[31] it is not unreasonable to expect that for PM2 DNA in 0.5 mM Tris, a portion of the molecule is denatured at 25°. Then the number of negative superhelical turns in native PM2 DNA at 25° in this buffer could be close to zero with no added ethidium. Addition of Eth^+ would then cause the formation of positive superhelical turns to give the observed reduction in the relaxation times.

Acknowledgments

The authors are indebted to J. W. Jost, M. C. Y. Kwan, and E. Haberkern for assistance with the on-line computer and the electro-optic data handling programs. We were happy to receive the PM2 samples from Professor J. C. Wang's laboratory.

REFERENCES

1. LERMAN, L. S. 1961. J. Mol. Biol. **3:** 18.
2. FULLER, W. & M. J. WARING. 1964. Ber. Bunsen. **68:** 805.
3. NEVILLE, D. M., JR. & D. R. DAVIES. 1966. J. Mol. Biol. **17:** 57.
4. CRAWFORD, L. V. & M. J. WARING. 1967. J. Mol. Biol. **25:** 23.
5. BAUER, W. & J. VINOGRAD. 1968. J. Mol. Biol. **33:** 141.
6. WANG, J. C. 1969. J. Mol. Biol. **43:** 25.
7. REVET, B. M. J., M. SCHMIR & J. VINOGRAD. 1971. Nature [New Biol.] **229:** 10.
8. HOUSSIER, C., B. HARDY & E. FREDERICQ. 1974. Biopolymers **13:** 1141.
9. RAMSTeIN, J., C. HOUSSIER & M. LENG. 1973. Biochem. Biophys. Acta **335:** 54–68.
10. HSIEH, T. S. & J. C. WANG. 1975. Biochemistry **14:** 527.
11. ESPEJO, R. T. & E. S. CANELO. 1968. Virology **34:** 738.
12. MAHLER, H. R., B. LINE & B. D. MEHROTA. 1964. J. Mol. Biol. **9:** 801.
13. ESPEJO, R. T., E. S. CANELO & R. L. SINSHEIMER. 1969. Proc. Nat. Acad. Sci. USA **63:** 1164.
14. O'KONSKI, C. T. 1968. Encyclopedia of Polymer Science and Technology. Vol. p: 551. John Wiley & Sons. New York, N.Y.
15. YOSHIOKA, K. & H. WATANABE. 1969. Physical Principles and Techniques of Protein Chemistry. Part A. S. J. Leach, Ed. Academic Press. New York, N.Y.
16. FREDERICQ, E. & C. HOUSSIER. 1973. Electric Dichroism and Electric Birefringence. Clarendon Press. Oxford, England.
17. O'KONSKI, C. T. & S. KRAUSE. 1976. Molecular Electro-Optics. Vol. 1, Chap. 3. Marcel Dekker, Inc.
18. JOST, J. W. & C. T. O'KONSKI. Chapter 15 in Volume 2 of Molecular Electro-Optics. Vol. 2, Chap. 13. In press.
19. O'KONSKI, C. T., J. W. JOST & M. KWAN. In preparation.
20. MORGAN, A. R. & D. E. PULLEYBLANK. 1974. Biochim. Biophys. Res. Com. **61:** 396.
21. LEPECQ, J. B. & C. PAOLETTI. 1967. J. Mol. Biol. **27:** 87.
22. SCATCHARD, G. 1949. Ann. N. Y. Acad. Sci. **51:** 660.
23. WANG, J. C. 1971. *In* Proceedings of Nucleic Acid Research. Vol. 2. Cantoni and Davies, Eds.: 407.
24. WARING, M. J. 1965. J. Mol. Biol. **13:** 269.
25. SANDER, C. & P. O. P. TS'O. 1971. J. Mol. Biol. **55:** 1.
26. VINOGRAD, J., J. LEBOWITZ & R. WATSON. 1968. J. Mol. Biol. **33:** 173.
27. LLOYD, P. H., R. N. PRUTTEN & A. R. PEACOCKE. 1968. Biochem. J. **107:** 353.
28. ARMSTRONG, R. W., T. KURUCSEV & U. P. STRAUSS. 1970. J. Am. Chem. Soc. **92:** 3174.
29. GLAUBIGER, D. & J. E. HEARST. 1967. Biopolymers **8:** 691.
30. WANG, J. C. 1974. J. Mol. Biol. **89:** 783.
31. PRITCHARD, A. Unpublished observations.

ELECTRIC DICHROISM AND BIREFRINGENCE OF DNA, CHROMATIN, AND THEIR COMPLEXES WITH CATIONIC DYES. THE STRUCTURE OF CHROMATIN

Claude Houssier, José Bontemps, Xavier Emonds-Alt, and Eugène Fredericq

Laboratoire de Chimie Physique
Université de Liège
Sart-Tilman
B-4000 Liège, Belgium

Introduction

The aim of this paper is to summarize the most essential information about the conformation of DNA and chromatin (the naturally occurring DNA-protein complex of the nucleus), in solution, gained through the use of electric dichroism and birefringence measurements. The study of the interaction of these biological molecules with divalent cations and with cationic dyes used as probes of the macromolecular structure will also help to draw the general picture of the chromatin structure that now emerges from these works in the light of the models recently described in the literature.

The techniques of electro-optical measurements and the procedures useful for the analysis of the data have been fully described in several recent reviews[1-6] and will not be considered here. The essential features of the interaction studies have also been reviewed in the case of the divalent cations interactions[7] as well as the dye-complex studies.[8-11]

The observations described here will be limited to the range of low ionic strengths ($I \leq 0.001$) for three main reasons: (1) the electro-optical technique is not easily applicable at higher ionic strengths because of electrolysis and heating effects during the application of electric pulses of high field strengths; (2) the chromatin precipitates in the range of $0.005 < I < 0.1$ and again becomes soluble for $I \simeq 0.3$–0.5 with partial dissociation of its proteins; (3) in some particular conditions, e.g. in the presence of divalent cations at concentrations as low as 3×10^{-4}M, chromatin and DNA adopt compact structures that could be similar to those encountered *in vivo* and in physiologic media.

Most of the results presented in this paper were obtained with sonicated samples (300,000–500,000 molecular weight), which are less polydispersed than the native samples and whose behavior approaches that of rigid particles, making easier the analysis of some of the data.

*Supported by the Fonds de la Recherche Fondamentale Collective. J. B. and X. E. are recipients of fellowships from the Institut pour l'Encouragement de la Recherche Scientifique dans l'Industrie et dans l'Agriculture.

ELECTRIC PROPERTIES OF DNA AND CHROMATIN

Orientation Mechanism

Two parameters easily accessible from pulsed electro-optical measurements are useful for elucidating the nature of this mechanism: 1) the depth of the minimum $\Delta_{min} = \Delta n_{min}/\Delta n_0$, of the electro-optical signal at the reversal of the field when consecutive pulses of reversed polarity are applied to the samples;[12] and 2) the ratio S_R/S_D of the rise and decay surfaces of the normalized birefringence or dichroism signal.[2] Both parameters yield the ratio $R = \beta^2/2\gamma$ of the permanent ($\beta = \mu_p E/kT$) to the induced ($\gamma = \Delta\alpha E^2/2kT$) dipole terms, provided that the measurements are made at low fields E (where the Kerr law is obeyed). A slow polarization mechanism, however, gives rise to a shape of signal similar to that of a permanent dipole contribution, and the two types of mechanism are thus very difficult to distinguish.[12]

TABLE 1

ORIENTATION MECHANISM AS REVEALED BY THE REVERSING PULSE TECHNIQUE

Samples	Medium	Δ_{min}	$R = \beta^2/2\gamma$	S_R/S_D
DNA				
native sample	1 mM NaCl	0.8–0.9	0.1–0.2	0.3
sonicated sample	1 mM NaCl	0.8–0.9	0.1–0.2	0.9
+ Mg^{2+}	1 mM NaCl	1.0	0	—
Chromatin				
native sample	H_2O pH 6.5	0.8	0.2	0.3
sonicated sample	H_2O pH 6.5	0.8	0.2	0.9
	1 mM NaCl	0.6	0.5	—
DNA-Ethidium Bromide (r = 0.1)	1 mM NaCl	0.9	0.1	—
Chromatin-Ethidium Bromide (r = 0.1)	H_2O pH 6.5	0.7–0.75	0.3	—
	1 mM NaCl	0.4	1.0	—
Chromatin-Proflavine (r = 0.1)	H_2O pH 6.5	0.7–0.75	0.3	—
	1 mM NaCl	0.3–0.4	1–1.5	—

The values of Δ_{min} and S_R/S_D observed for DNA, chromatin, and several of their complexes with dyes are given in TABLE 1. The values of S_R/S_D much smaller than 1 observed with the native samples are interpreted as being due to a polydispersity and/or a flexibility effect resulting in a segmental orientation, and a saturation of the electro-optical effect at low fields for the largest particles in the medium. It should be kept in mind that the ratio S_R/S_D decreases with increasing field strength out of the Kerr law region,[15] even for rigid monodispersed particles, and *a fortiori* for polydispersed systems in which comparatively smaller particles will be oriented to a larger degree as the field strength increases.

For DNA, Δ_{min} is always close to 1; the addition of traces of Mg^{2+} cations (~0.1 mM) makes the small transient at the pulse reversal disappear. This small transient may be interpreted as originating from a very small contribution of a permanent dipole, or a slow polarizability. We do not presently have any explanation for the slightly smaller Δ_{min} value observed by Hanss and Bernengo.[13] Greve and DeHeij[14] reported observations in very good agreement with ours. In the analysis of the field strength dependence of the anisotropy below, we shall thus consider a pure, induced polarization mechanism.

Several authors[13,16–18] claimed that DNA behaves as if it held a longitudinal permanent dipole moment, of the order of 10^4–10^6 D. The conclusions of Ding *et al.*[18] were based on the measurement of the initial slope of the electric dichroism signal found to be zero, and on the shape of the field strength dependence of the dichroism. The first argument appears to rest on inadequate experimental observations using oscilloscopic recordings showing a misalignment of the two traces of the double-beam oscilloscope. This problem has been carefully reexamined by Greve and DeHeij,[14] who reported conclusions in agree-

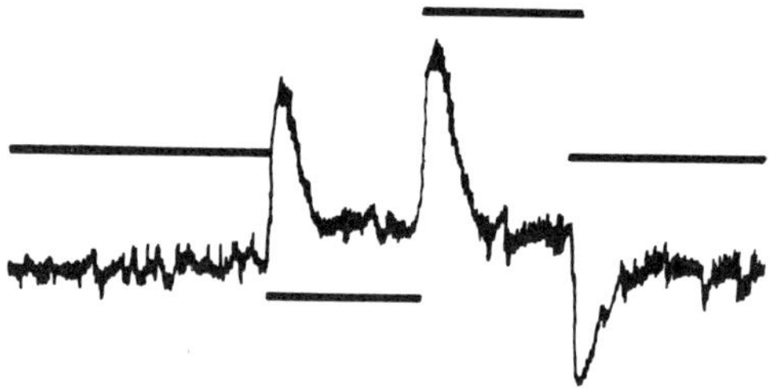

FIGURE 1. Electric birefringence signal of sonicated DNA at low fields in 1 mM NaCl. Field strength 1–1.5 kV/cm; DNA concentration = ~250 μg/ml.

ment with our findings. The second argument used by Ding *et al.*[18] seems to us an inappropriate indication of the orientation mechanism. The conclusions of the other reports are based on low-frequency dielectric[13] and electric-light scattering[17] dispersion measurements, but the possibility of interpreting the observations by an induced polarization mechanism was not totally excluded by these authors. We shall further discuss this problem in connection with a new interpretation of the dielectric dispersion properties of solutions of flexible polyelectrolytes recently developed by van der Touw and Mandel.[19]

At large DNA concentrations (c > 100 μg/ml) and low field strength ($E \sim$ 2–3 kV/cm), the sonicated DNA samples showed "anomalous" electro-optical signals characterized by the presence of a positive birefringence contribution (FIGURE 1) that disappeared upon the addition of Mg^{2+} (0.5×10^{-4} M[14,20]). The sonicated chromatin samples never showed this particular behavior, but the DNA extracted from these samples did. Recent observations of Wilkinson and Thurston,[21] using electric birefringence measurements in sinusoidal fields, also indicated a positive Kerr constant at low frequencies and very low fields for a shear-degraded DNA (M_w = 600,000). No satisfactory explanation has been found so far for these observations. In what follows, the DNA concentrations will be sufficiently low to make the contribution of this effect negligible.

In the case of chromatin, the depth of the minimum at the pulse reversal increased on going from distilled water as solvent to aqueous NaCl solution and on binding of intercalating dyes (TABLE 1). This effect could be explained by a partial loosening of some regions of the protein chains (thc Hl-histone fraction, for example), giving rise to a nonuniform charge distribution at the surface of the particle.

Field Strength Dependence of the Anisotropy

The analysis of the birefringence versus field-strength curves with the aid of a general multiparametric curve-fitting program[22] yielded the following results.

For DNA, a satisfactory fitting of the experimental data could be reached only if a sum of three orientation functions $\Sigma_i \omega_i \Phi(\gamma_i)$ with different polarizability terms was taken into account. The mean polarizability values $\overline{\Delta\alpha} = \Sigma \omega_i \Delta\alpha_i$ decreased only slightly with the molecular weight of the samples from $\sim 20 \times 10^{-32}$ Fm^2 for native DNA ($M_w \simeq (5\text{–}10) \times 10^6$) to $\sim 10 \times 10^{-32}\,Fm^2$ for sonicated DNA ($M_w \simeq 5 \times 10^5$), in 1 mM NaCl.

There are various possible approaches to the interpretation of the electric polarizability of polyelectrolytic macromolecules. Hornick and Weill[23] chose to calculate α on the basis of the various theories available, using length values determined by Kerr effect or light-scattering measurements. An average length of about 180 nm was obtained for sonicated DNA samples of 3.8×10^5 average molecular weight. These authors found that Mandel's theory gave polarizability values two orders of magnitude too large compared with the experimental observations. A better agreement was reached if the counterion repulsion was taken into account by using the Oosawa's theory.[24] However, the estimation of the correction factor introduced in this theory is subject to much uncertainty in the estimation of the local dielectric constant in the vicinity of the polyelectrolyte and in the radius of the layer of bound counterions.

Recent works by van der Touw and Mandel[19] and by Sakamoto *et al.*[25] seem to yield a better interpretation compatible with the electro-optical observations. Polyelectrolyte solutions display two dielectric dispersion regions. The low-frequency dispersion (in the kHz region) is considered as originating from the rotation of the whole molecule (behaving as a kinked chain, rather than as a rigid extended rod), which would hold a dipole moment arising from a very slow polarization, distinct from a free diffusion, of the counterion atmosphere over the whole chain with a relaxation time much longer than the dielectric relaxation time. The amplitude of this dispersion and the corresponding relaxation time are molecular weight-dependent. This low-frequency dispersion is not observed in the case of rigid polyelectrolytic molecules.[26] In the high-frequency range MHz region) the second dispersion observed would arise from a fast-polarization mechanism (relaxation time $< 1\ \mu$sec) resulting from the diffusion of the counterions along short segments of the flexible chain.

We believe that the electric birefringence and dichroism properties discussed in this paper proceed from the fast-polarization mechanism, and correspond to the orientation of segments within the macromolecular chain, a conclusion that had been reached from our previous studies.[27] The slow-polarization mechanism does not show up in the electric birefringence experiments, except perhaps in the range of very low field strengths;[21] this is because it corresponds to entities of much lower optical anisotropy, even though their electric polarizability is much larger.

On the basis of the theory of van der Touw and Mandel,[19] we can estimate the length b of the segments corresponding to the measured electric polarizability, using the equation:

$$\alpha = Z^2 e^2 N b^2 \phi / 12kT \tag{1}$$

where α is the polarizability along the long axis of the particle (perpendicular polarization assumed to be zero), e the unit electric charge (1.6×10^{-19} C), ϕ the fraction of bound counterions (estimated to 0.8 for DNA[28]) and N the number of charged sites.

For a length b of base pairs (two phosphate groups) at a distance h (= 3.4 Å), $N = 2b/h$, assuming a rigid cylinder. For a rigid particle (or a rigid extended macromolecular segment), we obtain:

$$b^3 = \alpha 6kTh/e^2\phi. \tag{2}$$

The observed polarizability values for DNA ranging from $(10\text{–}20) \times 10^{-32}$ F m^2 yield b values of 34–43 nm.

For sonicated DNA samples with intrinsic viscosity of the order of 3.4 dl g^{-1} (in 0.1 M NaCl), i.e. an average molecular weight of about 5×10^5,[29] the length in the fully extended configuration would be equal to ca. 260 nm. Such samples show mean relaxation times of about 20 μsec at high fields ($\geq$ 6 kV/cm) and about 40 μsec at moderate fields (1–2 kV/cm), which would correspond to lengths of 130–160 nm, assuming a rigid cylinder model and an axial ratio of about 100.

For native DNA samples of about 10×10^6 molecular weight (intrinsic viscosity of the order of 50–70 dl g^{-1} in 0.1 M NaCl), the length in the fully extended configuration would be 3200 nm, whereas the value obtained from the mean relaxation time at high fields (0.5–1 ms) is equal to 400–500 nm, assuming an axial ratio of 1000.

From all these pieces of evidence, we can tentatively draw the following conclusions: (1) the electric polarizability measured by the Kerr effect on the native and sonicated DNA samples is a segmental polarizability; (2) the optical anisotropy measured by the Kerr effect is also a segmental anisotropy, and this is why it does not vary appreciably with the molecular weight;[27] (3) the relaxation times correspond to the rotation of macromolecular segments hindered by their interconnection; the length deduced from $\bar{\tau}$ is thus related to the persistence length. The number of segments free to rotate in sonicated DNA is smaller, their length is shorter, and their length distribution is sharper than in native DNA. The distribution of relaxation times reflects the complex relaxation process of the joined segments in the partially flexible macromolecular chain and is not directly influenced by the polydispersity of the samples.

Similar conclusions can be drawn from the behavior of chromatin. Giving a complete interpretation of the electric polarizability of this molecule is, however, much more difficult, owing to the complex orientation mechanism revealed by the reversing pulse technique, in 1 mM NaCl and in the presence of intercalating dyes (TABLE 1). The mean polarizability of chromatin in 1 mM NaCl lies in the range of (3–6) × 10^{-32} Fm^2, which is about two to three times smaller than that of DNA. This is easily understandable on the basis of the neutralisation of the phosphate groups by the histone proteins.

INTERACTIONS WITH DYES

Various types of information may be obtained from the study of the optical and electro-optical properties of nucleic acid complexes with dyes.

Dichroism. By comparing the reduced dichroism values $\Delta A/A$ for the bases and for the bound dye molecules, at low binding ratios *r*, it is possible to determine the orientation of the dye chromophoric group, provided that at least two transitions with different orientations of their moments are accessible. As will be shown below, extrapolation to infinite field strength to obtain the intrinsic optical parameter is not required if the degree of orientation in the electric field is not affected by the binding, a reasonable assumption at very low amounts of binding.

The reduced dichroism of the bases and of the dye chromophoric group in their respective absorption bands may be expressed by the relations:

$$(\Delta A/A)_{\text{bases}} = 1.5\phi f_0(3\cos^2\alpha - 1) \tag{3}$$

$$(\Delta A/A)_{\text{dye}} = 1.5\phi f(3\cos^2\beta - 1) \tag{4}$$

where $0 \leq \phi \leq 1$ is the orientation function; f and f_0 are empirical parameters characterizing the kinked aspect of the chain in the oriented segments; α and β are the local angles between the transition moments (of the bases and of the dye chromophore, respectively) and the double helical axis.

If we consider that the purine and pyrimidine bases are oriented close to the perpendicularity with respect to the helix axis of the DNA backbone ($\alpha = 90°$), in DNA as well as in the secondary arrangement of DNA in chromatin, Equations 3 and 4 yield:

$$\frac{(\Delta A/A)_{\text{dye}}}{(\Delta A/A)_{\text{bases}}} = f(1\text{–}3\cos^2\beta)/f_0. \tag{5}$$

If the degree of kinking of the chain is not affected by the binding, then $f = f_0$ and Equation 5 allows the calculation of the angle β. Under this assumption, $(\Delta A/A)_{\text{dye}}$ can never be larger than $(\Delta A/A)_{\text{bases}}$. Values of the ratio $(\Delta A/A)_{\text{dye}}/(\Delta A/A)_{\text{bases}}$ ranging from 0.9 to 1.1 are considered as reflecting a parallelism between the base planes and the dye ring ($\beta = 90°$), the deviation from unity being of the order of the experimental accuracy for these determinations. For $(\Delta A/A)_{\text{dye}}/(\Delta A/A)_{\text{bases}}$ smaller than 0.9, we consider that the plane of the dye deviates significantly from the perpendicularity, and we estimate the β values given in TABLE 2, assuming $f/f_0 = 1$. β-values in the range of 50 to 60° indicate the absence of intercalation of the dye ring, and can be explained by an external binding in one of the grooves of the double helix that are inclined at an angle of about 60° with respect to the helix axis. When the ratio $(\Delta A/A)_{dye}/(\Delta A/A)_{bases}$ exceeds 1.1, as in the case of several complexes with chromatin, the dye is considered as being intercalated in accessible DNA regions having larger values of f; i.e., presenting a more regular parallel base stacking. This will be further discussed below in the light of the proposed chromatin structure.

For proflavine complexes, it has been shown[30] that the base composition and the methylation of DNA does not affect the intercalation arrangement of the dye,

TABLE 2

ELECTRIC DICHROISM OF DNA AND CHROMATIN COMPLEXES WITH DYES

Samples*	$(\Delta A/A)_{13}^{r\to 0}$ kV/cm: Bases‡ (260 nm)	$(\Delta A/A)_{13}^{r\to 0}$ kV/cm: Dye Chromophore (λ_{nm})	$(\Delta A/A)_{dye}$ / $(\Delta A/A)_{bases}$	$\beta(f_0/f)$ See Text	Ref.
DNA complexes with					
proflavine	−0.525	−0.535(460)	1.02	90°(0.98)	6
	−0.58	−0.55 (460)	0.95	90°(1.05)	30
acridine orange	−0.55	−0.60 (502)	1.09	90°(0.92)	32
		−0.40 (310)	0.73	72°(1.)	32
ethidium bromide	−0.57	−0.46 (510)	0.81	75°(1.)	31
		−0.38 (310)	0.67	70°(1.)	31
actinomycin-D	−0.55	−0.2 (460)	0.364	62°(1.)	§
dibutylproflavine	−0.55	+0.06 (500)	−0.109	52°(1.)	33
		−(0.05–0.1) (435)	(0.09–0.18)	57–59°(1.)	33
33258 Hoechst	−0.55	+0.3 (377)	−0.545	44° (1.)	34
Chromatin complexes with					
proflavine†	−0.43	−0.54 (460)	1.25	90°(0.8)	36
acridine orange†	−0.43	−0.635(502)	1.48	90°(0.685)	§
ethidium bromide	−0.30	−0.48 (510)	1.6	90°(0.625)	§
actinomycin-D†	−0.43	−0.18 (460)	0.42	64°(1.)	§
dibutylproflavine	−0.30	+0.05 (500)	−0.167	51°(1.)	33
33258 Hoechst	−0.30	+0.18 (377)	−0.6	43°(1.)	34

*All measurements in 1 mM NaCl, except for †.
†Measured in H_2O pH 6.5.
‡Measured with the uncomplexed macromolecular samples.
§Unpublished observations.

which remains closely parallel to the base planes. Ethidium bromide, as well as actinomycin-D, significantly deviates from this parallelism (TABLE 2) because of the incomplete insertion of their heterocyclic ring resulting from the presence of a hindering substituant (phenyl and oligopeptide chain, respectively).

In the strong binding range, the reduced dichroism and thus the orientation of the dye chromophore do not change with the binding ratio r (molar bound-dye concentration over molar nucleotide concentration) for the intercalating dyes. As soon as externally bound dye molecules are present, the dichroism decreases (FIGURE 2). If we consider that the free dye in the solution and the externally

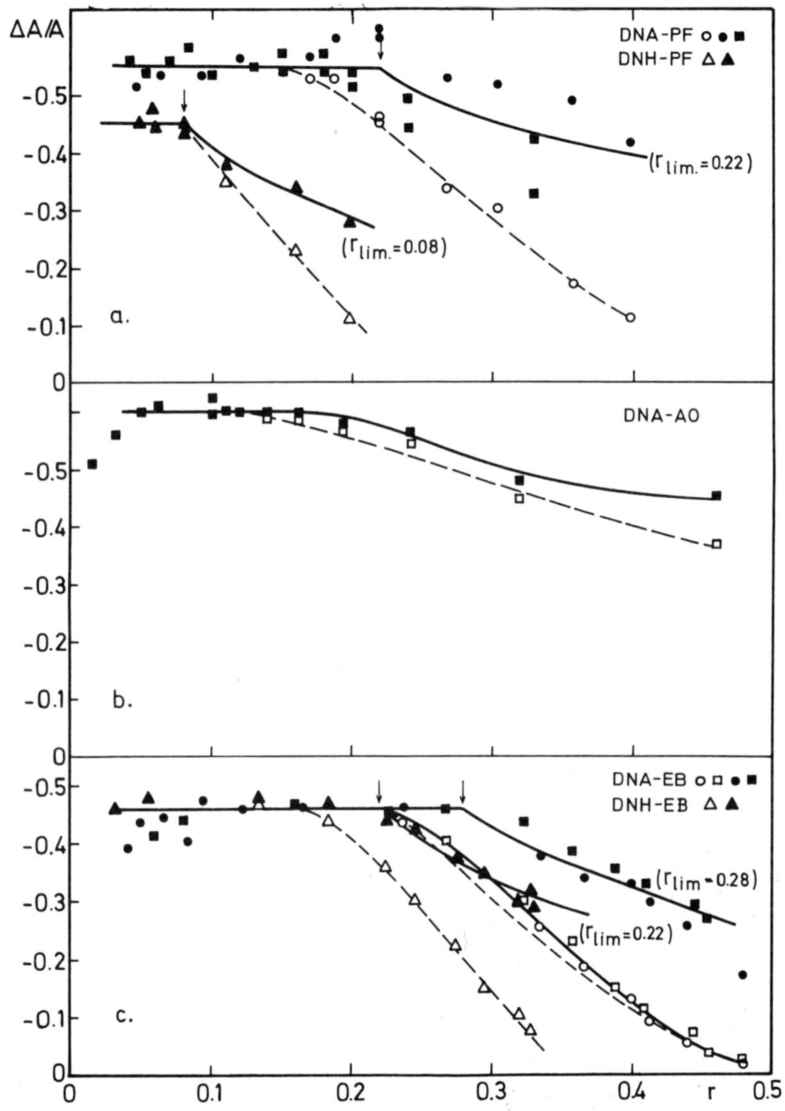

FIGURE 2. Electric dichroism versus r for complexes of proflavine (PF), acridine orange (AO), and ethidium bromide (EB) with DNA and chromatin (DNH), in 1 mM NaCl, at about 13 kV/cm. The full lines in parts a and c are calculated on the basis of Equations 6–9 with the appropriate adjustment of the r_{lim} value (indicated by the vertical arrow) to fit the experimental data satisfactorily. Open and filled symbols, respectively, refer to measured dichroism and dichroism corrected for free-dye contribution to the absorption.

bound dye molecules do not contribute to the measured dichroism, and that the degree of orientation of the complex particles is not modified by the binding, it can be shown that the reduced dichroism, corrected for the free-dye contribution to the absorption, is given by:[30]

$$(\Delta A/A)_{corr} = \frac{(\Delta A/A)_{int}}{1 + \frac{\epsilon_{ext}}{\epsilon_{int}} \frac{\alpha_{ext}}{\alpha_{int}}} \quad (6)$$

where $(\Delta A/A)_{int}$ is the constant reduced dichroism measured in the intercalation range, ϵ_{ext} and ϵ_{int} the extinction coefficients for the externally bound and intercalated molecules respectively, and $\alpha_{ext} = c_{ext}/c_{tot}$ and $\alpha_{int} = c_{int}/c_{tot}$ their respective bound fractions.

Since the values of ϵ_{ext}, ϵ_{int}, α_{ext}, and α_{int} have not been determined for all the set of data in FIGURE 2, we shall consider now some further simplification of the model. Assuming that intercalation can occur up to a limit or r_{lim}, the α_{int} and α_{ext} bound fractions will be related to the mononucleotide-to-dye molar ratio P/D, for $r > r_{lim}$ by the relations:

$$\alpha_{int} \approx r_{lim} \cdot (P/D);\ \alpha = \alpha_{int} + \alpha_{ext} = r \cdot (P/D) \quad (7)$$

so that Equation 6 becomes:

$$(\Delta A/A)_{corr} = \frac{(\Delta A/A)_{int}}{1 + \frac{\epsilon_{ext}}{\epsilon_{int}}\left(\frac{r}{r_{lim}} - 1\right)}. \quad (8)$$

If the measurements are made in a wavelength region where the absorption does not change too much with the binding ratio ("isosbestic region") Equation 8 gives:[31]

$$(\Delta A/A)_{corr} = (\Delta A/A)_{int} \cdot r_{lim}/r \quad (9)$$

For $r < r_{lim}$, $\alpha_{ext} \simeq 0$ and $(\Delta A/A)_{corr} = (\Delta A/A)_{int}$ = constant.

The full lines drawn in FIGURE 2, a and c, correspond to the calculated $(\Delta A/A)_{corr}$ using the above equations with the appropriate adjustment of the r_{lim} value. The agreement with the experimental observations is quite satisfactory. The limit of intercalation in DNA is found to be close to one dye molecule for two base pairs (exclusion of adjacent binding sites). For acridine orange, which has a high aggregation tendency, the externally bound molecules in the weaker binding processes seem to be present in a more organized arrangement than for the other dyes (FIGURE 2, b), because of the binding of dimer species.[32]

For the complexes with nonintercalating dyes, complex variations dependent on the wavelength of the measurement were encountered, and low positive and negative dichroism values were observed.[33,34]

Birefringence. By studying the variation of the birefringence Δn with the amount of binding, it is possible to see if the structure of the polynucleotide backbone is affected by the binding. The birefringence Δn_r measured in the visible range is the sum of the macromolecule birefringence Δn_0 plus the birefringence arising from the presence of new dispersion curves in the dichroic bands of the dye $\Delta n(\tilde{\nu})$ which can be estimated by the Kramers-Kronig relationships[35] and is given by:

$$\Delta n(\tilde{\nu}) = k\, \Delta A_{\max} / \ell \tag{10}$$

where $\Delta A_{\max}$ is the dichroism at the maximum of the band, k is a constant related to the position and bandwidth of the dichroic absorption band, and to the wavelength $\lambda = 1/\tilde{\nu}$ of the birefringence measurements; ℓ is the optical pathlength. Since the birefringence of the macromolecule is proportional to the macromolecular concentration P ($\Delta n_0 = \Delta n_M \cdot P$) expressed in mononucleotide, we obtain the following relationship for the relative change of birefringence upon dye binding at a concentration c_b of bound dye:

$$\begin{aligned}(\Delta n_r - \Delta n_0)/\Delta n_0 &= k(\Delta A/A)_{\max}\, \epsilon_{\max}\, c_b / \Delta n_M \cdot P \\ &= k(\Delta A/A)_{\max}\, \epsilon_{\max}\, r/\Delta n_M = K\, r \end{aligned}\tag{11}$$

or, for a set of i absorption bands:

$$(\Delta n_r - \Delta n_0)/\Delta n_0 = \frac{r}{\Delta n_M} \Sigma_i\, k_i (\Delta A/A)_{\max,i} \cdot \epsilon_{\max,i} \tag{12}$$

In the whole binding range, where the reduced dichroism $(\Delta A/A)_{\max}$ remains constant (intercalation range), the relative increase of birefringence should be proportional to r, the slope of the dependence being larger the bigger the extinction coefficient $\epsilon_{\max}$ of the dye, and the smaller the molar birefringence Δn_M of the macromolecule. We found, in the case of proflavine, that the ultraviolet absorption band of the dye (near 260 nm) accounted for about 20% of the calculated increase of birefringence at 550 nm. In the case of the ethidium bromide complexes, the near uv band, which has a much larger extinction coefficient than the visible band, contributed to more than 80% of the total calculated increase of birefringence at 600 nm.

FIGURE 3 compares the calculated relative increase of birefringence with the experimental observations for proflavine and ethidium bromide complexes with DNA and chromatin. In general, the calculated slope of the Δn_r versus r-dependence closely parallels the experimental findings. The deviation noticed for the DNA-ethidium bromide complexes may result from the lower accuracy of the determination in this case: the relative increase of birefringence does not exceed 10% for $r \leq 0.1$. We can thus conclude that no drastic alteration of the base-stacking arrangement of the polynucleotide backbone can be detected from these observations, although a modification of the orientation mechanism was clearly evidenced by the reversing-pulse technique for the intercalation complexes with chromatin.

Birefringence Relaxation. The intercalation of dye molecules in DNA-like structures is expected to produce a length increase corresponding to a base-pair separation per dye molecule added; i.e., the length of the complex molecules should vary with r according to the following relation:

$$L/L_0 = 1 + 2\,r. \tag{13}$$

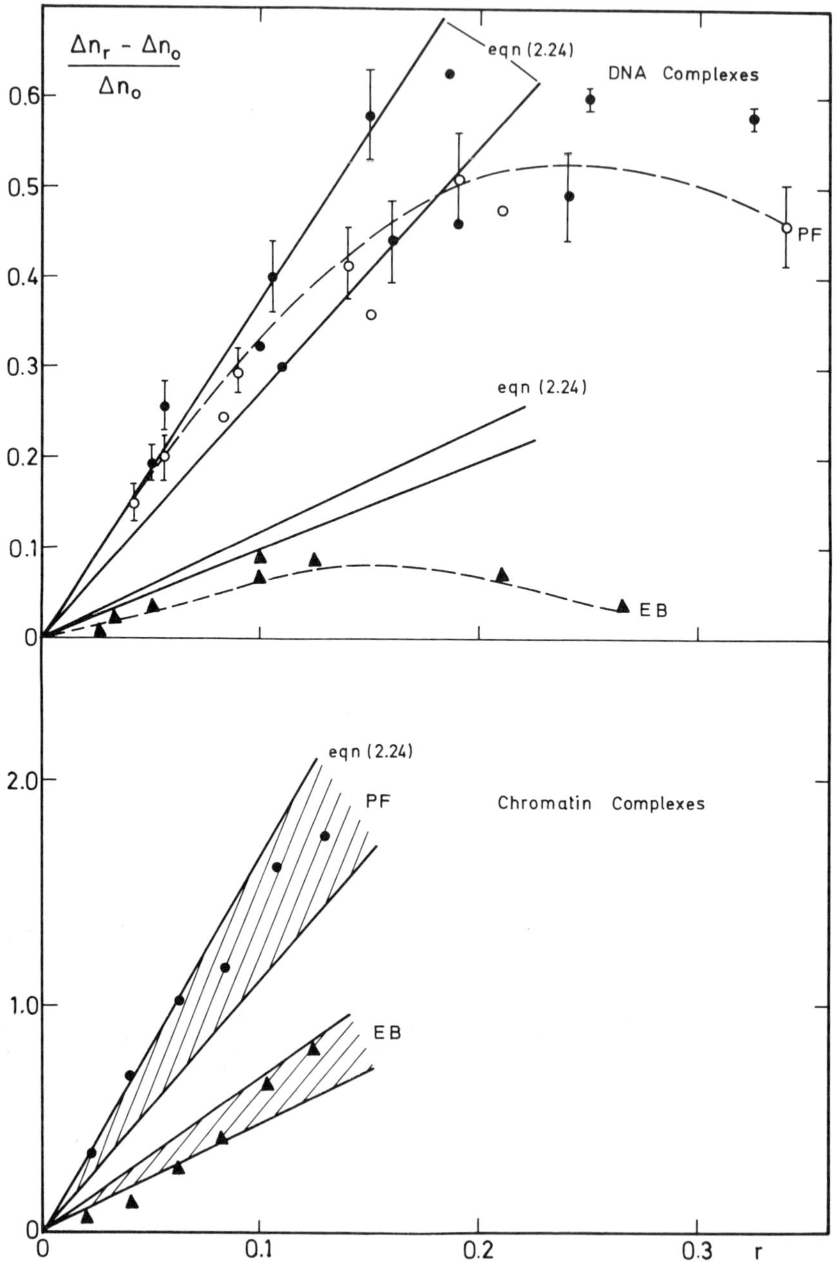

FIGURE 3. Relative increase of electric birefringence upon complexation of proflavine (PF) and ethidium bromide (EB) to DNA and chromatin, in 1 mM NaCl, at about 13 kV/cm. The two lines drawn for each complex on the basis of Equation 12 indicate the range of inaccuracy of the calculation resulting from the variations in all the parameters involved: Δn_M, dichroism values, . . .

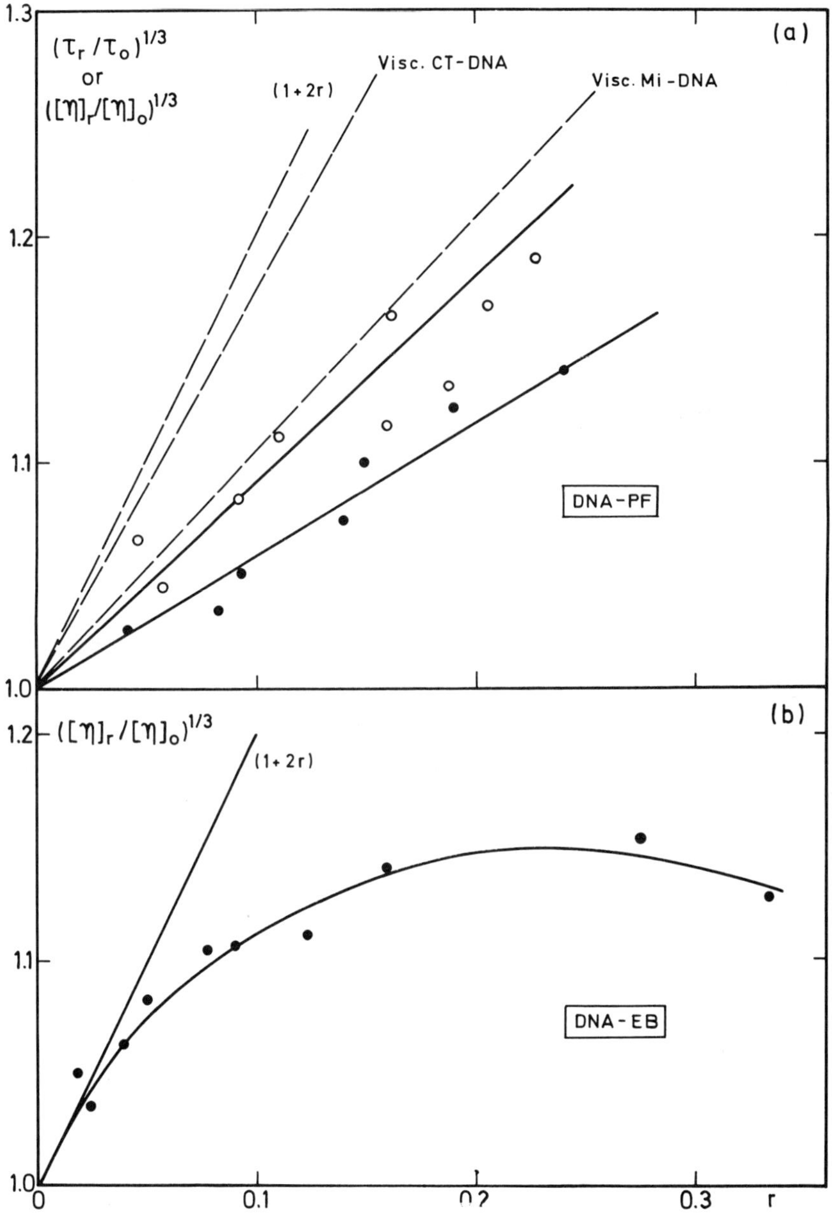

FIGURE 4. Increase of viscosity and birefringence relaxation time upon complexation of proflavine (PF) and ethidium bromide (EB) to DNA, in 1 mM NaCl, at about 13 kV/cm.

Although the increase of viscosity observed was compatible with this model, the increases of relaxation times $\bar{\tau}$ measured with sonicated samples were always much smaller (FIGURE 4), and in some cases (in particular, DNA-ethidium bromide complexes), no change of $\bar{\tau}$ was observed at all.[31] Sonicated chromatin complexes showed irregular variations of $\bar{\tau}$ with r, as well as native DNA complexes.[30] We believe that the interpretation of these observations can be sought in the segmental orientation mechanism discussed above. The dye-binding produces changes in the flexibility of the macromolecular chain at the same time as it produces increases of the segment lengths. We do not yet have any quantitative interpretation of these results.

THE CHROMATIN STRUCTURE

Electro-optical studies do not give direct access to accurate geometrical parameters characterizing a macromolecular structure. They mainly reflect the mean orientation of chromophoric groups with respect to the orientation axis (dichroism and, more indirectly, birefringence), the overall (rigid particles) or segmental (flexible particles) length of the oriented particles (obtained from the relaxation times), some useful additional information being obtained through the study of the interactions with dyes as regards the geometry, accessibility, and properties of the binding sites. The electro-optical measurements are also of invaluable help in the study of the dynamics of conformational changes.

Some of the most important electro-optical parameters of DNA, chromatin, and chromatin subunits in the presence of different cations are summarized in TABLE 3; they lead to the following observations.

1. The intrinsic optical anisotropy of chromatin is appreciably smaller than that of DNA. The removal of the proteins from chromatin progressively restores the DNA anisotropy,[37] but the difference between Hl-depleted and whole chromatin anisotropy is not significant (TABLE 3). According to x-ray diffraction results,[43b] histone Hl is not directly involved in the stabilization of chromatin tertiary structure; we can thus conclude that the optical anisotropy of chromatin determined in 1 mM NaCl is related to its tertiary structure.

2. The addition of Mn^{2+} cations produces a decrease of the optical anisotropy of DNA and of the extended chromatin subunit, and a reversal of the birefringence sign for whole chromatin and for its compact subunit. This change of birefringence sign is prevented by the removal of histone Hl, and is correlated to the formation of a quaternary structure.

3. Mg^{2+} ions do not appreciably affect the optical anisotropy of DNA and of the extended chromatin subunit but produce for whole chromatin the appearance of complex signals characterized by the presence of positive and negative optical anisotropy contributions.

We shall try to find a general interpretation of this set of experimental observations by considering the various models of chromatin structure recently described in the literature.

The basic subunit of the chromatin structure, which can be isolated by chromatography or sucrose gradient separation after partial nuclease digestion or ultrasonic degradation, is the nucleosome (or ν-body, or monosome or mono-nucleosome), having a spherical shape (10 nm diameter; 200,000 molecular weight) and containing 40 nm of a DNA-chain wound in a superhelical arrangement around a protein core (120,000 molecular weight) in which histone-histone interactions play a dominant role.[38-43]

In the native chromatin structure, these nucleosomes, interconnected by short internucleosome segments (40–60 base pairs) would be arranged in a more or less

TABLE 3

ELECTRO-OPTICAL ANISOTROPY OF DNA AND CHROMATIN (VALUES AT INFINITE STRENGTH OBTAINED FROM A FITTING OF THE EXPERIMENTAL DATA BY A THREE-TERM ORIENTATION FUNCTION*)

	Dichroism $(\Delta A/A)_s$		Birefringence $10^7 \times (\Delta n_s/A^{1\,cm}_{260\,nm})$ ‖		
Samples	In 1 mM NaCl	In Cacodylate Buffer .1 mM	In 1 mM NaCl	In .33 mM Mg^{2+} $(I/P > 2)$	In. .33 mM Mn^{2+} $(I/P > 2)$
DNA	−(.7–.8)	−1.35†‡	−26.6	−26.7	−22.4
Whole Chromatin	−0.35	−(.5–.6)† −(.7–.85)‡	− 7.4	− 2.1§ + 0.28¶	+ 2.8
Chromatin Without H1	—	—	− 7.7	—	− 2.8
Extended** Subunit	—	—	− 5.6	−5.6	− 2.1
Compact** Subunit	—	−0.2‡	− 3.5	(−)— + 0.915¶	+ 2.75

*See text. Birefringence and dichroism measured at 550 and 260 nm, respectively.
†Taken from Reference 18.
‡Taken from Reference 45.
§Negative components of a complex birefringence signal.
¶Positive components of a complex birefringence signal.
‖ The optical anisotropy factor $(g_a - g_b)$ may be obtained from the $(\Delta n_s/A^{1\,cm}_{260})$ values by the relation: $(g_a - g_b) = (\Delta n_s/A^{1\,cm}_{260}) \cdot n \cdot E_{260}{}^{1g/ml}/2\pi\bar{v}$ where n is the solution refractive index (1.33 for dilute aqueous solutions), $E_{260}{}^{1g/ml}$ the extinction coefficient in the (ml/g) cm^{-1} scale at 260 nm (~20,000 for DNA; ~(21,000/2.3) for chromatin at a Prot/DNA ratio of 1.3); the partial specific volume $\bar{v}$, however, was not determined in our experimental conditions, and the occurrence of important conformational changes in the presence of the divalent cations for chromatin may cause appreciable modifications of its value. The values of $\Delta n_s/A^{1\,cm}_{260}$ in the presence of Mg^{2+} and Mn^{2+} are given for ionic over nucleotide molar concentrations ratios $I/P > 2$.
**Prepared from sonicated chromatin by ECHTAM chromatography.[46]

compact helical array[42,43] or in a linear or flexible string.[41] The structural parameters of this "beads-on-a-string" model remain very vague, and it is most probable that in solution, after the shaking and sonication steps used to obtain homogeneous solutions, the quaternary arrangement depicted in FIGURE 5 will be partially unfolded.

It is possible to estimate the reduced dichroism $(\Delta A/A)_{sh}$ of such superhelical arrangements from the reduced dichroism of the initial double-helical structure

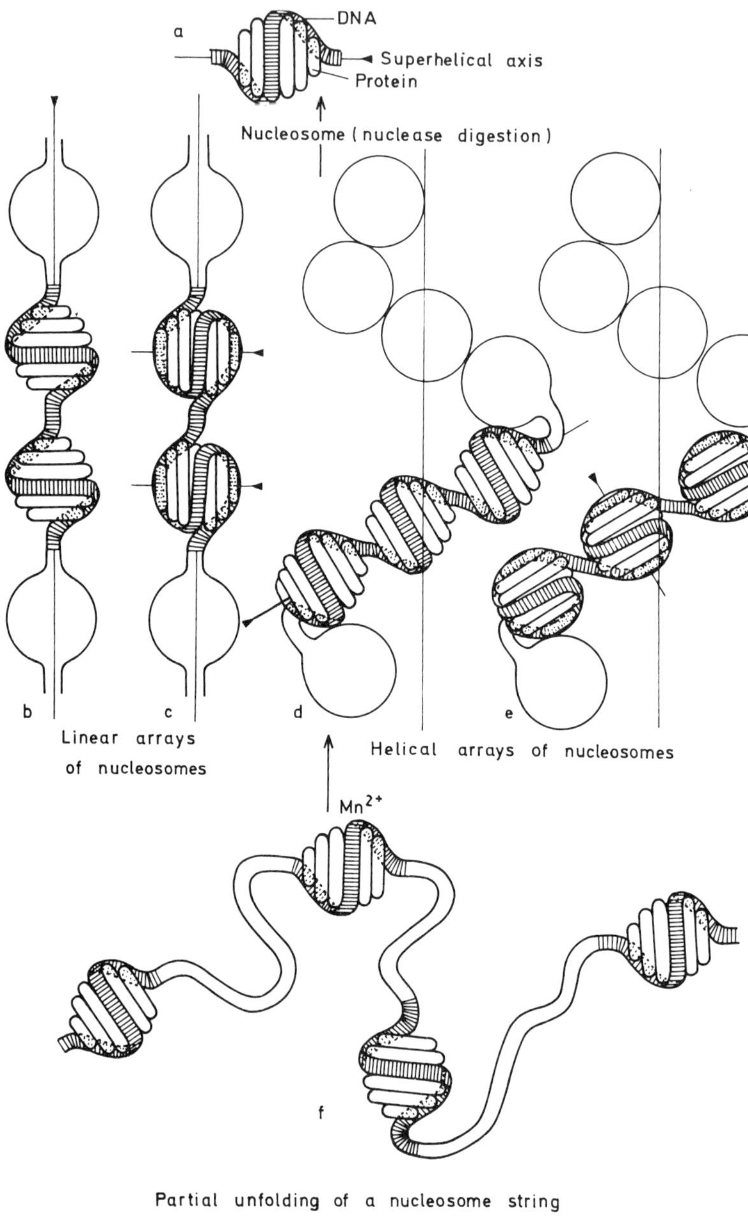

FIGURE 5. Tertiary and quaternary arrangements of chromatin.

using the relation derived by Rill[44] for a complete orientation of the particle with its superhelical axis parallel to the applied orientation force:[50]

$$(\Delta A/A)_{sh}/(\Delta A/A)_h = (3\cos^2\Theta - 1)/2 \quad \textbf{(14)}$$

where $\Theta = \pi d/p$ is the superhelix tilt angle, with d the diameter and p the pitch of the superhelix.

The dichroism values found on the basis of this relation for the various superhelical arrangements proposed in the literature (and which would be present in the nucleosomes) are summarized in TABLE 4, taking for $(\Delta A/A)_h$ a value of −0.8, i.e., the saturation value of the reduced dichroism found for DNA in our experi-

TABLE 4
REDUCED DICHROISM FOR VARIOUS STRUCTURAL ARRANGEMENTS OF CHROMATIN

Structures References	Geometrical Parameters Pitch (nm)	Diameter (nm)	Tilt Angle (Θ)	$(\Delta A/A)_{sh}$ / $(\Delta A/A)_h$	$(\Delta A/A)_s$
Secondary structure					
DNA B-form	3.4	2	0° (no superhelix)	1	−0.8*
Tertiary structure (DNA superhelical arrangement in the nucleosomes)					
47	4.5	5.6	76°	−0.41	+0.33
48	12.0	10.0	69°	−0.31	+0.25
39	5.3	13.0	83°	−0.475	+0.38
40	4.5	7.4	79°	−0.445	+0.36
Quaternary structure (helical arrangement of the nucleosomes)					
43	50.0	26.0	58°	−0.085	±0.030†
42	10.0	25.0 (20–30)	83°	−0.476	±0.170†

*Experimental value determined from the curve fitting of the electric dichroism versus field strength data, in 1 mM NaCl.
†Calculated for $(\Delta A/A)_{sh} \simeq 0.35$.

mental conditions (1 mM NaCl). They are all positive if we assume that the orientation axis is the superhelical axis, in which case Θ is larger than 54°45′. However, it is not evident that the axis of larger polarizability of the nucleosome particles will be the superhelical axis. The dichroism of the nuclease-resistant fragments of the chromatin was found to be negative,[45] as was the birefringence of the compact fraction obtained by the procedure of Simpson and Reeck[46] (see TABLE 3). These fractions are most probably not isolated nucleosome particles but arrays of nucleosomes of various lengths. We can thus envisage two explanations for the negative anisotropy of these nucleosome strings: (1) the superhelical axis of the nucleosome would be perpendicular to the long axis of the array, or (2) extended internucleosome fragments and/or unfolded nucleosomes with DNA in its normal double-helical arrangement (negative anisotropy) are present in sufficient amount (60–70%) to reverse the sign of the anisotropy.[50]

Equation 14 can also be used to calculate the reduced dichroism of the quaternary helical structure in which the nucleosomes are presumed to be arranged in native chromatin.[41–43] The sign of the calculated anisotropy will depend on the orientation of the nucleosome superhelical axis in this quaternary structure. If it is perpendicular to the backbone axis as in the linear array considered in (1) above, then the resulting dichroism for the quaternary arrangements proposed by Bram *et al.*[43] and by Carpenter *et al.*[42] would be positive. This would enable us to explain the reversal of the birefringence sign of the whole chromatin and of the compact fraction in aqueous solution in the presence of Mn^{2+} (TABLE 3), the helical arrangement of the nucleosomes being favored by the concerted action of the divalent cation and of the Hl-histone fraction. We also tentatively conclude that in this arrangement the nucleosome superhelical axis is perpendicular to the backbone axis (FIGURE 5, e). The above discussion rests on two important assumptions. The form contribution[6] to the measured anisotropy has not been taken into account; its estimation still remains too uncertain to allow a satisfactory appraisal of its importance. The contribution of the internucleosome segments to the overall anisotropy of the molecule could not be estimated because of the lack of knowledge of its structural arrangement.

The dye-interaction studies presented above make possible the following additional information about the chromatin structure in solution.

1. The intercalation sites in chromatin are DNA regions in which access to the dye is probably provided by one of the grooves of the DNA helix not covered by the proteins. Because of the high negative dichroism of the intercalated dye molecules, we believe that these regions are the internucleosome segments. The amount of such regions (60%) is compatible with the number of binding sites accessible for strong binding in chromatin compared to DNA. A study of the dye binding to isolated nucleosomes would help to clarify the question of the accessibility of DNA in the superhelical regions.

2. The intercalation of dye molecules displaces (or dissociates) some histone fractions (or segments of their chain), resulting in the appearance of a permanent dipole component on the chromatin molecule. The overall chromatin structure does not seem, however, to be greatly affected by this displacement, since no appreciable change of the bulk contribution to the birefringence is observed (FIGURE 3); this is understandable if the Hl histone fraction is at the origin of this effect.

3. The affinity of these "accessible" regions for the dye is found to be lower than that of free DNA.[49]

4. The closer proximity between the dye molecules in the accessible regions gives rise to more important dye-dye interactions yielding a larger induced optical activity (Bontemps *et al.*, in preparation).

Much additional work will be required to complete the picture of the chromatin structure. The main difficulty with the solutions studies is that we see only average behavior for a set of distinct structural arrangements which are continuously changing. It would be particularly useful to evidence the factors which are responsible for the induction of the compact arrangement of chromatin in the chromosomes.

References

1. O'Konski, C. T. 1968. Kerr effect. *In* Encyclopedia of Polymer Science and Technology. N. Bikales, Ed. John Wiley and Sons. New York, N.Y. Vol. **9:** 551–590.
2. Yoshioka, K. & H. Watanabe. 1969. Dielectric properties of proteins. II. Electric birefringence and dichroism. *In* Physical Principles and Techniques of Protein Chemistry. S. J. Leach, Ed. Part A, Chap. **7:** 335–367. Academic Press, New York, N.Y.
3. Charney, E. 1971. Linear dichroism with special emphasis on electric field induced linear dichroism. *In* Procedures in Nucleic Acid Research. G. L. Cantoni and D. R. Davies, Eds. Vol. **2:** 176–204. Harper and Row, New York, N.Y.
4. Stoylov, S. P. 1971. Colloid electro-optics. Electrically induced optical phenomena in disperse systems. Adv. Coll. Interface Sci. **3:** 45–110.
5. Hanss, M. & B. Roux. 1972. Biréfringence électrique. Application à l'étude des macromolécules. Ann. Phys. Biol. Méd. **3:** 133–163.
6. Fredericq, E. & C. Houssier. 1973. Electric Dichroism and Electric Birefringence. Clarendon Press. Oxford, England.
7. Daune, M. 1974. Interactions of metal ions with nucleic acids. *In* Metal ions in Biological Systems. H. Siegel, Ed. **3:** 1–43.
8. Blake, A. & A. R. Peacocke. 1968. The interaction of amino-acridines with nucleic acids. Biopolymers **6:** 1225–1253.
9. Gale, E. F., E. Cundliffe, P. E. Reynolds, M. H. Richmond & M. J. Waring. 1972. Inhibitors of nucleic acid synthesis. *In* The Molecular Basis of Antibiotic Action. Ch. **5:** 173–277. John Wiley & Sons. New York, N.Y.
10. Lochmann, E. R. & A. Micheler. 1973. Binding of organic dyes to nucleic acids and the photodynamic effect. *In* Physico-Chemical Properties of Nucleic acids. J. Duchesne, Ed. Vol. **1:** 223–267. Academic Press. New York, N.Y.
11. Bloomfield, V. A., D. M. Crothers & I. Tinoco, Jr. 1974. Physical Chemistry of Nucleic Acids. Harper & Row. New York, N.Y.
12. Tinoco, I., Jr. & K. Yamaoka. 1959. The reversing pulse technique in electric birefringence. J. Phys. Chem. **63:** 423–427.
13. Hanss, M. & J. C. Bernengo. 1973. Dielectric relaxation and orientation of DNA molecules. Biopolymers **12:** 2151–2159.
14. Greve, J. & M. E. DeHeij. 1975. Transient electric birefringence of sonicated DNA. Biopolymers **14:** 2441–2443.
15. Nishinari, K. & K. Yoshioka. 1969. A theory for the rise of the electric birefringence of rigid macromolecules at high field strength. Kolloid-Z. Z. Polymere **235:** 1189–1192.
16. Takashima, S. 1963. Dielectric dispersion of DNA. J. Mol. Biol. **7:** 455–467.
17. Jennings, B. R. & H. Plummer. 1970. Electro-optic scattering studies on deoxyribonucleic acid. Biopolymers **9:** 1361–1372.
18. Ding, D., R. Rill & K. E. Van Holde. 1972. The dichroism of DNA in electric fields. Biopolymers **11:** 2109–2124.
19. Van der Touw, F. & M. Mandel. 1974. Dielectric increment and dielectric dispersion of solutions containing simple charged linear macromolecules. I. Theory. Biophys. Chem. **2:** 218–230.
20. Colson, P., C. Houssier, E. Fredericq & J. A. Bertolotto. 1974. Electro-optical properties of nucleic acids and nucleoproteins VI. Reversal of the birefringence sign for sonicated DNA at low fields. Polymer **15:** 396–397.
21. Wilkinson, R. S. & G. B. Thurston. 1975. The optical birefringence of DNA solutions induced by oscillatory electric and hydrodynamic fields. Biopolymers **15:** 1555–1572.
22. Meites, L. 1973. The general multiparametric curve-fitting program CFT3. Computing Laboratory, Dept. of Chemistry. Clarkson College of Technology. Potsdam, N.Y. 13676.
23. Hornick, C. & G. Weill. 1971. Electro-optical study of the electric polarizability of rodlike fragments of DNA. Biopolymers **10:** 2345–2358.

24. OOSAWA, F. 1970. Counterion fluctuation and dielectric dispersion in linear polyelectrolytes. Biopolymers **9:** 677–688.
25. SAKAMOTO, M., H. KANDA, H. REINOSUKE & Y. WADA. 1976. Dielectric relaxation of DNA in aqueous solutions. Biopolymers **15:** 879–892.
26. VAN DER TOUW, F., J. W. H. BRIEDÉ & M. MANDEL. 1973. Electric permittivity of alfalfa mosaic virus in aqueous solution. Biopolymers **12:** 111–119.
27. COLSON, P., C. HOUSSIER & E. FREDERICQ. 1974. Electro-optical properties of nucleic acids and nucleoproteins. V. Study of sonicated DNA and nucleohistone solutions. Biochim. Biophys. Acta **340:** 244–261.
28. DAUNE, M. 1969. Interactions des ions métalliques avec le DNA. I. Comportement de polyelectrolyte et liaison des ions compensateurs. Biopolymers **7:** 659–670.
29. EIGNER, P. A. & P. DOTY. 1965. The native, denatured, and renatured states of deoxyribonucleic acid. J. Mol. Biol. **12:** 549–580.
30. RAMSTEIN, J., C. HOUSSIER & M. LENG. 1974. Electro-optical properties of nucleic acids and nucleoproteins. IV. Influence of base composition and methylation on the properties of DNA and DNA-proflavine complexes. Biochim. Biophys. Acta **335:** 54–68.
31. HOUSSIER, C., B. HARDY & E. FREDERICQ. 1974. Interaction of ethidium bromide with DNA. Optical and electro-optical study. Biopolymers **13:** 1141–1160.
32. FREDERICQ, E. & C. HOUSSIER. 1972. Study of the interaction of DNA and acridine orange by various optical methods. Biopolymers **11:** 2281–2308.
33. BONTEMPS, J., C. HOUSSIER & E. FREDERICQ. 1974. Optical and electro-optical properties of the complexes of dibutylproflavine with DNA and nucleohistone. Biophys. Chem. **2:** 301–315.
34. BONTEMPS, J., C. HOUSSIER & E. FREDERICQ. 1975. Physico-chemical study of the complexes of "33258 Hoechst" with DNA and nucleohistone. Nucleic Acids Res. **2:** 971–984.
35. HOUSSIER, C. & H. G. KUBALL. 1971. Electro-optical properties of nucleic acids and and nucleoproteins. III. Kramers-Kronig relationships in linear birefringence and dichroism. Application to a DNA-proflavine complex. Biopolymers **10:** 2421–2433.
36. HOUSSIER, C. & E. FREDERICQ. 1966. Electro-optical properties of nucleic acids and nucleoproteins. II. Study of the deoxyribonucleohistone-proflavine complexes. Biochim. Biophys. Acta **120:** 113–130.
37. FREDERICQ, E. & C. HOUSSIER. 1967. Physico-chemical properties of native and recombined calf-thymus nucleohistones. Eur. J. Biochem. **1:** 51–60.
38. VAN HOLDE, K. E., C. G. SAHASRABUDDHE & B. R. SHAW. 1974. A model for particulate structure of chromatin. Nucleic Acids Res. **1:** 1579–1586.
39. HYDE, J. E. & I. O. WALKER. 1975. A model for chromatin sub-structure incorporating symmetry considerations of histone oligomers. Nucleic Acids Res. **2:** 405–421.
40. PARDON, J. F., D. L. WORCESTER, J. C. WOOLEY, K. TATCHELL, K. E. VAN HOLDE & B. M. RICHARDS. 1975. Low-angle neutron scattering from chromatin subunit particles. Nucleic Acids Res. **2:** 2163–2176.
41. CARLSON, R. D. & D. E. OLINS. 1976. Chromatin model calculations: Arrays of spherical ν-bodies. Nucleic Acids Res. **3:** 89–100.
42. CARPENTER, B. G., J. P. BALDWIN, E. M. BRADBURY & K. IBEL. 1976. Organisation of subunits in chromatin. Nucleic Acids Res. **3:** 1739–1746.
43a. BRAM, S., G. BUTLER-BROWNE, P. BAUDY & K. IBEL. 1975. Quaternary structure of chromatin. Proc. Nat. Acad. Sci. USA **72:** 1043–1045.
43b. BRAM, S. 1975. A double coil chromatin subunit model. Biochimie **57:** 1301–1306.
44. RILL, R. 1972. The linear dichroism of oriented helical and superhelical polymers. Biopolymers **11:** 1929–1941.
45. RILL, R. & K. E. VAN HOLDE. 1973. Properties of nuclease-resistant fragments of calf-thymus chromatin. J. Biol. Chem. **248:** 1080–1083.
46. SIMPSON, R. T. & G. R. REECK. 1973. A comparison of the protein of condensed and extended chromatin fractions of rabbit liver and calf-thymus. Biochemistry **12:** 3853–3858.
47. BRAM, S. & H. RIS. 1971. On the structure of nucleohistone. J. Mol. Biol. **55:** 325–336.

48. PARDON, J. F. & M. H. F. WILKINS. 1972. A super-coil model for nucleohistone. J. Mol. Biol. **68:** 115–124.
49. BONTEMPS, J. & E. FREDERICQ. 1974. Comparative binding study of the interaction of quinacrine and ethidium bromide with DNA and nucleohistone. Biophys. Chem. **2:** 1–22.
50. RILL, R. & K. E. VAN HOLDE. 1974. Electric dichroism of chromatin. J. Mol. Biol. **83:** 459–471.

CHEMICAL TRANSITIONS OF BIOPOLYMERS INDUCED BY AN ELECTRIC FIELD AND THEIR EFFECTS IN DIELECTRICS AND BIREFRINGENCE*

Gerhard Schwarz

Department of Biophysical Chemistry
Biocenter of the University of Basel
Basel, Switzerland

INTRODUCTION

Certain biological regulation processes might be controlled by nonthermal effects of electric fields on the physicochemical state of the matter. Especially well-established phenomenological evidence for such conjectures has been presented regarding action potentials on excitable membranes. These are obviously associated with ionic permeabilities that depend on the transmembrane electric field.[1] The molecular mechanism of this phenomenon, however, is not so far known.

Rather obvious electric field effects to be considered in pertinent discussions are displacements of charged molecules or groups as well as rotation of molecular dipoles. On the other hand, it seems to be much less known that there is also a direct effect on chemical equilibria. In ordinary reactions involving only small molecules this can, however, usually be neglected. Once macromolecules participate, as in molecular biology, the situation might be quite different. Actually, such field-induced chemical transitions, particularly conformational changes, permit a rather specific control of biomolecular processes much better than simple electric-charge transfers would allow.

A survey of the fundamental theory and some pertinent experimental evidence will now be given.

THERMODYNAMIC FOUNDATION

If an electric field is applied to a chemical reaction system its equilibrium constant will generally be changed. The effect of an externally applied field strength E is quantitatively described by the relation

$$\left(\frac{\partial \ln K}{\partial E}\right)_{p,T} = \frac{\Delta M}{RT} \tag{1}$$

where M denotes the overall dipole moment parallel to the field and ΔM, the reaction moment, is the change of M which is induced at constant T,p,E by a one-molar conversion according to the stoichiometry of the reaction (analogous to the well-known reaction enthalpy ΔH). Regarding K, it must be noted that this

*Supported by grant no. 3.487–0.75 of the Swiss National Science Foundation.

quantity, strictly speaking, represents the actual thermodynamic equilibrium constant **K** divided by a specific factor, taking into account activity coefficients. For this reason K can usually be expressed in terms of equilibrium concentrations (instead of activities) even under nonideal conditions.

Let us consider a differential change of state of any system at given E. If this implies a change dM, a certain amount of electric work must be done on the system, namely EdM, as pointed out in basic electrostatics.[2] Under reversible conditions the change of internal energy will therefore be

$$dU = TdS - pdV + EdM$$

(S, p, V: entropy, pressure, volume of the system, respectively) as follows immediately from the first and second laws of thermodynamics. Introducing an appropriate free enthalpy

$$G = U + pV - EM - TS$$

and chemical potentials of any molecular species B_i in the system as

$$\mu_i = \left(\frac{\partial G}{\partial n_i}\right)_{T,p,E} = \mu_i^0\,(T,p,E) + RT \ln\,(f_i c_i)$$

(n_i, f_i, c_i: mole numbers, activity coefficients, concentrations) then yields

$$dG = -\,SdT + Vdp - MdE + \sum_i \mu_i dn_i.$$

In a closed system where a chemical reaction occurs according to the stoichiometry

$$\sum_i \nu_i B_i = 0 \qquad (\text{so that } dn_i = \nu_i d\xi)$$

we have

$$dG = -SdT + Vdp - MdE - Ad\xi \tag{2}$$

with ξ being the extent of reaction and A the affinity defined as

$$A = -\sum_i \nu_i \mu_i = -\sum_i \nu_i \mu_i^0 - RT \sum_i \nu_i \ln\,(f_i c_i).$$

Since the first term on the right equals $RT \ln \mathbf{K}$ with $\mathbf{K}$ being the (concentration independent) thermodynamic equilibrium constant we may write

$$A = RT \left\{ \ln K - \sum_i \nu_i \ln c_i \right\} \tag{3}$$

$$K = \mathbf{K} / \prod_i f_i^{\nu_i}.$$

Owing to the basic mathematical fact that second derivatives of the function G must not depend on the order of differentiation it follows from (**2**) that

$$\left(\frac{\partial A}{\partial E}\right)_{T,p} = \left(\frac{\partial M}{\partial \xi}\right)_{T,p,E} \equiv \Delta M.$$

If one inserts (**3**), our basic equation (**1**) is immediately obtained (taking the c_i as independent of E). We note that the K developed in this way is not rigorously identical with the expression $\Pi_i \bar{c}_i^{\nu_i}$ as postulated above (the bar indicating equilibrium values). However, there is usually no practical difference, as pointed out in a more detailed derivation of (**1**), which can be found elsewhere.[3]

Application to a Simple Chemical Transition

Let us consider a molecule that may occur in two chemical states so that a corresponding transition

$$B_1 \rightleftharpoons B_2 \qquad (4)$$

can be assumed. In this case we have

$$K = \bar{c}_2 / \bar{c}_1$$

and a so-called degree of transition

$$\Theta = \frac{\bar{c}_2}{c_0} = \frac{K}{1 + K} \qquad (5)$$

($c_0 = c_1 + c_2$ being the total concentration independent of state).

The reaction moment naturally depends on the dielectric properties of the system. One of the simplest situations is encountered if there are permanent dipole moments associated with the two molecular states whose components parallel to the field are fixed (i.e., the dipoles are not to rotate). If these contribute $\hat{\mu}_1$ and $\hat{\mu}_2$, respectively, to M we have

$$\Delta M = N_A \hat{\mu} \qquad \text{with } \hat{\mu} = \hat{\mu}_2 - \hat{\mu}_1.$$

From (**1**) it now follows

$$K = K_0 e^{\gamma E} \qquad \text{with } \gamma = \hat{\mu}/kT \qquad (6)$$

(K_0 being the "equilibrium constant" at zero field). A change of E thus implies a corresponding displacement of equilibrium concentrations. This, of course, requires a finite time determined by the chemical rate constants of the transition.

In order to estimate the extent of possible field-induced chemical transitions, we consider an E value of 100 kvolts/cm. This represents the maximum order of magnitude under realistic conditions (the field inside a biological membrane may actually assume such values). With a $\hat{\mu}$ of 1 Debye (D) one obtains a $K/K_0 = 1.01$. Accordingly, there is practically no effect as long as only small molecules are involved. This may be quite different for macromolecular transitions where hundreds of Debye units can, in principle, be encountered. Since a $\hat{\mu}$ of 480 D yields a $K/K_0 = 55$, almost complete transitions could indeed be induced here. Combining (**5**) and (**6**) leads to an expression for the degree of transition as a function of E, namely:

$$\Theta = \frac{e^{\gamma(E - E_m)}}{1 + e^{\gamma(E - E_m)}}.$$

The field strength associated with the midpoint of transition ($\Theta = 1/2$) is apparently given by

$$E_m = -\frac{1}{\gamma} \ln K_0.$$

It is therefore largely determined by the thermodynamic properties of the two molecular states at zero field. The dependence of Θ on E is graphically represented by a sigmoid curve which approaches unity (zero) for $E \to \infty$ if $\hat{\mu} > 0$ (<0, respectively). In other words, the state with the larger dipole moment parallel to the field is favored when E is increased. The slope of the curve turns out as

$$\left(\frac{\partial \Theta}{\partial E}\right)_{p,T} = \Theta(1 - \Theta)\gamma. \tag{7}$$

This demonstrates that the transition is most strongly affected at its midpoint (the above-mentioned $\hat{\mu}$ of 480 D yielding a $\delta\Theta$ of 1% per 1 kvolt/cm).

A system of some experimental interest concerns the helix-coil transition of polypeptides.[4] The α-helix is known to have an appreciable permanent dipole moment parallel to the helical axis, whereas in the randomly coiled state the individual dipoles of the peptide units cancel out each other. One would therefore expect that an applied field E tends to increase the amount of helical structure if not already at its maximum.

A pertinent theory for this phenomenon has been developed, but it is naturally more complicated than the simple example treated above.[5] The transition is a cooperative process involving a very large number of different intermediate states. In the transition range one has in principle a distribution of helical fragments with $N_0 = 1/\sqrt{\sigma}$ peptide units in them on the average at the midpoint of transition. The parameter σ (usually $\ll 1$) is defined by the reduced thermodynamic probability to form a new helical fragment instead of growing already existing ones. The elementary growth reaction is characterized by a corresponding equilibrium constant s. Owing to our basic equation (**1**) it must be affected by E according to

$$\frac{\partial \ln s}{\partial E} = \frac{\mu_0 \cos \Psi}{kT}$$

where μ_0 stands for the permanent dipole moment per helical peptide unit and Ψ is the angle between the field and the dipole. This means that the growth reaction will be affected quite differently, depending on the respective angle. This and the orientational distribution of dipoles by the field must be taken into account when the overall increase of helical structure is to be evaluated.

In aqueous solution of polyelectrolyte polypeptides an additional field effect on ionic dissociation complicates the issue. This could even lead to a decrease of the helix content.[6] Therefore nonionic polypeptides are more suitable for experimental tests of the field effect under consideration. An appropriate system proves to be the well-studied poly(benzyl-L-glutamate) (PBLG) in a mixture of ethylene dichloride and dichloroacetic acid. In this case ($\sigma = 3 \cdot 10^{-5}$, $\mu_0 = 1.5$ D, $T = 20°C$)[7] the theory implies that a field of 100 kvolts/cm may indeed induce an almost complete transition.

Dielectric Polarization Due to the Chemical Field Effect

The fact that changes of the applied external field induce corresponding variations of the overall dipole moment of a material system must in principle be reflected in conventional dielectric measurements. In these experiments small harmonic oscillations of E of a certain frequency are employed and a complex dielectric constant ϵ^* can be evaluated which is defined by

$$\epsilon_0(\epsilon^* - 1) = \frac{\partial P}{\partial E}$$

(ϵ_0: absolute permittivity of vacuum) where P stands for the dipole moment per unit volume.

Turning back to our simple transition (**4**) we can set

$$P = P_0 + N_A\{(1 - \Theta)\hat{\mu}_1 + \Theta\hat{\mu}_2\}c_0$$

(P_0 comprising all contributions in addition to those of the B_1 and B_2). Since Θ has been shown to depend on E, there will be a special contribution $\Delta\epsilon_{ch}{}^*$ to ϵ^* caused by the chemical field effect. Apparently we have

$$\epsilon_0 \Delta\epsilon_{ch}{}^* = N_A \hat{\mu} \frac{\partial\Theta}{\partial E} c_0 .$$

The relation (**7**) cannot generally be used here because it applies to equilibrium conditions only. An appropriate relaxation function $(1 + j\omega\tau)^{-1}$ must be added as a factor in order to include the phase shift between the instantaneous values of Θ and E ($j = \sqrt{-1}$, ω: angular frequency). The relaxation time τ characterizing this effect is that of the chemical transition. We then finally obtain

$$\epsilon_0 \Delta\epsilon_{ch}{}^* = N_A \Theta(1 - \Theta)\frac{\hat{\mu}^2}{kT} c_0/(1 + j\omega\tau).$$

This should be experimentally observable even with rather small dipole moments. However, in practice the condition of fixed dipoles is difficult to satisfy.

The problem could also be treated with freely rotatable dipoles.[3] Under these circumstances two relaxation processes are predicted, provided the chemical rate process is sufficiently fast compared with the rotational one. Then we obtain a chemically induced dielectric relaxation at higher frequencies, whereas at lower frequencies a purely rotational relaxation occurs. Fairly long rodlike polymers undergoing a very fast chemical transition should therefore reflect such dielectric behavior. Indeed, some pertinent experimental evidence for this has been obtained with helical polypeptides. One case concerns helix-coil transitions.[8,9] In another case, the diffusion-controlled proton exchange of polyproline helices with an appropriate acid-base system has been found to result in a chemical contribution to dielectric dispersion.[10,11]

Once the rotational processes are much faster than the chemical reactions, a chemically induced dielectric effect can no longer be expected at low field strengths. It may be encountered, however, if a high static field is applied.[12]

Electric Birefringence Measurements Reflecting a Field-Induced Chemical Transition

Various different ways to obtain experimental evidence of the chemical field effect are conceivable when a rather high field can actually be applied to the system. Measuring the resulting change of the equilibrium degree of transition by some means would then provide a comparatively direct experimental approach. In a case when the transition is associated with a considerable change of rodlike structure it should especially be reflected in optical birefringence. This has been examined with the PBLG system above and also with a similar poly(benzyl-L-aspartate) system.[7] The electric birefringence was measured at 15 kvolts/cm over the complete transition. The respective degrees of transitions at zero-field could easily be evaluated from optical rotation. The results clearly indicate that all birefringence is contributed by the helical state. In other words, the measured values of $\Delta n/\lambda$ ($\Delta n = n_{\parallel} - n_{\perp}$ being the difference of diffraction indices for light polarized parallel and perpendicular to the field, λ = wavelength of the light) finally become zero when the completely coiled state is assumed. The observed $\Delta n/\lambda$ thus must be a function of the helix content actually present. On the basis of the conventional theory[13] we may evaluate this according to the relation

$$\frac{\Delta n}{\lambda} = A_0 \sum_i C_i \left\{ 1 - \frac{3}{i\beta_0} \right\} L(i\beta_0)$$

where the optical anisotropy factor

$$A_0 = \frac{N_A}{2\,\epsilon_0\, n\lambda\, \mathbf{M}} (\alpha_1 - \alpha_2)$$

(n: diffraction index of the system, $\mathbf{M}$ molecular weight of a helix fragment, α_1 and α_2 optical polarizabilities parallel and perpendicular to the helix axis, respectively) can be taken as being independent of the number of peptide units, i, in a given helical fragment. The mass concentration of i-fragments is indicated by C_i, L means the Langevin function and

$$\beta_0 = \frac{\mu_0 E}{kT}$$

with μ_0 defined as before. The parameters A_0 and μ_0 were determined experimentally from the $\Delta n/\lambda$ for completely helical polypeptide chains (the corresponding number i being known from the molecular weight). The distribution of fragments in the transition range can be calculated by means of an appropriate helix-coil transition theory.[14,15] This is comparatively easy if infinite chain lengths are assumed. Starting out from the distribution at zero-field, $\Delta n/\lambda$ can indeed be derived as a function of the corresponding degree of transition. Strictly speaking, we should take into account the fact that our chains are actually not infinitely long. For this reason somewhat smaller values of $\Delta n/\lambda$ may turn out if insufficiently long polypeptides are employed. In the study under consideration this effect could be shown to be negligible in first approximation.

In fact, the measured birefringence proved to be much more pronounced than to be expected according to the above approach, the deviation being clearly beyond experimental inaccuracy. This can only mean that there is more helical structure than there had been at zero field strength. Indeed, once we take into account the field effect on the helix-coil transition (as pointed out before) we arrive at birefringence data that agree very well with the observed ones. Thus the latter can be taken as a quantitative confirmation of the field-induced transition.

It may be noted that the actual overall change of the degree of transition at 15 kvolts/cm is only about 1% in the above system. However, the electric birefringence is mainly determined by the long helical fragments, which are much more affected. For this reason the underlying chemical effect is appreciably amplified in electric birefringence and quite apparent already at comparatively low field strength.

Summary

An electric field tends to displace any chemical equilibrium toward a greater overall electric dipole moment of the system parallel to the field. This can be quantitatively derived from basic thermodynamics. It is found that only with macromolecules can an appreciable transition be induced that could possibly have some relevance to biological control processes.

The variation of the overall dipole moment associated with a chemical field effect implies a corresponding contribution to dielectric polarization. This can be measured even at low field strength if the dipole rotation is sufficiently slow compared with the chemical reaction rate. Experimental evidence of this kind has actually been obtained with helical polypeptides. Another approach takes advantage of the technique of electric birefringence. In the ranges of two appropriate α-helix-coil transitions the actual birefringence turned out to be clearly above that corresponding to a transitional state unperturbed by the field. Very good agreement is achieved, however, if the chemical field effect is taken into account.

References

1. Hodgkin, A. L. & A. F. Huxley. 1952. A quantitative description of membrane current and its application to conduction and excitation in nerve. J. Physiol. (London) **117:** 500–544.
2. Frank, H. S. 1955. The thermodynamics of a fluid substance in the electrostatic field. J. Chem. Phys. **23:** 2023–2032.
3. Schwarz, G. 1967. On dielectric relaxation due to chemical rate processes. J. Phys. Chem. **71:** 4021–4030.
4. Engel, J. & G. Schwarz. 1970. Cooperative conformational transitions of linear biopolymers. Angew. Chem. Int. Ed. **9:** 389–400.
5. Schwarz, G. 1975. On the displacement of cooperative transitions of linear biopolymers by an electric field. Biopolymers **14:** 1173–1179.
6. Kikuchi, K. & K. Yoshioka. 1976. Electric field-induced conformational change of poly(L-lysine) studied by transient electric birefringence. Biopolymers **15:** 1669–1676.
7. Schwarz, G. & U. Schrader. 1975. The effects of cooperativity, finite chain length, and field induced changes of the conformational equilibrium on the electric birefringence of polypeptides in the range of the helix-coil transition. Biopolymers **14:** 1181–1195.

8. Schwarz, G. & J. Seelig. 1968. Kinetic properties and the electric field effect of the helix-coil transition of poly(γ-benzyl-L-glutamate) determined from dielectric relaxation measurements. Biopolymers **6:** 1263–1277.
9. Wada, A., T. Tanaka & H. Kihara. 1972. Dielectric dispersion of the α-helix at the transition region to random coil. Biopolymers **11:** 587–605.
10. Schwarz, G. 1970. Acid-base catalysis of dielectric relaxation of zwitterions. J. Phys. Chem. **74:** 644–658.
11. Schwarz, G. & P. J. Bauer. 1974. Structural flexibility and fast proton transfer reflected by the dielectric properties of poly-L-proline in aqueous solution. Biophys. Chem. **1:** 257–265.
12. Bergmann, K., M. Eigen & L. De Maeyer. 1963. Dielektrische Absorption als Folge chemischer Relaxation. Ber. Bunsenges. Physik. Chem. **67:** 819–826.
13. O'Konski, C. T., K. Yoshioka & W. H. Orttung. 1959. Electric properties of macromolecules. IV. Determination of electric and optical parameters from saturation of electric birefringence in solution. J. Phys. Chem. **63:** 1558–1565.
14. Zimm, B. H. & J. K. Bragg. 1959. Theory of the phase transition between helix and random coil in polypeptide chains. J. Chem. Phys. **31:** 526–535.
15. Schwarz, G. 1968. General theoretical approach to the thermodynamic and kinetic properties of cooperative intramolecular transitions of linear biopolymers. Biopolymers **6:** 873–897.

FIELD INTERACTION WITH BIOLOGICAL MATTER

H. P. Schwan

Department of Bioengineering
University of Pennsylvania
Philadelphia, Pennsylvania 19174

INTRODUCTION

The interaction of electrical fields with macromolecules and biological membranes continues to be of interest to many biophysicists. It is the dominant topic of this meeting. Work on the giant squid axon as summarized by the Hodgkin Huxley Equations and their more recent modifications suggest that electrical currents will cause excitation if the induced membrane potential change is of the order of 10 mV. Orientation of polar macromolecules by electrical fields appears to require kilovolts/cm to become noticeable. Corresponding current densities in biological fluids are of the order of 10–100 amperes/cm^2 and cause rapid heating.

A third group of biophysically significant electrical phenomena is caused by electromechanical forces. These field-induced forces may cause orientation and "pearl chain" formation of cells, cellular deformation, or destruction or the movement of cells in inhomogeneous fields as they exist in tissues. These phenomena require a threshold field strength which is inversely related to particle size. This field strength is rather high for biopolymers and of the order of volts/cm for large cellular size.

These modes of field interactions do not provide for explanations of more subtle phenomena that have been reported in recent years. We shall first discuss the electrical properties of tissues and cells, since these properties contain relevant information.

TISSUE PROPERTIES

Suppose a step function potential is applied to a sample of matter. Then the time dependence of the accumulated charge can be frequently characterized by an exponential of the form a + b (1-exp($-t/T$). The corresponding behavior in the frequency domain is given by the Debye expression

$$\epsilon^* = \epsilon_\infty + \frac{\epsilon_0 - \epsilon_\infty}{1 + j\omega T} \qquad (1)$$

with $\epsilon^* = \epsilon' - j\epsilon''$ the complex dielectric constant; T the time constant that characterizes the exponential behavior in the time domain; $\omega = 2\pi f$ angular frequency; the subscripts 0 and ∞ indicate limit values of ϵ' for low and high frequencies. If the time dependence of the charge accumulation is not a simple exponential of time, it may be due to several processes characterized by several time constants. In all generality, any response may be written as a sum of exponentials and the corresponding frequency behavior in the frequency domain as a sum of "dispersions" of the type given in Equation 1.

The phenomena of dielectric dispersion is exhibited by practically all tissues following relaxation equations of the type:

$$\epsilon_0 - \epsilon_\infty + \frac{\epsilon_0 - \epsilon_\infty}{1 + (f/f_c)^2} \tag{2}$$

$$\kappa = \kappa_0 + (\kappa_\infty - \kappa_0)\frac{(f/f_c)^2}{1 + (f/f_c)^2}. \tag{3}$$

These equations are obtained from (**1**) by separation and introducing the conductivity $\kappa = \omega\epsilon''$ and allowing for a low-frequency limit conductivity κ_0

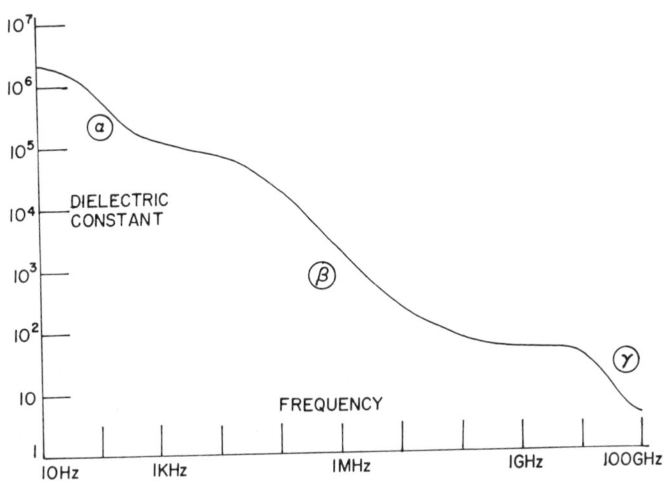

FIGURE 1. Dielectric constant of muscle tissue as a function of frequency. Three dispersion effects are centered near 0.1 kHz, 100 kHz, and 20 GHz. The data are characteristic for most tissues with high water content, even though detailed values for characteristic frequencies and dispersion magnitudes may vary. Room temperature. Data from Ref. 1.

different from zero. In (**2**) and (**3**) the "characteristic frequency" is $f_c = \frac{1}{2}\pi T$, ϵ stands for relative dielectric constant, and κ for conductivity. The dispersion magnitudes $\epsilon_0 - \epsilon_\infty$ and $\kappa_\infty - \kappa_0$ interrelated by

$$\kappa_\infty - \kappa_0 = (\epsilon - \epsilon_\infty)\epsilon_r \cdot 2\pi f_c \tag{4}$$

which results from the Kronig-Kramer relationships for the case of one time constant and with ϵ_r dielectric constant of free space.

The electrical properties of tissues are characterized by three major dispersions labeled, α, β and γ in FIGURE 1. Each dispersion is fairly well defined by either a single relaxation time $T = \frac{1}{2}\pi f_c$ or a small spectrum of relaxation times. Major mechanisms responsible for these dispersions have been fairly well identified and include[1,2] the following:

γ-dispersion: Dielectric relaxation of free water (f_c near 20 GHz);

β-dispersion: Maxwell-Wagner type relaxation resulting from the charging of cell membranes:

α-dispersion: Variability with frequency of the apparent outer-cell membrane capacitance. This membrane capacitance frequency dependence may result from

1) a frequency-dependent access to inner membrane systems that connect with the outer membrane;[3]
2) a frequency-dependent surface admittance tangential to the cell membrane caused by counter ion displacement about the charged cell and, possibly, fixed charge sites near the outer membrane;
3) a boundary potential-related capacitance element in series with the membrane capacitance;
4) a frequency-dependent membrane capacitance *per se* that results from ionic gating currents.[1,4,5]

Not entirely masked by these major mechanisms are secondary dispersions resulting from

1) Maxwell-Wagner relaxation effects resulting from the charging of membranes of subcellular organelles;
2) Debye type relaxations caused by the polar moments of biopolymers, primarily proteins;
3) relaxation of primarily protein-bound water occurring at some hundred MHz.[1,6-8]

These secondary effects may well be responsible for deviations from a single time-constant response for the β region, particularly at its high frequency tail. Other reasons relate to the complexity of cellular shape and density of packing. The relative contribution of all these secondary mechanisms is not yet entirely clarified.

A considerable amount of work has been done in order to identify the various mechanisms summarized above, particularly by Cole[9] and Schwan.[1] It is apparent that cellular and tissue properties are largely determined by the properties of biological membranes and tissue water. The contribution of biological macromolecules is probably largely masked by the strong Maxwell-Wagner polarization effects that result from the charging of membrane systems through the extracellular and intracellular fluids.

Water and Biopolymers

The interaction of alternating electrical fields with biological macromolecules has been extensively studied. (For summaries see Oncley,[10] and Takashima and Minakata.[11] Much of this work was done with macromolecules suspended in water or electrolyte. It has been suggested that tissue water may be differently structured than is free water.[12] It is, therefore, of interest to discuss the electrical properties of tissue water and compare them with normal water. Free-water estimates may be derived from microwave conductivity data obtained well above 1 GHz and the dielectric constant between β- and γ-dispersion. In addition, the mobility of ions may be compared with those in normal water from an analysis of tissue conductivity data obtained at frequencies between the β- and γ-dispersion ranges. We shall briefly discuss these three possibilities and conclude that in the few tissues studied so far water appears to be "free."

The dielectric properties of protein bound water have been investigated.[1,6–8] FIGURE 2 indicates the dispersions of the dielectric constant for bound water, ice and normal water. Ice relaxes at audio frequencies, bound water between 100 and 1000 MHz, and free water near 20 GHz. Normal or free water is for all practical purposes characterized by a single time constant.[13] Ice and particularly bound water relax over a broader frequency range, possibly integrating over various states.

The magnitudes of the dielectric constant dispersions $\epsilon_0 - \epsilon_\infty$ for ice, bound water, and free water are similar, as indicated in FIGURE 2. Hence the Debye con-

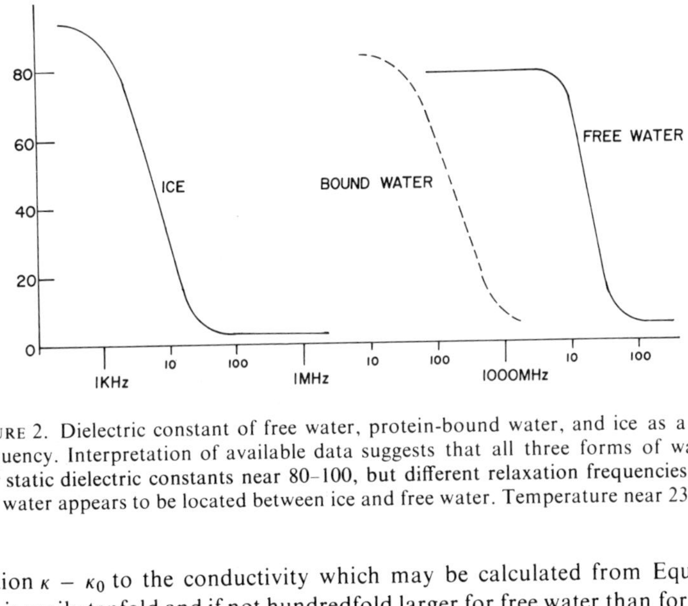

FIGURE 2. Dielectric constant of free water, protein-bound water, and ice as a function of frequency. Interpretation of available data suggests that all three forms of water have similar static dielectric constants near 80–100, but different relaxation frequencies. Protein-bound water appears to be located between ice and free water. Temperature near 23°C.

tribution $\kappa - \kappa_0$ to the conductivity which may be calculated from Equations 2 and 3 is easily tenfold and if not hundredfold larger for free water than for protein-bound water. This strong contribution increases sharply with frequency for f-values below and approaching 20 GHz, as may be seen from the appropriate reduction of Equations 3 and 4.

$$\kappa - \kappa_0 = (\epsilon_0 - \epsilon_\infty) 2\pi \epsilon_r f^2/f_c. \quad (5)$$

This provides for a possibility to determine the amount of free water in tissue from microwave conductivity. FIGURE 3 shows the well-established conductivity of free water, that of bound water, and a few values obtained for liver tissue. The 10 GHz value for liver is about 30% lower than that of free water, indicating a free-water content in liver above 60%. A more detailed analysis of this sort has been carried out recently for muscle tissue.[14] The analysis yields a free water content equal to most muscle water.

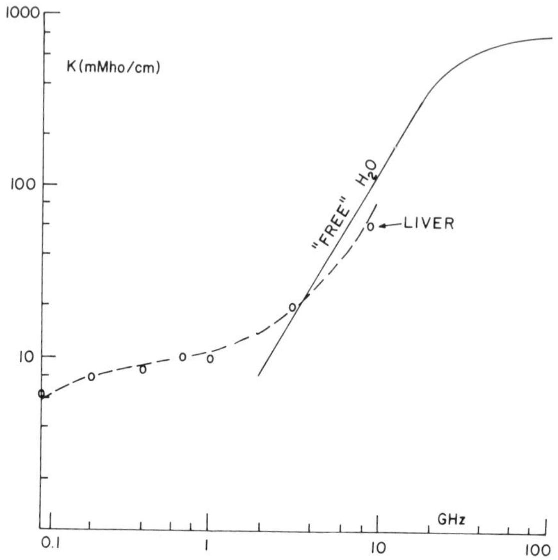

FIGURE 3. Conductivity of liver tissue is compared with that of water at microwave frequencies. The presence of substantial amounts of free water in liver tissue is indicated by the strong increase of κ (liver) above 2000 MHz. Liver data as quoted by Schwan.[1] The water conductivity is calculated by a relaxation frequency of 20 GHz and a static dielectric constant of water of 78 and the Debye Equation.[1] The very slight deviation of the behavior of water from that characterized by a single relaxation type[13] does not noticeably affect this presentation.

FIGURE 4 is chosen to demonstrate how the dielectric constant may be used to estimate the amount of free water from the relationship

$$\epsilon = \epsilon(H_2O) - \delta C \qquad (6)$$

where C is the volume fraction occupied by biopolymers, membranes, and bound water. This relationship applies between the end of the β-dispersion and beginning of the γ-dispersion. The dielectric decrement δ is primarily caused by the protein fraction and near 1, as indicated by theoretical estimates from mixture equations and experimental work with protein and larger particles.[1] The flat part of the muscle curve near 1000 MHz indicates $\epsilon = 50$, whereas free water has a value $\epsilon(H_2O) = 78$ at room temperature. A water content of 100–28—i.e., 72%—is indicated.

Ionic mobility in tissues and blood can be judged from microwave measurements near 100 MHz.[15] These mobilities compare with those in water fairly well if the inner frictional effect of macromolecular surfaces on the mobility is duly considered. Tissue water appears, therefore, largely of the normal type for most tissues investigated so far. However, an analysis of some initial results[16] indicates that the simple relationships that appear valid for most tissues with high water content may not apply to brain tissues, with a possibility for more structured water.

A major fraction of all biopolymers may therefore be anticipated in a normal nonstructured water environment, and selective interactions at moderate or small field strength values are not indicated by these bulk dielectric data. But further refinements are recommended to elaborate on this approach.

Membrane Properties and Excitation

Membrane capacitance values are available for many cell types from bulk measurements. Here the applied current to a tissue or cell suspension is varied with frequency and analyzed with use of appropriate spherical, ellipsoidal, or cylindrical shape approximations for cells. Measurements are also available with electrode systems impaled into the cells and directly measuring across the membrane. Although this technique is restricted to larger cell size, it has provided membrane capacitance values in agreement with those obtained from the bulk measurements. Briefly, the capacitance is of the order of 1 $\mu F/cm^2$ with a range extending from 0.6 to 1.3 $\mu F/cm^2$ and an accuracy of about 20% or 30% in most cases. This capacitance appears frequency-independent for the frequency range from about 1 KHz to 100 MHz. It corresponds to a membrane dielectric constant of about 10 relative to free space, and this dielectric constant has been rationalized to reflect a mixture of values from the lipid and protein contributions to the membrane.[1]

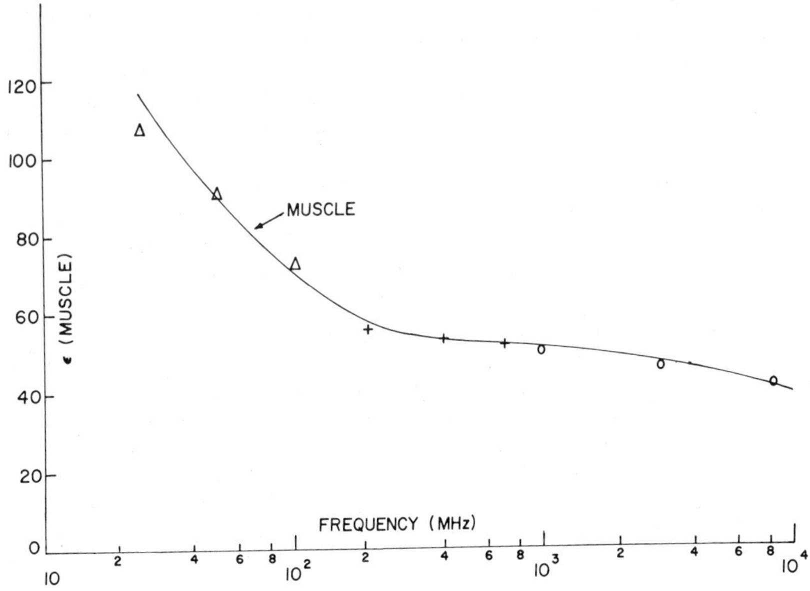

FIGURE 4. Dielectric constant of muscle at microwave frequencies at room temperature. The increase below 100 MHz is due to a Maxwell-Wagner type β-dispersion effect.[1] The decrease above 3000 MHz is due to water and the slow gradual change between 300 and 3000 MHz at least partially due to the δ-dispersion of protein-bound water.[1,6-8] The value between 1000 and 3000 MHz may be used to estimate the amount of free tissue water. The data are taken from Schwan.[1]

The membrane capacitance is polarized by an applied external field with cytoplasmic and extracellular tissue fluids serving as access impedance elements in series with the membrane capacitance. Thus the evoked membrane potential is given by the applied field E sampled over the cell dimension

$$\Delta V_m = 1.5RE \text{ (for spherical shape)} \tag{7}$$

provided that the applied frequency is low enough to make the membrane impedance large in comparison with that of the access impedance elements; i.e., f must be lower than the characteristic frequency $f_c = \frac{1}{2}\pi T$ where T is the time constant given by membrane capacitance, cellular size, and access impedance. For example, for cellular spherical shape[1]

$$T = RC_m(\rho_i + \tfrac{1}{2}\rho_a) \tag{8}$$

where R radius in cm, C membrane capacitance per cm^2, ρ_i internal and ρ_a external resistivity in Ohm-cm. It follows then that larger cells are more sensitive to an external field than smaller ones. This principle of sampling the field over a distance given by the cellular dimensions can explain the electrical sensitivities demonstrated by some large cells.[17,18] It appears also to be used effectively by the receptor organs (Ampullae Lorenzini) of electrosensitive fishes including sharks and rays.[19]

Membrane conductances cannot be extracted from bulk data of tissues or cell suspensions, since minor variations in the extracellular shunt path correspond to major membrane conductance values. Thus, precise knowledge of membrane conductances is limited to larger cells and presently excludes cellular organelles. The data available for large plant cells, nerve cells, and muscle indicate values of the order of 10 $mMho/cm^2$, but over a wide range. The membrane time constant $T = C_m/G_m$ is of the order of msec; i.e., at frequencies above about 1 KHz the membrane admittance is capacitive.

Above-indicated principles have been extensively tested on a large number of membrane systems. These include all sorts of sea eggs, muscle and most other tissues, erythrocytes, subcellular organelles, bacteria, synaptosomes, a variety of vesicles, PPLO, and bilayer membranes. They are responsible for the β-dispersion, which terminates at the 1 to 100 MHz frequency range.

The mechanism of the β-dispersion demands that at its center frequency the membrane impedance per unit area is equal to that of the access impedance $R(\rho_i + \frac{1}{2}\rho_a)$. At the high frequency tail of the β-dispersion the membrane impedance becomes smaller; i.e., the membranes are at frequencies corresponding to the end of the β-dispersion increasingly transparent. At these high frequencies, cell membranes therefore no longer receive the field strength sampled over cellular dimensions. Applied fields result in membrane potentials of the order of R/D times, 1000-fold smaller than at low frequencies (D membrane thickness). Rapidly increasing current densities are therefore needed to sustain any given imposed AC membrane potential. From this we would conclude that at high frequencies the likelihood of any direct interaction of electrical fields with biological membranes is small.

If it is assumed that induced membrane potential changes must be a noticeable fraction of the resting potential, 1–10 μV, bulk membrane current densities needed for excitation can be calculated. Such calculations have been carried out,[20] and indicate values ranging from 0.1 to 10 mA/cm^2, depending on cellular size, membrane conductance values, and extracellular space. Consider, for example, Equation 7, $\Delta V_m = 1.5RE$. The field strength E is related to the average tissue current density $j = E/\rho$ with the tissue resistivity ρ typically of the order of 600–700 Ohm-cm.[21] Hence

$$\Delta V_m = Rj\,(mA/cm^2). \tag{9}$$

Thus for $j = 1 mA/cm^2$, a radius of 10 μm would correspond to $\Delta V_m = 1$ mV and a radius of 100 μm to $\Delta V_m = 10$ mV. A value of very approximately 1 mA/cm^2 is therefore indicative of the threshold of excitation with low-frequency currents of about 100 Hz or less. More sophisticated models can be based on the cable core conductor approach, using either the HH equations for the membrane properties or newer revisions of the HH model.

Considerable experimental evidence on current threshold for stimulation exists, including muscle and heart muscle tissues. More recent data have been stimulated by the development of the cardiac pacemaker field.[22] These data are supportive of the mA/cm^2 threshold for excitation but depend, of course, in detail on electrode configuration and size and the mode of excitation.

Considerable experience has also been assembled by those interested in electrical accidents. The extensive body of this knowledge is not summarized here, but it is supportive of above-stated threshold figures. (See for example Ref. 20.)

Many data are furthermore available about currents needed to evoke cerebral responses such as sleep, anesthesia, and narcosis. Currents are usually in the range between 10 and 100 mA, with current densities probably an order of magnitude lower. For details, see the report "An Evaluation of Electroanesthesia and Electrosleep," Div. Med. Sci. Nat. Res. Council.

All available knowledge from the more basic principles, literature on hazards, and more clinically oriented work appears internally consistent. But this internal consistency does not necessarily exclude the possibility of more subtle responses, particularly if they are masked by the more strong and readily apparent excitation phenomena discussed above.

Indeed, weaker interactions corresponding to membrane potentials of less than 1 mV have been reported. Terzuolo and Bullock[23] estimated the voltage gradient in the saline solution surrounding a neuron (stretch receptor of the crayfish) just sufficient to cause a noticeable change in the firing frequency to be of the order 0.01 V/cm. This corresponds to a current density in the external medium of about 0.02 mA/cm^2 and is almost one order of magnitude lower than the above postulated range of 0.1 to 10 mA/cm^2. It is not very different from the current densities which can significantly interact with cardiac tissues if applied with large electrodes.[22] Schmitt *et al.*[24] have recently pointed out that interactions in the central nervous system are mediated by graded electronic potential changes that are propagated through highly sensitive synapses. Corresponding membrane potentials are quoted at the submillivolt threshold level. There appears but little doubt that significant effects other than the propagation of spike potentials occur at current densities and corresponding potentials across membranes about tenfold lower than indicated by the above-quoted range extending down to 0.1 mA/cm^2

and 1–10 mV across the membrane. It must be pointed out, however, that the translation of membrane potentials into bulk current densities in the surrounding tissue or extracellular fluid is difficult and only approximate estimates can be made, as stated above. The real significance of the reported membrane sensitivities at the fractional mV level is that these values are at least tenfold lower than suggested by axonology so far.

Some Extraordinary Sensitivities to Electrical Fields

Probably the highest confirmed sensitivities to electrical fields have been reported for a variety of electrosensitive fish species. For sharks and rays, fields as low as 0.01 μV/cm can be perceived.[19,25] How can such an extraordinary sensitivity be achieved? Two principles come to mind. Sampling of the electrical field as suggested by cellular structures and expressed in Equation 7 and narrow bandwidth and, related, sampling over corresponding long periods of time. Equation 7 would suggest that large cellular structures or corresponding devices that sample the field strength over appreciable distances be utilized. Fish tissues have a higher resistivity than the surrounding sea water. Hence external field strength levels are of the same order of magnitude as *in situ* internal tissue values. One device used by electrosensing fish is long tubular structures, the ampullae of Lorenzini. These organs integrate the field strength over their total length at sufficiently low frequencies and apply this potential to their end receptor epithelium. These structures extend over more than 10 cm. Therefore, above-quoted field strength values result in receptor potentials of the order of 0.1 μV. Detection of these exceedingly small membrane potentials in the presence of a typical noise level of about 1 μV across the membrane is aided by the low band pass characteristics of the ampullae Lorenzini. The ability to detect the quoted low field values decreases rapidly and more than linearly with increasing frequencies above 8 or 10 Hz. The receptor potential of 0.1 μV may therefore correspond to sensitivities of the order of 0.01 mV on a broad band basis extending to 1000 Hz. This potential is about ten times less than potentials quoted above as involved in electrotonic conduction processes in the brain. The evolution of these highly sensitive field-integrating tubular structures, in combination with very sensitive receptor membrane structures, may well have been aided by the high conductivity of water in comparison to that of air, favoring adequate "coupling" of external fields. The receptor membrane structures operate more closely to noise level. Induced potentials are not buried in it and hence, in principle, detectable. By contrast, axons operate at a level about 1000-fold above noise.

A large number of reports have been published claiming that fairly low external fields of the order of a volt per cm in air can be perceived. Field-strength values in air "couple" fairly ineffectively into tissues. Consider, for example, a sphere of tissue surrounded by air. Then for low frequencies, the ratio of internal field E_i to external field in air E_a is

$$E_i/E_a = f_\rho/6 \cdot 10^{11} \qquad \textbf{(10)}$$

(f (frequency in Herz, ρ tissue resistivity in Ohm-cm). For a tissue slab exposed to a vertical air field

$$E_i/E_a = f_\rho/18 \cdot 10^{11}. \tag{11}$$

Thus, for 60 Hz and a tissue resistivity of about 1000 Ohm-cm, E_i is about 10^7 times weaker than the air field strength E_a. External field levels of the order of 1 V/cm therefore result in internal field values about 10^{-7} V/cm, even though local fields may be higher because of tissue differences in electrical properties. Corresponding current densities are of the order of 10^{-7} mA/cm^2 and corresponding membrane potentials for 10-μm sized cell 10^{-7} mV or 10^{-4} μV. This is about 1000-fold less than the sensitivity of the end receptor membranes in the sharks, and 10,000 times below membrane noise. Clearly, exceedingly large distance sampling of the field strength such as achieved by the shark's tubular structures and very limited frequency response characteristics are needed to perceive signals of the order of 1 V/cm in air. Sharks perceive *in situ* tissue field-strength levels of 10^{-8} V/cm. But they sample the field over distances of 10 cm not demonstrated so far in other than fish species.

Most of the evidence for subtle effects of ELF fields on biological systems appears subject to inquiry. However, in recent years evidence in favor of such effects has been increasing and is well discussed in a text edited by Adey and Bawin[26] containing many pertinent references. A direct interaction with the central nervous system has been suggested by Adey. But the possibility of subtle peripheral transduction effects occurring at the surface of the body should also be considered.[27] The possibility is suggested by the fact that short-wave fields of a frequency near 150 MHz as well as ELF fields near 10 Hz have been noted as biologically effective by Bawin *et al.*[28,29] In both cases, similar external field-strength levels have been used, but internal field levels consequently differ by a factor of about 10^7. There appears to be agreement that if the reported sensitivities of nervous tissues can be confirmed and are not caused by surface transduction effects, exceedingly effective spatial and temporal integration mechanism must be involved. An example for spatial field integration was discussed above as it is utilized by the shark receptor organs. Another principle that concerns field-generated forces acting on particles is discussed below. It also demands larger size for smaller field threshold values to overcome thermal noise. In addition, several proposals have been made stating that highly cooperative molecular phenomena and metastable giant dipole effects may well provide a biophysical explanation of the claimed unusual sensitivities. These cooperative models either apply to exceedingly high frequencies above 100 GHz, where water loses its dampening properties,[30] or are so far rather qualitative and therefore not subject yet to verification.[31] Most of these models have one element in common. They demand interactions over extended regions or large volumes reacting to fields. There appears only little possibility that low-frequency fields of the mentioned low magnitudes can interact with individual macromolecules or membranes.

Some of the strongest evidence for subtle field effects, as reported by Adey and his associates and Wever's work on circadian rhythm as described in Reference 26, remains to be confirmed. If these findings can be confirmed and if it can be shown that they are caused by direct interaction with the central nervous system, entirely new modes of interaction of electrical fields with biological systems are indicated.

FIELD-GENERATED FORCE EFFECTS AND MEMBRANE DESTRUCTION

Two classes of effects are less subtle than some of the previously discussed phenomena and occur at similar field levels. They will be discussed, since they probably account for a large number of experimental observations. The first deals with field-generated forces of sufficient magnitude to affect cellular characteristics, and the second deals with fields high enough to apply to membrane potentials that are sufficient to cause temporary or permanent damage.

Electrical fields can directly interact with matter and create forces that can act on molecules as well as on cellular and larger structures. Most of these interactions are reversible and do not necessarily yield demonstrable biological effects. An example is the movement of ions in an alternating field, which is inconsequential provided that the field is sufficiently weak to prevent undue heating from molecular collisons (i.e., below about 1 V/cm corresponding to 1 mA/cm^2 in a physiological medium). Another example is the orientation of polar macromolecules. For field-strength values of interest here, only a very minor preferential orientation with the field results. Complete orientation and consequent dielectric saturation require field strength values of the order of KV/cm. Changes of this magnitude occur in membranes upon depolarization. Hence field-induced orientation or changes in orientation of membrane molecules appear entirely possible. Corresponding current tissue current densities are of the order of mA/cm^2, as discussed above.

But electrical fields can just as well interact with nonpolar cells and organelles in the absence of any net charge. The application of the electrical field will create induced dipole moments in suspended cells or particles. These dipole moments can interact with each other or they may be responsible for orientation effects or cellular shape changes. Reported field interactions at the cellular level have been summarized by Schwan & Sher[32] and include a variety of effects such as:

- Cellular and particle movement in inhomogeneous field;
- orientation of particles and cells with or perpendicular to the field;
- "pearl chain" formation;
- cellular-shape changes;
- AC-induced cytoplasmic streaming; and
- erythrocyte and other cellular destruction in a nonthermal manner.

Common to all these effects appear to be field-induced mechanical forces which drive the system involved toward a state of lowered electrical energy. These forces are in competition with random thermal collisional forces. The response to the field emerges when the electrical potential energy or its change with the response (i.e., upon orientation, etc.) becomes larger than the thermal energy KT (K Boltzmann constant, T absolute temperature). Thus there exists a threshold field strength needed for the phenomena of interest to occur. Since the electrical potential energy is proportional to E^2 and the volume V of the responding particle,[33] the threshold field strength E_{th} must be given by the equation

$$E_{th}^2 V \sim aKT = A \rightarrow E_{th}^2 \sim 1/R^3 \qquad \textbf{(12)}$$

where $A = aKT$ is a constant for a given temperature. The constant a depends on the particular phenomena of interest, electrical properties of suspended and suspending phase, shape of particles, and so on. Fairly abundant experimental evidence confirms that the principle and threshold field values for responses of 10-μm-sized cells are of the order of 10 V/cm. FIGURE 5 summarizes some of these results. But for 100-Å macromolecules they are about 10 KV/cm, and comparable to the fields needed for complete orientation due to the existence of a typical dipole moment of the order of 100 Debye units.

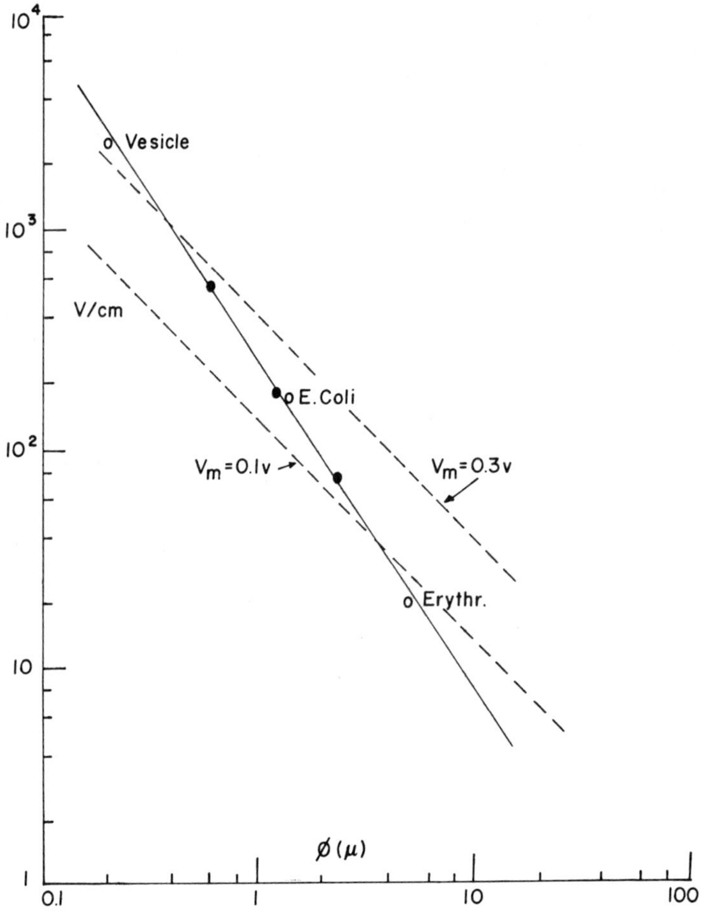

FIGURE 5. Threshhold of field effects on particles and cells. The solid curve gives the threshold for field-generated force effects. The dashed curves give the threshhold for dielectric membrane breakdown, assuming two different values for the membrane breakdown potential. Some typical membrane breakdown potentials have been listed by Schwan[27] and are of the order of 200 mV. For erythrocytes Zimmermann *et al.*[43] give a breakdown potential above 1000 mV. Data given are for permeability changes in chromaffin granule vesicles (Neumann & Rosenheck[42]), orientation and pearl chain formation effects in *E. coli* and erythrocytes (Sher & Schwan[32,38], and pearl-chain formation threshold data for silicone particles (solid points, Sher & Schwan[32,38]), axis: particle diameter.

The field effects may demonstrate themselves as movement or deformation of cells in an inhomogenous field, destructive effects on the cells, orientation, and alignment phenomena.[32] Particularly, the movement and deformation or destruction of cells in highly inhomogenous fields (for example, near the tip of a small electrode) can be very dramatic.[34] Cellular movement in inhomogeneous field (dielectrophoresis) has been used for purposes of cell sorting.[35] However, in physiological environments heat development is often pronounced, since high field-strength values are needed.

The usual random movement of a population of protozoa *Euglena* is replaced by movement in the direction of an applied alternating RF field or perpendicular to it, depending on frequency.[36] Orientation effects on *Fucus* eggs and *E. coli* have been observed by Novak and Bentrup[37] and by Sher,[38] respectively. The orientation of nonspherical particles has been treated theoretically by Schwarz *et al.*[39] and by Saito *et al.*[40] Further theoretical and experimental work is summarized by Schwan and Sher.[32] Experimental and theoretical evidence indicates that pulsed fields cannot be more effective than continuous fields of the same average power. Hence, modulation effects are not to be expected.[41] Field forces due to the induced dipole moment of the field have sometimes been listed as evidence of nonthermal action of electrical fields on biological systems. However, the effects require fairly large field-strength levels, frequently above those which give rise to undue heat development or stimulation of excitable tissues. The field forces are also strongly dependent on the electrical properties of the particle considered and its environment. Hence the threshold above noise is a strong function of frequency and has been proposed for purposes of cell classification.[35] In general, available evidence and present understanding indicate that significant effects with field-evoked forces above thermal noise require field-strength values above 1 V/cm in the medium unless cellular dimensions are well above 100 μm.

A second principle which relates to cellular responses to fields is also indicated in FIGURE 5. At sufficiently low frequencies the total potential applied across a cell by an external field is applied to the outer membranes, since the cytoplasmic phase is at an equipotential, as indicated by Equation 7. Let it be assumed that dielectric breakdown of membrane systems occur at a given field level V_m above that typical for the resting state. Then a threshold field strength E_{th} for cellular effects resulting from dielectric breakdown in membrane system is given by

$$E_{th} = V_m/1.5\mathrm{R} \tag{13}$$

where V_m is the induced membrane breakdown potential. The curves in FIGURE 5 demonstrate that both membrane breakdown as well as field-generated force cellular effects require external *in situ* field-strength levels of the order of 10–100 V/cm or more for cellular sizes and that the field strength for smaller dimensions is accordingly higher. On the other hand, large cells such as investigated by Friend *et al.*[17] can respond at field levels of some V/cm in the medium. FIGURE 5 also indicates that membrane-breakdown effects and effects caused by fields acting on the total cell can be expected at comparable field-strength levels in the medium. This makes it somewhat difficult to establish clearly what mechanism may be involved in any particular situation. But in any case, a perusal of the studies so far conducted on fields acting on cells and vesicles reveals experimental findings in accord with the classical principles stated.

Summary

Electric field interactions with biological matter can be summarized as follows.

1. Rather strong field values of the order of KV/cm are probably needed to orient polar macromolecules significantly.

2. Various interactions at the cellular level depend strongly on size and may occur at field levels as low as 1 V/cm or less for unusually large unicellular organisms. They appear to be largely caused by field-generated forces and the dipole moments induced by the electrical field. These forces emerge above the thermal level at a threshold field-strength that is inversely related to size. For usual cellular size, required field values produce significant heating resulting from ionic movement.

3. Strong membranc interactions leading to membrane breakdown correspond to the order of some hundred mV across the membrane. Corresponding field-strength values in the medium are comparable to those which yield significant field-generated forces.

4. Classical membrane excitation phenomena occur at field levels of the order of 1 V/cm and current densities of very approximately 1 mA/cm^2. Corresponding membrane potentials are in the range of 1–10 mV.

5. Weak membrane interactions corresponding to a fraction of a mV have been reported for CNS tissues. Some fish species display still higher sensitivities corresponding to values below 1 μV/cm, with a highest sensitivity near 0.01 μV/cm in sea water. These highest sensitivities are achieved by a sampling of the field over large distances with special organs and use of narrow bandwidth characteristics. Corresponding membrane sensitivities are below 0.1 μV across the membrane and above noise level over a rather narrow frequency range of a few Hertz.

6. Extraordinary weak field effects have been reported corresponding to tissue field gradients of the order of 10^{-7} V/cm. These effects, if shown to be related to more complex interactions at the membrane level, would require the development of new concepts.

References

1. Schwan, H. 1957. Electrical properties of tissue and cell suspensions. *In* Advances in Biological and Medical Physics. J. H. Lawrence and C. A. Tobias, Eds. Vol. **V:** 147–209. Academic Press. New York, N.Y.
2. Schwan, H. P. 1965. Biological impedance determinations. J. Cell. Comp. Physiol. **66:** 5–12.
3. Fatt, P. 1964. An analysis of the transverse electrical impedance of striated muscle. Proc. Royal Soc. B. **159:** 606–651.
4. Takashima, S. & H. P. Schwan. 1974. Passive electrical properties of squid axon membrane. J. Membrane Biol. **17:** 51–68.
5. Takashima, S., K. S. Cole & H. P. Schwan. 1975. Membrane impedance of squid axon during hyper- and depolarization. Biophys. J. **15** (no. 2, part 2); 39a.
6. Schwan, H. P. 1965. Electrical properties of bound water. Ann. N. Y. Acad. Sci. **125:** 344–354.
7. Pennock, B. and H. P. Schwan. 1969. Further observations on the electrical properties of hemoglobin bound water. J. Phys. Chem. **73:** 2600–2610.
8. Grant, E. H. 1965. The structure of water neighboring proteins, peptides and amino acids as deduced from dielectric measurements. Ann. N. Y. Acad. Sci. **125:** 418–427.

9. Cole, K. S. 1968. Membranes, Ions and Impulses, parts I and II. Univ. California Press. Berkeley, Calif.
10. Oncley, J. L. 1943. Chapter 22 *In* Proteins, Amino Acids and Peptides as Ions and Dipolar Ions. E. J. Cohn and J. T. Edsall, Eds. Reinhold. New York, N.Y.
11. Takashima, S. & A. Minakata. 1975. Dielectric properties of biological macromolecules. *In* Digest of Dielectric Literature **37:** 602.
12. Ling, G. N., C. Miller & M. M. Ochsenfeld. 1973. The physical state of solutes and water in living cells according to the association induction hypothesis. Ann. N. Y. Acad. Sci. **204.** C. F. Hazlewood, Ed.
13. Schwan, H. P., R. J. Sheppard & E. H. Grant. 1976. Complex permittivity of water. J. Chem. Phys. **64:** 2257–2258.
14. Schwan, H. P. & K. R. Foster. 1977. Microwave dielectric properties of tissues: Some comments on the rotational mobility of tissue water. Biophys. J. **17:** 193–197.
15. Pauly, H. & H. P. Schwan. Dielectric properties and ion mobility in erythrocytes. Biophys. J. **6:** 621–639. 1966.
16. Stoy, R. D. & H. P. Schwan. 1976. Dielectric properties of rabbit cerebral cortex in the radio frequency range. Physiologist **19**(3).
17. Friend, W. A., E. D. Finch & H. P. Schwan. 1974. Low frequency electric field induced changes in the shape and motility of amoebae. Science **187:** 357–359.
18. Goodman, E. M., B. Greenbaum & M. T. Marrow. 1976. Effects of ELF radiation on *Physarum polycephalum*. Technical Report N00014-76-C-0180. See also Nature. 1975. **254:** 66–67.
19. Kalmijn, A. J. 1966. Electroperception in sharks and rays. Nature **212:** 1232–1233.
20. Schwan, H. P. 1972. Biological hazards from exposure to ELF electrical fields and potentials. NWL Technical Report TR-2713. US Naval Weapons Lab. Dahlgren, Va.
21. Schwan, H. P. & C. F. Kay. 1957. Conductivity of living tissues. Ann. N. Y. Acad. Sci. **65:** 1007–1013.
22. Roy, O. Z., J. R. Scott & G. C. Park. 1976. 60 Hz ventricular fibrillation and pump failure thresholds versus electrode area. IEEE Trans. Biomed. Eng. **BME-23:** 45–48.
23. Terzuolo, C. A. & T. H. Bullock. 1956. Measurement of imposed voltage gradient adequate to modulate neuronal firing. Proc. Nat. Acad. Sci. USA **42:** 687–694.
24. Schmitt, F. O., P. Dev & B. H. Smith. 1976. Electronic processing of information by brain cells. Science **193:** 114–120.
25. Kalmijn, A. J. 1974. The detection of electrical fields from inanimate and animate sources other than electric organs. Chapter 5. *In* Handbook of Sensory Physiology. Vol. III/3. A. Fessard, Ed. Springer Verlag. New York, N.Y.
26. Adey, W. R. & S. M. Bawin. 1977. Brain interactions with weak electric and magnetic fields. Neurosci. Res. Prog. Bull. 15. MIT Press. Cambridge, Mass.
27. Schwan, H. P. 1977. Tissue determinants of interactions with electrical fields. *In* Brain Interactions with Weak Electric and Magnetic Fields. W. R. Adey and S. M. Bawin, Eds. Neurosci. Res. Prog. Bull. 15. MIT Press. Cambridge, Mass.
28. Bawin, S. M. & W. R. Adey. 1976. Sensitivity of calcium binding in cerebral tissue to weak environmental electric fields oscillating at low frequency. Proc. Nat. Acad. Sci. USA **73:** 1999–2003.
29. Bawin, S. M., L. K. Kaczmarek & W. R. Adey. 1975. Effects of modulated VMF fields on the central nervous system. Ann. N. Y. Acad. Sci. **247:** 74–81.
30. Fröhlich, H. 1977. Possibilities of long- and short-range electric interactions of biological systems. *In* Brain Interactions with Weak Electric and Magnetic Fields. W. R. Adey and S. M. Bawin, Eds. Neurosci. Res. Prog. Bull. 15. MIT Press. Cambridge, Mass.
31. Grodsky, I. T. 1977. Biophysical basis of tissue interactions. *In* Brain Interactions with Weak Electric and Magnetic Fields. W. R. Adey and S. M. Bawin, Eds. Neurosci. Res. Prog. Bull. 15. MIT Press. Cambridge, Mass.
32. Schwan, H. P. & L. D. Sher. 1969. Alternating current field induced forces and their biological implications. J. Electrochem. Soc. **116:** 170–174.
33. Schwarz, G. 1962. A theory of the low frequency dielectric dispersion of colloidal particles in electrolyte solution. J. Phys. Chem. **66:** 2636–2642.

34. ELUL, R. 1967. Fixed charge in the cell membrane. J. Physiol. **189:** 351–365.
35. POHL, H. A. 1973. Biophysical aspects of dielectrophoresis. J. Biolog. Phys. **1:** 1–16.
36. TEIXEIRA-PINTO, A. A., L. L. NEJELSKI, J. L. CUTLER & J. H. HELLER. 1960. Exp. Cell Res. **20:** 548.
37. NOVAK, B. & F. W. BENTRUP. 1973. Orientation of Fucus egg polarity by electric ac and dc fields. Biophysik **9:** 253–260.
38. SHER, L. D. 1963. Mechanical effects of AC fields on particles dispersed in a liquid. Biological implications. Ph.D. Thesis, Univ. Pennsylvania. Philadelphia, Pa.
39. SCHWARZ, G., M. SAITO & H. P. SCHWAN. 1965. On the orientation of non-spherical particles in an alternating field. J. Chem. Phys. **43:** 3562–3569.
40. SAITO, M., H. P. SCHWAN & G. SCHWARZ. 1966. Response of non-spherical biological particles to alternating electric fields. Biophys. J. **6:** 313–327.
41. SHER, L. D., E. KRESCH & H. P. SCHWAN. 1970. On the possibility of nonthermal biological effects of pulsed electromagnetic radiation. Biophys. J. **10:** 970–979.
42. NEUMANN, E. & K. ROSENHECK. 1972. Permeability changes induced by electric impulses in vesicular membranes. J. Membrane Biol. **10:** 279–290.
43. ZIMMERMANN, U., G. PILWAT & R. RIEMANN. 1974. Dielectric breakdown of cell membranes. Biophys. J. **14:** 881–889.

DISCUSSION

J. L. Oncley* and A. Wada,† *Moderators*

**Biophysical Research Laboratory*
Institute of Science and Technology
University of Michigan
Ann Arbor, Michigan 48109

†Department of Physics
University of Tokyo
Tokyo, Japan

DR. MANDEL: Prof. O'Konski, does the average relaxation time, which you measured for the superhelical structure, really correspond to the rotational relaxation time?

DR. O'KONSKI: In terms of a flexible model, yes. That is, the times are within the range that I would expect. If you would like a segmental model, one way of looking at it is to consider that there are segments of various lengths and inertial effects are negligible, so you can calculate the frictional coefficient of rodlike segments and from that you can get a distribution of relaxation time. If you look at the shortest relaxation time, about ten microseconds, that would correspond to a length of about 2,000 Å for the shortest persistence length.

Now that corresponds quite nicely with the expectation of a persistence length on the order of a couple of thousand angstrom for DNA. That also raises a point with respect to Prof. Houssier's calculation using your model where the discrepancy between the persistence lengths from this measurement and from others and B values he calculates is quite appreciable.

DR. MANDEL: One has to be very careful if one uses persistence length, because I think the wormlike chain is only well defined at the dilution.

DR. O'KONSKI: We are at a very low concentration of the DNA, down to the order of micrograms per mil, and in that region even with the longer calf thymus DNA, the relaxation times go asymptotically to a constant value.

DR. MANDEL: If the molecules are very large, one has to be very careful to define what infinite dilution is. One parameter we can use is the overlap volume, supposing every molecule must turn more or less freely without touching each other at infinite dilution. That should define the region of infinite dilution, and I have the feeling that with very large molecules, this low concentration where no overlap occurs must be really low.

DR. O'KONSKI: I have done these calculations years ago, and I am confident that we are at the essentially infinite dilution value for the PM-2 DNA.

DR. HOUSSIER: We should not compare the values that were presented by Professor Mandel from the dielectric dispersion and the one I presented from electric birefringence, for two reasons.

First of all, the ionic strength was not the same in the two measurements. Second we are discussing average values and the type of averages that we are obtaining from relaxation processes, and from electric polarizability measurements we do not get the same type of average, so we cannot strictly make the comparison this way.

DR. O'KONSKI: Dr. Houssier, you gave for DNA a length of 32,000 Å and an L zero value of 4,000–5,000 Å, which you did not explain. Could you tell me what that value of L-zero corresponds to?

DR. HOUSSIER: I think all the L zero values correspond to the equivalent length calculated from a rigid model.

DR. O'KONSKI: That seems to suggest a very long persistence length, about ten times longer than the B value you calculated from Mandel's theory, which seems to me to be a very serious discrepancy.

DR. HOUSSIER: You don't obtain the persistence length if you use a rigid-rod model. If you analyze the data in terms of a wormlike chain model, you always obtain much lower values of L.

DR. MANDEL: The persistence length does not measure the length of equivalent chains which can represent the same end-to-end distance, for instance. It is just a measure of the rigidity of the chain. The persistence is therefore defined clearly, mathematically or physically, so that there is not direct relation between the persistence length and the length of any model one uses to present the wormlike chain.

DR. M. NADAUTA (*Yale University, New Haven, Conn.*): Dr. Houssier, did you ever try to study the birefringence with *H*-1-depleted chromatin?

DR. HOUSSIER: Yes, I mentioned the results on this. The general observation is that the birefringence is not very much affected by the removal of the *H*-1 fraction. But the reversal of the sign that we observed in the presence of manganese is not observed if *H*-1 is removed.

DR. L. E. CHURNEY (*NIH, Bethesda, Md.*): It seems to me that the only way in which a relaxation time corresponding to persistence length will show up in one of these measurements is if it happens to coincide with one of the normal modes of the flexible molecule. Otherwise, you just get any relaxation times corresponding to the overall motion of the molecule as well as possibly to some of the normal modes in the more flexible ones. I could just add that we made measurements on very small fragments of DNA and on polyadenylic acid fragments. Within the accuracy of a measurement, we see only a single relaxation time of the order of 3 or even 4 log units in the decay time.

DR. O'KONSKI: Similar measurements were made by Dr. Stellwagen on sonicated DNA years ago, and they behave like rigid rods. I want to make a general remark now. Certainly, one needs to use a distribution of relaxation time to represent the data. The three discreet values we use are only a mathematical modeling of the actual relaxation process for a flexible macromolecule. You can get good agreement even though there may be many modes involved. One has to do a transformation of the relaxation data to calculate the distribution function to get at those modes.

The point I am making with respect to the discrepancy is this. The rigid rod has a relaxation time proportionate to the cube of the length, so that relaxation time is a sensitive measure of the length. If you have a discrepancy on the order of ten in length, that corresponds to a discrepancy on the order of a thousand relaxation time. I am inclined to think that the parameter of a few hundred angstroms calculated from dielectric relaxation measurement might correspond to the mean distance a counterion has to travel in a diffuse layer.

DR. GOLDBERG (*University of Pittsburgh, Pittsburgh, Pa.*): I wanted to ask Dr. Schwarz if the 15 kilovolts per centimeter he spoke of was an internal electric field or an external electric field, or whether there is no significant difference between them in this experiment.

DR. SCHWARZ: The fifteen kilovolt per centimeter is an externally applied electric field, but in this case, because we have organized structures (dipole moments along the helical axis), there is no difference between internal and external field strengths.

DR. STOCKMAYER: I just wanted to say a word about the theoretical situation with respect to the wormlike chain. Actually, if it were frozen into a rigid wormlike conformation with no change of shape, the calculation of the rotatory diffusion would be quite feasible, with appreciable hydrodynamic details. However, the real DNA-type chain probably is also flexing itself, and it is not a bead and spring model, but not a rigid rod. It is something in between. There are many modes of motion, but they are not independent normal modes, so I think the theoretical problem in that case is really formidable. I think the experimenters are going to be forced for a while to deal with effective L-zero values such as Professor Houssier displayed, having a precise theoretical label to put on them.

MECHANISMS OF RAPID OPTICAL CHANGES OF POTENTIAL SENSITIVE DYES*

A. S. Waggoner

Department of Chemistry
Amherst College
Amherst, Massachusetts 01002

A. Grinvald

Department of Physiology
Yale University School of Medicine
New Haven, Connecticut 06510

INTRODUCTION

The electrical potential is an important property of many cell and organelle membranes. Nerve impulses are propagated as changes in nerve cell membrane potential. Energy transformation, nutrient uptake, and sensory transduction are all cell processes that are integrally dependent on the magnitude and time course of the membrane potential. In many situations the use of microelectrodes to follow the membrane potential is either impossible or inconvenient, and recently developed molecular membrane potential probes have found a variety of applications. One new technique is based on the observation that a change in potential across membranes stained with certain dyes is accompanied by changes in their optical properties, usually absorption or fluorescence (see FIGURE 1). Optical detection of membrane potential has been accomplished with single cells and with suspensions of cells (FIGURE 2). For example, simultaneous monitoring of electrical activity in 14 individual neurons in the central nervous system of an invertebrate has been done optically.[1] Other biological preparations that have been studied by this technique are erythrocytes,[2,3] mitochondria,[4] single muscle fibers,[5,6] several species of bacteria,[7,8,11] heart cells,[9] a variety of vesicle preparations,[3,10] ascites cells,[12] and chloroplasts.[13] Several review articles are available on these subjects.[14,15]

The most sensitive dyes (largest optical change per unit potential change) fall into three classes: merocyanine, cyanine, and oxonol. Often both absorption and fluorescence properties of these dyes are sensitive to the membrane potential.[17] For studying preparations of single cells the best dyes discovered so far are merocyanines.[17] For example, a merocyanine oxazolone (NK 2367 of Ref. 48) used with the squid giant axon produces transmittance changes, $\Delta I/I$, of $5 \times 10^{-4}/100$ mV change. The response time for these dyes is usually less than 5–100 μsec; therefore, we call them "fast" dyes. Potential dependent changes in the fluorescence of stained cells can be observed as well. For example, merocyanine 540 gives fast fluorescence changes with single cells.[16] Although these signals are relatively small

*Supported by NIH grants NS10489 and EY-00930 to A. S. W. A. G. is a long-term EMBO fellow. Address correspondence to A. S. W.

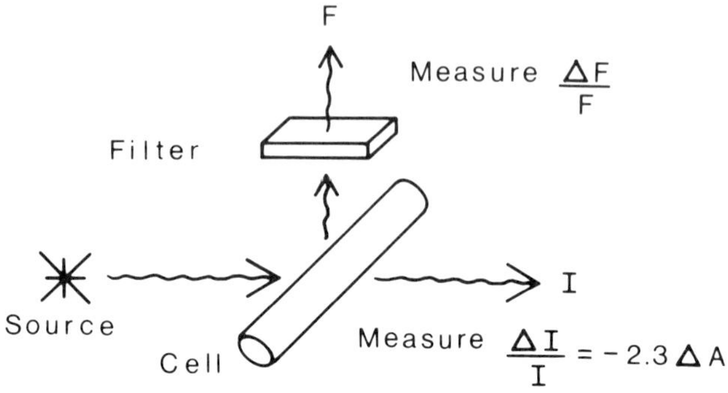

FIGURE 1. Schematic diagram of optical experiment involving measurement of a transmittance change, ΔI, and a fluorescence change, ΔF.

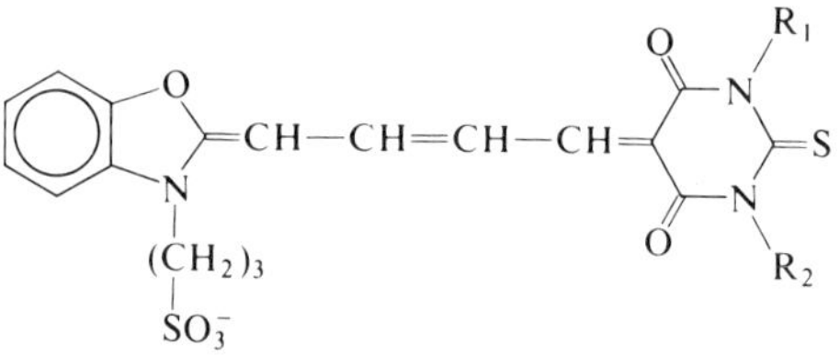

merocyanine 540

$R_1 = R_2 =$ butyl

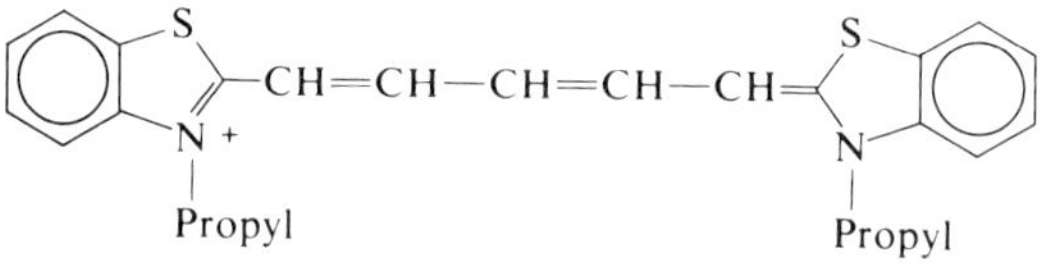

cyanine dye [diS—C_3—(5)]

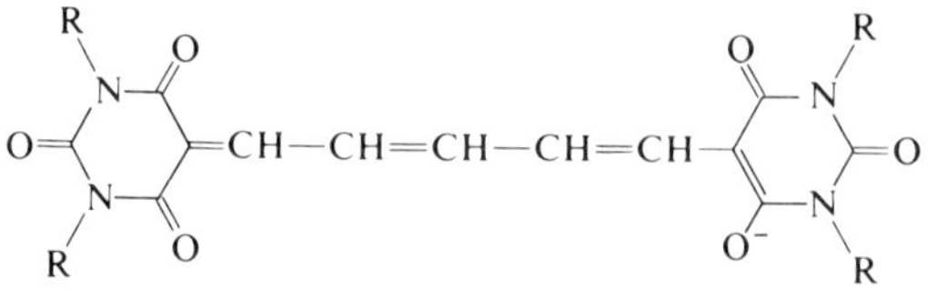

oxonol dye [diBa—C_4—(5)]

R = butyl

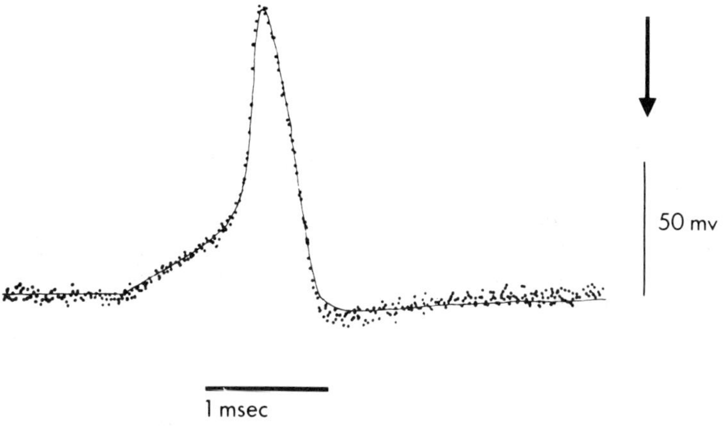

FIGURE 2. Changes in absorption (dots) of a giant axon stained with a merocyanine rhodanine dye (dye XVII of Ref. 17) during a membrane action potential (smooth trace) recorded simultaneously. The change in absorption and the action potential had the same time course. The direction of the vertical arrow to the right of the trace indicates the direction of an increase in transmission; the size of the arrow represents a change in transmission of 10^{-4}. The transmission change was measured at 750 nm; 32 sweeps were averaged. The response time constant of the light measure system was 5 μsec. (Taken from Ross *et al.*, Ref 40.)

($\Delta F/F \sim 10^{-3}$ per 100 mV), they can be detected with a signal-to-noise ratio of greater than 10 without signal averaging in squid axon experiments.[17] Furthermore, "fast" dyes (derivatives of merocyanine 540) were recently used to measure potential changes in suspensions of bacteria[18] and synaptosomal ghosts.[19] The fluorescence changes in these systems were as large as 20%, two orders of magnitude larger than those observed in squid axon experiments. Some cyanine and oxonol dyes also give fast, but somewhat smaller, absorption and fluorescence changes when used with single-cell preparations.[17]

For studies of suspensions of cells and vesicles, the cyanine dyes have proved to be the most useful, in particular, diS—C_3—(5).[3] These positively charged dyes are membrane-permeant and distribute across the membrane according to the membrane potential. When accumulated by the cell or vesicle to high concentrations, these dyes form nonfluorescent aggregates.[3,52] Thus, when the cells are hyperpolarized the dye is accumulated and the suspension fluorescence decreases. The aggregates formed are probably unbound in the interior of vesicles and cell ghosts that contain no internal protein constituents, but may be substantially associated with internal constituents in whole cells, as in the case with intact red blood cells.[52] The fluorescence changes that are seen with dyes that behave according to an accumulation mechanism, such as diS—C_3—(5), are *large* (80%) and relatively *slow* (1–20 sec). Thus, the mechanism underlying the response of the "slow" dyes is fairly well understood.[3,14,52]

An insight into the mechanisms of dyes that show fast optical changes would probably accelerate the design of more sensitive probes that would make optical experiments with single cells, and even cell suspensions, technically easier to perform. Several possible mechanisms for these changes can be suggested.

1. *Electrochromism:* Changes in the extinction coefficient of chromophores that result directly from a change in the electric field rather than from movement of the chromophores.

2. *Dye Movement,* rotational or translational, in a changing electric field:

Rotational movement of dipolar or charged dye molecules in the membrane electric field may produce optical changes that result from changes in the microenvironment of the dye, or rotation of the absorption transition moment of the dye into a more or less favorable orientation for absorption of light.

Translational movement of dye may produce optical changes that result from alteration of the polarity, polarizability, or rigidity of the microenvironment of the dye, or changes in steric constraints that orient dye in different oppositions to the membrane.

3. *Dye Aggregation:* Changes in the membrane electric field may perturb dye monomer-aggregate equilibria if, for example, the different species involved have different dipole moments. Rotational or translational movement of dye could be involved in this process.

4. *Indirect:* A change in electrical potential across the membrane may affect some property of the membrane that in turn affects the orientation or binding of dye to the membrane. Among the possible examples of these effects are electrostriction and reorientation of lipid dipoles at the membrane surface.

Below is a discussion of the relative importance of these possible mechanisms. The discussion will be based partly on our[17] experience with more than 750 dyes tested on the squid axon and bilayer lipid membranes, partly on experiments and theory found in the literature (mainly on the work of Tasaki and Warashina[21] and Conti[22]), and partly on theoretical expressions we have developed from a thermodynamic model of the movement of dye molecules in a membrane electric field. From the theoretical expressions we will point out parameters that suggest how dye sensitivity might be improved, and we will try to estimate the maximum possible optical change than can be expected for each mechanism under optimal conditions. Finally, we will discuss an important practical consideration; optimization of signal-to-noise ratios in absorption and fluorescence experiments with single cells.

Absorption Changes

A list of the absorbance changes, ΔA, that have been observed for a number of the more sensitive dyes used in single nerve cells and in bilayer lipid membrane experiments are shown in Table 1. The values of ΔA in this table range from 10^{-6} to 10^{-4} per hundred mV potential change. Hundreds of other dyes have absorbance changes ranging from $\sim 10^{-7}$ (the limit of detection) to 10^{-4}. In some experiments the dye remains in the bulk aqueous solution during the optical experiment. In other experiments the axon is stained and rinsed before illumination. The dye concentration in the bathing medium was usually in the 10–200 μM range. Let us now consider the source of these absorbance changes.

The total absorbance change, ΔA_T, for a dye associated with a membrane during a potential change is given by the sum of the absorbance changes that take place by the four different mechanisms listed earlier. Thus $\Delta A_T = \Delta A_{\text{electrochrom}} + \Delta A_{\text{rot}} + \Delta A_{\text{transl}} + \Delta A_{\text{aggreg}} + \Delta A_{\text{other}}$. The signs, magnitudes, and wavelength dependence of ΔA for each mechanism may be quite different. The absorbance change for any mechanism can be given by a form of Beer's law, $\Delta A(\lambda) = 1000 \, \Delta\epsilon(\lambda)\Delta n$ where $\Delta\epsilon(\lambda)$ is the change in "effective" extinction (to be discussed later) of the dye that results from the potential change and Δn is the number of moles of dye/cm^2 membrane that are involved in the extinction change. This absorbance change is related to the change, ΔI, in light intensity, I, that passes through the membrane preparation and reaches the photodector by

$$\Delta A = \frac{-1}{2.3}\left(\frac{\Delta I}{I}\right) = 1000\,\Delta\epsilon\Delta n \qquad (1)$$

when $\Delta I/I$ is small ($<10^{-2}$). In the discussion, we will occasionally interchange ΔA and $\Delta I/I$. When talking about absorption properties of dyes, ΔA is often

TABLE 1

ABSORBANCE CHANGES AND ESTIMATED FRACTIONAL ABSORBANCE CHANGES FOR DYES STUDIED IN SQUID AXONS AND BLMS

Dye	System	V(mV)*	ΔA†	$\Delta A/A_{max}$‡	Reference
diS—C_2—(5)	BLM	75	1×10^{-5}	1×10^{-2}	20
diI—C_2—(5)	BLM	75	2×10^{-6}	2×10^{-3}	20
diI—C_1—(5)	squid	75	2×10^{-5}	8×10^{-3}	21
Cyanine XVIII	squid	100	5×10^{-5}	2×10^{-2}	17
Merocyanine I	squid	100	9×10^{-5}	3×10^{-2}	17
Merocyanine XVII	squid	100	17×10^{-5}	6×10^{-2}	17

*These membrane potentials correspond to membrane electric field strengths of 2×10^5 V/cm if the thickness of the BLM and squid axon membranes are 40 and 50 Å, respectively.

†The observed ΔA for squid experiments was divided by 3 to facilitate comparison with BLM data. No correction for the geometry of the membrane was used. The data obtained from Ref. 17 were transformed to the units used in this paper.

‡In estimating A_{max} it is assumed that the surface density of dye on the membrane is 10^{-11} moles/cm^2 and $\epsilon_{max} = 10^5$. Then $A_{max} = 1000\,\epsilon_{max} n$.

more convenient. On the other hand, when talking about the size of electrical signals generated at photodetectors, we more often use $\Delta I/I$. The quantity ΔI is usually termed the "transmittance change of the sample" and $\Delta I/I$ is the "fractional transmittance change."

A glance at Equation 1 tells us that in order to understand the mechanism of voltage-dependent absorbance changes and to learn how to increase the sensitivity ($\Delta A/\Delta$mV) of the technique, we must understand the factors upon which $\Delta\epsilon$ and Δn depend. In the discussion that follows, we will discuss mechanisms that can produce large changes in $\Delta\epsilon$. We will begin with electrochromism, then move on to changes in $\Delta\epsilon$ that result from dye movement. Later, we will discuss Δn and develop an equilibrium thermodynamic treatment that will enable us to estimate its magnitude as a function of the size of the membrane potential change and several other important parameters.

ELECTROCHROMISM

Electrochromism (the Stark effect) is the change in absorption properties of a chromophore in an electric field that results from a frequency shift of the absorption peak. This frequency shift is given by

$$\Delta\nu = \frac{-1}{h}\Delta\vec{\mu}\cdot\vec{F} - \frac{1}{2h}\Delta\alpha F^2 \quad (2)$$

where $\Delta\mu$ is the change in dipole moment of the chromophore upon electronic excitation, $\Delta\alpha$ is the change in polarizability of the chromophore upon excitation, and F is the strength of the electric field.[23-25] The first term describes frequency shifts that depend on the first power of the electric field and are therefore called linear electrochromic shifts. Linear electrochromic changes are strongly dependent on the relative orientation of the chromophore and the electric field. The shift may be red or blue, depending on whether the direction of the dipole change is oriented parallel or antiparallel to the direction of the electric field. If the chromophore structure is symmetrical, $\Delta\vec{\mu}$ is zero, and there can be no linear electrochromism. The second term in Equation 2 describes quadratic electrochromic shifts that occur even for randomly oriented dyes with symmetrical structures. Under optimal conditions (large $\Delta\vec{\mu}$) the linear term is expected to produce the largest frequency shifts. The theory for electrochromism has been more extensively discussed elsewhere.[23-28]

Electrochromic changes have been experimentally observed in solutions, gases, polystyrene matrices, evaporated films, and oriented lipid multilayers.[29-36] They have been postulated to account for the "cartenoid shift" that has been observed in chloroplasts and chromatophores.[37-39] Since the number of membrane layers the light passes through before detection varies with the different preparations studied, it is difficult to make direct comparisons of the ΔA values for these systems. In an attempt to get around this problem we have normalized the absorbance changes by dividing them by the total absorbance of the membrane system at the wavelength of maximum absorption. The normalized absorbance change, $\Delta A/A_{max}$, we call the "fractional absorbance change." It is assumed in this procedure that most of the dye in the membrane system being studied is localized on the membrane where the potential change occurs. An advantage of using $\Delta A/A_{max}$ is that for dyes with the same mechanism, this value ought to be somewhat independent of the membrane system studied, and we can make order-of-magnitude comparisons of fractional absorbance changes in bilayer lipid membranes and single nerve fibers and absorbance changes in suspensions of cells, organelles, and vesicles.

The fractional absorbance changes for a number of membrane preparations have been tabulated in TABLE 2. It can be seen that in most of the multilayer experiments, the $\Delta A/A_{max}$ values with electric fields of 10^5 V/cm are two to three orders of magnitude smaller than the fractional changes observed with the dyes shown in TABLE 1. The dye in TABLE 2 that we expect to be most similar to the

TABLE 2

FRACTIONAL ABSORBANCE CHANGES THAT ARE PRESUMED TO RESULT FROM ELECTROCHROMIC PEAK SHIFTS

Chromophore	System	Voltage (V/cm)	$\Delta A/A_{max}$*	$\Delta A/A_{max}$ Norm, 10^5 V/cm	Reference
Methyl red	polystyrene	3×10^{-5}	5×10^{-3} (quadr?)	5.5×10^{-4}	29
Porphyrin	polystyrene	10×10^{5}	1.5×10^{-3} (quadr?)	1.5×10^{-5}	30
4,4-dimethylamino-nitrostilbene	evaporated film	2×10^{-6}	2×10^{-2} (linear)	1×10^{-3}	31
Merocyanine	multilayer	1×10^{5}	1×10^{-4} (linear) 6×10^{-6} (quadr)	←	32
Rhodamine B	multilayer	1×10^{5}	6×10^{-5} (linear) 2×10^{-5} (quadr)	←	33 Figure 3,4
Lutein	multilayer	1×10^{5}	2×10^{-5} (quadr)		34 Figure 5
Clorophyll a	multilayer	1×10^{5}	3×10^{-5} (linear) 5×10^{-6} (quadr)	←	35 Figure 3
Chlorophyll b	multilayer	1×10^{5}	4×10^{-4} (linear) 1×10^{-5} (quadr)		36 Figure 1,2
Chlorophyll	*R. spheroides*	?	7×10^{-3} (?)	?	37 Figure 1
Chlorophyll	*R. spheroides*	?	4×10^{-3} (?)	?	38 Figure 1
Chlorophyll	*R. rubrum*	?	3×10^{-3} (?)	?	39 Figure 2

*In some cases A_{max} had to be calculated from data in the methods section of the referenced paper.

dyes mentioned in TABLE 1, in terms of chromophore structure and orientation, is trans-4-dimethylamino-4′-nitrostilbene. This dye was especially selected by Fischer *et al.*[31] to demonstrate a very large linear electrochromic effect. The fractional change in absorbance for this dye normalized to a field of 10^5 V/cm is 10^{-3} (Figure 5 of Ref. 31). But this value is still an order of magnitude smaller than the largest value in TABLE 1.

It is worth while at this point to try to place an upper limit on the size of an electrochromic absorbance change that can occur under optimal conditions in a biological membrane during a potential change of around 50–60 mV. Consider a merocyanine-like dye with a large $\Delta\vec{\mu}$, say 20 debye (thought to be one of the largest known),[31] that is directed along the conjugated structure of the chromophore. Assume that the dye is oriented with $\Delta\vec{\mu}$ parallel to the membrane electric field (and with no dye oriented antiparallel) and that we are observing the absorbance change with light that is polarized parallel to the chromophore. By modifying the equation of Reich *et al.*[23] to accomodate wavelength rather than wave number, we get

$$\Delta\epsilon_{\text{linear}} = \frac{-\Delta\vec{\mu}}{h}\cdot\vec{F}\,\frac{\partial\epsilon}{\partial\lambda}\,\frac{\lambda^2}{c}\times 10^{-7}\left(\frac{\text{liters}}{\text{mole-cm}}\right). \qquad (3)$$

In this equation $\partial\epsilon/\partial\lambda$ is the slope of the absorption peak (in ϵ-units/nm) at the wavelength λ (nm).† It is clear that dyes with narrow absorption peaks will give the largest electrochromic absorption changes. For the merocyanine, cyanine, and oxonol dyes, the steepest slopes, which occur on the red side of the absorption maximum, are usually near 10^4 ϵ/nm.‡ This slope is nearly 40 times steeper than that of dimethylaminonitrostilbene. From Equation 3 we see that the hypothetical merocyanine-like dye, which absorbs at 600 nm and has a $\Delta\bar{\mu}$ of 20 debye, will have $\Delta\epsilon \simeq 10^4$ for an electric field change of 10^5 V/cm. If ϵ_{max} for this dye is about 10^5, then $\Delta\epsilon/\epsilon_{max} = 0.1$ and, if all the chromophores in the membrane were involved in the electrochromic shift, $\Delta A/A_{max} = \Delta\epsilon/\epsilon_{max} = 0.1$. Therefore, under ideal conditions, with an appropriate dye, and with an electric field change of 10^5 V/cm, a $\Delta A/A_{max}$ could be generated that is two times larger than the fractional absorbance change observed in the squid axon with the merocyanine rhodanine dye (XVII)[17] shown in TABLE 1.

Whether the rapid absorption changes observed in nerve fibers are actually occurring because of an electrochromic shift seems inconclusive according to the above discussion. We can, however, present some observations that suggest that electrochromism is not the predominant mechanism for most of the optical changes observed. (1) Electrochromic changes are fast. They are expected to lag potential changes by less than nanoseconds. Cohen *et al.*[16] have determined the time constant for the absorbance change of merocyanine 540 in the squid axon to be 30 μsec. Many other merocyanine, cyanine, and oxonol dyes show optical changes several μsec to msec slower than the time that was required for the axon

†$\Delta\mu$ must be converted from debye by multiplying by 2.1×10^{-9} electron-cm/debye, $h = 4.1\times10^{-15}$ eV-sec, F = V/cm, and $c = 3\times10^{10}$ cm/sec. Alternately, V can be inserted in stat volts with the other constants in CGS units.

‡For ANS-type dyes $\partial\epsilon/\partial\lambda$ is 20 × smaller.

membrane potential to change. However, some absorbance changes (e.g. WW 375) are too fast to measure ($< 5\ \mu$sec).[17] (2) According to the theory of linear electrochromism, $\Delta\epsilon$ is proportional to $\partial\epsilon/\partial\lambda$.[23] Therefore, the difference spectrum for an electrochromic change should resemble the first derivative of the absorption spectrum of the chromophore. The shapes of absorption-difference spectra that result from potential changes in stained axons generally do not resemble the first derivative of the absorption spectra.[17] (3) Many cyanine and oxonol dyes that have been studied have symmetrical or nearly symmetrical chromophore structures and should therefore not exhibit linear electrochromism. Nevertheless, a number of these dyes show substantial absorbance changes in the squid axon and the BLM.[17,20] (4) Dichroism spectra obtained by the illumination of the center and the edge of axons with polarized light are inconsistent with predictions based on an electrochromic mechanism (Grinvald, Kamino, Salzberg and Cohen, unpublished). Thus, although electrochromic signals might be very large, we think that the signals found in axons and BLM's cannot be explained by this mechanism.

$\Delta\epsilon$ that Results from Dye Movement and Changes in Aggregation

Several possible "microenvironments" that a potential-sensitive dye might encounter are given below.

1. Hydrocarbon region: low polarity, low polarizability, high viscosity, may preferentially orient dye molecules.
2. Interface: localized phospholipid charges and dipoles, exact polarizability and polarity unknown, probably viscous, may preferentially orient dye.
3. Aqueous region: high polarity, low viscosity, allows random dye orientation.
4. Other dye molecules: coupling of electronic transitions yields new adsorption spectra, fluorescence might be quenched.

As a specific example let us consider a $\Delta\epsilon$ that results from a change in polarity of dye microenvironment. In FIGURE 3a the absorption spectra of merocyanine 540 in 500 mM NaCl and bound to phosphatidylcholine (PC) vesicle membranes are shown. The dye that produced the salt-water spectrum was a mixture of monomers and aggregates (the degree of aggregation is uncertain, but dimers and other small aggregates are probably the predominant species). The spectrum of the membrane-bound dye shown in this figure is similar to the spectrum obtained for this dye in hydrocarbon solvents such as octanol and pyridine. The difference spectrum formed by subtracting the aqueous solution spectra from the spectrum of merocyanine 540 in membranes is shown in FIGURE 3b. It is clear from the difference spectrum that when this dye molecule moves from the salt solution into the membrane as a monomer, extinction changes as large as 10^5 can be expected. There is a general resemblance between the difference spectrum of FIGURE 3b and the squid axon absorption difference spectrum obtained by Ross *et al.* (FIGURE 5d). Although this resemblance might be taken as evidence to support an On–Off mechanism, in which the dye undergoes potential-dependent movement between the membrane and the aqueous phase near the membrane, the dissimilarity of the difference spectra in the region of 500–530 nm, together with evidence to be presented below, weigh against this mechanism. The On–Off mechanism may still explain the BLM difference spectra obtained with cyanine dyes.[20]

Let us now consider absorbance changes that result from changes in the state of dye aggregation. Ross *et al.*[40] were the first to suggest that the fluorescence and absorption signal of merocyanine 540 in squid axon experiments are different expressions of a single mechanism and that the *membrane-bound* dye exists in a potential-dependent equilibrium between monomers and dimers having different absorption maxima and different quantum yields. The membrane-associated monomer absorbs at 570 and is highly fluorescent, whereas the dimer absorbs at 520 and is only slightly fluorescent. Aggregation phenomena of this sort are not unusual. Large differences in spectral properties of aggregated dyes have been observed for more than 30 years.[49–51] Additional support for the membrane-localized monomer-dimer mechanism was recently published by Tasaki and Warashina.[21] And recent experiments of Ross *et al.*[17] rule out electrochromism, and an On–Off mechanism as major contributors to the signal observed with the squid axon. In the experiments to be discussed below, we have determined the spectral properties of the membrane-bound species of merocyanine 540 by titrating phosphatidylcholine vesicles with the dye. The titration experiments provide direct evidence of the existence of membrane-bound dimers at micromolar dye concentrations.

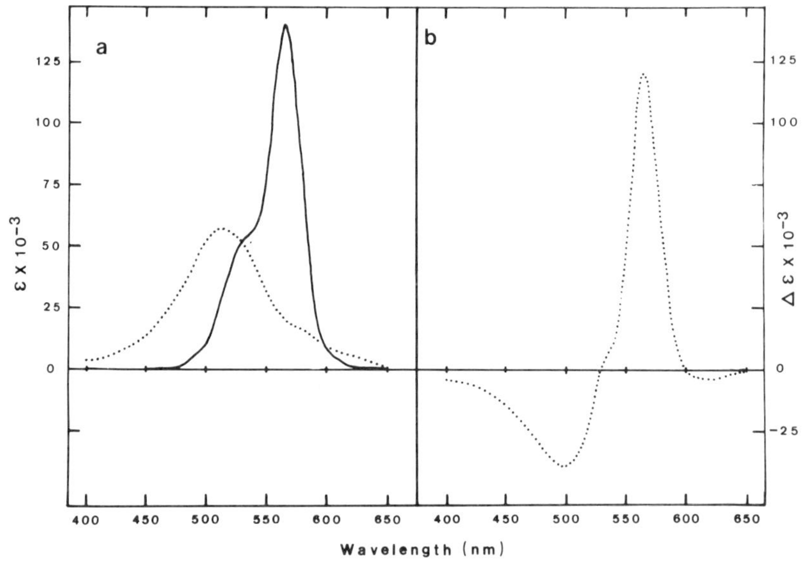

FIGURE 3. a: Absorption spectra of merocyanine 540. Solid spectrum, 2.7 μM dye plus 620 μM phosphatidylcholine in distilled water; dotted spectrum, 21 μM dye in 0.5 M NaCl. b: Difference absorption spectrum calculated from spectra in a: phospholipid spectrum minus 0.5 M NaCl spectrum.

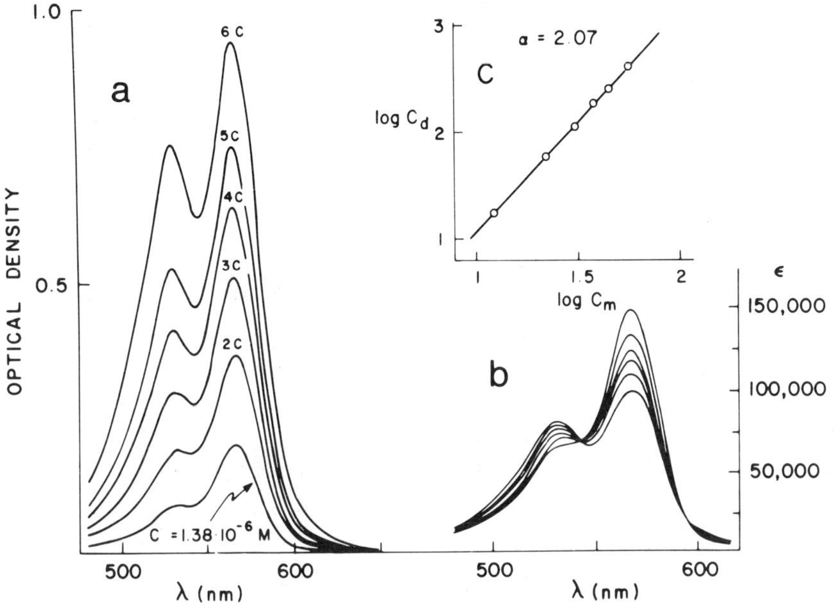

FIGURE 4. Titration of phosphatidylcholine (PC) vesicles with merocyanine 540. Vesicles were prepared according to Waggoner *et al.*[20] The absorption cuvet contained 2.1 × 10^{-4} M PC and 0.5 M NaCl. a: Absorbance of the vesicle-dye suspension after successive additions of dye. In each step the dye concentration was increased by 1.38 μM. b: Normalized absorption spectra obtained by dividing each spectrum of a by the dye concentration at which it was obtained. The normalized spectra, which are plotted as extinction vs. λ, show that only two absorbing species exist in the concentration range studied. c: If a monomer-dimer equilibrium exists on the vesicles, the concentrations of monomer dye, c_m, and dimer dye, c_d, should obey the relationship $c_m^2/c_d = K$. As one varies the total concentration of dye participating in the equilibrium, a plot of log c_d vs. log c_m should be linear with a slope of two. By following the procedure of West and Pearce[49] we have determined c_m and c_d from the absorption spectra of a. The slope of the log-log plot of these data is 2.07. This indicates that one species is a monomer and other is a dimer.

FIGURE 4a shows the result of the titration. Each of the absorption spectra was divided by the dye concentration at which it was obtained to yield a family of spectra expressed in terms of extinction (FIGURE 4b). The spectra have isosbestic points at 542 and 593 nm. This observation favors the existence of only two different species in this concentration range (at higher dye concentrations the spectra no longer cross the isosbestic point). The data of this experiment were analyzed on a computer, assuming that the two species involved obey a simple monomer-dimer equilibrium law.[49] The individual spectra of the monomer and dimer calculated by this procedure are shown in FIGURE 5a.§ The calculated spectrum for the monomer was then compared with the monomer absorption spectra observed at a very large vesicle to dye-concentration ratio. The two

§The data were also analyzed by a procedure that does not assume *a priori* that the two species are monomers and dimers. Similar results were obtained.

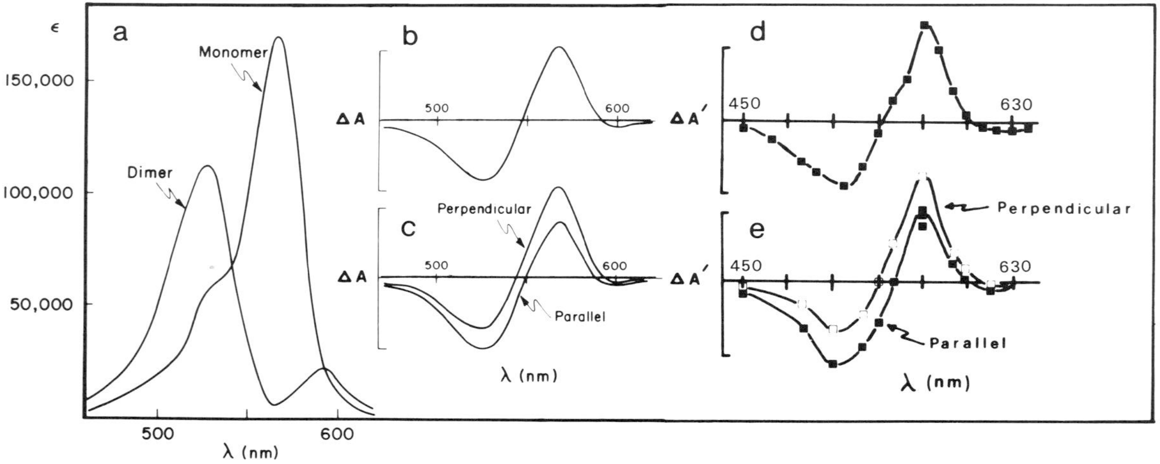

FIGURE 5. a: Calculated spectra of phosphatidylcholine (PC) vesicle-bound merocyanine 540 monomers and dimers. The spectra were calculated from the data in FIGURE 4a essentially by the procedure of West and Pearce,[49] assuming that only two species exist and that their concentrations obey a monomer-dimer equilibrium law. b: Difference spectrum for conversion of membrane-bound dimers to monomers (calculated by subtracting the dimer spectrum from the monomer spectrum of FIGURE 5a. c: Calculated difference spectra predicted for a membrane-localized monomer-dimer shift assuming that the monomer absorption moments are more perpendicular to the surface of the membrane than the absorption transition moments of dimer dye. More specifically, to fit the squid axon spectra (FIGURE 5d) obtained with light polarized parallel(∥) and perpendicular (⊥) to the axon the following equations were used: $\Delta A_{11} = K(\alpha_{11}\epsilon_m + \beta_{11}\epsilon_d)$ and $\Delta A_{\perp} = k(\alpha_{\perp}\epsilon_m + \beta_{\perp}\epsilon_d)$, with $\alpha_{11} = 0.64$, $\beta_{11} = 1.44$, $\alpha = 1.0$, $\beta = 1.26$. d: Relative changes in absorption, $\Delta A'$, for squid giant axon stained with merocyanine 540. (Data from Ref. 17) e: Relative changes in absorption, $\Delta A'$, for squid axons stained with merocyanine 540 and illuminated with light polarized parallel and perpendicular to the axis of the axon.

spectra were almost identical, suggesting that the assumption used for the spectral decomposition is correct. The insert in FIGURE 4c shows the log-log relation between the concentrations of the two species. The slope of the line is 2.08, suggesting that the second species is indeed a dimer. Since a tenfold dilution of the stained vesicles did not appreciably alter the spectra, it is clear that the monomer and dimer are membrane-bound.

The individual absorption spectra of the membrane-associated monomer and the dimer obtained by the procedure described above were used to calculate the absorption difference spectra shown in FIGURE 5b,c. These calculated difference spectra are qualitatively similar to the *experimental* absorption difference spectra obtained with the squid axon that are shown in FIGURE 5d,e.

Although the vesicle experiments described above with merocyanine 540 support the two-state monomer dimer hypothesis for squid axon optical changes, additional axon experiments have shown that the absorption spectra and the fluorescence excitation spectra are concentration-dependent. Furthermore, the excitation spectra are dependent on the emission wavelength used. Experiments with phosphatidylcholine vesicles and merocyanine 540 showed that at high concentrations of bound dye a new fluorescence peak appears that is probably due to weak emission from the dimer molecules. Thus the mechanism for the squid axon optical change is more complicated than the two-state monomer-dimer mechanism.

Recently, Dragsten and Webb[41] studied the mechanism of merocyanine 540 fluorescence changes in polarization experiments on hemispherical lipid bilayer membranes. Their results indicate that at low dye concentrations the optical response results mainly from a reorientation of the membrane associated chromophore in the electric field. At higher dye concentrations (in the relevant range for squid axon experiments) effects from the formation of nonfluorescent aggregates were seen. Their observations at higher dye concentrations are compatible with the monomer-dimer mechanism of Ross *et al.*[17]

One more mechanism for changes in $\Delta\epsilon$ must be considered. It is possible to have absorbance changes take place without any changes in the wavelength or extinction as dye moves to a different environment. In other words, the "intrinsic" extinction coefficient of the dye might remain constant, but changes in the orientation of the dye transition moment can produce changes in the "effective" extinction of dye.¶ Consider an example where dye highly oriented in a membrane is illuminated by a beam normal to the plane of the membrane. If the transition moment of membrane-bound dye is perpendicular to the plane of the membrane, no light will be absorbed by this population of dye molecules. If, however, a fraction of the membrane-bound dye is rotated, this fraction will be capable of light absorption. Also, if a fraction of the membrane-bound dye is driven into the water by a potential change, one of every three randomly oriented aqueous dye molecules will be capable of absorbing light. Thus there will be an absorbance change even if there is no solvent shift of the absorption maximum. Studies of ΔA using polarized light and different geometric optical arrangements (i.e., edge vs. center of an axon) of nerve fibers and bilayer lipid membranes may help de-

¶Fluorescence can also be highly dependent on dye orientation in the membrane.[17,41–43]

termine the mechanism of the optical changes. These dichroism spectra often show large differences among the different types of merocyanine, cyanine, and oxonol dyes that have been studied.[17,21] Dichroism studies with naphthalene sulfonate dyes have been discussed by Conti[22] and by Carbone *et al.*[44]

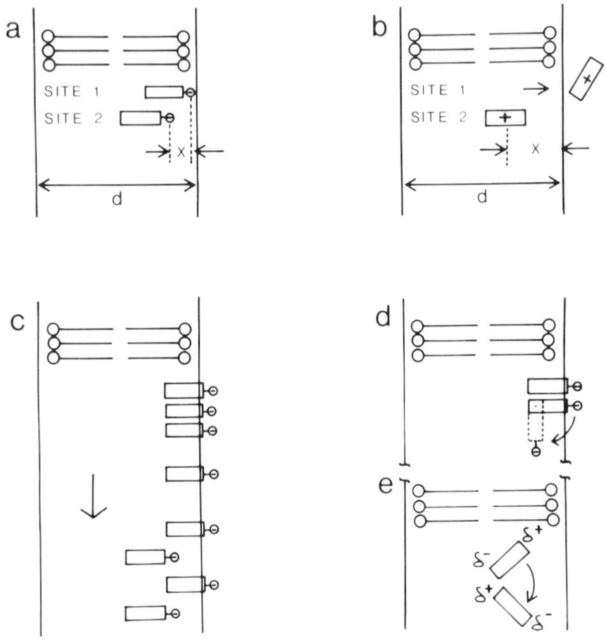

FIGURE 6. Models of possible mechanisms for optical changes that involve potential dependent changes in the position of dye in the membrane. a: A chromophore (symbolized by rectangle) with an attached negative charge (analogous to a mericyanine dye) shifts to a new depth in the membrane. The charge moves through a distance X in moving from site 1 to site 2. b: A dye with a positive charge delocalized on the chromophore structure (analogous to a cyanine dye) moves a distance X from site 2 on the membrane to site 1 off the membrane (an "On–Off" mechanism). c: As the potential changes, a fraction of the dye that was initially aggregated is driven into the membrane in a monomeric form. d: As the electric field rotates the charged dye into the plane of the membrane, aggregates are broken up. e: An uncharged but dipolar dye rotates in the changing electric field.

Movement of Charged or Dipolar Dyes in the Membrane Electric Field: An Equilibrium Thermodynamic Expression for Δn

We will now consider the extent by which a change in membrane potential can move a charged or dipolar dye molecule within or to and from a membrane. Models for some possible modes of potential dependent movement are shown in FIGURE 6. The discussion and the quantitative treatment is restricted to the simplest case possible, movement between only two sites or states, labeled 1 and 2, under the influence of the changing electric field. We realize, of course, that dye molecules associated with a membrane may often assume a distribution of positions and orientations. Nevertheless, the simple two-state model provides easily obtainable estimates of the degree of dye movement that can be expected under certain conditions. Models *a* and *b* of FIGURE 6 involve only translational movement of the dye charge to a different depth in the membrane. As the charge on the dye moves, the chromophore is also shifted to a new environment. In the new environment the dye absorbs light differently because the polarity of the dye microenvironment is different or because the orientation or state of aggregation of the dye has changed. Both sites may be localized within the membrane or one of the sites may be in the aqueous phase (On–Off model). In models *d* and *e* the dipolar dye rotates to a new orientation in the membrane, in which light is absorbed to a greater or lesser degree.

The general equation for the number of dye molecules that move between sites 1 and 2 is developed in the APPENDIX and given below. The development assumes that the dye is in rapid equilibrium between the two sites. A shift in equilibrium between the two sites results from a change in electrical free energy of the charge on the dye in the two sites as the membrane potential changes. Thus Equation 4 results from an equilibrium thermodynamic treatment, and in the approximate form presented is accurate to better than 10% for $Z\mathfrak{F}/RT\ \alpha V$ less than 0.2. For large αV equation A5 in the APPENDIX should be used.

$$\Delta n = n_T \frac{Z\mathfrak{F}}{RT} \alpha V/(2 + 1/K + K). \tag{4}$$

In Equation 4, n_T is the total concentration of dye involved in the equilibrium (in moles/cm^2), K ($= n_2/n_1$) is the equilibrium constant for the dye in the two sites before the transmembrane electrical potential is changed by an amount V, and α ($= x/d$, see FIGURE 6) is the fraction of the transmembrane potential change that the charge on the dye moves through in going from site 1 to site 2. $Z\mathfrak{F}/RT$ has the value 0.04 mV^{-1}. For a noncharged but dipolar dye, μ/ed $(\cos\theta_2 - \cos\theta_1)$ can be substituted for α (μ is the dipole moment of the dye in esu-cm, e is 4.8×10^{-10} esu, and d is the membrane thickness in cm). In this case, the dye rotates within the membrane between two states characterized by θ_1 and θ_2, which are the angles formed by the dipole moment of the dye and a vector perpendicular to the plane of the membrane.

Equation 4 can now be incorporated into Equation 1 to get the absorbance change expected, according to this model, during a potential change.

$$\Delta A = 10^3\, \Delta\epsilon\, n_T \frac{Z\mathfrak{F}}{RT} \alpha V\Big/\left(2 + \frac{1}{K} + K\right). \tag{5}$$

This equation assumes small αV. For large αV, Equation A5 (APPENDIX) should be used in conjunction with Equation 1. Let us now consider, one at a time, the parameters upon which ΔA depends.

1. The absorbance change is linear in n_T. Thus the greater the amount of dye involved in the equilibrium, the larger the signal. There are practical limits on n_T. It seems likely that the surface charge and permeability properties of the membrane may be severely altered if the dye concentration on the membrane rises above 10^{-11} moles/cm^2 (1 dye/25 phospholipid molecules). In BLM optical

experiments and in binding experiments with phospholipid vesicles, most of the dyes that give large signals reach surface concentrations near 10^{-11} moles/cm^2 when the aqueous dye concentration is in the micromolar range.[20] Even higher surface concentrations have been observed with merocyanine 540 on stable phosphatidyl ethanolamine bilayer membranes (Dr. S. Krasne, personal communication).

2. The signal varies nearly linearly with αV. It is therefore important for the dye charge to move between sites as far apart as possible. In other words, the dye must "sense" a large fraction of the membrane potential change. It is hard to predict how deeply the charged portion of a dye molecule might penetrate into the hydrocarbon region of a membrane. Because such penetration is energetically unfavorable, there would have to be strong forces on other parts of the dye that tend to force the charge into the membrane. Cyanine and oxonol dyes with distributed or "delocalized" charges may partition more deeply into the membrane than merocyanine dyes with "localized" or point charges. Certainly for impermeant dyes like the merocyanines, α must be less than 0.5. In the case of the neutral dipolar molecule, where α is replaced by μ/ed (cos θ_2 − cos θ_1), it is desirable that the dipole moment be as large as possible and that the rotation be as large as possible (180° is the optimum).

3. The factor in the denominator has an intuitive explanation. If K is excessively large or small, this signifies that the dye is effectively locked into site 1 or 2. In such a case the equilibrium is not easily shifted during normal potential changes, and ΔA will be small. Absorbance changes are expected to be optimal when each site is nearly equally populated, in other words, when K is near 1.‖

The Maximum Absorbance Change Possible

The factors that can be modified by appropriate synthetic modification of the dyes are α, Z, K, n_T and $\Delta\epsilon$ (refer to (**5**)). For the merocyanine dyes it is possible that α can be increased by delocalizing the charge on the dye so that it can be driven deeper into the membrane. In doing this there is danger that the dye will become permeant. K can be adjusted by alteration of the length of hydrocarbon chains attached to the chromophore. TABLE 3 demonstrates how this can markedly affect the size of an optical signal. No doubt one of the reasons why merocyanine, cyanine, and oxonol dyes are such good optical probes of potential is that they have very large extinction coefficients (~0.5–2.5 × 10^5). This fact might explain why dyes that have longer conjugated chains, and thus larger extinction coefficients, usually show greater absorbance changes in squid axon experiments. Of course, other factors may contribute as well to their greater sensitivity. Thus, $\Delta\epsilon$ can be made large by developing dyes with very large intrinsic extinction coefficients and chromophore absorption spectra that are very solvent-sensitive or

‖ K and n_T are more complicated parameters in the case of an On–Off mechanism. One must decide the size of the pool of aqueous dye that is involved in the equilibrium. The size of this pool will depend on the time it takes to diffuse through the aqueous regions to and from the membrane when the potential changes.

very sensitive to the state of dye aggregation. The factor n_T can be increased by increasing the concentration of dye in the bathing medium.

We are at a point now where we can estimate the maximum ΔA possible for a dye associated with a single bilayer membrane. We will assume $\Delta\epsilon = 10^5$, $n_T = 2 \times 10^{-11}$, $K = 1$, $\alpha = 1$ (a shift of dye charge entirely across the membrane), and that the potential change is 100 mV. With these values substituted into Equation 5 we get $\Delta A \simeq 10^{-3}$. It is remarkable that several of the most sensitive merocyanine dyes shown in TABLE 1 are already within a factor of 10 of the estimated maximum. It seems unlikely that absorbance signal can be increased by more than a factor of 3 unless the concentration of membrane-associated dye can be raised higher than 2×10^{-11} moles/cm^2.**

FLUORESCENCE CHANGES

Potential-dependent fluorescence changes have been observed with more than 300 dyes in squid axon experiments.[17] The fractional changes in fluorescence, $\Delta F/F$, of a number of the more sensitive dyes are listed in TABLE 4. In this table

TABLE 3

EFFECT OF ALYKL GROUP SUBSTITUTION ON THE SIGNAL-TO-NOISE RATIO (S/N) OF THE MEROCYANINE 540 FLUORESCENCE CHANGES IN THE SQUID AXON (FROMREF. 14,16)

R_1*	R_2*	S/N
Methyl	methyl	0.02
Ethyl	ethyl	1.0
Butyl	butyl	5.0
Pentyl	pentyl	4.9
Hexyl	hexyl	4.5
Ethyl	octyl	4.7
Octyl	octyl	0.7
Ethyl	Octadecyl	0.00

*R_1 and R_2 are the *N*-substituted alkyl groups on the merocyanine 540 structure.

we have *not* listed results of experiments using the permeant cyanine dyes in suspensions of cells, organelles, and vesicles, since in these systems the cyanine dyes show large, slow fluorescence changes that take place by an accumulation mechanism.[3,14] Instead, we will restrict the discussion to the small, fast fluorescence changes that take place when dyes, such as those in TABLE 4, are added to only one side of a membrane. For these experiments we shall again assume that the optical change originates from a potential-dependent shift in the equilibrium of dye molecules that can exist in two different sites (or states). The fractional fluorescence change is given by

**In most experiments, however, the signal-to-noise ratio is more important than the actual magnitude of the optical change. As we shall see later, improvements can yet be made in S/N.

$$\frac{\Delta F}{F} = \frac{(F_{\text{site 2}} - F_{\text{site 1}})\Delta n}{n_2 F_{\text{site 2}} + n_1 F_{\text{site 1}} + F_{\text{nonspec}}} = \frac{(\epsilon_2 Q_2 - \epsilon_1 Q_1)\Delta n}{n_2 \epsilon_2 Q_2 + n_1 \epsilon_1 Q_1 + F_{\text{nonspec}}}.$$

In this equation, ϵ_1 and ϵ_2 are the effective extinction coefficients of the dye in sites 1 and 2, Q_1 and Q_2 are the quantum efficiencies of dye in sites 1 and 2, n_1 and n_2 are the concentrations of dye in sites 1 and 2, and Δn is the amount of dye that moves between the two sites during the potential change. F_{nonspec} is the background fluorescence that arises from all dye that is not in the two sites that are involved in the potential change (inner filter effects are neglected in this equation). As was the case for absorption experiments, we expect that Δn will be given by equation (4) for potential-dependent movement of charged dye and for rotation of uncharged dipolar dyes.

Optimal conditions for a large $\Delta F/F$ should occur with three conditions satisfied: (1) when dye in one of the two sites is not fluorescent ($\epsilon_1 Q_1$ or $\epsilon_2 Q_2 = 0$), (2) when all of the fluorescence from the sample is arising from the membrane

TABLE 4

FRACTIONAL FLUORESCENCE CHANGES

Dye	Membrane System	$\Delta F/F$	Reference
Merocyanine 540	*S. lactis* cell suspension	0.2	18
Merocyanine 540	squid axon	10^{-3}	17
Merocyanine 540	skeletal muscle	10^{-1}	5
Merocyanine 540	heart muscle	2×10^{-2}	9
diI—C_1—(5) (cyanine XIX)[17]	squid axon	10^{-4}	17
diI—C_1—(5) (cyanine XIX)[17]	skeletal muscle	3×10^{-2}	5
diBA—C_4—(5) (oxonol XXI)[17]	squid axon	10^{-3}	17

at which the potential change is taking place ($F_{\text{nonspec}} = 0$), and (3) when the two sites are nearly equally occupied (recall a similar argument for optimization of ΔA). Under these conditions

$$\frac{\Delta F}{F} = \frac{2\Delta n}{n_T} \tag{7}$$

where $n_T = n_1 + n_2$. Let us now compare the relative sizes of $\Delta F/F$ and $\Delta I/I$ ($= -2.3\,\Delta A$) in an experiment in which we have a dye that will show both an optimal fluorescence change and an optimal absorbance change ($\Delta\epsilon \sim 10^5$).

$$\frac{\frac{\Delta F}{F}}{\frac{\Delta I}{I}} = \frac{\frac{2\Delta n}{n_T}}{2300\,\Delta\epsilon\,\Delta n} = \frac{10^{-8}}{n_T}. \tag{8}$$

Since n_T can be no larger than 10^{-9} moles/cm^2 (a bilayer of pure dye), it is clear that under the conditions mentioned above, fluorescence measurements will provide a superior approach for potential measurements. Even with a reasonably high amount of membrane-associated dye, say $n_T = 2 \times 10^{-11}$ moles/cm^2, the fractional change in fluorescence is predicted by this equation to be 500 times larger than the fractional transmittance change. This characteristic advantage of fluorescence measurements is well known, and the relative size of the changes in fluorescence *vs.* absorption has been evaluated for different applications.[46] Unfortunately, in many single-cell experiments the advantage of the fluorescence approach has not been realized, probably due to nonspecific fluorescence from dye bound to Schwan cell membrane and possibly due to fluorescence of dye at potential-insensitive sites on the axon membrane. Even with high background fluorescence, $\Delta F/F$ in the squid axon is 10× larger than $\Delta I/I$.[16] In spite of this fact it is frequently more profitable to measure *transmittance* changes with the probes presently available. The reason for this is explained below.

Signal-to-Noise Ratios

The optical signals from single cells and BLMs are small, typically in the range of 10^{-4}/100 mV to 10^{-3}/100 mV. This fact makes these experiments technically difficult, because the light reaching the photodetector must be stabilized to about 10^{-5} of the resting light level. Several precautions, including stabilization of all optical components against vibration, must be taken in order to reach this performance. Even if all these disturbances are eliminated, shot noise, which originates from statistical fluctuations in the photon flux reaching the photodiode, remains.[47]

The shot noise level is proportional to the square root of the light intensity reaching the photodector. Thus for an absorption measurement $(S/N)_A$ is proportional to $\Delta I/\sqrt{I}$. The complete expression for the signal-to-noise is

$$\left(\frac{S}{N}\right)_A = \left(\frac{\Delta I}{I}\right)(2\tau q)^{1/2}(I)^{1/2} \qquad \mathbf{(9)}$$

where q is the quantum efficiency of the photodiode, and τ is the rise time of the detector circuit ($\tau = 1/4\,\Delta f$; Δf is the power bandwidth).

For fluorescence-measurement experiments, the light intensity radiated from the illuminated region of the sample is denoted by F. This light contains the signal fluorescence, background fluorescence, and scattered light. The detector optics will capture a fraction, g, of F. In the apparatus used by Cohen and his colleagues to study squid axons, the efficiency of the light collection is fairly high, $\sqrt{g} = 0.38$.[17] For fluorescence experiments

$$\left(\frac{S}{N}\right)_F = \left(\frac{\Delta F}{F}\right)(2\tau q)^{1/2}(gF)^{1/2}. \qquad \mathbf{(10)}$$

It is clear from this equation that the more efficient the light collection and the greater the amount of fluorescence from the dye that is involved in the potential change (compared with the nonspecific fluorescence), the greater the signal-to-noise ratio will be.†† The total fluorescence can be increased by increasing the total dye concentration; however, eventually the point will be reached where energy transfer between dye molecules and fluorescence quenching will prevent further increases in fluorescence.

We shall now compare the signal-to-noise ratios obtainable from fluorescence and absorption experiments.

$$\frac{\left(\frac{S}{N}\right)_F}{\left(\frac{S}{N}\right)_A} = \frac{\left(\frac{\Delta F}{F}\right)}{\left(\frac{\Delta I}{I}\right)} \times \frac{(gF)^{1/2}}{(I)^{1/2}}. \qquad \textbf{(11)}$$

In most experiments with single cells the transmitted light, I, will be more than 1000 times as intense as the captured fluorescence light, gF. Thus the apparent advantage in measuring fluorescence (where $\Delta F/F \gg \Delta I/I$) largely disappears in optical experiments with single cells. And S/N is the critical factor in these experiments! This phenomenon is abundantly clear in squid axon experiments in which $\Delta F/F$ is 10 times larger than $\Delta I/I$ but $(S/N)_F$ is only half $(S/N)_A$. Of course, the way to improve $(S/N)_F$ is to design dyes with larger $\Delta F/F$. If this can be done (mainly by reducing nonspecific background fluorescence), fluorescence methods may be the better choice for detecting potential changes in single cells. For example, (**6**) and (**11**) can be combined to show that under ideal conditions

$$\frac{\left(\frac{S}{N}\right)_F}{\left(\frac{S}{N}\right)_A} \simeq 16.$$

In this estimate $\sqrt{g}$ was taken to be 0.4 and the other values were optimized as before (that is, $\epsilon_1 Q_1$ or $\epsilon_2 Q_2 = 0$, $F_{\text{nonspec}} = 0, K = 1, \Delta\epsilon = 10$, and $n_T = 2 \times 10^{-11}$ moles/cm^2).

SUMMARY

We have tried to quantitate the sizes of absorption and fluorescence changes of the "fast" potential sensitive dyes in terms of the spectral properties of the dyes and the electrical forces in membranes that can affect the dye molecules. The following conclusions were made.

1. In principle, dye molecules with appropriate absorption properties (narrow absorption peaks and large absorption dipole moment changes) that are properly oriented in a membrane can show very large linear electrochromic

††If the fluorescence intensity is low, the photodiode dark current may become the limiting noise factor. Under this circumstance it will be advantageous to use a photomultiplier, even though photomultiplier tubes usually have a lower value for q and their shot noise is somewhat larger.

changes. However, the potential-dependent optical changes of the cyanine, merocyanine, and oxonol dyes studied so far do not seem to occur by this mechanism.

2. The size of potential-dependent absorption changes, ΔA, is proportional to the product of the extinction change of the dye, $\Delta\epsilon$, and the number of dye molecules involved in the extinction change, Δn. There is strong experimental evidence that for merocyanine 540 at high concentrations $\Delta\epsilon$ results from a potential-dependent change in the state of dye aggregation localized on the membrane. For some cyanine and oxonol dyes it was suggested that $\Delta\epsilon$ occurs because of potential-dependent movement of dye between membrane-binding sites and an aqueous region near the membrane.
3. A simple two-state equilibrium thermodynamic model was used to quantitate the dependence of Δn on the membrane potential (V), the fraction of V that the dye senses in moving between the two states (α), the total amount of dye involved in the equilibrium (n_T), and the equilibrium constant (K) for the dye in the two sites in the absence of a membrane potential. By using estimates of the maximum amount of dye that can bind to a membrane (to get n_T) together with the maximum value of $\Delta\epsilon$ for existing dyes, we were able to propose an upper limit for ΔA for nonelectrochromic mechanisms. Our estimate of $\sim 10^{-3}$ is only $10\times$ larger than the absorbance changes seen in squid axon with presently available dyes.
4. Fluorescence changes, ΔF, also depend on Δn and $\Delta\epsilon$, but in addition depend on the quantum efficiencies of the dye in the two sites and on the nonspecific fluorescence from dye not involved in the potential change. The fractional fluorescence change, $\Delta F/F$, in principle should be much larger than the fractional transmittance change, $\Delta I/I$, measured in an absorption experiment ($\Delta I/I = -2.3\ \Delta A$). However, in practice $\Delta F/F$ is considerably reduced because of nonspecific binding of dye to sites where potential does not change (especially in single-cell experiments). Future work should concentrate on reducing $F_{nonspec}$ by localizing dye only at membrane-binding sites that can be affected by potential changes.
5. Since optimization of signal-to-noise (S/N) is a critical part of all single cell optical experiments, we have compared S/N ratios that can in principle be obtained by fluorescence vs. absorption measurements. Because shot noise is the predominant noise source in single-cell experiments, absorption experiments, in which intensity of light reaching the photodetector is high, often have larger S/N ratios. If $F_{nonspec}$ can be reduced, we show that $(S/N)_F$ can be made greater than $(S/N)_A$. Hence, proper design of fluorescent potential probes could lead to substantial improvements in the optical technique for measuring membrane potentials.

ACKNOWLEDGMENTS

We thank R. Tolles for technical assistance, and L. Cohen for helpful discussions. Special thanks to W. Ross who participated in early vesicle experiments with merocyanine 540.

References

1. Salzberg, B. M., A. Grinvald, L. B. Cohen, H. V. Davila & W. N. Ross. 1977. Simultaneous optical recording of electrical activity in several neurons in an invertebrate central nervous system. J. Neurophysiol. In press.
2. Hoffman, J. F. & P. C. Laris. 1974. Determination of membrane potentials in human and *Amphiuma* blood cells by means of a fluorescent probe. J. Physiol. London **239:** 519.
3. Sims, P. J., A. S. Waggoner, C.-H. Wang & J. F. Hoffman. 1974. Studies on the mechanism by which cyanine dyes measure membrane potential in red blood cells and phosphatidylcholine vesicles. Biochemistry **13:** 3315.
4. Laris, P. C. & H. A. Pershadsingh. 1974. Estimations of membrane potentials in *Streptococcus faecalis* by means of a fluorescent probe. Biochim. Biophys. Acta **376:** 415.
5. Oetliker, H., S. M. Baylor & W. K. Chandler. 1975. Simultaneous changes in fluorescence and optical retardation in single-muscle fibers during activity. Nature **257:** 693.
6. Vergara, J. & F. Bezanilla. 1976. Fluorescence changes during electrical activity in frog muscle stained with merocyanine. Nature (Lond). **259:** 684.
7. Kashket, E. R. & T. H. Wilson. 1974. Protonmotive force in fermenting *Streptococcus lactis* 7962 in relation to sugar accumulation. Biochem. Biophys. Res. Commun. **59:** 879.
8. Renthal, R. & J. K. Lanyi. 1976. Light induced membrane potential and pH gradient in *Halobacteriumhalobium* envelope vesicles. Biochemistry **15:** 2136.
9. Salama, G. & M. Morad. 1976. Merocyanine 540 as an optical probe of transmembrane electrical activity in the heart. Science **191:** 485.
10. Blaustein, M. P. & J. M. Goldring. 1975. Membrane potentials in pinched-off presynaptic nerve terminals monitored with a fluorescent probe: Evidence that synaptosomes have potassium diffusion potentials. J. Physiol. London **247:** 589.
11. Laris, P. C. & H. A. Pershadsingh. 1974. Estimations of membrane potentials in *Streptococcus faecalis* by means of a fluorescent probe. Biochem. Biophys. Res. Commun. **57:** 620.
12. Laris, P. C., H. A. Pershadsingh & R. M. Johnstone. 1976. Monitoring membrane potentials in Ehrlich ascites tumor cells by means of a fluorescent dye. Biochim. Biophys. Acta. **436:** 475.
13. Pick, U. & M. Avron. 1976. Measurement of transmembrane potentials in *Rhadospirillum rubrum* chromatophores with an oxacarbocyanine dye. Biochim. Biophys. Acta **440:**189.
14. Waggoner, A. S. 1976. Optical probes of membrane potential. J. Membr. Biol. **27:** 317.
15. Cohen, L. B. & B. M. Salzberg. Optical measurement of membrane potential. Rev. Physiol. Biochem. Pharm. In press.
16. Cohen, L. B., B. M. Salzberg, H. V. Davila, W. N. Ross, D. Landowne, A. S. Waggoner & C.-H. Wang. 1974. Changes in axon fluorescence during activity: Molecular probes of membrane potentials. J. Membrane Biol. **19:** 1.
17. Ross, W. N., B. M. Salzberg, L. B. Cohen, A. Grinvald, H. V. Davila, A. S. Waggoner & C.-H. Wang. 1977. Changes in absorption, fluorescence, dichroism and birefringence in stained giant axon: Optical measurement of membrane potential. J. Membrane Biol. **33:** 141.
18. Carbone, D. P. & A. S. Waggoner. 1977. Kinetics of membrane potential changes in *Streptococcus lactis* cells following addition of glucose. Biophys. J. **17:** 35a.
19. Kamino, K. 1977. Measurement of transmembrane potential in synaptosomal ghosts using merocyanine 540. Submitted for publication.
20. Waggoner, A. S., C.-H. Wang & R. L. Tolles. 1977. Mechanism of potential-dependent light absorption changes of lipid bilayer membranes in the presence of cyanine and oxonol dyes. J. Membrane Biol. **33:** 109.

21. TASAKI, I. & A. WARASHINA. 1976. Dye membrane interaction and its changes during nerve excitation. Photochem. Photobiol. **24:** 191.
22. CONTI, F. 1975. Fluorescent probes in nerve membranes. Ann. Rev. Biophys. Bioeng. **4:** 287.
23. REICH, R., R. SCHEERER & K.-U. SEWE. 1976. Electrochromic spectra of dyes in lipid layers. Ber. Bunsenges. Phys. Chem. **80**(3): 245.
24. REICH, R. & S. SCHMIDT. 1972. Über den einfluss elektrischer felder auf das absorptionsspektrum von farbstoffmolekülen in lipidschichten: I. Theorie. Ber. Bunsenges. Phys. Chem. **76**(7): 589.
25. CHENG, W. K. 1975. Electrochromic properties of membrane probes. *In* Biomembrane. Vol. 7. H. Eisenberg, E. Katchalski-Katzir, and Lionel A. Monson, Eds.:56–60. Plenum Press. New York, N.Y.
26. LIPTAY, W. 1976. Optical absorption of molecules in liquid solutions in an applied external electric field (electrochromism). Ber. Bunsenges. Phys. Chem. **80**(3): 207.
27. PLATT, J. R. 1961. Electrochromism, a possible change of color producible in dyes by an electric field. J. Chem. Phys. **34**(3): 862.
28. LABHART, H. 1976. Applications of electrochromism. Ber. Bunsenges. Phys. Chem. **80**(3): 240.
29. KUMAMOTO, J., J. C. POWERS, JR. & W. R. HELLER. 1962. Electrochromism (stark effect) in solids due to isolated absorbing centers. J. Chem. Phys. **36**(11): 2893.
30. MALLEY, M., G. FEHER & D. MAUZERALL. 1968. The stark effort in porphyrins. J. Mol. Spectry. **25:** 544.
31. FISCHER, J. K., D. M. VON BRÜNING & H. LABHART. 1976. Light modulation by electrochromism. Appl. Opt. **15**(11): 2812.
32. BÜCHER, H., J. WIEGAND, B. B. SNAVELY, K. H. BECK & H. KUHN. 1969. Electric field induced changes in the optical absorption of a merocyanine dye. Chem. Phys. Lett. **3**(7): 508.
33. SCHMIDT, S. & R. REICH. 1972. Über den einfluss elektrischler felder auf das absorptionsspektrum von farbstoffmolekülen in lipidschichten: II. Messungen an rhodamine B. Ber. Bunsenges. Phys. Chem. **76**(7): 599.
34. SCHMIDT, S. & R. REICH. 1972. Über den einfluss elektrischer felder aug das absorptionsspektrum von farbstoffmolekülen in lipidschichten: III. Elektrochromie eines carotinoids (lutein). Ber. Bunsenges. Phys. Chem. **76**(11): 1202.
35. KLEUSER, D. & H. BUCHER. 1969. Elektrochromie von chlorophyll-a und chlorophyll-b in monomolekularen filmen. Z. Naturforsch. Sect. B **24:** 1371.
36. REICH, R. & R. SCHEERER. 1976. Effect of electric fields on the absorption spectrum of dye molecules in lipid layers. IV. Electrochromism of oriented chlorophyll b. Ber. Bunsenges. Phys. Chem. **80**(6): 542.
37. JACKSON, J. B. & A. R. CROFTS. 1971. The kinetics of light induced carotenoid changes in *Rhodopseudomonas spheroides* and their relation to electrical field generation across the chromatophore membrane. Eur. J. Biochem. **18:** 120.
38. SAPHON, S., J. G. JACKSON & H. T. WITT. 1975. Electrical potential changes, H^+ translocation and phosphorylation induced by short flash excitation in *Rhodopseudomanas sphaeroides* chromatophores. Biochim. Biophys. Acta **408:** 67.
39. BARSKY, E. L. & V. D. SAMUILOV. 1973. Absorption changes of carotenoids and bacteriochlorophyll in energized chromatophores of *Rhodospirillum rubrum*. Biochim. Biophys. Acta **325:** 454.
40. ROSS, W. N., B. M. SALZBERG, L. B. COHEN & H. V. DAVILA. 1974. A large change in dye absorption during action potential. Biophys. J. **14:** 983.
41. DRAGSTEN, P. R. & W. W. WEBB. 1977. Mechanism of membrane potential sensitivity of merocyanine 540. Biophys. J. **17:** 215a.
42. YGUERABIDE, J. & L. STRYER. 1971. Fluorescence spectroscopy of an oriented model membrane. Proc. Nat. Acad. Sci. USA **68:** 1217.
43. FREHALAND, E. 1976. Dynamical theory of fluorescence polarization in a planar array of oriented pigment molecules. Biophysical Chem. **4:** 65.
44. CARBONE, E., F. MALERBA & M. POLI. 1976. Orientation and rotational freedom of fluorescent probes in lecithin bilayers. Biophys. Struct. Mech. **2:** 251.

45. KRASNE, S. 1977. Cyanine dye-induced electrical and fluorescence effects in neutral and negative bilayer membranes. Biophys. J. **17:** 214a.
46. RIGLER, R., C. R. RABL & T. M. JOVIN. 1974. A temperature jump apparatus for fluorescence measurements. Rev. Sci. Instr. **45:** 581.
47. WILLIAMS, S. W. & O. A. BECKLUND. 1972. Optics, a Short Course for Engineers and Scientists. 294. John Wiley & Sons. New York, N.Y.
48. GRINVALD, A., B. M. SALZBERG, L. B. COHEN, K. KAMINO, A. S. WAGGONER, C.-H. WANG & D. TI. 1977. Simultaneous recording from 14 neurons in the supraesophageal ganglion of the barnacle using a new potential-sensitive dye. Nature. In press.
49. WEST, W. & S. PEARCE. 1965. The dimeric state of cyanine dyes. J. Phys. Chem. **69:** 1894.
50. RABINOWITCH, E. & L. F. EPSTEIN. 1941. Polymerization of dystuffs in solution. Thionine and methylene blue. J. Am. Chem. Soc. **63:** 69.
51. SHEPPARD, S. E. 1942. The effects of environment and aggregation on the absorption spectra of dyes. Rev. Mod. Phys. **14:** 303.
52. HLADKY, S. B. & T. J. RINK. 1976. Potential difference and the distribution of ions across the human red cell membrane: A study of the mechanism by which diS-C_3-(5) reports membrane potential. J. Physiol. (London) **263:** 287.

APPENDIX

Consider a charge in equilibrium between two sites, 1 and 2 (as in FIGURE 6), with concentrations n_1 and n_2 in the two sites. The equilibrium constants with and without a potential difference across the membrane are, respectively,

$$K^v = \frac{n_2^v}{n_1^v} \qquad \text{and} \qquad K^0 = \frac{n_2^0}{n_1^0} \tag{A1}$$

The standard free-energy changes for moving the charge from site 1 to site 2 take the usual form:

$$\Delta G^v = -RT \ln K^v \qquad \Delta G^0 = -RT \ln K^0. \tag{A2}$$

The differences between these two free energies is assumed to be linearly dependent on the electrical potential change the dye experiences in moving between the two sites. Thus,

$$\Delta G^v - \Delta G^0 = -Z\mathfrak{F}\alpha V \tag{A3}$$

where $\alpha = x/d$ is fraction of the total transmembrane potential drop, V, sensed by the dye. The equilibrium constant is therefore related to the membrane potential by

$$K^v = K_0 e^{-Z\mathfrak{F}/RT\alpha V}. \tag{A4}$$

We wish to find the number of charges that move between the two sites when the potential changes from 0 to V. Using $\Delta n = n_2^V - n_1^0$ and assuming that the total amount of charge involved in the equilibrium, n_T, remains constant during the potential changes ($n_1 + n_2^0 = n_T = n_1^v + n_2^v$), we get the following:

$$\frac{\Delta n}{n_T} = [e^{-Z\mathfrak{F}/RT\alpha V} - 1]/[1 + \frac{1}{K_0} + (1 + K_0)e^{-Z\mathfrak{F}/RT\alpha V}]. \tag{A5}$$

For small $Z\mathfrak{F}/RT\,\alpha V$ this reduces to

$$\frac{\Delta n}{n_T} = \frac{Z\mathfrak{F}}{RT}\alpha V/[2 + K^0 + \frac{1}{K^0}]. \quad \textbf{(A6)}$$

For rotation of a dipolar but uncharged chromophore, the change in the depth of the charged ends of the dipole in the membrane is given by $\ell/2(\cos\theta_2 - \cos\theta_1)$, where θ_1 and θ_2 are the initial and final angles of the dipole relative to a vector normal to the membrane and ℓ is the length of the dipole in cm. The dipole moment, μ, of the molecule in esu-cm is given by $\mu = \ell q$, where q is the charge of an electron in esu. The fraction, α, of the total potential drop that the two charges on the dipole move through upon rotation of the dipole is given by

$$\alpha = 2\frac{\ell}{2d}(\cos\theta_2 - \cos\theta_1) = \frac{\mu}{qd}(\cos\theta_2 - \cos\theta_1). \quad \textbf{(A7)}$$

DISCUSSION

Hans Meves, *Moderator*

Laboratory of the Marine Biological Association
Plymouth, England

DR. HOUSSIER: Have you looked at the possible influence of the polarization of light beam on your change of absorption?

DR. WAGGONER: Yes, these experiments have been done by Dr. Larry Cohen and his associates at Woods Hole, Mass., with the squid axon. You see rather striking differences when the light is polarized with the electric vector parallel to the length of the axon contrasted with the electric vector oriented perpendicular to the axon. The changes in dichroism during the potential change could be explained in several ways. The dye could rotate in the membrane under the influence of the electrical potential change. Alternatively, the dye could move from a highly oriented state in the membrane into the water, where it will be randomly oriented. The first mechanism might explain dichroism results with merocyanine dyes, whereas the second mechanism might explain the results with cyanine dyes.

DR. GUTTMAN: Dr. Waggoner, you showed one very noisy optical trace over a smooth one, and that smooth one was the action potential. What percentage of the optical change was the maximum of that curve?

DR. WAGGONER: The third slide showed a squid axon experiment with merocyanine 540 taken from Cohen *et al.* (1974, J. Membrane Biol., **19:** 1). The fluorescence change was one thousandth of the fluorescence background. Even with these small changes you can see single action potentials without signal averaging. With some of the better dyes, the signal-to-noise ratios are about five to ten times larger than those with merocyanine 540.

DR. ROBERT S. EISENBERG: Do you know why these dyes are, on the one hand, not toxic, and on the other hand give such large signals in the heart in Martin Morad's experiments?

DR. WAGGONER: Martin Morad uses a relatively low light intensity in his experiments with the heart muscle. With the intense beam of light used by Dr. Cohen with the squid, the axon is killed in less than 20 seconds.

There are several new absorption dyes that produce considerably less photodynamic damage in the squid axon.

DR. O'KONSKI: Did you consider electrometachromasie? Metachromatic dyes are dyes that shift their electronic spectra a great deal by electronic coupling between identical resonating units, and it's explained in terms of the exiton theory of spectra. Upon dimerization, the metachromasie is associated also with a loss of fluorescence, so I think what you should look for is metachromatic dyes that you can concentrate on one side of the membrane by changing the electric field. That should give a very large effect.

DR. WAGGONER: In fact, we like the idea of changes in the state of aggregation very much. Virtually all of the cyanine and merocyanine dyes show metachromasie. We see H bands, J bands, dimer bands, and so on all forming in aqueous solution and on membranes when we raise the dye concentrations high enough. We are just not sure we can identify changes in the state of aggregation as being responsible for all the optical changes we see.

PART VI. ELECTRICAL AND TRANSPORT PHENOMENA OF ARTIFICIAL MEMBRANES

STUDIES OF THE PHYSICAL CHEMISTRY OF PLANAR BILAYER MEMBRANES USING HIGH-PRECISION MEASUREMENTS OF SPECIFIC CAPACITANCE*

Stephen H. White†

Department of Physiology
University of California
Irvine, California 92717

INTRODUCTION

A solution is, by definition, any phase containing two or more components. The cell membrane is a phase(s) distinct from the cytoplasm and interstitium containing lipids and proteins, and is therefore a solution. Simply calling the membrane a solution, however, provides few new insights into membrane architecture. My laboratory is concerned with developing quantitative methods for describing the solution properties of membrane systems that we believe will lead to a deeper understanding of membrane organization. Toward this end, we have been studying the physical chemistry of the planar bilayer membrane, which is a particularly simple membrane solution consisting of alkane molecules dissolved in a lipid bilayer. This paper describes the rationale for and the results of some of our experiments.

THE CELL MEMBRANE AS A LIPID-PROTEIN SOLUTION

The basic structural element of cell membranes is the lipid bilayer,[1-3] which is believed to act as a two-dimensional "solvent" for hydrophobic proteins.[4-7] A highly pictorial description of the membrane as a two-dimensional solution is shown in FIGURE 1. Accepting this scheme as a working hypothesis, the next logical step is to construct a quantitative physicochemical theory. This is a difficult step, however, because of our limited knowledge of solutions in volumes of molecular dimensions. Describing the solution properties of membranes is related to the problem of describing the solution properties of a very thin slice of an ordinary bulk solution. Both problems are equivalent to the notoriously intractable one of describing the behavior of solutions at the molecular level.

The complexity of describing the membrane as a lipid-protein solution can be appreciated more easily by constructing a Corey-Pauling-Koltum model of part of the hydrophobic segment[9,22] of the major glycoprotein ("glycophorin") of human red blood cell membranes and comparing it (FIGURE 2) with a CPK model

*Research supported by research grants from the NIH (NS-10837), the NSF (GM-40054 and PCM76-20691), and gifts from the Guggenheim Multiple Sclerosis Foundation.

†Recipient of a Research Career Development Award from the National Institute of Neurological and Communicative Disorders and Stroke (1 KC4 NS00146).

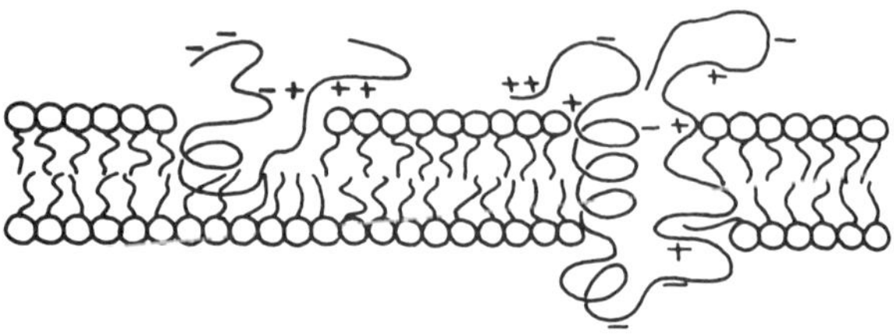

FIGURE 1. A highly schematic representation of the cell membrane as a two-dimensional solution in which the lipid bilayer acts as a "solvent" for hydrophobic proteins. Based on Singer and Nicolson.[8]

of a phospholipid. FIGURE 2 suggests that we must be cautious in describing the membrane as a simple two-dimensional solution. For one thing, at the level of a single amino acid residue, the bilayer thickness is sufficient to attribute a third dimension to the system. Perhaps the term *quasi-two-dimensional* is more appropriate. It may be reasonable to treat the membrane as a two-dimensional solution at the level of the polypeptide or protein, but then a problem arises because of the mass of the polypeptide. The approximate molecular weight of the decapeptide shown in FIGURE 2 is 1,300, and this accounts for less than half of the 23 residue hydrophobic segment[22] of glycophorin. Hexadecane, which has a molecular weight of only 226 and is ideally hydrophobic, is only slightly soluble in planar bilayers.[10,11] Heneicosane ($C_{21}H_{44}$, MW297) is probably completely insoluble (FIGURE 3). Thus, hydrophobicity alone does not assure solubility in bilayers. In one sense this comparison is unfair, because solubility of proteins will be determined in part by tertiary structure, which probably renders them amphiphilic (surface active). Nevertheless, the low solubility of the longer alkanes in planar bilayers raises a significant conceptual problem with regard to describing the membrane as a simple solution of protein in lipid bilayer. Further examination of FIGURE 2 reveals an additional problem that is usually ignored in discussions of hydrophobic proteins. The polypeptide contains numerous carbonyl oxygens that must make the polypeptide more polar than an alkane of equivalent mass. Organizing the polypeptide into a helical configuration[22] could, however, partially alleviate this problem.

Thus, the membrane is likely to be an exceptionally complex solution. It is probably proper to view the membrane as a quasi-two-dimensional solution of proteins or parts of proteins dissolved (or at least dispersed) in the bilayer. Viewed in this way, a major problem of membrane biophysics is to develop quantitative methods for describing the solubility of proteins or polypeptides in bilayers. Where shall we start? In thinking about this problem, I realized that we are largely ignorant of how even the simplest molecules interact with the bilayer. Consider, for example, hexane. This molecule should, and apparently does (S. H. White and M. Yafuso, unpublished) readily dissolve in the membrane bilayer and lipid dispersions.[41] What kind of environment does it find? Is the interior of the bilayer like a simple alkane liquid, as many people believe, or does the interior have properties not normally observed in ordinary bulk solutions? The answers to these questions are important for understanding lipid-protein interactions in membranes. For this reason, my laboratory is engaged in studies of the interaction of alkanes with bilayers and cell membranes. The studies I describe here are concerned with the simplest possible membrane solution (compare FIGURES 2 and 8) consisting of alkane dissolved in planar lipid bilayers. Several specific questions motivated the study. 1) Is the interior of the bilayer equivalent to a bulk alkyl liquid? 2) What structural features of the lipid affect the solubility of alkanes in the bilayer? 3) What structural properties of the alkanes determine solubility? Partial answers to these questions are beginning to emerge and will be described below. I believe the results provide exciting and provocative insights into the nature of the bilayer.

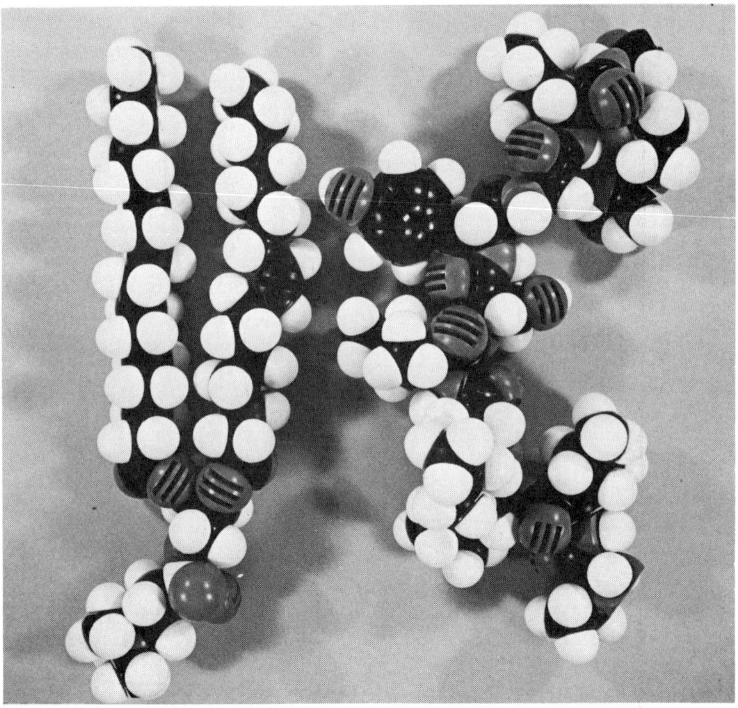

FIGURE 2. Corey-Pauling-Koltum models of a lecithin molecule (left) and a membrane polypeptide (right) consisting of a sequence of ten amino acids (numbers 87–96) of the 23 amino acid hydrophobic segment of the major red blood cell membrane glycoprotein (glycophorin).[9,22] This decapeptide has a molecular weight of about 1,300 and is far too bulky to be soluble in the bilayer in any simple sense because the 21 carbon *n*-alkane heneicosane (MW 297) is probably completely insoluble in bilayers. Relative to a single amino acid, the thickness of the bilayer is significant. It is perhaps more reasonable to refer to the membrane as a *quasi-two-dimensional* solution.

The Potential or Interactive Volume of Bilayers

The planar bilayer membrane first described by Mueller *et al.*[12] is a natural system for examining the alkane-bilayer interaction since an alkane or equivalent solvent is essential[13] for bilayer formation. Various techniques used in the formation and examination of these bilayers have been well described in two recent publications.[14,15] A measurement of particular importance is that of specific capacitance,[16] which yields information on the thickness and composition of the bilayer. The hydrophobic core of the bilayer has a specific geometric capacitance (C_g) given by

$$C_g = \frac{\epsilon_0 \epsilon_B}{\delta_B} \tag{1}$$

where $\epsilon_0 = 8.854 \times 10^{-14}$ Farads/cm, ϵ_B is the dielectric coefficient and δ_B the thickness of the alkyl interior. ϵ_B can be estimated satisfactorily[18] and C_g can be accurately measured with high precision[17] by A.C. bridge techniques, with appropriate corrections for the electrolyte-bilayer dispersion[17] and polarization-charge capacitance.[34] In my laboratory, C_g can be measured with a precision of better than 0.3% which means that changes in δ_B of about 0.2A or less can be reliably detected.

Thickness, determined from Equation 1, is a direct measure of bilayer composition.[10,23] A planar bilayer of thickness δ_B and area A will have a volume given by

$$\delta_B A = N_{AC} V_{AC} + N_A V_A$$

where N_{AC} and N_A are the total number of acyl chains and alkane molecules, respectively, and V_{AC} and V_A are the molecular volumes. In terms of the number (n) of acyl chains and alkanes per unit area of membrane, Equation 2 becomes

$$\delta_B = n_{AC} V_{AC} + n_A V_A. \tag{2}$$

For monoglyceride and phospholipid bilayers, the number of stabilizing lipid molecules per unit area of bilayer appears to be independent of the type and size of the alkane used to form the bilayer. Therefore, $\delta_B \propto n_A$, so that thick membranes have a larger volume fraction of alkane than do thin ones.[10,11]

Figure 3 shows the results of recent measurements in my laboratory of the thickness of bilayers formed from glycerol monooleate and *n*-alkanes at 30°C in 0.1 M NaCl solutions. The maximum thickness (48.3 Å) is obtained for *n*-decane. As the size of the alkane increases, the thickness of the bilayer decreases and approaches the theoretical limiting value of 24.9A (calculated from data in Ref. 18). This decrease is striking and, I believe, profoundly significant. First, *n*-alkanes are not equally soluble in the bilayer, for if they were, thickness would be independent of alkane type. This indicates that the interior of the bilayer is not completely equivalent to a simple bulk alkyl liquid because we know that liquid alkyls are miscible in one another. Second, the thickness of the bilayer can be changed by almost a factor of two without changing the number of stabilizing lipid molecules per unit area. This suggests that the bilayer has a *potential* volume available for accommodating proteins or other molecules that is approximately equal to the volume of the acyl chains of the stabilizing lipid molecules composing the bilayer;

that is, given a fixed amount of lipid in a bilayer, the membrane thickness can adjust to accept additional constituents without necessarily changing the interfacial organization. I like to call this potential volume the *interactive* volume of the bilayer because it is the volume which can be made available for the interaction of solute molecules with the bilayers. I emphasize that in the alkane-free bilayer this volume does not exist. It is a potential volume that is revealed only when other constituents (such as alkanes) are introduced.

Presumably, biological membranes have an interactive volume partially occupied by membrane proteins or polypeptides. The introduction of alkanes into biological membranes should reveal the available interactive volume and allow one to examine the competition between alkane and protein for the interactive volume. It may be possible in this way to quantitate the interaction of the proteins with the bilayer.

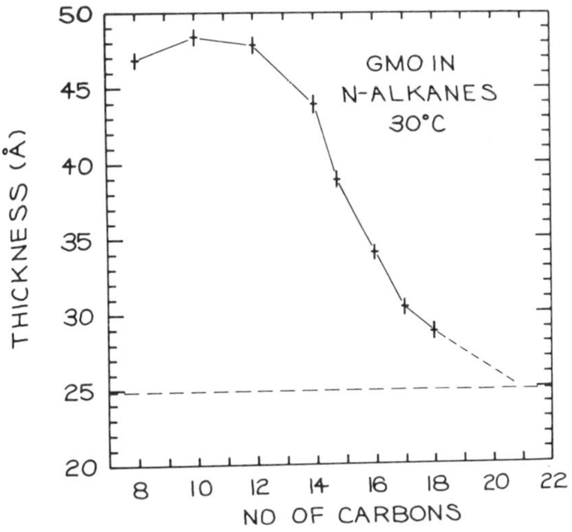

FIGURE 3. The thickness of planar bilayer membranes formed from glycerol monooleate (GMO) dispersed (10 mg/ml) in *n*-alkanes, as a function of the number of carbons in the alkane. The aqueous phase was unbuffered 0.1M NaCl (pH $\simeq$ 6); T = 30.0°C. Thickness was determined from high-precision capacitance measurements corrected for the electrolyte-membrane dispersion and the polarization-charge capacitance. Since the number of GMO molecules per unit area of membrane is independent of alkane size, thickness is a direct measure of the amount of alkane in the bilayer. Note that as the size of the *n*-alkane increases, its solubility in the bilayer decreases. The horizontal dashed line represents the thickness of a hypothetical alkane-free bilayer. The extrapolation of the curve connecting the data points shows that heneicosane ($C_{21}H_{44}$) is probably insoluble in the bilayer.

How Alkanes Occupy the Interactive Volume of the Bilayer

Why is it that the interactive volume of the bilayer cannot be occupied as easily by, say, *n*-hexadecane as by *n*-decane? That is, why is *n*-decane more soluble in the

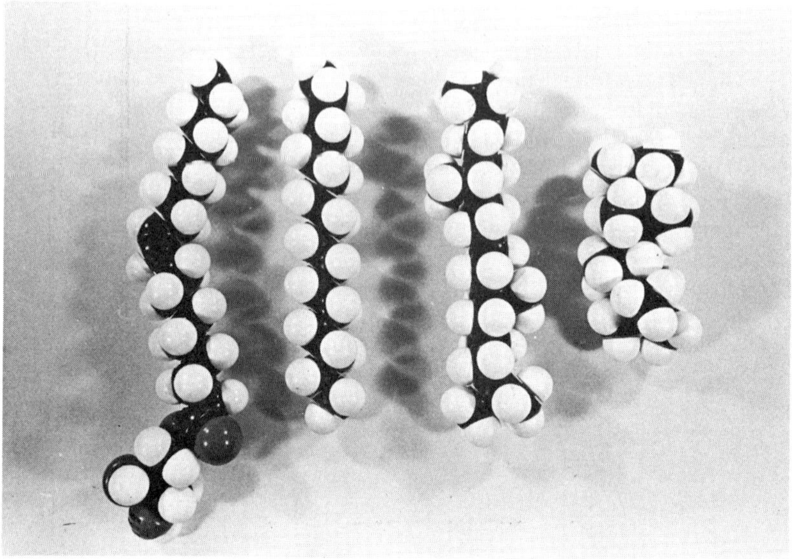

FIGURE 4. Corey-Pauling-Koltum models of (from left to right) glycerol monooleate, *n*-hexadecane, 2,6,11,15-tetramethylhexadecane, 2,2,4,4,6,8,8-heptamethylnonane. Heptamethylnonane has about the same molecular volume as hexadecane, but its length is the same as that of *n*-decane. Tetramethylhexadecane has about the same molecular volume as eicosane but the length of hexadecane. Length is the primary variable determining alkane solubility (see FIGURE 5).

bilayer than *n*-hexadecane? Before this question can be answered, we must first establish what structural property of the alkane is important in determining solubility. In FIGURE 3, thickness was plotted against the number of carbons in the alkane. Structurally, both the molecular volume and the length of the alkane vary with the number of carbons. Which of these variables is important in setting the solubility? This question can be answered by examining bilayers formed from isomers of *n*-alkanes (FIGURE 4). The results of measurements made on bilayers

TABLE 1

PHYSICAL PROPERTIES OF PLANAR BILAYERS FORMED FROM GLYCEROL MONOOLEATE AND SEVERAL ALKANES AT 30°C

Alkane	Bilayer Thickness (Å)	ΔH‡ (kcal/mol)	ΔS‡ (cal/deg/mol)
n-Hexadecane	34.0 ± 0.2	4.03 ± 0.10	13.25 ± 0.36
Tetramethylhexadecane*	32.5 ± 0.2	3.20 ± 0.25	10.65 ± 0.86
Heptamethynonane†	45.0 ± 0.2	0.37 ± 0.01	1.23 ± 0.03

*2,6,11,15-tetramethylhexadecane.

†2,2,4,4,6,8,8-heptamethylnonane.

‡ΔH is the enthalpy of transfer and ΔS the entropy of transfer of the alkane from bulk (microlenses and annulus) to the bilayer at saturation equilibrium (ΔH = TΔS).

formed from *n*-hexadecane, heptamethylnonane (a hexadecane isomer), and tetramethylhexadecane (an eicosane isomer) are shown in TABLE 1 and FIGURE 5. These data strongly suggest that the primary variable is alkane *length,* with molecular volume having a second-order effect. Heptamethylnonane has approximately the same molecular volume as *n*-hexadecane, but the bilayer has a thickness of 45.0 Å rather than 34.0 Å. A short extrapolation of the data in FIGURE 3 shows that bilayers formed from eicosane (n-$C_{20}H_{42}$) should be essentially alkane-free and have a thickness of about 25 Å. If molecular volume were the primary

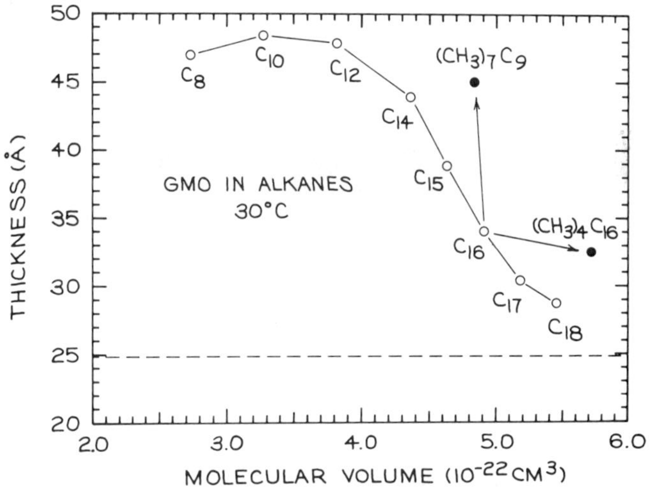

FIGURE 5. The thickness of planar bilayer membranes formed from glycerol monooleate, dispersed (10 mg/ml) in various alkanes at 30°C, as a function of the molecular volume of the alkane. The open circles (○) are values for the *n*-alkanes (C_N) from FIGURE 3, and the shaded circles (●) are values for heptamethylnonane [$(CH_3)_7C_9$] and tetramethylhexadecane [$(CH_3)_4C_{16}$]. Note that the membranes formed from $(CH_3)_7C_9$] are much thicker than from C_{16}, even though the molecular volumes of the two alkanes are about the same. $(CH_3)_4C_{16}$ gives membranes of about the same thickness as C_{16}, even though the molecular volume is about the same as C_{20}. The thickness of C_{20} membranes should theoretically give a value of thickness close to that shown by the dashed line, which represents the hypothetical alkane-free thickness of the bilayer. These data indicate that alkane length determines solubility. Molecular volume has a second-order effect.

variable, membranes formed from tetramethylhexadecane (i-$C_{20}H_{42}$) should give membranes of about this thickness. However, the observed thickness is 32.5 Å—close to the value of *n*-hexadecane. The length of tetramethylhexadecane is about the same as *n*-hexadecane, whereas the length of heptamethylnonane is about the same as *n*-decane (see FIGURES 4 and 8). FIGURE 5 shows bilayer thickness plotted against the molecular volume of the alkanes. The points for heptamethylnonane [$(CH_3)_7C_9$] and tetramethylhexadecane [$(CH_3)_7C_9$] are out of place in a way most easily explained by alkane length being the primary variable. Note, however, that the thicknesses are smaller than expected if length were the only vari-

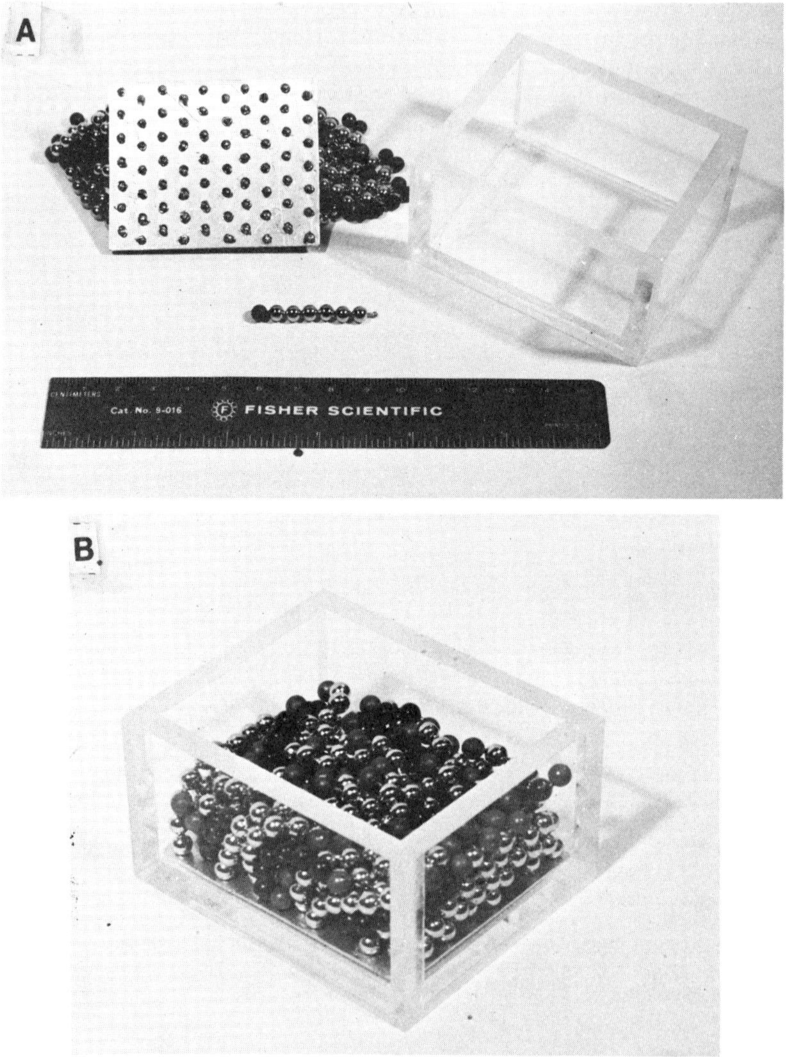

FIGURE 6, A. & B. A model for one monolayer of a bilayer constructed from number 10 nickel-plated beaded chain. A: Disassembled model showing the "acyl chains" attached to a $\frac{1}{16}''$ aluminum plate in a hexagonal pattern. The spacing is scaled to give an area/acyl chain of $40A^2$, taking the cross-sectional area of a single bead as $20A^2$. The terminal "methyl groups" are painted red but appear gray in the black-and-white photo. B: Model assembled in a Plexiglass® box. The height of the box corresponds to the fully extended length of the "acyl chain."

C

D

FIGURE 6, C. & D. C: Side view of the assembled model (see legend for FIGURE 6, A & B). A few chains (20%) were painted black. Note that most of the chains are bent so that the terminal methyl half of the chain *tends* to run parallel to the bilayer surface. D: Top view of the model. Note the high density of "methyl groups."

able. These deviations are probably caused by a molecular volume effect which is second order to the length effect.

Andrews *et al.*[29] have suggested that the alkane solute resides largely (but not exclusively[21]) in the central portion of the bilayer. As discussed by White,[11] this is reasonable for several reasons. First, magnetic resonance studies of phospholipid multilayers and vesicles (see e.g. Refs. 30 and 31) indicate a significant gradient of fluidity within bilayers. The methylene groups ($—CH_2—$) closer to the terminal methyl group ($—CH_3$) have more degrees of freedom (fluidity) than those near the polar group. The alkanes should fit more easily into the less constrained (more fluid) central zone than into the outer zones where the lipid acyl chains have fewer degrees of freedom. A graphic illustration of the fluidity gradient is shown in FIGURE 6, in which a model of one half of a bilayer has been constructed from beaded chain of the type used in pull-string electric lamp sockets. The model is a gross approximation to reality, but it does have the following essential constraints of lipid acyl chains in bilayers:[3] 1) The polar ends of the "molecules" are closely packed on an approximately hexagonal two-dimensional lattice. 2) The average distance between $—CH_2—$ groups on neighboring chains is constant throughout the bilayer (i.e., the density of the bilayer is uniform with respect to thickness). 3) Neighboring chain segments run parallel to one another. These constraints cause the bilayer to have a much smaller thickness (δ_B) than expected from the fully extended length (ℓ) of the acyl chain. Approximately, $\delta_B = 2\ (A_0/A)\ell$ where A_0 is the cross-sectional area of an acyl chain ($\sim 22A^2$, Ref. 3) and A is the area per acyl chain in the interface ($\sim 39A^2$ for glycerol monooleate, Ref. 29). For glycerol monooleate, $\ell = 22.3A$ (Ref. 11) and $\delta_B = 25A$. As a result (as the model clearly shows), each chain at any instant of time tends to be "bent" so that the segments of the chain near the terminal methyl group tend to run *parallel* to the plane of the bilayer. This means that as time runs the terminal methyl groups move about in a central zone parallel to the bilayer and sweep out much larger areas than the polar groups; hence, the increased fluidity near the center of the bilayer. In glycerol monooleate bilayers containing little solvent $\delta_B \simeq \ell$ and on the average the acyl chains will be bent near their centers, causing the chain near the polar end to tend toward the normal of the bilayer and the chain near the $—CH_3$ group to tend toward the parallel of the bilayer.

Second, the alkane molecules will tend to align themselves parallel to the acyl chains (see discussions in Refs. 32 and 33) just as neighboring acyl chains tend to align themselves parallel to one another. If the alkane molecules were located exclusively near the polar surfaces, they could align themselves parallel to the acyl chains only if the area per polar group increased significantly. Such an area increase would also increase the free energy (surface tension) of the bilayer and would therefore not be favored. If, on the other hand, the alkane molecules were located mostly in the center of the bilayer, they could simultaneously occupy volume and be parallel to the acyl chains without changing the area per polar group (FIGURES 7 and 8). This, coupled with the idea that the chains must be bent much of the time, explains why decane is more soluble than hexadecane (and in general why alkane length and not molecular volume is the important structural parameter). As will be explained shortly, the terminal methyl groups of the opposing monolayers must be free to make frequent contacts with one another across the bilayer midplane. All of the hexadecanes cannot simply span the distance from the polar group to the $—CH_3$ group without causing an energetically unfavorable increase in the area per polar group, and they therefore shift toward the fluid center of the bilayer. But, if the hexadecanes occupied a volume fraction of about 0.50, as *n*-decane does,[29] the terminal methyl groups could not come into contact at the midplane. Since this contact must occur, some hexadecanes must be excluded from the structure. The bent terminal methyl ends of the acyl chains probably swing out toward the bilayer midplane somewhat to form "slanted pockets" into which the hexadecanes can fit (FIGURE 7). In this way, the hexadecanes can be parallel to the acyl chains without occupying interfacial area, and the terminal methyls can interact. This treatment assumes that the monolayers function independently of one another; that is, a given hexadecane does not often span the midplane so that it is simultaneously mixed in both monolayers. Cross-monolayer mixing requires that space for part of the hexadecane be available in both monolayers at the same point in the midplane simultaneously. The probability of this is low unless the motions of the acyl chains in the two monolayers are highly correlated, which also seems unlikely. If the hexadecanes could span the midplane, it would be possible to have them occupy a large volume fraction (0.50) and give membranes of the same thickness as decane membranes. Since the volume fraction is actually low (0.27) and the membranes are thin, it appears that spanning does not occur, and that the monolayers act independently. This is consistent with the

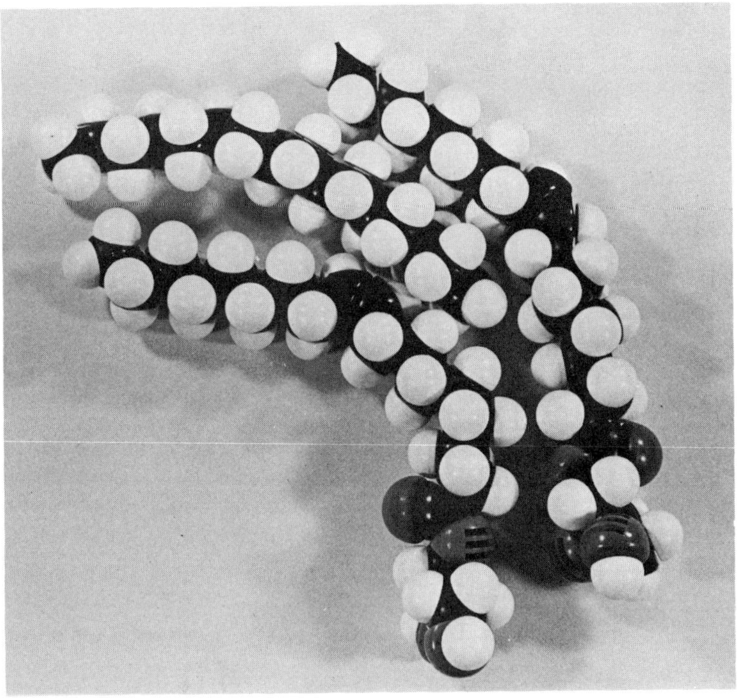

FIGURE 7. Corey-Pauling-Koltum models of *n*-hexadecane and glycerol monooleate showing how hexadecane might fit into the central zone of the bilayer. Some alkane must also be located in the interfacial zones near the polar groups but only in very limited amounts, because small increases in area per polar group lead to large increases in the interfacial free energy. It is suggested that the "bent" terminal halves of the acyl chains swing out toward the midplane slightly to generate space for accommodating the alkanes. Note that the hexadecane must be in frequent contact with $-CH_3$ groups from the opposed monolayer. The packing of a long alkane within the acyl chains (between planes defined by the terminal methyl groups and the double bonds) is probably not as tight as in bulk, so that some free volume is generated.

small free-energy difference between $1 cm^2$ of alkane-saturated bilayer and $2 cm^2$ of monolayer adsorbed at the alkane/water interface[36] and with the suggestion of Andrews *et al.*[29] that the volume fraction of alkane in the monolayer at the alkane/water interface is the same as in the bilayer saturated with alkane.

Now consider decane (FIGURE 8). The same general considerations apply,

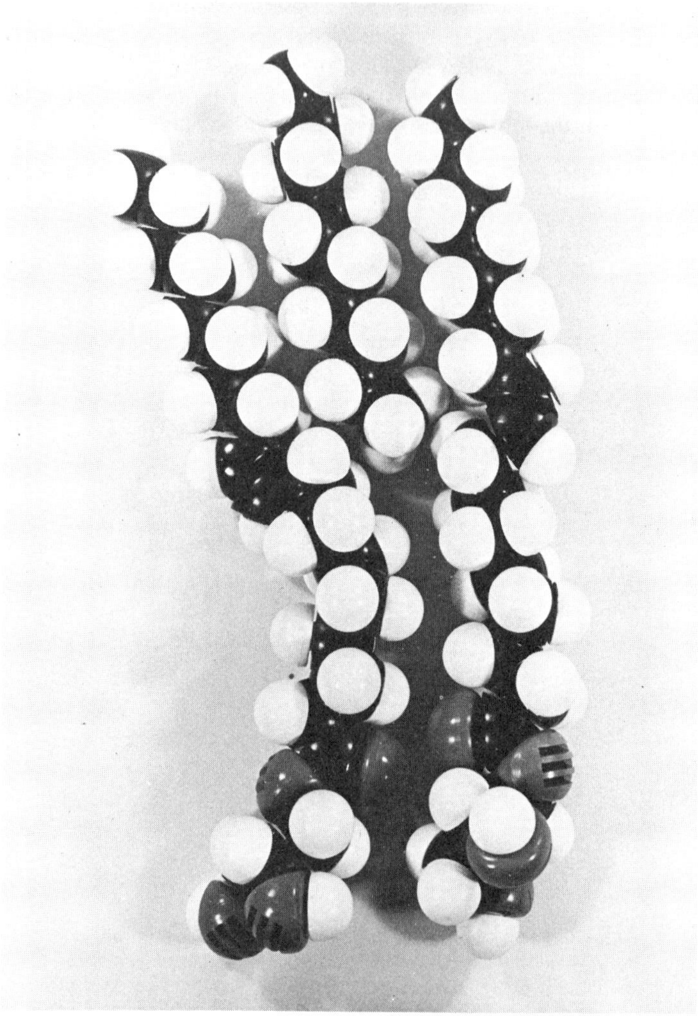

FIGURE 8. Corey-Pauling-Koltum models of *n*-decane and glycerol monooleate showing how decane might fit into the central zone of the bilayer. Because decane is shorter than *n*-hexadecane, full extension of some of the acyl chains is possible and provides enough space to accommodate a large volume fraction (about 0.5) of decane. The thickness of the bilayer is about twice the fully extended length of the acyl chain.

except that the alkane chains are shorter and can occupy a large volume fraction without occupying interfacial area or causing the terminal methyls to lose contact. The larger volume fraction (interactive volume) is achieved by having the acyl chains spend much more time in fully extended configurations. Decane bilayers, in fact, have a thickness about equal to twice the fully extended length of the acyl chain. Andrews *et al.*[29] point out that only a small percentage of the acyl chains need to be extended at any instant of time to maintain this thickness. It can be imagined that the acyl chains are whipping back and forth between contracted and extended states. As this whipping occurs, there will be a continual exchange between space occupied by acyl chains and space occupied by alkanes.

Now consider why the terminal methyls on opposing monolayers must make frequent contact at the bilayer midplane. That is, why cannot the membrane absorb arbitrarily large amounts of alkane by growing indefinitely in thickness? As discussed by White[20] and Andrews *et al.*,[29] a compressive force acts across the bilayer due to the attractive van der Waals force between the separated aqueous phases (see Ref. 35 for an excellent discussion of this force). If the bilayer were much thicker than 2ℓ, this compressive force would cause the excess alkane to flow between the monolayers toward the annulus and microlenses until the acyl chains of the opposing monolayers came into contact and presented a significant repulsive force. Until the acyl chains meet, the only opposing force is the transient one of viscous flow of the excess alkane between the monolayers. This is precisely why the planar bilayers thin from an initially thick film in the first place.

The Position of the Acyl Chain Double Bond Affects the Interactive Volume

The acyl chains of the bilayer are, on the average, bent, causing the methyl end of the chain to have a tendency to run parallel to the bilayer. A bilayer containing little solvent has a thickness slightly greater than the fully extended length of a single acyl chain. This means that the statistical bend will occur in approximately the middle of the chain. It was proposed above that the alkanes interact predominantly with the terminal methyl half of the chain. The double bond of the glycerol monooleate is located at carbon number 9, which places it directly in the center of the chain. Since the double bond is *cis*, it favors nicely the statistical bend in the center of the chain. Thus, the alkane molecules probably spend most of the time in the central region of the bilayer defined by the double bonds of the acyl chains of the opposed monolayers. This suggests that the location of the double bond may have an effect on the way the alkanes can occupy the interactive volume of the bilayer. The data of Figures 3 and 5 indicate that glycerol monooleate bilayers have a maximum thickness when *n*-decane is used to form the bilayer. Bilayers formed from *n*-octane are slightly thinner. Requena *et al.*[36] have made a similar observation. Preliminary measurements in my laboratory indicate that cyclodecane and 2,2,3,3, tetramethylhexane also give thinner membranes than *n*-decane. Requena and Haydon[18] report that membranes formed from 2,2,4 trimethylpentane, cyclohexane, and *cis*-5-decene also yield thinner membranes. It thus appears that glycerol monooleate bilayers have a *maximum* thickness for *n*-decane. For alkyl solutes longer or shorter than decane, the membranes decrease in thickness. This effect might be due to the length of the alkane relative to the double-bond position. Note that in Figure 8 the decane molecule precisely spans the distance between carbon 9 (the double bond) and the methyl group. A reasonable hypothesis is that if the alkane can precisely span the double bond-methyl group distance, it is possible for the acyl chain to spend some time fully extended in the presence of the alkane without generating free volume. If the

TABLE 2

PHYSICAL PROPERTIES OF PLANAR BILAYERS FORMED FROM ISOMERS OF GLYCEROL MONOOLEATE AND HEXADECANE* AT 30°C

Isomer	Bilayer Thickness (Å)	ΔH (kcal/mol)	ΔS (cal/deg/mol)
18:1(Δ6)†	31.9 ± 0.2	3.01 ± 0.19	10.02 ± 0.57
18:1(Δ9)‡	34.0 ± 0.2	4.03 ± 0.10	13.25 ± 0.36
18:1(Δ11)§	32.0 ± 0.2	3.66 ± 0.22	12.15 ± 0.65

*ΔH and ΔS have same meaning as in TABLE 1.
†Glycerol monopetroselinin.
‡Glycerol monooleate.
§Glycerol monovaccenin.

alkane is shorter than the double bond-methyl distance, the acyl chain cannot spend much time fully extended without generating energetically unfavorable free volume. Therefore, the acyl chains do not extend as fully with the shorter alkanes.

To examine the role of double-bond position further, membranes were formed from *n*-hexadecane and positional isomers of glycerol monooleate. TABLE 2 shows a comparison of glycerol monooleate (18:1, Δ9) with glycerol monovaccenin (18:1, Δ11) and glycerol monopetroselinin (18:1, Δ6). When the double bond is shifted to either position 6 or 11, the membrane becomes thinner. It seems that the maximum interactive volume is achieved when the double bond is located in the center of the acyl chain. This is consistent with the work of Barton and Gunstone,[37] who have shown that the bilayer phase transition temperature (T_c) of dioctadecenoyl lecithins is minimum when the double bond is at position 9. For example, $T_c = -21$°C for Δ9, but for Δ3 and Δ15, $T_c = +35$°C. Maximum internal fluidity is obtained when the double bond is in the center of the chain. This makes it easier to insert alkanes into the interior.

THERMODYNAMICS OF ALKANE SOLUBILITY

Not only does the thickness (and therefore the composition) of the bilayer depend upon alkane structure, it also depends strongly on temperature.[19,20] This fact is indicative of a nonideal bilayer solution and permits important thermodynamic information about the interaction of alkanes with the bilayer to be obtained.[21]

The temperature-dependence of the solubility of solid, immiscible liquid and gaseous solutes in liquids provides a means of calculating the entropy (ΔS) and enthalpy (ΔH) of solution.[25] Consider, for example, a saturated solution in which excess solid solute is present. At equilibrium, the free energy of solute in the pure solid must be the same as the partial molal free energy (chemical potential) of the solute in the saturated solution; that is, the free energy of transfer from solid to solution is zero at equilibrium. Under these circumstances, the ΔH and ΔS of transfer from solid to solution are easily calculated.[25] The planar bilayer is completely equivalent to this system; it is saturated with alkane because the bilayer is surrounded by an annulus (Plateau-Gibbs border) of the bulk solution of lipid in alkane. The alkane is present in great excess in the annulus and can be treated for our purposes as a reservoir of pure alkane. The proof that the bilayer is saturated comes from the observation that numerous microlenses (dia. ~1 μM) of excess solvent reside in the plane of the bilayer.[26,27] Immediately after formation, the "black" film or bilayer appears completely smooth and structureless in reflected light (FIGURE 9A). Within a few minutes, however, bright "pinpoints" appear which become fully developed in 10–20 minutes (FIGURE 9B). These "pin points" are the microlenses, which are thick relative to the bilayer and scatter incident light. Exceedingly careful measurements of specific capacitance show a slight thinning of the membrane that follows the time course of appearance of microlenses. In general, however, the microlenses have less than a one-percent effect on the measured specific capacitance.[18,23,28]

Thus, the planar bilayer is saturated with alkane. The microlenses and annulus are excess solute and are completely equivalent to the excess solid found in saturated solutions of solid in liquid described earlier. If the alkane of the microlenses (M) and annulus (A) is in equilibrium with the bilayer (BI), then the enthalpy and entropy of transfer of alkane to the bilayer can be calculated by the following standard equations:[25]

$$\Delta H = \overline{H}_{BI} - H_{M,A} = -R\left[\frac{\partial \ln X_s}{\partial(1/T)}\right]_{SAT,P} \tag{3}$$

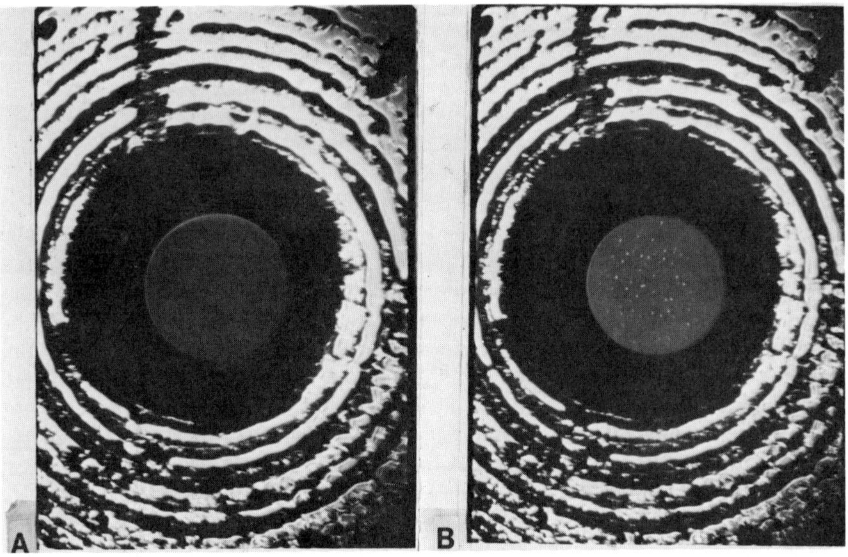

FIGURE 9. Reflected-light photographs of a planar bilayer formed from glycerol monooleate in heptamethylnonane. A: The membrane immediately after thinning. Note the smooth structureless surface. B: The membrane about twenty minutes after thinning. Numerous microlenses of excess alkane can be seen in the surface. These microlenses indicate that the bilayer is saturated with the alkane. The bright concentric rings surrounding the bilayer are due to the end-mill used in the machining of the septum.

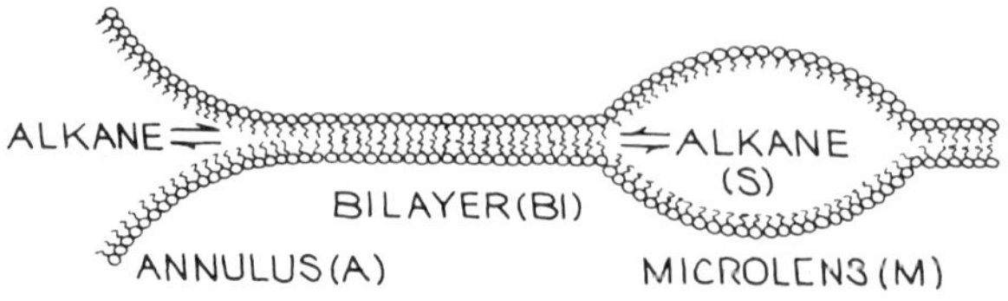

AT SATURATION EQUILIBRIUM, $\bar{F}_{BI} = F_{M,A}$

$$\Delta H = \bar{H}_{BI} - H_{M,A} = -R\left[\frac{\partial \ln X_S}{\partial(1/T)}\right]_{SAT,P}$$

$$\Delta S = \bar{S}_{BI} - S_{M,A} = R\left(\frac{\partial \ln X_S}{\partial \ln T}\right)_{SAT,P}$$

$X_S \equiv$ MOLE FRACTION OF ALKANE IN BILAYER

FIGURE 10. Schematic drawing summarizing the structure and thermodynamics of planar bilayers saturated with alkane. At equilibrium, the partial molal free energy of the alkane in the bilayer (F_{BI}) must be the same as in the bulk alkane of the annulus and microlenses ($F_{M,A}$). The equations give the enthalpy (ΔH) and entropy (ΔS) of transfer *from* bulk *to* bilayer, assuming the activity of the alkane is a linear function of concentration, which is probably a reasonable assumption at saturation. The mole fraction of alkane solute (X_S) can be calculated from measurements of specific geometric capacitance and a knowledge of the interfacial concentration of the stabilizing lipid molecule (see text).

$$\Delta S = \bar{S}_{BI} - S_{M,A} = R\left(\frac{\partial \ln X_S}{\partial \ln T}\right)_{SAT,P} \tag{4}$$

The derivation of these equations assumes that the activity of the alkane solute in the bilayer is a linear function of concentration. In these equations, R is the gas constant, T the temperature in °K, and X_S is the mole fraction of solute in the saturated bilayer. $\bar{H}_{BI}$ and $\bar{S}_{BI}$ are, respectively, the partial molal enthalpy and entropy of the alkane solute in the bilayer, whereas $H_{M,A}$ and $S_{M,A}$ are the values in the bulk phases of microlenses and annulus. ΔH and ΔS give, therefore, the enthalpy and entropy of transfer from bulk to bilayer. At saturation equilibrium, the free energy of transfer $\Delta F = F_{BI} - F_{M,A}$ is zero, so that $\Delta H = T\Delta S$. FIGURE 10 summarizes the thermodynamics of the alkane-bilayer interaction. The beauty of this approach is that a *direct* thermodynamic comparison of the interior of the bilayer with the interior of the bulk alkane is obtained.

Procedurally, ΔS and ΔH are found from the slopes of plots of $\ln X_S$ vs. $\ln T$ and $(1/T)$. The only problem is to determine X_S, which is easily done from the measurements of specific geometric capacitance and knowledge of the concentration of the surface active lipid in the bilayer.[10,11,18,21] It is absolutely essential that the planar bilayer-annulus-microlens system (FIGURE 10) be in chemical equilibrium. That equilibrium exists can be easily verified, because the composition of the bilayer, and therefore the thickness and specific capacitance, will be invariant with time at equilibrium. Measurements as a function of time of films

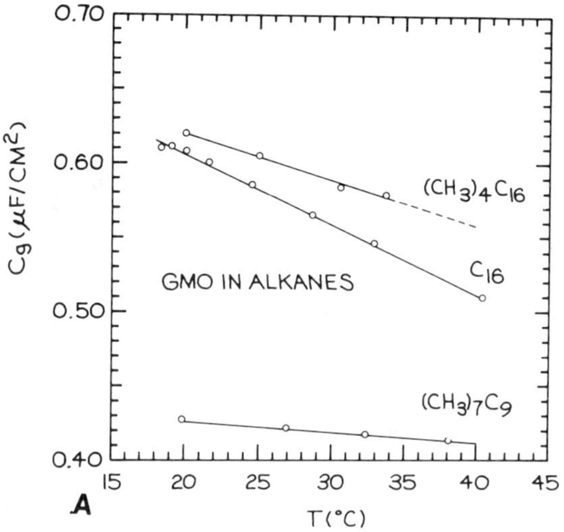

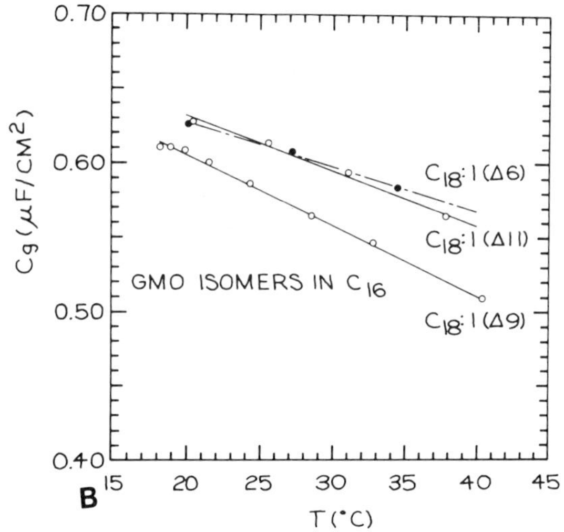

FIGURE 11. Plots of the specific geometric capacitance (C_g) as a function of temperature (T). The diameters of the data points represent the standard error of the mean ($\simeq$0.3%) of triplicate determinations on each of three membranes at each temperature. A: C_g of bilayers formed from 2,6,11,15-tetramethylhexadecane [$(CH_3)_4C_{16}$], *n*-hexadecane [C_{16}], and 2,2,4,4,6,8,8-heptamethylnonane [$(CH_3)_7C_9$]. B: C_g of bilayers formed from positional isomers of glycerol monooleate [C_{18}:1(Δ9)] and *n*-hexadecane.

formed from various lipids indicate that equilibrium is easily achieved in monoglyceride films[21] but not necessarily with films formed from lecithins.[28] The monoglycerides and shorter alkanes are slightly soluble in water,[21,29] and to have equilibrium the aqueous phase must be saturated with the bulk solution; otherwise the values of specific capacitance will be too high and time-dependent changes will be observed

Measurements of specific geometric capacitance (C_g) of bilayers as a function of temperature for glycerol monooleate in several related alkanes and for positional isomers of glycerol monooleate in *n*-hexadecane are shown in FIGURE 11. From these data, plots of $\ln X_S$ against $(1/T)$ and $\ln T$ were prepared as shown in FIGURE 12 for glycerol monooleate in various alkanes. A linear regression analysis on these plots, using Equations 3 and 4, yield the enthalpies and entropies of transfer (TABLES 1 and 2). The results are interesting. Consider first the data for glycerol monooleate in *n*-hexadecane (TABLE 1). The enthalpy of transfer from bulk hexadecane to saturated bilayer is +4.03 Kcal/mol and the entropy is +13.25 cal/mol/°K.‡ These values are exceptionally large. As pointed out elsewhere,[21] if the interior of the bilayer were equivalent to a simple alkyl liquid, the enthalpy would be only a few calories per mol. Therefore, the immediate and unequivocal conclusion is that the hexadecane molecules find themselves in an environment markedly different from a simple alkyl liquid. The large values of ΔH and ΔS mean that the hexadecane molecules are not bound as tightly in the bilayer as in bulk, and that they have more degrees of freedom. One explanation is that, in general, the cohesive forces in the bilayer are not as large as in bulk. The work of Gershfeld and Pagano[38] and Katz[39], however, indicates that the cohesive forces between acyl chains in monolayers and bilayers are the same as in bulk. The x-ray diffraction data of Wilkins *et al.*[3] are consistent with this conclusion. An alternative explanation was offered by White[21] and depends upon the idea outlined earlier that the alkane molecules are located largely (but not exclusively, see below) in the central zone of the bilayer. Numerous x-ray diffraction studies of bilayers (e.g. Ref. 3) reveal a zone of low electron density in the bilayer mid-plane due to the methyl (—CH_3) groups (FIGURE 6D illustrates this point). There is thus little interdigitation between the apposed monolayers, and a well-defined interface must exist between them. Because the methyl groups have a lower electronic polarizability than methylene groups, molecules located chiefly in the vicinity of the mid-plane should experience smaller van der Waals cohesive forces. Support for this hypothesis is given by Dean and Hayes,[40] who measured the heat of sorption of hexane vapor on close-packed stearic acid monolayers at the vapor-water interface. At low partial pressures of hexane (that is, low surface coverage by adsorbed hexanes), the heat of sorption is 4.5 Kcal/mol. At high partial pressures (high-surface coverage approaching closest packing of hexane on the stearic acid surface) the heat of sorption becomes identical to the heat of vaporization of liquid hexane (7.5 Kcal/mol). This indicates that the exposed —CH_3 surface of the steric acid does not attract hexanes as strongly as the fully covered hexane surface or bulk liquid.

‡These values are slightly different than previously reported[21] because of slightly different assumptions about how the area per acyl chains in the interface changes with temperature.

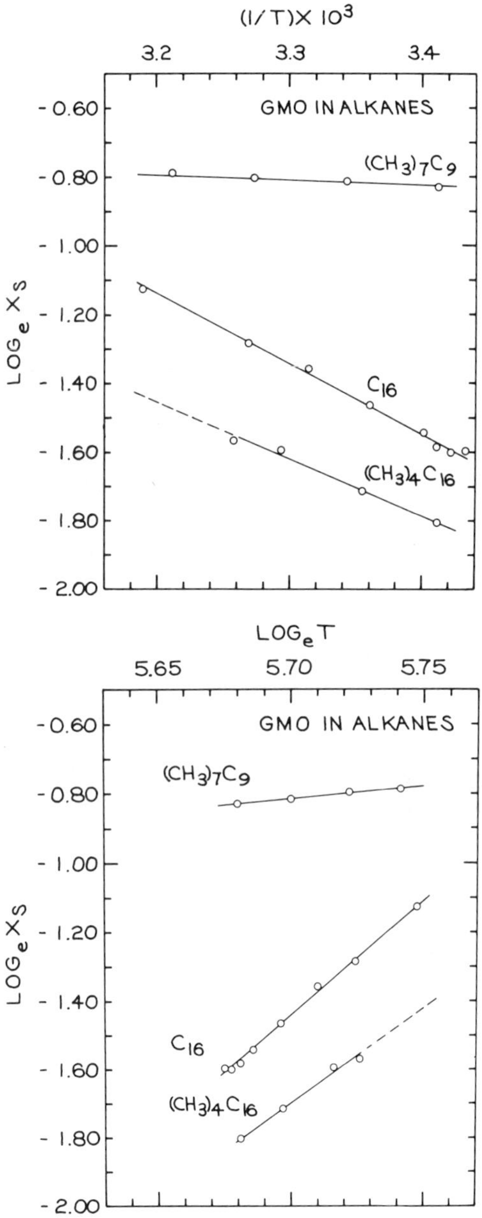

FIGURE 12. Plots from data of FIGURE 11A of $\log_e X_s$ against (A) 1/T and (B) $\log_e$T. The enthalpy and entropy of transfer, respectively, can be calculated from the slopes of these curves using Equations 3 and 4 (see also FIGURE 10).

An additional contribution to the large positive ΔH for hexadecane may come from the generation of free volume in the central zone. The insertion of hexadecane into the "slanted pockets" between the terminal methyl halves of the acyl chains probably creates some free volume because the acyl chains cannot flex enough to pack tightly about end of the hexadecane facing the interface (see FIGURE 7). This generation of free volume would increase the fluidity and therefore the entropy of the central zone. Thus, compared to bulk hexadecane where the alkane-alkane packing is tighter than the alkane-acyl chain packing of the bilayer, the cohesive energy density would be decreased. But this decrease would be compensated for by the increased entropy.

The enthalpy of transfer of heptamethylnonane into glycerol monooleate (TABLE 1) is only 0.37 Kcal/mol, compared to 4.03 Kcal/mol for hexadecane. This means that the heptamethylnonane finds an environment in the bilayer much closer to the bulk environment. This is consistent with measurements I have made on the shorter *n*-alkanes (C_{15} through C_8). The results of these measurements will be reported elsewhere, but the general conclusion is that as the length of the alkane decreases, ΔH decreases and, in fact, reverses sign at about *n*-decane or *n*-hendecane. One reason for this is the larger volume fraction (F) of the bilayer occupied by the shorter alkanes (e.g., for heptamethylnonane at 30°, F = 0.45 while for *n*-hexadecane, F = 0.27), which means more nearest neighbors of the alkanes will be alkanes rather than acyl methyl groups. This does not fully explain the difference, however, since tetradecane with F = 0.43 has a ΔH of 1.02 Kcal/mol (S.H. White, unpublished), about 3 times the heptamethylnonane value. The difference is probably due to the large number of $—CH_3$ groups on the heptamethylnonane. In bulk, this molecule is already in a $—CH_3$ rich environment and does not pack as tightly as tetradecane in bulk. Therefore the environment of the bilayer is closer to that of the bulk environment. There is also likely to be a free-volume effect for this molecule. Because the heptamethylnonane is more spherical than the neighboring acyl chains, it cannot pack as compactly in the bilayer as in bulk and consequently free volume is produced resulting in a lower cohesive energy density.

Now compare the tetramethylhexadecane data with the hexadecane data in TABLE 1. Note that ΔH is reduced to 3.20 Kcal/mol. The volume fraction of the bilayer occupied by the tetramethylhexadecane is 0.23 compared to 0.27 for hexadecane, so the lower ΔH cannot be explained by the volume fraction effect. The difference is probably due to the presence of the additional $—CH_3$ groups on the tetramethylhexadecane, which cause looser packing in bulk compared to hexadecane. The lower polarizability of the additional $—CH_3$ groups should also lower the bulk cohesive energy density compared to hexadecane. The fact that the melting point of tetramethylhexadecane is far lower than that of hexadecane supports this point of view.

The data of TABLE 2 for the bilayers formed from the glycerol monooleate positional isomers in *n*-hexadecane is more difficult to interpret. ΔH is significantly smaller for the Δ6 and Δ11 isomers, even though the volume fractions for these are smaller (0.22) than for the double bond at Δ9. The situation is not completely clear, but it may be due to a bimodal distribution of alkane. Even though the alkane is located largely in the central zone of the bilayer, there must also be some in the monolayers near the aqueous interface. Since the area per glycerol monooleate is about $39A^2$ (Ref. 10) and the cross-sectional area of the acyl chain is $22A^2$, it is possible to pack a small number of alkanes into the bilayer parallel to the acyl chain beginning at the interface, rather than at the center of the acyl chain. To accomplish this, it is only necessary for an occasional chain to be fully extended. These few alkane molecules would find an environment much more like the bulk liquid and would consequently have a lower ΔH and ΔS of transfer. I will report elsewhere that the maximum positive ΔH is achieved for *n*-hexadecane with the value decreasing slightly for *n*-heptadecane and *n*-octadecane. I believe this is because the length of the longer alkanes starts to favor insertion near the interface, rather than into the central zone. Not much can be packed in this way, of course, without increasing the free energy of the system due to an increase in the area per acyl chain. Essentially, in the general case, the central zone of the bilayer has degrees of freedom not found in bulk alkane, which favors the entry of alkane because of a positive ΔS paid for at the expense of an unfavorable positive ΔH. I propose that as the alkane chain length increases, its presence in the center of the bilayer becomes more unlikely stericly, and a more favorable enthalpy is achieved at the expense of entropy by having the alkane line up with acyl chains beginning at the interface. This cannot long continue without an unfavorable free-energy increase in the bilayer because of the interfacial changes. The amount of alkane at the interfacial zone is probably small, but it is not until the amount of alkane in the central zone is reduced that its effect on the enthalpy can be seen.

Now consider the problem of changing the location of the double bond. As the double bond is shifted to either end of the acyl chain, the acyl chain behaves more and more like an unsaturated chain, as evidenced by the thermal studies of Barton and Gunstone.[37] I believe this decreases the interactive volume and therefore favors the shifting of the alkane chains toward the interface. The increased fraction of alkane in the interfacial zone may account for the decrease in ΔH and ΔS observed for hexadecane in the two isomers of glycerol monooleate. This effect will probably not be observed for the short alkanes, because the amount of alkane in the central zone must be reduced far enough to permit the lowered ΔH of the alkane in the interfacial zone to be observed. Also, the alkane chain must be long enough to achieve a favorable enthalpy of interaction with the acyl chain. Requena *et al.*[36] report that the surface pressure of glycerol monooleate at the alkane/water interface increases as the length of the alkane increases. The insertion of a few alkanes into the absorbed monolayers should cause such an increase,[24] provided the area per glycerol monooleate in the interface does not increase. The increase in surface pressure with increasing alkane chain length is consistent with the idea that the thermodynamics favor the entrance of the longer alkanes into the interfacial zone.

Conclusions

This research represents the first attempt of my laboratory to explore systematically the solution properties of bilayers using alkane molecules as solutes. The data clearly support the notion that the interior of the bilayer is not simply equivalent to a bulk alkyl hydrocarbon, since the enthalpy of transfer from bulk to bilayer for hexadecane is greater than 3 Kcal/mol. The solubility of alkanes depends primarily upon the length of the alkane and only secondarily upon molecular volume. The variation in solubility suggests that the bilayer has an interactive or potential volume available for interactions with other molecules roughly equivalent to the volume of the acyl chains. This potential volume can become occupied without significant changes in interfacial organization. It appears that the interactive volume is maximized by having the double bond in the center of the octadecenoyl acyl chain. It is suggested that the alkane molecules occupy primarily the central zone of the bilayer but may also occupy the interfacial zone to a limited extent, with the longer alkanes having a greater likelihood of occupying the interfacial zone than the shorter alkanes. In the central zone, it is suggested that the presence of the terminal methyl groups of the acyl chains and the greater disorder found there give rise to the positive enthalpy and entropy observed for alkanes of lengths greater than eleven carbons. The qualitative analysis of the data is consistent with having the two monolayers of the bilayer act independently; it appears unlikely that a single alkane molecule spans the bilayer midplane and mixes simultaneously in both monolayers.

Perhaps the single most important conclusion from this work is that alkane molecules do not distribute uniformly through the thickness of the bilayer. That is, the bilayer is physicochemically anisotropic normal to the plane of the bilayer. Therefore, membrane solutions are unlikely to be true two-dimensional solutions in which thickness is an irrelevant variable. The bilayer also has properties unobserved in three-dimensional solutions, and it therefore seems reasonable to refer to membranes as quasi-two-dimensional solutions.

Acknowledgments

I am pleased to acknowledge the outstanding technical help of Mr. John Lowe and the support of my wife, Buff, and our six children.

References

1. Branton, D. 1966. Proc. Nat. Acad. Sci. USA **55:** 1048–1056.
2. Steim, J. M., M. E. Tourtellotte, J. C. Reinert, R. N. McElhaney & R. J. Roder. 1969. Proc. Nat. Acad. Sci. USA **63:** 104–109.
3. Wilkins, M. H. F., A. E. Blaurock & D. M. Engleman. 1971. Nature [New Biol.] **230:** 72–76.
4. Wallach, D. F. H. & P. H. Zahler. 1966. Proc. Nat. Acad. Sci. USA **56:** 1552–1559.
5. Frye, L. D. & M. Edidin. 1970. J. Cell Sci. **7:** 319–335.
6. Bretscher, M. S. 1971. Nature [New Biol.] **231:** 229–232.
7. Lenard, J. & S. J. Singer. 1966. Proc. Nat. Acad. Sci. USA **56:** 1828–1835.
8. Singer, S. J. & G. L. Nicolson. 1972. Science **175:** 720–731.
9. Tomita, M. & V. T. Marchesi. 1975. Proc. Nat. Acad. Sci. USA **72:** 2964–2968.
10. Fettiplace, R., D. M. Andrews & D. A. Haydon. 1971. J. Membrane Biol. **5:** 277–296.
11. White, S. H. 1975. Biophys. J. **15:** 95–117.
12. Mueller, P., D. O. Rudin, H. T. Tien & W. C. Wescott. 1962. Circulation **26:** 1167–1170.

13. WHITE, S. H., D. C. PETERSEN, S. SIMON & M. YAFUSO. 1976. Biophys. J. **16:** 481–489.
14. TIEN, H. T. 1974. Bilayer Lipid Membranes (BLM). Marcel Dekker. New York, N.Y.
15. FETTIPLACE, R., L. G. M. GORDON, S. B. HLADKY, J. REQUENA, H. P. ZINGSHEIM & D. A. HAYDON. 1975. Techniques in the formation and examination of "black" lipid bilayer membranes. *In* Methods in Membrane Biology. Vol. **4**. E. D. Korn, Ed.: 1–75. Plenum Press. New York, N.Y.
16. HANAI, T., D. A. HAYDON & J. TAYLOR. 1964. Proc. R. Soc. (London) **A281:** 377–391.
17. WHITE, S. H. & D. N. BLESSUM. 1975. Rev. Sci. Instrum. **46:** 1462–1466.
18. REQUENA, J. & D. A. HAYDON. 1975. Proc. R. Soc. Med. **A347:** 161–177.
19. WHITE, S. H. 1970. Biochim. Biophys. Acta **196:** 354–357.
20. WHITE, S. H. 1970. Biophys. J. **10:** 1127–1148.
21. WHITE, S. H. 1976. Nature **262:** 421–422.
22. SEGREST, J. P., R. L. JACKSON & V. T. MARCHESI. 1972. Biochem. Biophys. Res. Commun. **49:** 964–969.
23. WHITE, S. H. 1974. Biochim. Biophys. Acta **356:** 8–16.
24. FOWKES, F. M. 1962. J. Phys. Chem. **66:** 385–389.
25. HILDEBRAND, J. H., J. M. PRAUSNITZ & R. L. SCOTT. 1970. Regular and Related Solutions. Van Nostrand Reinhold Co. New York, N.Y.
26. ANDREWS, D. M. & D. A. HAYDON. 1968. J. Molec. Biol. **32:** 149–150.
27. HENN, F. A. & T. E. THOMPSON. 1968. J. Molec. Biol. **31:** 227–235.
28. WHITE, S. H. & T. E. THOMPSON. 1973. Biochim. Biophys. Acta **323:** 7–22.
29. ANDREWS, D. M., E. D. MANEV & D. A. HAYDON. 1970. Spec. Discus. Faraday Soc. No. **1:** 46–56.
30. HUBBELL, W. L. & H. M. MCCONNELL. 1969. Proc. Nat. Acad. Sci. USA **64:** 20–27.
31. LEVINE, Y. K., N. J. M. BIRDSALL, A. G. LEE & J. C. METCALFE. 1972. Biochemistry **11:** 1416–1421.
32. VANDENHEUVEL, F. A. 1968. Chem. Phys. Lipids **2:** 372–395.
33. PHILLIPS, M. C., R. M. WILLIAMS & D. CHAPMAN. 1969. Chem. Phys. Lipids **3:** 234–244.
34. WHITE, S. H. 1973. Biochim. Biophys. Acta **323:** 343–350.
35. PARSEGIAN, V. A. 1973. Ann. Rev. Biophys. Bioengineer. **2:** 221–255.
36. REQUENA, J., D. F. BILLETT & D. A. HAYDON. 1975. Proc. R. Soc. Med. **A347:** 141–159.
37. BARTON, P. G. & F. D. GUNSTONE. 1975. J. Biol. Chem. **250:** 4470–4476.
38. GERSHFELD, N. L. & R. E. PAGANO. 1972. J. Phys. Chem. **76:** 1231–1237.
39. KATZ, Y. 1976. Biophys. J. **16:** 52a.
40. DEAN, R. B. & K. E. HAYES. 1952. J. Am. Chem. Soc. **74:** 5982–5984.
41. SIMON, S. A., W. L. STONE & P. BUSTO-LATORRE. 1976. Biophys. J. **16:** 137a.

ELECTRICAL CHARACTERISTICS OF ION TRANSPORT IN LIPID BILAYER MEMBRANES*

Gabor Szabo

Department of Physiology and Biophysics
University of Texas Medical Branch
Galveston, Texas 77550

INTRODUCTION

Bilayer membranes, routinely formed from a variety of lipids in pure aqueous solutions of common electrolytes such as Na^+ or Cl^-, behave as insulators ($R_m > 10^7\ \Omega cm^2$) despite their molecular thinness (~ 50 Å). This is because the low dielectric constant membrane interior, composed of apposing hydrocarbon tails of lipid molecules, imposes a large energy barrier against the movement of small hydrated ions into the membrane interior.[1,2] Selective translocation of ions across the bilayer may nevertheless occur by one of three well-known mechanisms.[4] The simplest of these operates for large hydrophobic ions for which the energy of transfer across the membrane interior is not exceedingly unfavorable.[5] These hydrophobic ions, which often adsorb at the surface of the bilayer, can therefore move across the bilayer directly, without the assistance of carriers or polar channels in the membrane. The somewhat more complex carrier-mediated transport is similar to direct ion transport in that a hydrophobic ion, formed by association of a hydrophobic carrier molecule with a hydrophilic ion, is the permeant species. However, the ion-carrier reaction, which allows the selective transport of otherwise impermeant ions,[6] introduces complexities into the behavior of this transport mechanism.[7] The presence of polar heterogeneities or channels through the bilayer has also been observed to enhance ionic permeability by providing an energetically more favorable pathway fixed within the membrane for the movement of ions.

Detailed electrical properties of ion-conductive lipid bilayers depend strongly on the type of transport mechanism operative in the membrane.[8] This paper concentrates on examining some of the salient features of the conductive properties induced in lipid bilayer membranes by the hydrophobic anion tetraphenylborate (TFB) not only because unambiguous interpretation of the results is possible for this simple conduction mechanism, but also because direct ion translocation appears to be a good model for gating currents observed in nerve and muscle membranes.[9,10,11] It is predicted and experimentally verified that the small-signal TFB-induced electrical response varies as a function of the electrical field set up in the membrane by applying a steady holding potential (bias). These small signal properties, as well as the nonlinear large-signal properties of the membrane, are predicted by an extension of a simple kinetic model that previously was found to describe the electrical behavior of the membrane in the absence of an electrical field.[12-14]

*Supported by USPHS grant HL-19639.

The last section extends these conclusions to spontaneous current fluctuations elicited in bilayers by TFB. Characteristics of these current fluctuations are compared with fluctuations seen for channel and carrier mediated ion transport.

Kinetics of Hydrophobic Ion Transport

Hydrophobic ions such as TFB, dipicrylamine, and cyanine dyes induce in lipid bilayer membranes an electrical response characterized by an exponential decay of the membrane current in response to a voltage jump under potentiostatic conditions.[12,15] A kinetic scheme in which the permeant ion is adsorbed at the membrane surface and moves through the membrane interior with voltage-dependent rate constants k' and k'' was found to describe the step-change response.[12] This scheme can be summarized as follows:

$$I' \longleftrightarrow I_0' \underset{k_d}{\overset{k_a}{\rightleftarrows}} I_0'^{*} \underset{k''}{\overset{k'}{\rightleftarrows}} I_d''^{*} \underset{k_a}{\overset{k_d}{\rightleftarrows}} I_d'' \longleftrightarrow I''. \qquad (1)$$

diffusion adsorption translocation adsorption diffusion

The rate constants k' and k'' may be related to the physical properties of the membrane by deriving these in terms of electrodiffusion across the nonpolar energy barrier of the membrane interior.[3,12,13] Such derivations, as well as available experimental data, indicate that the voltage dependence of k' and k'' may be described for symmetrical membranes by the following equations:

$$k' = kf(u)\exp(-\mathrm{nu}) \qquad k'' = kf(u)\exp(\mathrm{nu}) \qquad (2)$$

where $u = zFV/RT$ is the normalized potential, n is an experimentally determined constant, $f(u)$ is an experimentally determined function, while k is the translocation rate for the permeant ion in the limit of small applied potentials.[8,13,14]

Strongly hydrophobic permeant ions are expected to partition at the membrane-solutions interface. For symmetrical membranes bathed by solutions of identical composition, the concentrations N' and N'' (moles/cm^2) of the hydrophobic ion adsorbed at surfaces $'$ and $''$ of the membrane are expected to be equal and related to their aqueous concentration by the adsorbtion coefficient β.

$$N'_{\mathrm{eq}} = N''_{\mathrm{eq}} = \frac{k_a}{k_d}c = 2\beta c. \qquad (3)$$

β may be viewed as the thickness (in cm) of the aqueous layer that contains an amount of permeant ion equal to N_{eq}.

For a strongly partitioned permeant ion such as TFB, absorption and desorption are limited by diffusion within the unstirred aqueous layers that adhere to the membrane surfaces.[8,14] Thus, the total amount, $N_T = N' + N''$, of TFB in the membrane remains unaltered for short times (e.g., $t < 10$ msec) despite the fact that an applied potential displaces the surface concentrations N' and N'' from their equilibrium value. Thus TFB is virtually trapped at the membrane surfaces so that

$$N'_{(t)} + N''_{(t)} \simeq N'_{\mathrm{eq}} + N''_{\mathrm{eq}} = \beta c. \qquad (4)$$

It should be noted that, at longer times, when the system approaches steady state, $N'_{(t)} + N''_{(t)}$ need no longer be time-independent. We will not concern ourselves with such slow (e.g. time scale of seconds) kinetic processes described elsewhere.[8,16]

Small-Signal Electrical Response

The influence of the initial holding potential, V_h, on TFB-induced electrical properties of bilayer membranes has not been explored, V_h being usually held at its equilibrium value ($V_h = 0$ for membranes separating identical aqueous solutions). Can V_h, (e.g., the electrical field within the membrane) alter the magnitude and time course of TFB induced membrane currents?

Intuitively, V_h should have an effect on membrane response, since large V_h's would transfer much of the negatively charged TFB to that side of the membrane

TABLE 1
INFLUENCE OF A HOLDING POTENTIAL ON THE SMALL-SIGNAL SQUARE-PULSE RESPONSE INDUCED IN LIPID BILAYER MEMBRANES BY 2×10^{-7} M TFB, $\Delta V = 10$ mV*

V_H mV	I_0 $\mu A/cm^2$	τ msec	Q nC/cm^2	R $k\Omega\ cm^2$	C $\mu F/cm^2$	R^{CAL} $k\Omega\ cm^2$	C^{CAL} $\mu F/cm^2$
0	8.2	0.66	5.4	1.2	0.54	1.3	0.48
10	8.4	0.57	4.8	1.2	0.48	1.3	0.45
25	8.0	0.50	4.0	1.3	0.40	1.4	0.41
50	5.7	0.41	2.3	1.7	0.24	1.8	0.26
75	3.7	0.33	1.2	2.7	0.12	2.7	0.13
100	2.3	0.25	0.58	4.3	0.070	4.2	0.064
125	1.4	0.20	0.28	6.9	0.028	6.8	0.029

*R^{CAL} and C^{CAL} were calculated by equation 10A with $\beta = 1.5$ mm; $k = 808\ sec^{-1}$; $n = 0.42$; and $\xi = 0.009$.

surface which is positive. A small test pulse ΔV would then result in a relatively small TFB current, since nearly all of the TFB is held at the positive membrane surface by the large holding potential and is thereby unable to move. This intuitive expectation is observed experimentally. TABLE 1 shows the effect of different V_h's on membrane current elicited by a $\Delta V = 10$ mV displacement of the membrane potential for lipid bilayers formed from cholesterol-monoolein mixtures (5:1 molar ratio; lipids dissolved in decane at 25 mM total concentration). The aqueous solutions contained 2×10^{-7} M TFB with 0.1 M NaCl serving as the indifferent electrolyte. Normally, the membrane potential was kept at $V_0 = 0$. Holding potentials V_h were applied only for sufficiently long times (typically 100 msec) to insure complete decay of current transients but not long enough to disturb the concentration profile of TFB in the aqueous layers near the membrane surface.†

†See Reference 8 for a comprehensive discussion of these conditions.

Test pulses $\Delta V = 10$ mV were superimposed upon V_h while recording digitally the resulting current transient. The initial time course of the current transient $I(t)$ was always exponential. This was evidenced by the experimental data falling on a straight line on an ln $I(t)$ vs. time plot. The membrane response could be characterized from such plots by an initial current I_0, calculated from the zero-time intercept, and by a time constant τ, calculated from the slope. I_0's and τ are listed respectively in the second and third column of TABLE 1 for different values of V_h. Relatively small ($< 10^{-2} I_0$ for $\Delta V = 10$ mV) but significant residual TFB currents were observed following the decay of the main exponential current transient. These residual decays had a slow time course (half time of 10 seconds) and followed a $1/\sqrt{t}$ dependence previously ascribed to diffusion of TFB within the aqueous unstirred layers. During the time course of the holding (V_h) and test (ΔV) pulses, typically shorter than 100 msec, residual currents were practically constant. The residual currents were routinely subtracted from the current records. This procedure greatly improved the quality of exponential fit at longer times ($t > 2\tau$) without altering significantly I_0's and τ's obtained from initial segments of $I(t)$ records for which exponential fit was good without any correction.

TABLE 1 shows that I_0, τ, as well as the transferred charge $Q = I_0\tau$, decrease with increasing holding potential. This effect is symmetrical about $V_h = 0$ for positive and negative V_h's. Because of the exponential response, the membrane could be formally represented by an electrical equivalent composed of a resistor, R, and a capacitor, C, connected in series. Equivalent R's and C's are listed in columns 5 and 6 of TABLE 1. The dependence of R and C on V_h is closely predicted by scheme 1, for which APPENDIX A calculates R's and C's as a function of four parameters: β = the coefficient of adsorption for TFB at the membrane surface; k = the intrinsic translocation rate of TFB across the membrane interior, and the two empirical constants n and ξ, which simply prescribe the voltage dependence of the translocation rates k' and k'', as specified by Equation 7A of the APPENDIX. When the translocation rates k' and k'' are derived within the framework of the Nernst-Planck diffusion equation, $1 - 2n$ can be interpreted as the fraction of the voltage drop that is effective in driving TFB across the membrane interior. Similarly, ξ may be related to the expected effect of image forces acting on TFB during its transit through the membrane interior.[3,8,14] A nonlinear least-square procedure gave $\beta = 0.0015$, $k = 808$, $n = 0.42$, $S = 0.009$ as the parameters that yield the best fit of Equations 10A to the experimental R's and C's. Inspection of TABLE 1 shows good agreement between experimental and fitted R's and C's. Thus the simple model, based on Equations 2 through 5, adequately describes the small-signal step response induced in bilayers by TFB.

It is shown in the APPENDIX that the small-signal series RC equivalent of the membrane response is expected to model accurately the membrane response to an arbitrary but small ($|V(t)| < 25$ mV) time-variant membrane potential, $V(t)$, superimposed on a constant holding potential, V_h. Such linear membrane response is best characterized by the electrical admittance, Y, of the membrane.

The right-hand side of FIGURE 1 shows admittance expected for the electrical analog of a bilayer, composed of a TFB-induced series RC element connected in parallel with the intrinsic membrane capacitance of the membrane, observable in the absence of TFB. The calculated admittances also take into account the

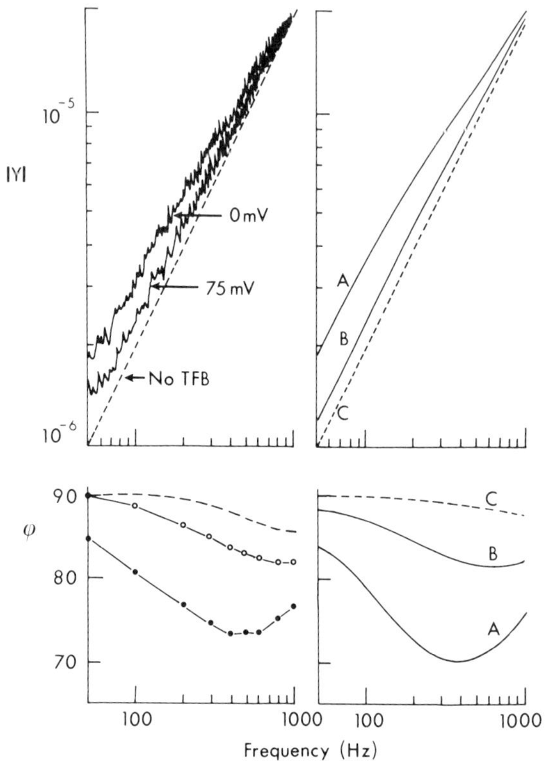

FIGURE 1. Magnitude and phase of the electrical admittance for lipid bilayers in the presence of 2×10^{-7} M TFB. *Left:* Experimental data. *Right:* Theoretical expectations calculated from data of TABLE 1. Membrane area: 6.1 mm^2.

series electrolyte resistance ($R_s = 2 \times 10^3\ \Omega$ for 0.1 M NaCl and a membrane of $6.1 \times 10^{-3}\ cm^2$ area). The upper part shows the magnitude $|Y|$ and the lower part shows the phase (ϕ) of the admittance calculated for $V_M = 0$ and 75 mV using Equations 10A and β, k, n, ξ determined for TABLE 1.

The left-hand pannels of FIGURE 1 compare the observed magnitude (upper part obtained by applying a white-noise signal to the bilayer) and phase (lower part obtained by measuring the phase of the membrane current, relative to sinusoidal $V(t)$) of the admittance for $V_h = 0$ and 75 mV. Admittance and phase in the absence of TFB for the geometrical membrane capacitance and electrolyte series resistance are shown by dashed lines. An agreement is seen between the predicted and observed admittance of the membrane, despite the fact that no adjustable parameters were used in the calculation of the expected membrane response.

Both the square-pulse and admittance measurements indicate that the linear small-signal response of the membrane depends strongly on the holding potential, V_h, and that the effect of V_h can be understood in terms of the simple translocation scheme 1. The next section shows that similar conclusions can be reached for the large-signal response of the membrane as well.

Large-Signal Electrical Response

Although the electrical response is no longer linear, measurements reported here and elsewhere[14] show that TFB-mediated membrane currents decay exponentially not only for small- ($\Delta V < 25$ mV), but also for large-step changes of the membrane potential ($25 \text{ mV} < |\Delta V| < 200$ mV). The time constant of the decay, however, changes as a function of ΔV; furthermore, I_0 is not proportional to ΔV so that the response of the membrane is strongly nonlinear. TABLE 2, which shows I_0's and τ's measured for large ΔV's under conditions identical to those of TABLE 1, illustrates typical nonlinear large-signal responses for a holding potential $V_h = -100$ mV. The time constant, τ, the initial current, I_0, as well as the charge transferred across the membrane, $\Delta Q = I_0 \tau$, are seen to vary steeply with ΔV. Thus, the apparent large-signal membrane capacitance, $C = Q/\Delta V$, varies as a function of ΔV, as illustrated in the fifth column of TABLE 2.

It is shown in the APPENDIX that the large-signal step response, as well as the small-signal response, should be directly predictable from the four transport parameters k, β, n and ξ, provided that the transport model of scheme 1 is valid. Since the four transport parameters have been determined from the small-signal membrane properties, the large-signal response should be completely predictable from these. The agreement between observed I_0', τ's and ΔQ's and those predicted from the small-signal membrane properties of TABLE 1, using Equation 11A (last three columns of TABLE 2), shows that the kinetic model accurately describes the membrane response even for large signals and large holding potentials.

The reasons for the nonlinear change transfer are not difficult to understand. A large holding potential will transfer nearly all of the negatively charged, adsorbed TFB ions to that surface of the membrane which is positive. A large test pulse of the same polarity will therefore not be able to transfer any more TFB, so that $\Delta Q \sim 0$. A large test pulse of the opposite polarity, however, will be able to transfer much of the TFB to the opposite surface of the membrane now made

TABLE 2

LARGE-SIGNAL SQUARE-PULSE RESPONSE INDUCED IN LIPID BILAYER MEMBRANES BY 2×10^{-7} M TFB AT A HOLDING POTENTIAL OF -100 MV*

ΔV mV	I_0 $\mu A/cm^2$	τ msec	Q nC/cm^2	C $\mu F/cm^2$	I_0^{CAL} $\mu A/cm^2$	τ^{CAL} msec	Q^{CAL} $\mu F/cm^2$
25	6.6	0.38	2.5	0.10	6.8	0.36	2.4
50	14	0.51	7.3	0.15	15	0.47	7.1
75	25	0.61	15	0.20	27	0.58	16
100	40	0.66	27	0.27	44	0.62	27
125	61	0.63	38	0.30	67	0.58	39
150	93	0.51	48	0.32	99	0.47	47

*I_0^{CAL}, τ^{CAL}, Q^{CAL} were calculated by Equation 11A with $\beta = 1.5$ mm; $k = 808$ sec^{-1}; $n = 0.42$; and $\xi = 0.009$.

positive by the test pulse, so that $\Delta Q \simeq 2\beta c$. In contrast, for $V_h = 0$ identical responses of opposite polarity are expected for test pulses of opposite polarity.

The potential dependence of the TFB-induced ionic current is qualitatively similar to that of gating currents observed in excitable membranes such as the squid giant axon and skeletal muscle.[9-11] In particular, both the time course and magnitude of the current in TABLE 2 are analogous to those seen for the squid giant axon.[10] Thus, the translocation of a surface-adsorbed permeant ion through lipid bilayer membranes appears to be a useful model for the gating currents, at least insofar as large-signal electrical properties are concerned. If the analogy could be extended to small signals, the bilayer model would predict maximal charge transfer as well as the appearance of a series *RC* element at holding potentials for which the gating particles are equally distributed between the two membrane surfaces; that is, at zero field across the membrane. This expectation appears to be supported by data available for the squid axon membrane.[17,18]

However, Fishman *et al.*[19] have not been able to detect the appearance of a series *RC* element in admittance measurements on the squid giant axons. Inactivation of the gating current has been advanced to explain the failure to observe admittance changes when the holding potential is varied.

It should be noted that in the particular circumstances of the present experiments, the phase of the admittance is a much more sensitive indicator of a parallel *RC* branch than the magnitude. If there is a quantitative similarity between the data presented here and gating currents in squid axon, then phase measurements of axonal admittance are expected to be more sensitive indicators of gating phenomena than the magnitude of these quantities.

INFLUENCE OF DIPOLAR SURFACE POTENTIALS ON MEMBRANE ELECTRICAL PROPERTIES

A comparative study of cationic and anionic permeability has shown that the electrostatic potential at the surface of monoolein bilayers can be increased by increasing the cholesterol content of the membrane-forming lipids.[20] This surface potential, also seen in monolayers spread at the air-water interface,[21] must originate from oriented dipolar groups, since the lipids have no net charge. "Dipolar" potential changes may be large (approximately 100 mV for membranes of high cholesterol content relative to pure monoolein membranes), and their variation as a function of lipid cholesterol content has been characterized for bilayers.[20] Therefore, these calibrated bilayers were used to assess the influence of a dipole potential on electrical properties of ion-transporting bilayer membranes.

Translocation rates k and adsorbtion coefficients β were determined for monoolein membranes of increasing cholesterol content from the small-signal square-pulse response at $V_h = 0$. At low concentration of TFB ($C_{\text{TFB}} < 10^{-6}$ M), I_0's were proportional to the aqueous TFB concentration, while τ's remained unaltered within the limits of experimental error.[13] Furthermore, k's and β's were also independent of the concentration of the indifferent electrolyte (0.01 M or 0.1 M NaCl) indicating the absence of charged, screenable groups at the membrane surface.[22] TABLE 3 shows the extent to which increasing the cholesterol mole fraction (X_{chol}) in the membrane-forming lipids increase k's without altering β's.

TABLE 3

THE INFLUENCE OF CHOLESTEROL CONTENT ON TFB TRANSPORT PARAMETERS IN MONOOLEIN BILAYERS

X_{chol}	k	β
0	20	0.0015
0.5	81	0.0021
0.75	613	0.0018
0.83	675	0.0016

The increase of the k's can be related to the changes of the dipolar surface potential $\Delta\psi$, determined by independent methods for the same lipids.[22] This is shown in FIGURE 2, which plots k's (filled circles, logarithmic scale) as a function of $\Delta\psi$. There is a strong correlation between $\log_{10} k$ and $\Delta\psi$, as indicated by the regression line ($\Delta\log k/\Delta\psi = 0.015$; $\log k = 1.41$ for $\Delta\psi = 0$), which has a correlation coefficient of 0.96. The slope of the regression line implies that a 66 mV increase in $\Delta\psi$ results in a tenfold increase in k. This value is close to a tenfold change in k for a 59 mV increase in $\Delta\psi$ (at $T = 22°C$) that would be expected if all of the dipole potential $\Delta\psi$ would contribute to decrease the energy barrier for the translocation of the hydrophobic ion from one membrane surface to the other; that is, if k's would obey the following relationship:[13]

$$k' = k \exp(F\Delta\psi/RT). \tag{5}$$

The dashed line in FIGURE 1 illustrates this expectation. The most parsimonious explanation of the results of FIGURE 2 locates the plane of TFB adsorbtion at the aqueous side of the dipolar groups from which $\Delta\psi$ originate, so that $\Delta\psi$'s become part of the energy barrier for TFB translocation. The fact that the adsorbtion coefficient β is practically unaltered by $\Delta\psi$ is consistent with this interpretation.

It should be noted that surface potentials that originate in the region between the plane of TFB adsorbtion and the aqueous phase are expected to alter the absorbtion coefficient β, but not the translocation rate k. For example, charged

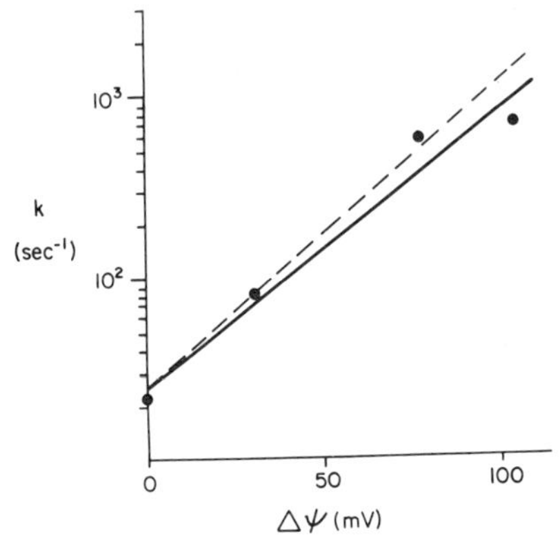

FIGURE 2. Correlation between translocation rates for TFB (K, logarithmic scale) and dipole potential ($\Delta\psi$) for cholesterolmonoolein bilayers.

groups at the bilayer membrane surface are found to alter primarily β and not k (unpublished observations). Thus $\Delta\psi$'s need not always alter k's; a careful study of the changes in β's and k's as a function of lipid composition may lead to the localization of the adsorbtion plane of hydrophobic permeant ions.

Comparison of Current Fluctuations for Different Permeation Mechanisms

The process of TFB-conduction through lipid bilayer membranes is expected to be associated with fluctuations of the net TFB movement and therefore the electrical current through the membrane. Under the simplest circumstances, these fluctuations would originate from thermal fluctuations of the TFB fluxes in opposite directions through the membrane. Within the framework of the previously developed electrical equivalent for TFB-induced electrical properties of bilayer membranes, such equilibrium thermal noise can be represented, in accordance with Nyquist's theorem, by inserting a white-noise source e_n in series with the series RC electrical equivalent. For a purely thermal noise source, e_n should have a spectral density $S_v(\omega) = 4\ \mathrm{kTR}$. The spectral density of the electrical current, $S_I(\omega)$, is predicted for this equivalent by Nyquist's theorem as:

$$S_I(\omega) = S_v(\omega)\,|\,H(j\omega)\,|^2 \tag{6}$$

where $H(j\omega)$ is the system function.[25]

It was possible to observe in carefully designed experiments TFB-induced current fluctuations for lipid bilayer membranes held at 0, 50, 75, and 100 mV potentials. The results are shown in FIGURE 3 as solid circles which show spectral density of the current noise calculated from the difference of S_I observed in the

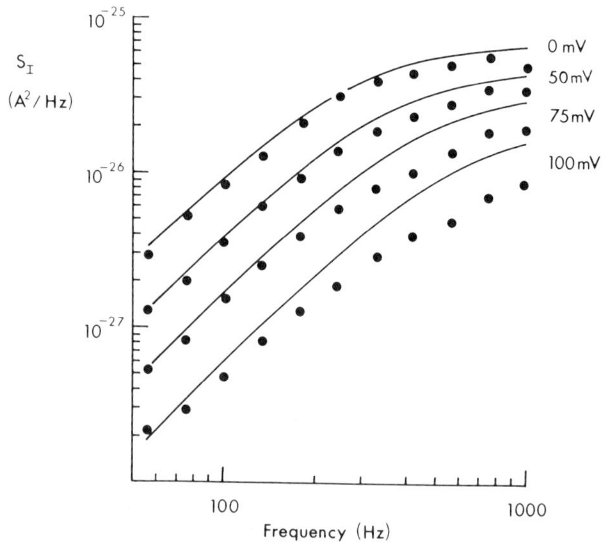

FIGURE 3. Spectral density of current fluctuations induced in lipid bilayers by the presence of 2×10^{-7} M TFB for different values of a holding potential. Membrane area: 5.5 mm^2; 0.1 M NaCl.

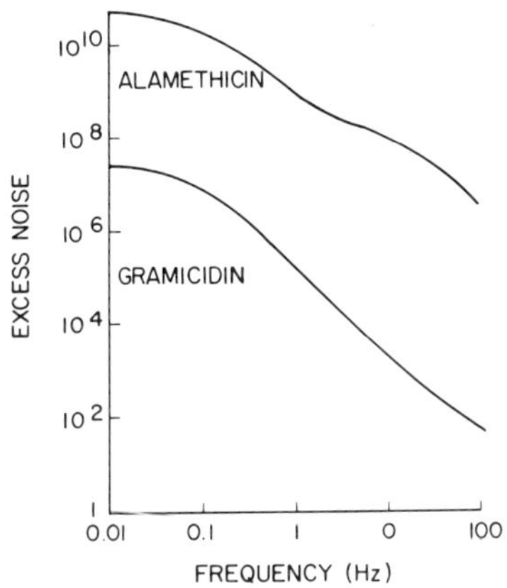

FIGURE 4. Excess current noise for the channel formers gramicidin A (monoolein membranes, 0.2 M NaCl, $V = 20$ mV, $G = 7.6 \times 10^{-7}$ S) and alamethicin (bacterial phosphatidyl ethanolamine membranes, 1 M KCl, $V = 100$ mV, $G = 10^{-8}$ S).

presence and absence of TFB. The solid lines are drawn to the theoretical expectations of Equation 6 in which

$$|H(j\omega)|^2 = |Y|^2 = \frac{1}{R^2} \frac{(2\pi fRC)^2}{1 + (2\pi fRC)^2} \tag{7}$$

is the equivalent TFB-induced admittance determined from the small-signal response for bilayers formed under the same conditions (see TABLE 1). Since R increases and C decreases with holding potentials that are increasingly different from 0 mV, the holding potential is expected to reduce the TFB-induced noise. This is seen both experimentally and theoretically. Furthermore, the excellent agreement for small holding potentials ($V_h = 0$ and 50 mV) between the observed and predicted current noise indicates that the TFB-induced current noise is accounted for entirely by equilibrium thermal noise filtered by the series-equivalent capacitance of the conduction process. In particular, the high-frequency asymptotes of S_I are close to 4 kTG expected for thermal current noise for a conductance G.

Thus it appears that there are only thermal fluctuations in the translocation of TFB across the interior of the bilayer membrane. At high frequencies ($f > 1/2\pi$RC), these current fluctuations are observed directly. At lower frequencies, however, the fluctuations are reduced by TFB translocation, which is capable of maintaining the concentration of TFB at the membrane surface at these slow time scales.

In contrast to the small, thermal current fluctuations seen for TFB, channel formers such as gramicidin or alamethicin give rise to large current fluctuations under potentiostatic conditions. Spectral density of such current noise is shown in FIGURE 4 for gramicidin and alamethicin in monoolein/decane bilayer membranes. For a meaningful comparison the data is plotted as excess noise, calculated by dividing the density of the current noise S_I by the constant S_j = 4 kTG, where G is the membrane conductance calculated as the ratio between the mean membrane current $\bar{I}$ and the transmembrane potential. The current noise is thus normalized to the thermal (Johnson) noise S_j expected for conductance G.

The excess noise generated by gramicidin and alamethicin can be understood in terms of the known magnitude and open-close kinetics of conductive channels induced by these compounds in lipid bilayers. Extensive studies[23,24] have shown that gramicidin, which forms a conductive dimer according to the reaction

$$Gr + Gr \underset{k_D}{\overset{k_R}{\rightleftharpoons}} Gr_2, \tag{8}$$

gives rise to a Lorentzian spectral density of current fluctuations relatable to the single-channel conductance, g, and the rate constants k_D and k_R. A simple expression, in terms of g, and characteristic channel lifetime τ_c can be derived for the excess noise.[23]

$$\frac{S_I}{S_j} = \frac{V^2}{kT} g \frac{\tau_c^2 k_D}{1 + (2\pi\tau_c f)^2} \tag{9}$$

For independent channels, one may further simplify Equation 1, since $\tau_c \sim 1/k_D$ for this case.[23]

Equation 9 shows that large single-channel conductance is the single most important variable that gives rise to large excess noise, provided that the characteristic channel lifetime is not too small for practical measurement of the fluctuations; e.g., $\tau_c < 1$ second.

A qualitatively similar but much more complex expression could also be derived for alamethicin, for which the excess noise shows at high frequencies an extra hump that results from conductance fluctuations within an active conductive structure.[26,27]

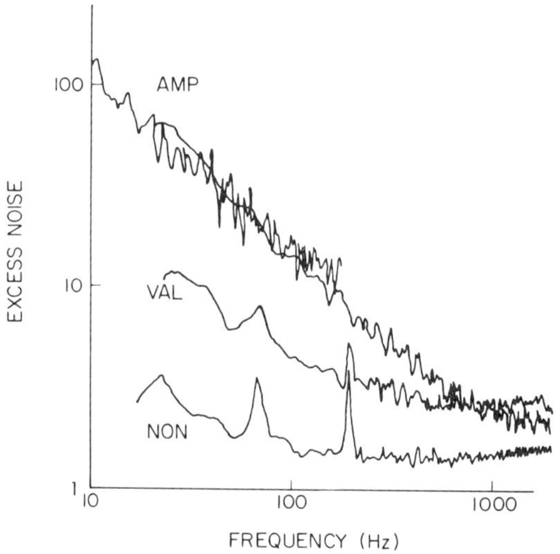

FIGURE 5. Excess noise for amphotericin B (monoolein-cholesterol membranes 1:1 molar ratio, 1 M KCl, $V = 100$ mV, $G = 1.4 \times 10^{-6}$ S), valinomycin (monoolein membranes, 1 M KCl, $V = 100$ mV, $G = 4.3 \times 10^{-7}$ S) and nonactin (monoolein membranes, 1 M KCl, $V = 100$ mV, $G = 1.5 \times 10^{-6}$ S).

In contrast to the large gramicidin- and alamethicin-induced excess noise, FIGURE 5 shows that much smaller but significant excess noise is observed for bilayers made conductive by amphotericin *B*, nonactin, or valinomycin. Why is the excess current noise so much smaller for these membranes? There are at least three possibilities. First, long-lived, stable conductive pores may be created in the membrane. Under these circumstances, the noise resulting from opening and closing of the channels would not be accessible on a practical measurement time scale. Amphotericin *B* appears to form such long-lived pores in the membrane, since both direct electrical detection of conductive channels[28,29] and indirect permeability measurements[30] point to the existence of large ($g \sim 10^{-10}$ S) conductive channels. Second, weakly conducting (e.g., small g) and fast (e.g., small τ_c) channels may mediate the ionic permeability. Third, a relatively noise-free permeation mechanism may mediate the ionic permeability in the membrane. This is expected to be the case for nonactin and valinomycin, which are well documented neutral carriers of ions[6] and should exhibit noise properties similar to those seen for TFB. The carrier mechanism may generate some excess noise arising from fluctuations of the carrier concentration in the membrane, which is the likely origin of the low-frequency excess noise seen for nonactin and valinomycin in FIGURE 5. It would appear that under ideal circumstances the electrical properties of the membrane, and particularly the electrical noise, may be a useful indicator of the permeation mechanism operating in the membrane.

REFERENCES

1. MUELLER, P. & D. O. RUDIN. 1969. Translocators in bimolecular lipid membranes. *In* Current Topics in Bioenergetics. D. S. Sohdai, Ed. Vol. **3**: 157. Academic Press. New York, N.Y.
2. PARSEGIAN, A. V. 1975. Ion-membrane interactions as structural forces. Ann. N.Y. Acad. Sci. **264**: 161.
3. HALL, J. E., C. A. MEAD & G. SZABO. 1973. A barrier model for current flow in lipid bilayer membranes. J. Membr. Biol. **11**: 75.
4. EISENMAN, G. 1968. Ion permeation in cell membranes and its models. Fed. Proc. **27**: 1249.
5. MUELLER, P. & D. O. RUDIN. 1967. Development of K^+–Na^+ discrimination in experimental bimolecular lipid membranes by macrocyclic antibiotics. Biochim. Biophys. Res. Comm. **26**: 398.
6. SZABO, G., G. EISENMAN & S. CIANI. 1969. The effects of the macrotetralide actin antibiotics on the electrical properties of phospholipid bilayer membranes. J. Membr. Biol. **1**: 346.
7. LÄUGER, P. & G. STARK. 1970. Kinetics of carrier-mediated ion transport across lipid bilayer membranes. Biochim. Biophys. Acta **211**: 458.
8. HAYDON, D. A. & S. B. HLADKY. 1972. Ion transport across thin lipid membranes: a critical discussion of mechanisms in selected systems. Quart. Rev. Biophys. **5**: 187.
9. SZABO, G. 1977. Admittance and step-clamp response for ion redistribution in bilayers: a simple model for gating currents. Biophys. J. **17**: 11a.
10. ARMSTRONG, C. M. & F. BEZANILLA. 1975. Currents associated with the ionic gating structures in nerve membrane. Ann. N.Y. Acad. Sci. **264**: 265.
11. CHANDLER, W. K. & M. P. SCHNEIDER. 1976. Effects of membrane potential on the capacitance of skeletal muscle fibers. J. Gen. Physiol. **67**: 125.
12. KETTERER, B., B. NEUMCKE & P. LÄUGER. 1971. Transport mechanism of hydrophobic ions through lipid bilayer membranes. J. Membr. Biol. **5**: 225.
13. SZABO, G. 1976. The influence of dipole potentials on the magnitude and kinetics of

ion transport in lipid bilayer membranes. *In* Extreme Environments. M. R. Heinrich, Ed. Academic Press. New York, N.Y.
14. ANDERSEN, D. S. & M. FUCHS. 1975. Potential energy barriers to ion transport within bilayers: studies with tetraphenylborate. Biophys. J. **15:** 795.
15. WAGGONER, A. S., C. H. WANG & R. L. TOLLES. 1977. Mechanism of potential dependent light absorption changes of lipid bilayer membrane in the presence of cyanine and oxonol dyes. J. Membr. Biol. **33:** 109.
16. LEBLANC, O. 1969. Tetraphenylborate conductance through lipid bilayer membranes. Biochim. Biophys. Acta **193:** 350.
17. KEYNES, R. D. & E. ROJAS. 1974. Kinetics and steady state properties of the charged system controlling sodium conductance in the squid giant axon. J. Physiol. **239:** 393.
18. MEVES, H. 1976. The effect of zinc on the late displacement current in squid giant axons. J. Physiol. **254:** 787.
19. FISHMAN, H. M., L. E. MOORE & D. POUSSART. 1977. Charge movements and admittance in squid axon. Biophys. J. **17:** 11a.
20. SZABO, G. 1974. Dual mechanism for the action of cholesterol on membrane permeability. Nature **252:** 47.
21. HLADKY, S. B. & D. A. HAYDON. 1973. Membrane conductance and surface potential. Biochim. Biophys. Acta **318:** 464.
22. SZABO, G, G. EISENMAN, S. G. A. MCLAUGHLIN & S. KRASNE. 1972. Ionic probes of membrane structures. Ann. N.Y. Acad. Sci. **195:** 273.
23. KOLB, H. A. & E. BAMBERG. 1977. Influence of membrane thickness and ion concentration on the properties of the gramicidin A channel. Autocorrelation, spectral power density, relaxation and single-channel studies. Biochim. Biophys. Acta **464:** 127.
24. ZINGSHEIM, H. P. & E. NEHER. 1974. The equivalence of fluctuation analysis and chemical relaxation measurements: a kinetic study of ion pore formation in thin lipid membranes. Biophys. Chem. **2:** 197.
25. PAPOULIS, A. 1965. Probability, Random Variables and Stochastic Processes. McGraw Hill. New York, N.Y.
26. BOHEIM, G. 1974. Statistical analysis of alamethicin channels in black lipid membranes. J. Membr. Biol. **19:** 277.
27. GORDON, L. G. M. & HAYDON, D. A. 1972. The unit conductance channel of alamethicin. Biochim. Biophys. Acta **255:** 1014.
28. EMIRSHKIN, L. N., K. M. KASUMOV & V. M. POTZELUYEV. 1976. Single ionic channels induced in lipid bilayers by polyene antibiotics amphotericin B and nystatine. Nature **262:** 698.
29. BUNOW, M. & I. W. LEVIN. 1977. Vibrational raman spectra of lipid systems containing amphotericin B. Biochim. Biophys. Acta **464:** 202.
30. FINKELSTEIN, A. & R. HOLZ. 1973. Aqueous pores created in thin lipid membranes by the polyene antibiotics nystatin and amphotericin B. *In* Membranes. G. Eisenman, Ed. Marcel Dekker. New York, N.Y.

APPENDIX

The Time Course of Ionic Currents Through Bilayers

The objective of this appendix is to derive the electrical properties of a lipid bilayer membrane in which ionic conduction takes place by translocation of permeant hydrophobic ion such as tetraphenylborate. Following the original derivation of Ketterer *et al.*[12] and according to scheme 1 of the text, the ionic current I through the bilayer is given by

$$\frac{I(t)}{ZF} = \Phi(t) = k'N'(t) - k''N''(t) = -\frac{dN'(t)}{dt} \tag{1A}$$

where $\Phi(t)$ is the net flow of permeant ions i of valence Z from side$'$ to side$''$ of the membrane. If the total amount of i at the membrane surfaces is assumed to be constant, one can write, using Equation 3,

$$N'(t) + N''(t) = N_T = 2\beta c. \tag{2A}$$

For a membrane that is held initially at a potential V_h, a steady state is reached for which

$$N'_{(0)} = N'_h;\, N''_{(0)} = N''_h;\, \Phi = k_h N''_h - k''_h N''_h \tag{3A}$$

where k'_h and k''_h are the translocation rate constants at V_h. It is convenient to define incremental surface concentrations ΔN:

$$\Delta N' = N'(t) - N'_h; \qquad \Delta N'' = N''(t) - N''_h. \tag{4A}$$

Equations 1A, 2A, 3A, and 4A can be combined to give:

$$\frac{I(t)}{ZF} = \Phi(t) = -\frac{dN'}{dt} = (k' + k'')\Delta N' + 2\beta c \left\{\frac{k''_h(k' + k'')}{k'_h + k''_h} - k''\right\}. \tag{5A}$$

Defining $Q = ZF\Delta N$ as the change in the charge density at the membrane surface, Equation 5A can be written in the form

$$I(t) = ZF\Phi(t) = -\frac{dQ(t)}{dt} = (k' + k'')Q(t) + 2ZF\beta c\,\frac{k''_h k' - k'_h k''}{k'_h + k''_h}. \tag{6A}$$

The voltage dependence of translocation rates k' and k'' have been derived by solving the diffusion equation for the translocation of an ion across the low dielectric constant interior of the bilayer membrane[3,8,13,14]

$$k' = kf(u)\exp(-nu);\; k'' = kf(u)\exp(nu) \tag{7A}$$

where $u = FV/RT$ is the normalized membrane potential, k is the intrinsic translocation rate; e.g., the translocation rate in the limit of $u \rightarrow 0$. The function $f(u) \simeq e^{-\xi u^2}$ is an empirical relation derived to take into account image forces acting on the ion i in the bilayer,[8] ξ being determined experimentally. The parameter $0 \leq n \leq 0.5$ takes into account the possibility that the membrane potential may not be entirely applied to drive the ion i across the membrane.[14]

Incorporating equation 7A into equation 6A yields

$$I(t) = -\frac{dQ}{dt} = k\,f(u)\cdot \left\{(\exp(nu) + \exp(-nu))Q + 2ZF\beta c\,\frac{\exp n(u_h - u) - \exp n(u - u_h)}{\exp(nu_h) + \exp(-nu_h)}\right\} \quad \textbf{(8A)}$$

For arbitrary but small variations of the membrane potential about the holding potential V_h we set

$$u = F(V_h + V(t))/RT.$$

For $|V(t)| < RT/F \simeq 25$ mV, Equation 8A simplifies to

$$-\frac{dQ}{dt} \simeq [2k\,f(u_h)\cosh nu_h]\,Q + \frac{2\,Z^2F^2\,k\,f(u_h)n\beta c}{RT\cosh(nu_h)}\,V(t). \quad \textbf{(9A)}$$

This differential equation is formally equivalent to that describing the behavior of a series RC circuit:

$$-\frac{dQ}{dt} = \frac{1}{R_{el}C_{el}}\,Q + \frac{1}{R_{el}}\,V$$

Therefore, it is possible to describe the small-signal behavior of the bilayer by that of a well-known series RC circuit, provided that one sets

$$R_{el} = \frac{RT}{2nF^2\beta ck}\;\frac{\cosh(nu_h)}{f(u_h)}$$

$$C_{el} = \frac{nF^2\beta c}{RT}\;\frac{1}{\cosh^2(nu_h)} \quad \textbf{(10A)}$$

The small-signal square-pulse response and electrical admittance of the bilayer thus become immediately calculable with use of results of elementary electrical circuit analysis.

Equation 8A is nonlinear for large signals, since the coefficients are voltage-dependent. There is a simple solution, however, for step changes ΔV of the membrane potential, since under these conditions the coefficients have a constant value. For a ΔV from a holding potential V_h to $V_h + \Delta V$, the solution of Equation 8A is exponential. This exponential decay of the membrane current can again be represented by that of an RC circuit with

$$R_{el} = \frac{RT}{2nF^2\beta ck}\;\frac{\cosh(nu_h)}{f(v_h + \Delta u)}$$

$$C_{el} = \frac{nF^2\beta c}{RT}\;\frac{1}{\cosh(nu_h)\cosh n(u_h + \Delta u)} \quad \textbf{(11A)}$$

where $u_h = FV_h/RT$ and $\Delta u = F\Delta V/RT$, as defined previously. It must be noted that the large-signal RC equivalent holds only for the square-pulse response of the membrane; the current response is nonlinear for other waveforms of applied potential.

CHANNELS ACROSS BLACK LIPID MEMBRANES

Moisés Eisenberg and Michael E. Kleinberg

Department of Pharmacological Sciences
Health Sciences Center
State University of New York
Stony Brook, New York 11794

Joel H. Shaper

The Oncology Center and the Department of Pharmacology and
Experimental Therapeutics
The Johns Hopkins University Medical Center
Baltimore, Maryland 21205

Lipid bilayer membranes are highly impermeable to small inorganic ions, as evidenced by their low electrical conductivity ($\sim 10^{-8}$ mho cm^{-2} in 0.1 M sodium chloride). Biological membranes, on the other hand, may have conductivities many orders of magnitude higher. In particular, the plasma membranes of nerve and muscle cells have conductivities of 10^{-3} to 10^{-1} mho cm^{-2}. This is mainly due to their high permeability to sodium and potassium ions. It is fairly well accepted that these ions pass through those biomembranes through specialized pores or channels.[1-3]

A variety of molecules, among them several antibiotics, are known to interact with lipid bilayer membranes and increase their permeability to small ions to levels comparable with nerve membranes.[4-8] Some of these (e.g. valinomycin, nonactin) facilitate the passive diffusion of ions via a carrier mechanism. That is to say, the carrier molecule complexes with the ion and shuttles across the membrane down the electrochemical gradient.[7] Another class of membrane-modifying molecules (e.g. gramicidin A, alamethicin) facilitate the passive diffusion of ions via the formation of structures that can bridge across the membrane and create a pathway through which ions can move down their electrochemical gradient. This latter class is known as pore- or channel-formers. In most instances these ion passages are only transient.

Most direct evidence for the existence of antibiotic-mediated pores in black lipid membranes comes from experiments in which the electrical current through the membrane is monitored as a function of time, while the voltage across the membrane is kept constant. Under these conditions, discrete conductance changes can be seen in the current-time traces. These quantized conductance changes range from $\sim 10^{-11}$ to $\sim 2 \cdot 10^{-9}$ mho for a variety of pore-formers and correspond to fluxes of $\sim 10^{7}$ to $\sim 2 \cdot 10^{9}$ univalent ions per second at 100 mV of applied membrane potential, respectively. Such large ion fluxes cannot be due to a carrier molecule shuttling across the membrane, since diffusion through the membrane phase limits the maximum conductance to about three orders of magnitude lower than is observed in these discrete jumps.[8]

The polyene antibiotics nystatin and amphotericin B are pore-formers in lipid bilayers.[9-12] Early indirect evidence for this[9-11] stemmed from the difference between tracer and osmotic permeability, the cutoff radius of the permeability to nonelectrolytes at the size of glucose, and the high power (4th–12th) dependence of conductance on antibiotic concentration. Recent observation of discrete conductance changes[12] ($2.5 \cdot 10^{-12}$ mho in 2 M KNO_3 for amphotericin B and $1–3 \cdot 10^{-12}$ mho in 3 M KCl for nystatin) directly demonstrates that they form channels

Gramicidins A, B, and C interact with lipid bilayers and form pores that are almost perfectly cation-selective for the alkali metal, ammonium, and hydrogen chlorides.[3] The conductance of a single pore ($4 \cdot 10^{-11}$ mho in 1.0 M potassium chloride solution) can be distinctly resolved. The characterization of the gramicidin/black lipid membrane interaction has been carried out quite extensively. This includes selectivity measurements, kinetic and statistical analyses,[13,14] and the use of fluorescent analogs,[15] to mention a few. These studies indicate that the conducting channel is a gramicidin dimer.

EIM (excitability-inducing material),[16] monazomycin,[17] hemocyanin,[18] alamethicin,[19] and suzukacillin,[20] induce the formation of pores across lipid bilayers. The conductance of black lipid membranes doped with these antibiotics is strongly voltage-dependent. This is not the case for the gramicidins or nystatin. Whereas EIM and hemocyanin induce a conductance that is maximal at zero-applied potential and diminishes with increasing voltage, monazomycin, alamethicin, and suzukacillin show a marked exponential increase in conductance with increasing voltage. For all these pore-formers the voltage dependence is a result of a change in the average number of pores in the open state, and is not due to a voltage-dependent change in the single-channel conductance.

A purified toxin from black widow spider venom has been shown recently to form pores in lipid bilayers.[21] These pores are ideally selective for alkali cations relative to anions and are significantly permeable to calcium. The conductance of the single channel is about $3.6 \cdot 10^{-10}$ mho in 0.1 M KCl. These studies suggest a possible mechanism of action of the venom in the induction of neurotransmitter release at the neuromuscular junction.[22]

Direct evidence for the existence of pores in biological membranes has been reported by Neher and Sakmann.[3] Discrete conductance changes of $2 \cdot 10^{-11}$ mho were recorded from the plasma membranes of denervated frog muscle. Their response to specific pharmacological agents indicates that these channels are part of the postsynaptic nicotinic acetylcholine receptors. The statistical analysis of the amplitude and duration of these receptor channels is consistent with previous noise fluctuation analysis data.[23]

Our efforts at present are mostly directed toward the molecular mechanisms of action of the voltage-dependent conductance induced by alamethicin in black lipid membranes. We shall concern ourselves exclusively with alamethicin and hope that it will serve to illustrate some of the approaches used in the study of channels in black lipid membranes. First, we shall briefly summarize the current views and controversies on this antibiotic's mode of action.[19,24–35] Subsequently we shall present some new data of a preliminary nature to be discussed in greater detail elsewhere.

The primary structure of alamethicin was at first thought to be a cyclic eighteen amino acid polypeptide.[36] More recent studies, however, have shown that it is a linear molecule.[37] The N-terminus is acetylated and a phenylalanino (Pheol) is linked to the side-chain carboxyl group of the Glu(17). The amino acid sequence is: Acetyl - [Aib(0)] - Pro(1) - Aib(2) - Ala(3) - Aib(4) - Ala(5) - Gln(6) - Aib(7) - Val(8) - Aib(9) - Gly(10) - Leu(11) - Aib(12) - Pro(13) - Val(14) - Aib(15) - Aib(16) - Glu(17)(Pheol) - Gln(18). (Aib stands for α-amino isobutyric acid.) The free carboxyl group of Gln(18) is probably responsible for the antibiotic's pK_a of 5.5.[38] Martin and Williams postulated that a ring structure formed between Pro(1), Aib(0), and acetyl might yield a positive charge at the N-terminus,[37] making the overall molecule dipolar at neutral pH.

The addition of alamethicin to the aqueous phase on one side only of a black lipid membrane induces a rectifying steady state conductance which can be closely described by the expression:

$$G = G_0 + [\text{ala}]^{9\pm1}[\text{salt}]^{4.5\pm1}(g_+e^{(6\pm1)V/kT} - g_-e^{-(6\pm1)V/kT}) \tag{1}$$

where G_0 is the conductance of the unmodified membrane, [ala] and [salt] are the concentrations of alamethicin and salt (for any alkali or alkaline earth chloride), respectively, V is the voltage across the membrane (in millivolts, as measured on the side in which the alamethicin has been added relative to the opposite side), g_+ and g_- are constants that depend on the type of lipid, the type of salt, and the area of the membrane, k is the Boltzmann constant, and T the absolute temperature (kT in meV).

For all membranes studied, $g_+ > g_-$. In particular, for black lipid membranes made from bacterial phosphatidyl ethanolamine (PE) in *n*-decane, $g_+ \gg g_-$, and the negative branch of the exponential conductance-voltage curve is not experimentally observed. If the total diffusion of alamethicin from one bulk aqueous phase to the other is limited both by the unstirred layers and the lipid phase[7] in comparable magnitude, then at steady state the difference in the concentration of the antibiotic at both membrane-water interfaces could account for the assymmetry in the conductance vs. voltage relationship, relative to voltage polarity. On the other hand, if the diffusion of alamethicin across the membrane is much lower than that through the unstirred layers, then the antibiotic will be effectively present only at one of the membrane-water interfaces at steady state. It must be argued, therefore, that pores formed at voltages of opposite sign reflect two different ways in which the alamethicin molecules interact with the membrane potential.[29] We shall present some additional indirect evidence that favors the latter proposition.

In PE membranes, the kinetics of the onset of conductance following the application of a voltage pulse is first order and can be described by[26]

$$G = \frac{\mu}{\lambda}(1 - e^{-\lambda t}) \tag{2}$$

where both μ and λ are exponential functions of the voltage across the membrane. In membranes made with other lipids it is possible to see more complex kinetic behavior, including sigmoidal rise and inactivations.[32] In addition, pretreatment of alamethicin-doped lecithin membranes with an applied voltage for a prolonged time introduces a fast component (in the milliseconds) prior to the first-order kinetics,[41] which is typically in the seconds range.

By monitoring the current through the alamethicin-doped black lipid membrane as a function of time at constant potential such that the steady state conductance is slightly above that of the unmodified membrane, it is possible to resolve the discrete conductance fluctuations characteristic of the formation of single channels. The most salient features that can be observed are as follows:

1. The fluctuations occur in bursts. The current is either at the level of the unmodified membrane or it shifts into the fluctuations mode and jumps among five to seven discrete levels for a few seconds to subsequently return to the bare membrane conductance level and randomly repeat the cycle. The fluctuations are typically in the millisecond range at room temperature.
2. Each level has a characteristic amplitude which is reproducible over long periods of time and between different membranes.
3. All transitions occur from adjacent levels.
4. The amplitude of any of these five to seven levels is not a linear combination of smaller levels with integral coefficients. In addition, the differences in amplitude of consecutive levels increases monotonically with a characteristic pattern for each different kind of salt.
5. The probabilities of the different conductance levels distribute in a characteristic way, the middle levels being more probable than either lower or higher ones.
6. The main effect of the voltage is to increase the probability of occurrence of fluctuating conductance bursts. Indeed, under conditions in which more than the first seven levels are resolved, the amplitudes and probability distributions of the higher levels can be fit accurately by the superposition of two or three fluctuating bursts occurring simultaneously. As more levels build up by the increase in voltage, the fluctuations become too fast to be resolved. The analysis of the noise fluctuations in the current at large average number of open pores, furthermore, is consistent with the picture emerging from the single-channel conductance studies. A pore has therefore been defined operationally as the compound fluctuation burst described in 1, and the five to seven discrete levels are defined as the conductance states of the pore.
7. The conductance states of the pore are nearly ohmic.
8. The conductance states of the pore closely follow the same dependence on ionic concentration and viscosity as that of the bulk solution conductivity.
9. A secondary effect of the voltage is a directly proportional slight skew of the probability distribution of the conductance states of the pore. This effect, however, is insufficient in explaining the overall exponential voltage dependence of the induced conductance.

There are several current speculations on the molecular mechanisms of the formation of alamethicin pores. We shall briefly mention two. Boheim,[27] and Baumann and Mueller[28] have proposed that alamethicin monomers at the water-membrane interface can respond to the applied potential by flipping into the membrane phase, as it were, and aggregating into a cylindrical structure similar to the staves of a barrel. The discrete conductance fluctuations would reflect the addition or subtraction of a monomer to or from the aggregate.[32] On the other hand, it has been suggested[26,29,31] that an aggregate may already exist at the membrane-water interface and that some sort of conformational change may take place in response to a voltage change, perhaps involving some lipid molecules as well, which leads to the opening of the pores.

The monomer aggregation model is consistent with the observation that the two lower levels of the pore are more ion-selective than the higher levels, and that the amplitude of the difference between consecutive levels increases. As discussed below, however, it fails to predict the observed cutoff radius of the conductance levels to ions of various sizes.

The preaggregate model can be interpreted either as a cluster of almost identical parallel pores (with the exception of the smallest level)[29] or as a single pore of variable effective cross-sectional area.

Electrical measurements of alamethicin-doped black lipid membranes may not be sufficient to distinguish clearly between the different proposed models. Lateral diffusion in the plane of the membrane, for example, is rapid enough to allow consistency between the monomer aggregation scheme and first-order kinetics. We have considered, therefore, the use of alamethicin analogs in an attempt to sort out some of these questions.

Materials and Methods

All of the salts used were ultrapure grade from Alfa Inorganics, except for the substituted ammonium salts, which were obtained from Eastman and recrystallized. Water was of low conductivity ($<10^{-7}$ mho cm^{-1}) from a "Milli-Q" system (Millipore Corp.). Alamethicin was a kind gift from Dr. Whitfield of Upjohn, and it was used without further purification.

The methyl- and dansyl- analogs of the antibiotic were synthesized by condensation of alamethicin with methanol and dansyl-cadaverine, respectively. The major products were purified by repeated TLC on silica gel plates. The final purified analogs migrated as a single component in a variety of solvent systems. Unmodified alamethicin in the analog preparations was undetectable as visualized by the chlorine peptide spray on TLC.[39]

Black lipid membranes were formed from 2.5% w/v solutions of bacterial PE (Supelco) in *n*-decane (Gold Label, Aldrich). The membrane chamber was made of Teflon® with symmetric compartments of 15 ml capacity. The orifice in the septum separating the two compartments measured half a millimiter in diameter. Membranes typically 200 μm in diameter were in equilibrium with the torus of membrane-forming solution attached to the rim of the orifice. One pair of Ag/AgCl electrodes were used to apply the voltage and monitor the current across the membrane. The voltage source was simply a 1.35 V mercury battery divided through a 1 Kohm ten-turn potentiometer. This voltage was applied between one of the electrodes and ground. The other electrode was connected to the input of a Keithley 427 current amplifier (referred to the same ground). The output of the amplifier was monitored with a Tektronix 5103 storage oscilloscope, from which Polaroid® pictures were taken and analyzed.

Alamethicin and its analogs were always added from a 0.1 mg/ml solution in water to the aqueous phase after the black lipid film had been formed. The aqueous phases were then mixed for at least 15 minutes with Teflon-coated magnetic bars driven remotely by a magnet spinning on a motor. All experiments were carried out between 22 and 24°C.

RESULTS AND DISCUSSION

Selectivity and Cutoff Radius

The single-pore conductance induced by alamethicin in PE membranes was measured in a variety of chloride salts of cations of increasing size. In all cases the salt concentration was 1.0 M on both sides of the membrane. Single-channel conductance was measured for each salt in the range of 20–250 mV. The conductance was very linear up to 125 mV and was the value used in making up TABLE 1. For each salt, the ratio of the conductance of each of the levels relative to the same level in KCl was calculated. At the same time, the ratio of the bulk-solution conductivity to that of KCl was also obtained. A + sign on TABLE 1 indicates that the ratio for the corresponding conductance level is within 10% of that of the ratio of the bulk-solutions' conductivities. A − sign indicates that the ratio for the conductance level is much lower than that in bulk and could be explained to a first approximation by only the chloride ion moving through the pore. A ± sign indicates an intermediate situation.

TABLE 1

CATION SELECTIVITY OF THE CONDUCTANCE STATES OF AN ALAMETHICIN PORE RELATIVE TO K^+ FOR A VARIETY OF CHLORIDE SALTS

Cation	Radius	Conductance State					
	(Å)	0	1	2	3	4	5
K^+	1.8	+	+	+	+	+	+
Cs^+	1.8	+	+	+	+	+	+
NH_4^+	1.8	+	+	+	+	+	+
Na^+	3.3	−	±	±	+	+	+
$N(CH_3)_4^+$	3.5	−	−	−	±	±	±
Li^+	3.7	−	+	+	+	+	+
$(EtOH)NH_3^+$	2.5 × 3.0	−	−	−	±	+	+
$(EtOH)_2NH_2^+$	2.5 × 4.5	−	−	−	±	±	±
$(EtOH)_3NH^+$	4.0 × 4.5	−	−	−	±	±	±
Tris H^+	4.0	−	−	−	±	±	±
$N(CH_2CH_3)_4^+$	4.0	−	−	−	−	−	−
$N(CH_2CH_2CH_3)_4^+$	4.5	−	−	−	−	−	−

Our data is suggestive of a somewhat gradual cutoff radius, consistent with a picture of a pore whose effective minimal radius increases with increasing conductance level. Previously, it had been reported for substituted quaternary ammonium ions that the cutoff radius was uniform for all conductance levels (except the smallest).[29] The difference between both results, however, is not a large one. Our data were taken for some different ions in the same size range and in a different lipid bilayer membrane. It is clear, in both results, that cations larger than 4.5 Å in radius are excluded from the pore.

If one considers the pore as an aqueous cylindrical channel approximately 30 Å long and with a conductivity equal to that of the bulk aqueous solution, then from the relationship $G = (\gamma A)/1$ (where G is conductance, γ = specific conductivity, A = area of the cross section and l = length), one can predict that, to a first approximation, the area of the cross section of the pore ranges from about 10 $Å^2$ for the lowest level to about 240 $Å^2$ for the fifth conductance state. These correspond to radii of ≈ 2 Å and ≈ 9 Å, respectively. This approach is admittedly oversimplified. Further refinements to the approximation, however, serve only to increase the effective radius. The large calculated values are hard to reconcile with the experimentally determined cutoff radii, and this difficulty casts some doubts on whether the cylindrical barrel model for the alamethicin channel "holds water." The fact that the selectivity decreases gradually with increasing conductance state is not fully consistent with the idea of identical channels in parallel, to a first approximation.

Alamethicin Analogs

Although qualitatively both the methyl ester of alamethicin (me-ala) and the dansyl derivative (d-c-ala) behave in a similar manner to regular alamethicin, the following quantitative differences can be observed in PE membranes:

1. Both me-ala and d-c-ala show the positive and the negative induced conductance. The probability distribution of the conductance states of the single pore, furthermore, is skewed toward higher levels for the negative channels in the same way as reported by Gordon and Haydon[29] for regular alamethicin in monoolein-cholesterol membranes.

2. Under identical salt, membrane, and voltage conditions, it requires about 600 times less me-ala and about eight times less d-c-ala to reach the same average number of open pores than regular alamethicin.

3. The absolute value of the conductance states of me-ala are practically identical to the regular alamethicin, but the dansyl derivative has only about 43%, 43%, 58%, 66%, 71%, and 75% of the conductance of the natural compound for the first six levels of the pore, respectively, in 1.0 M KCl (FIGURE 1).

Both of the analogs' functional groups are covalently bound to the terminal carboxyl group of alamethicin. The fact that both analogs show the same overall exponential voltage-dependent conductance as regular alamethicin strongly suggests that the negative charge of the terminal carboxyl group of natural alamethicin is not directly involved in the interaction of the molecule with the electric field.

The calculated ratios of the conductances of corresponding conductance states of d-c-ala and regular alamethicin are not consistent with the model of parallel identical channels.

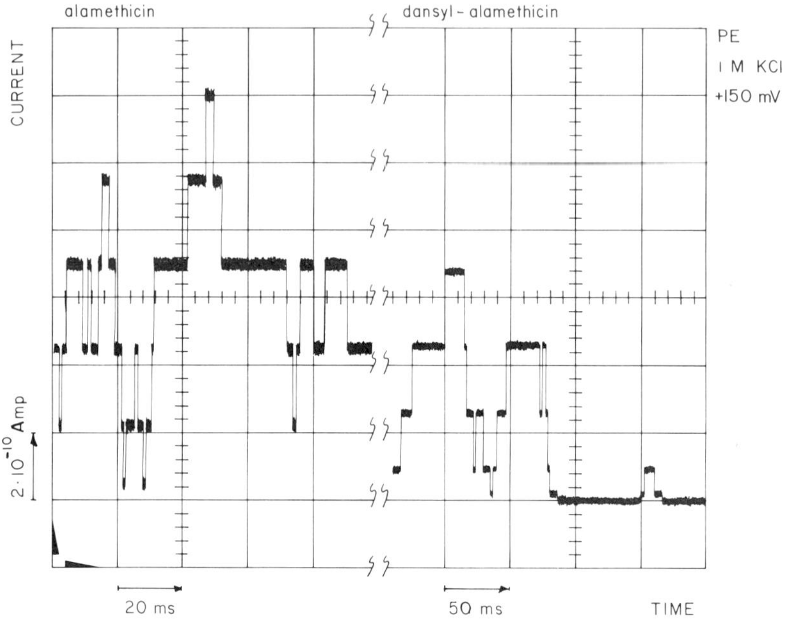

FIGURE 1. An artist's retracing of two oscillographs. On the left are the six conductance states of alamethicin. On the right, up to five conductance states of d-c-ala are shown for the same voltage, 150 mV, and KCl concentration, 1 M. The current of the unmodified membrane is one major division from the bottom of the screen, and it is practically indistinguishable from zero current in this scale. The concentrations of alamethicin and d-c-ala are 2.5×10^{-8} and 1.2×10^{-8} g/ml, respectively.

The asymmetry of the probability distribution for positive and negative channels strengthens the suggestion that channels opening from opposite polarities both originate from molecules on the same side of the membrane.[29]

Hybrid Channels

Since the conductance states of d-c-ala have amplitudes different and distinguishable from those of regular alamethicin, it is possible to add corresponding amounts of both molecules on the same side of a lipid bilayer and ask whether pores are formed that show conductance states of intermediate amplitudes. This is indeed the case, as shown in FIGURE 2. These homoside hybrids (so called because they originate from molecules added on the same side of the membrane) show statistical distributions that will be discussed in greater detail elsewhere. Given a pore in a conductance state, it is not possible to predict to which level the conductance is going to change (recall that in the case of homogeneous pores formed from only one kind of molecule, the conductance states have reproducible amplitudes, and transitions occur only among consecutive levels). We have seen up to 17 different conductance levels within a fluctuating hybrid pore. It is quite clear that the pore is composed of several molecules. These results are consistent with both the monomer aggregation scheme and the preaggregate of variable conformation.

It is also possible to ask whether one may form heteroside hybrids by the addition of regular alamethicin and d-c-ala on opposite sides of the membrane in quantities such that positive alamethicin channels and negative d-c-ala channels have the same probability of being open to a given membrane potential. Preliminary data show that these heteroside hybrids exist, but are different from the homoside hybrids, further strengthening the hypothesis that alamethicin pores opened by negative voltages form by a different mechanism than those opened by positive voltages.

It seems clear to us that alamethicin pores correspond to polymeric aggregates of the molecule. We think that rather than a cylindrical pore of varying diameter or a set of identical parallel channels, it is possible that the pore might adopt an extended linear conformation in the plane of the membrane, to form a "slit" (not necessarily straight) which would account for the increase in the effective cross-sectional area without a large increase in the effective cutoff radius.

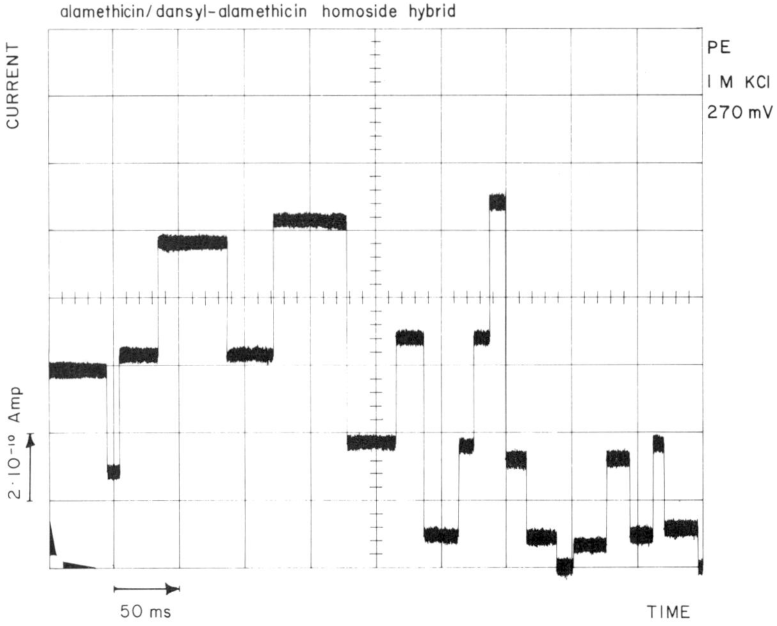

FIGURE 2. An artist's retracing of an oscillograph of the current of a homoside hybrid pore formed by the addition of alamethicin and d-c-ala on the same side of a PE membrane at 270 mV and 1 M KCl. The current through the unmodified membrane corresponds to the bottom of the screen, and it is undistinguishable from zero current in this scale.

In a three-dimensional isotropic water solution, many amphipathic molecules, such as several detergents, will form micelles, one of whose dimensions would be limited to a bimolecular width.[40] Because of the amphipathic nature of alamethicin, it is possible that the aggregates formed in the membrane (which is a two-dimensional isotropic fluid as far as lateral diffusion is concerned) are likewise limited to a bimolecular width in one of their dimensions. This concept is, of course, speculative and assumes that the alamethicin molecule spans the width of the membrane, that it has a hydrophobic and a hydrophillic face, and that the alamethicin molecules in the pore aggregate so as not to expose the hydrophillic faces to the hydrophobic interior of the membrane.

REFERENCES

1. HILLE, B. 1970. Prog. Biophys. Mol. Biol. **21:** 1–32.
2. EHRENSTEIN, G. & H. LECAR. 1972. Ann. Rev. Biophys. Bioeng. **1:** 347–368.
3. NEHER, E. & B. SAKMANN. 1976. Nature **260:** 792–802.
4. MUELLER, P. & D. O. RUDIN. 1969. Curr. Top. Bioenerg. **3:** 157–249.
5. FINKELSTEIN, A. 1972. Arch. Intern. Med. **129:** 229–240.
6. HAYDON, D. A. & S. B. HLADKY. 1972. Q. Rev. Biophys. **5:** 187–282.
7. MCLAUGHLIN, S. & M. EISENBERG. 1975. Ann. Rev. Biophys. Bioeng. **4:** 335–366.
8. LÄUGER, P. 1972. Science **178:** 24–30.
9. HOLZ, R. & A. FINKELSTEIN. 1970. J. Gen. Physiol. **56:** 125–145.
10. MARTY, A. & A. FINKELSTEIN. 1975. J. Gen. Physiol. **65:** 515–526.
11. FINKELSTEIN, A. & R. HOLZ. 1973. *In* Membranes, a Series of Advances. G. Eisenman, Ed. **2:** 161–177. Dekker. New York, N.Y.
12. ERMISHKIN, L. N., KH. M. KASUMOV & V. M. POTZELUYEV. 1976. Nature **262:** 698–699.
13. HLADKY, S. B. & D. A. HAYDON. 1970. Nature **225:** 451–453.
14. BAMBERG, E. & P. LÄUGER. 1973. J. Membrane Biol. **11:** 177–194.
15. VEATCH, W. R., R. MATHIES, M. EISENBERG & L. STRYER. 1975. J. Mol. Biol. **99:** 75–92.
16. EHRENSTEIN, G., H. LECAR & R. NOSSAL. 1970. J. Gen. Physiol. **55:** 119–133.
17. MULLER, R. J. & A. FINKELSTEIN. 1972. J. Gen. Physiol. **60:** 285–306.
18. LATORRE, R., O. ALVAREZ, G. EHRENSTEIN, N. ESPINOZA & J. REYES. 1975. J. Membrane Biol. **28:** 3–56.
19. MUELLER, P. & D. O. RUDIN. 1968. Nature **217:** 713–719.
20. BOHEIM, G., K. JANKO, D. LEIBFRITZ, T. OOKA, W. A. KONIG & G. JUNG. 1976. Biochim. Biophys. Acta **433:** 182–199.
21. FINKELSTEIN, A., L. I. RUBIN & M.-C. TZENG. 1976. Science **193:** 1009–1011.
22. CLARK, A. W., W. P. HURLBUT & A. MAURO. 1972. J. Cell Biol. **52:** 1.
23. ANDERSON, C. R. & C. F. STEVENS. 1973. J. Physiol. (Lond.) **235:** 655–691.
24. GORDON, L. G. M. & D. A. HAYDON. 1972. Biochim. Biophys. Acta **255:** 1014–1018.
25. HAYDON, D. A., S. B. HLADKY & L. G. M. GORDON. 1972. *In* Fed. Europ. Biochem. Soc. Mitochondria/Biomembranes. **28:** 307–316. North Holland. Amsterdam, The Netherlands.
26. EISENBERG, M., J. E. HALL & C. A. MEAD. 1973. J. Membrane Biol. **14:** 143–176.
27. BOHEIM, G. 1974. J. Membrane Biol. **19:** 277–303.
28. BAUMANN, G. & P. MUELLER. 1974. J. Supramolec. Struct. **2:** 538–557.
29. GORDON, L. G. M. & D. A. HAYDON. 1975. Philos. Trans. R. Soc. Lond. [Biol. Sci.] **270:** 433–447.
30. GORDON, L. G. M. & D. A. HAYDON. 1976. Biochim. Biophys. Acta **436:** 541–556.
31. HALL, J. E. 1975. Biophys. J. **15:** 934–939.
32. MUELLER, P. 1976. Ann. N.Y. Acad. Sci. **264:** 247–265.
33. MUELLER, P. 1976. *In* MTP International Review of Science, Biochemistry Series One.

3: 75–120. Energy Transducing Mechanisms. E. Racker, Ed. Butterworth. London, England.
34. Roy, G. 1975. J. Membrane Biol. **24:** 71–85.
35. Gisin, B. F., S. Kobayashi & J. E. Hall. 1977. Proc. Nat. Acad. Sci. USA **74:** 115–119.
36. Payne, J. W., R. Jakes & B. S. Hartley. 1970. Biochem. J. **117:** 757–766.
37. Martin, D. R. & R. J. P. Williams. 1976. Biochem. J. **153:** 181–190.
38. Pressman, B. C. 1968. Fed. Proc. **27:** 1283.
39. Nitecki, D. E. & J. W. Goodman. 1967. Biochemistry **5:** 5994.
40. Tanford, C. 1973. The Hydrophobic Effect: The Formation of Micelles and Biological Membranes. John Wiley & Sons. New York, N.Y.
41. Boheim, G. 1977. J. Membrane Biol. In press.

DISCUSSION

Robert E. Taylor, *Moderator*

Section on Molecular Biophysics
National Institutes of Health
Bethesda, Maryland 20014

DR. ARTHUR PILLER (*Columbia University, Physicians and Surgeons College, New York, N.Y.*): Dr. Szabo, if you make a detailed analysis of the impedance of electrodes or membranes, you would not see a simple exponential behavior, but instead at least two components. Moreover, at the concentrations of the ions that you are using, one would expect to see aqueous diffusion. I do not see all of these showing up in your particular approach.

DR. SZABO: Let me concentrate on tetraphenylborate because that is a very simple case, and I think it will also apply to the carriers.

First of all, as to the aqueous diffusion. When you take a trace of the membrane current that results from a square pulse of potential, you find the relaxation process I have shown, and for a few time constants it is really exponential. Then it deviates from exponentiality, and that deviation can be shown to be a result of aqueous diffusion of tetraphenylborate in the unstirred layers across the membrane. As to the absorption capacitance, you find that membrane current even at short times deviates from the exponenitality. Anderson and others have interpreted this as being an absorption capacity. I deliberately stay away from these concentrations because I am interested in the membrane properties alone.

DR. PILLER: In the case where you can see absorption, do you have techniques that allow you to go fast enough to isolate dielectric capacitance or geometric capacitance of the membrane?

DR. SZABO: Yes. Again, as you have noticed, I extrapolated to time zero. It is necessary to extrapolate because of the geometrical capacitance of the membrane. Usually, the speed of our measurements is limited by the series resistance of the electrolyte. You can still observe the relaxation because it is so slow. For carriers it is much more difficult, but you can still do it. Normally the capacitance of the membrane, which is a dielectric relaxation, is over in a few microseconds. What we are looking at is of the order of between 20 microseconds and a few milliseconds, so it is not a problem.

DR. KRONICK: The research done by Dr. Szabo concentrates on the bilayer and leaves out the properties of the water medium. The properties of the water do not seem to be important to the behavior you observed. I want to throw in an observation I made with the bilayer prepared in glycerol. In this medium, the effects of valinomicin and other polar molecules are completely abolished.

DR. SZABO: There is no conductance whatsoever when you have a potassium chloride solution?

DR. KRONICK: As you increase the concentrations of glycerol relative to the water, the increase in conductance due to the valinomicin in the presence of potassium diminishes until about 80% glycerol. In pure glycerol, there is no effect at all.

DR. SZABO: The way I would look at it is that as you add glycerol, you decrease the partition coefficient between the ion-carrying complex which is in the aqueous phase and the membrane surface. Now if you add glycerol, the glycerol presumably dissolves the ion carrier complex from the membrane surface.

DR. KRONICK: I do not have any direct evidence against it, except that there have been similar experiments in nonaqueous solvents, with acetone I think, in which the complex—not valinomicin, but some other compounds—was stable.

DR. SZABO: It is very important to have the water there. That is our reference phase, and even the dipole potentials that I am talking about come from the water.

DR. MURRAY LAMBERT: I think there is an obvious way in which water would have come into Dr. Szabo's talk. It seemed to me that the image force is clearly playing a role. Water has such a big dielectric constant that you had a very large rounding off, and that was purely an image force effect. It is one way in which the water could come in.

DR. SZABO: Conceivably the answer to that is yes, and that is why the energy barrier was drawn the way it was. That would be in accordance with some simple image force calculation.

DR. IBERAR: Dr. White, within your extremely exciting model, is there any difficulty that you envisage with regard to diffusion within the plane of the membrane?

DR. WHITE: Not really. The acyl chains are, of course, in constant motion, and they are whipping back and forth. In the model, I showed that sometimes you see a terminal methyl down at the interface and the chains migrate back and forth. As that happens, the volume left by acyl chain is then occupied by the alkane, so you have the shifting back and forth. I would not guess that there would be any problem with diffusion taking place.

DR. RICHARD PRIEST (*Naval Research Lab., Bethesda, Md.*): Dr. White, you demonstrated that the molecular length is the most important consideration for molecular solubility in the GMO bilayers. I wonder if you have any information on the role of molecular rigidity on the same question?

DR. WHITE: No. It takes a long time to do these experiments, and I have not examined that. Dennis Haydon has looked at a monoglyceride that has three double bonds in it, for example. When all those double bonds are inserted between position 9 and position 18, the membranes do become thinner and the acyl chain bends over; that is a rigidity factor. We have also looked at 5-decane, which is a decane with a double bond in the middle. You get somewhat thinner membranes. I do not know how to relate that to rigidity.

DR. ALAN DRAWSON: Dr. M. Eisenberg, the rectification characteristics you showed suggest that of a semiconductor metal junction or PN junction. Is there any relation, or does one add insight to the other?

DR. M. EISENBERG: It does look like a diode. However, in semiconductors you have holes and electrons, and there is a single interface. In this case the rectification is simply due to the fact that the number of conducting elements is increasing with the voltage very steeply, as though you are changing the conformation of units that are going to form these pores in such a way as to favor the equilibrium toward the formation of pores. I would say that I do not like to think in terms of analogies of semiconductors and this rectification.

DR. O'KONSKI: Dr. White, what value of dielectric constant of the bilayer did you use to calculate the thickness?

DR. WHITE: I will not go into the details of the calculation of dielectric coefficient. The pure acyl chain is 2.2, on the average. If you take all the values from octane through octodecane and average them up, you will find 2.13 plus or minus 0.02, so you get a pretty good estimate just taking 2.13.

DR. JOSHUA ZIMMERBERG: Dr. White, in considering why the thickness of the membrane without any alkane is about on the length of a lipid molecule, how did you rule out interdigitation of the lipid molecule?

DR. WHITE: When you do an x-ray diffraction study of dispersion of lipids, you get an electron density versus thickness, and there is a large decrease in electron density in the center where there is nothing but hydrogen at low electron density. There is absolutely no evidence of interdigitation.

DR. BASH: Dr. M. Eisenberg, have you looked at the effect of transmembrane potential on insertion of alamethicin monomers? You showed that conductance varies to a high power of the impressed voltage across the membrane. Have you looked at impressing voltage first and then trying to get the alamethicin monomers into the membrane?

DR. M. EISENBERG: If you form a membrane in the absence of alamethicin and you apply steady-state voltage and then add alamethicin in three or four minutes, which is the time it takes for alamethicin to reach the membrane, you will arrive at the same steady state as you did before. There is one noticeable difference: if you have some membranes, you are at equilibrium and well stirred. Now you apply a voltage pulse, and you see the first-order kinetics. You reach the steady-state conductance of large numbers of open pores. You leave them for a long time and then shut off the voltage. Now you wait a very short time, two or three seconds, and try to turn it on again. The conductance will no longer start from zero but will start from a steady state. This effect is, to some extent, irreversible with some lipids, in the sense that it would take a long time to wash out.

DR. PILLER: Dr. Eisenberg, is it possible for you to make reproducible measurements of the initial hop to your channel, and then successive hops? Can you examine the shape of the curve between plateaus?

DR. EISENBERG: The time resolution to see the single channels is limited in our system to about a tenth of a millisecond, so within that range all you see is a square rise, or an almost triangular one. This is due to the fact that in order to see single channels you have to start with a bare membrane with a parallel circuit of 10^{10} ohms, roughly speaking, with at best 5×10^{-5} picofarads and that limits the rise time. You cannot see much more detail than a hundred microseconds.

DISCUSSION PAPER: ELECTRICAL PROPERTIES OF LIPID BILAYERS IN NONAQUEOUS MEDIA*

Paul Kronick

Chemistry Department
The Franklin Institute Research Laboratories
Philadelphia, Pennsylvania 19103

Although synthetic bilayers in aqueous media appear to be a good model for biological membranes in their electrical properties, previous attention has usually been focused on the bilayer rather than on the water structure at the membrane-medium interface. We have found that when nonaqueous solvents are substituted for water, several of the "lifelike" responses of the model systems disappear.

In these experiments, membranes were prepared from 1% oxidized cholesterol in *n*-decane, or from 50% tocopherol (Calbiochem)-sphingomyelin (Analabs, Inc.) (25:1) in chloroform-methanol (3:2). In ethylene glycol or glycerol solutions more concentrated than 40 mole %, they were formed on a 0.01-cm^2 round hole in the side of an open polyethylene tube of 3 mm internal diameter thinned to 0.1 mm wall thickness. The lower end of the tube is closed by dipping into a shallow cup filled with tetrachloroethylene fixed to the bottom of a dish filled with the test medium (glycerol or ethylene glycol solution). Additions of various agents to the medium were made from a pipette held within 3 mm of the membrane. By observing the schlieren lines and the movement of small bubbles, we could establish that the agents were fully dispersed by the stirrer in a few seconds. Convection of the agents to the membrane was therefore much more rapid than the membrane reactions observed.

The capacitances of the high-resistance membranes were in the range of 3 to 7×10^{-7} farad/cm^2; breakdown voltages, 500–1000 mV, which is reasonable. Breakdown in all media occurred with a rapid (0.1 s) decrease of electrical resistance of the cell to 10^6 ohm (or less, in highly aqueous solutions), a change of three orders of magnitude. This also shows that the speed of the conductance and potential responses observed were limited by changes in the membranes, not in the solutions or at the electrodes.

Ethylene Glycol. Resistance and potential measurements were made on membranes of tocopherol-sphingomyelin under 24 mole % aq. glycol. The starting medium contained 1.67 mM potassium chloride and 1.6×10^{-6} g/ml valinomycin. When each increment of potassium chloride was added, the membrane potential slowly increased over 5 to 15 minutes to a stable value. Although the response was sluggish and there was considerable scatter, the final potentials followed the theoretical (Nernst) line. In 50 mole % glycol containing valinomycin (3×10^{-6} g/ml) and potassium chloride, the potentials across oxidized cholesterol membranes were far less than theoretical. With a concentration ratio as high as 310 mM/0.834 mM across the membrane, the potential was only 24 mV instead of the

*Supported by the National Heart Institute under Contract NIH-69-2237.

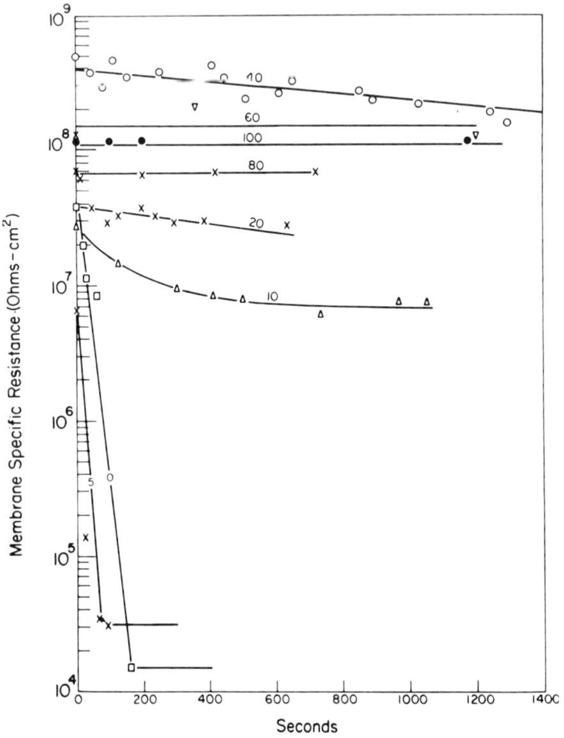

FIGURE 1. Lowering of membrane resistance on addition of 4×10^{-7} g/ml valinomycin to glycerol-water media containing 0.1 M KCl, 0.005 M histidine hydrochloride, pH = 7.0. Membranes are from oxidized cholesterol-decane. Parameter is mole-% glycerol. Measurements are made intermittently after the addition of valinomycin at time = 0 to each membrane in its solvent mixture.

TABLE 1

MEMBRANE POTENTIALS ACROSS OXIDIZED CHOLESTEROL MEMBRANES IN 40 MOLE % AQUEOUS GLYCEROL*

Glycerol Mole %	Potential
20 mole %	72 mV
10	15
5	28

*Valinomycin concentration 4×10^{-7} g/ml, KCl concentration gradient from 1 mM to 100 mM across membranes.

TABLE 2
MEMBRANE RESISTANCE OF TOCOPHEROL-SPHINGOMYELIN MEMBRANES IN 40 MOLE %
GLYCEROL CONTAINING 0.1 M NACL

Valinomycin	Potassium Chloride	Specific Resistance
0 g/ml	0.001 M	6.1×10^7 ohm cm^2
2.7×10^{-7}	0.001	5.3×10^7
2.7×10^{-7}	0.007	5.3×10^7

TABLE 3
RESISTANCE OF OXIDIZED CHOLESTEROL MEMBRANES IN POTASSIUM TRIIODIDE
SOLUTIONS IN WATER AND IN 40 MOLE % GLYCEROL

Potassium Triiodide Concentration	Specific Resistance in Water	Specific Resistance in 40 Mole % Glycerol
0 mM	1.1×10^7 ohm cm^2	—
0	—	4.2×10^7 ohm cm^2
0.11	1.5×10^6	—
0.23	1.5×10^6	—
0.34	9.7×10^5	—
0.58	—	3.2×10^7
0.92	1.4×10^6	—
1.15	—	3.2×10^7

theoretical 152 mV. This 300-fold increase in potassium concentration lowered the membrane resistance by only an order of magnitude, from 10^9 ohms to 10^8 ohms.

Glycerol. Only 10 mole % glycerol greatly diminished the response to valinomycin and made its onset very sluggish (FIGURE 1). In media of 60 mole % glycerol and above, there were no electrical effects of potassium chloride or sodium chloride in the presence of valinomycin or alamethicin at times of up to at least an hour after addition of antibiotic or salt. Although with valinomycin, a 100-fold potassium-ion gradient across oxidized cholesterol membranes in the more aqueous glycerol media was accompanied by a membrane potential, it was poorly reproducible and much smaller than the theoretical 118 mV. Some typical results are shown in TABLE 1, determined after waiting 15 minutes after the addition of antibiotic, when apparent asymptotic values had been reached. The data in FIGURE 1 were obtained on oxidized cholesterol films, but films of tocopherol-sphingomyelin in 40 mole % glycerol were also insensitive (TABLE 2), with no membrane potential greater than 15 mV with a 100-fold potassium concentration gradient. Similar experiments were done with alamethicin, which promotes the transport of both sodium and potassium ions through membranes in aqueous media. For example, an oxidized cholesterol membrane in 100% glycerol containing 1 mM NaCl and 100 mM KCl had a resistance of 5.5×10^9 ohm. Fifteen minutes after addition of 2×10^{-6} g/ml alamethicin, the membrane resistance was 4.9×10^9 ohm. The difference between two such large values of resistance is within experimental error. In 40 mole % glycerol, moreover, no electrical effects were observed (TABLE 3) on addition of triiodide; nor did an oxidized cholesterol membrane in glycerol medium show any change in electrical properties for ten minutes when crystals of dinitrophenol were added to saturate the medium.

ELECTRICAL IMPEDANCE OF EXCITABLE MEMBRANES

Robert E. Taylor

Section on Molecular Biophysics
Laboratory of Biophysics
National Institute of Neurological and Communicative Disorders and Stroke
National Institutes of Health
Bethesda, Maryland 20014

INTRODUCTION

I have been asked to make some comments about studies of the electrical impedance of excitable cell membranes. There are good sources for the early history,[1-3] and I will make no attempt to review the recent literature thoroughly. Most of my remarks will be in reference to the membrane of the squid giant axon. The geometry is relatively simple, and Dr. Eisenberg will be discussing muscle.

Nerve membrane impedance can be measured by noise and correlation techniques, as Guttman and Feldman[31] have demonstrated. I will not discuss this topic, since we shall hear about it from Dr. H. M. Fishman, Dr. D. Poussart, and Dr. L. E. Moore.

Measurements of the electrical and osmotic properties of various cells and tissues have led to the concept that all cells are bounded by a superficial layer with very special properties,[4] and I will not use the expression "cell membrane" to imply more than this. The word "membrane" is probably more commonly used to refer to a thin layer composed of many cells and/or noncellular material.

Some properties are common to the membranes of many different cell types, such as a rather high permeability to water and to lipid soluble substances and the ability to transport ions, sugars, and amino acids up or down electrochemical gradients. For electrical signals above a few thousand cycles per second (kHz) there appears to be nothing exceptional about excitable cell membranes. Excitable membranes, such as those of nerve and muscle, differ from others in that they respond to changes in the electrical potential across them in an active fashion. This activity seems to occur in specialized regions that provide imperfectly selective ionic channels such that the spatial average of an individual ionic current (sodium, potassium, calcium, and others) is a nonlinear function of the potential, and the nonlinearity is associated with a time delay. As K.S. Cole discussed long ago,[5,6] a voltage-dependent conductance with a delay manifests itself as a linear reactance when measured with very small signals. (For further discussion and a generalization by R. FitzHugh see Ref. 1, p. 170). The frequency ranges where these reactances are important depend on the magnitudes of the time delays.

The spatial average of an individual ionic conductance is almost certainly determined in a statistical fashion by the number of units which are either open or closed.[7] The potential dependence presumably results from the movement of charges or dipoles confined to the membrane. The delay in the movement of this charge would be the first step in the process resulting in delays in the voltage-dependent conductances and results in a frequency dependence of the measured capacitance associated with these "gating" currents in the same range as the reactances associated with the conductance changes. This charge movement saturates within plus or minus 100 millivolts (mV) (the external medium is taken as ground). It would seem that it is this saturation which is the first step in the process leading to the voltage-dependence of the ionic conductances. We shall hear more about these matters from Dr. Takashima and Dr. Meves.

Those who attempt to measure the impedance of natural membranes can well envy Dr. White[8] for the relative simplicity of the lipid bilayer preparations. It is not an accident that most impedance measurements have been made on suspensions of red blood cells, marine eggs, and the squid giant axon.[1,2] In those cases where the geometry is least troublesome, the low frequency capacitance is in the neighborhood of 0.8–1.3 microfarads per square centimeter ($\mu F/cm^2$). In 1923 H. Fricke[9,1] obtained a value of 0.81 $\mu F/cm^2$ for red blood cells. If the surface layer was composed of oil with a dielectric constant of 3, then the thickness would be about 33 Angstroms. In 1925 Gorter and Grendel[10] determined that the amount of lipid present was that for a bilayer. Davson and Danielli[11] wanted to put the charged groups on the outside, but for various reasons it was not thought for many years that such a structure would be stable without a surface layer of protein, a conclusion which I still feel is necessary to explain the mechanical properties. Because of the varieties of physiological behavior it was clear that the membrane had to be heterogeneous, and penetrating proteins were obvious candidates.

There seems to be no way to be certain of the surface area involved from optical measurements alone. I will not discuss this subject, and will only mention that there are cases where the surface membrane optically expands on stretch and the apparent capacitance decreases. It hardly seems likely that a flat membrane would be elastic (cf. Ref. 32).

Series Impedance

In their first paper on the transverse electrical impedance of the squid axon in 1938, Curtis and Cole observed a dispersion at frequencies above 200 kHz which Cole[12] has now ascribed to the presence of the Schwann cell layer. From his analysis of this early data he arrives at the equivalent of 12.5 membranes in series (.084 $\mu F/cm^2$ at 1.0 kHz) and a resistance of 1.6 ohm cm^2, both plausible results for Woods Hole squid axons.

The reactance contributed by the Schwann cell layer would thus not interfere significantly with measurements at frequencies below 100 kHz, but the resistance in series with the membrane does cause problems. Hodgkin *et al.*[13] discussed the effects of the series resistance in some detail, as have many authors since. I will only comment that the value for the axons from the squid obtained in Plymouth, England seems to be higher than that for the Woods Hole squid axons.

Components of Membrane Impedance

As I see the situation at the present time, the impedance can be dissected into five parallel components. Leakage conductance, linearized impedance of the ionic conductances, and impedance due to gating currents are strikingly potential-dependent; a well-behaved static capacitance, presumably contributed by a background lipid bilayer, and a lossy part due to penetrating proteins or lipoprotein complexes are not irresistably compelling but appear quite likely.

The last two, plus perhaps some leakage, are probably common to all cell membranes and many subcellular organelles. The lipid-to-protein-weight ratio varies from about 1:4 for the inner membrane of some mitochondria to about 4:1 for the modified Schwann cell membranes that comprise the myelin sheath of medullated axons.

Definitions

The impedance $Z(j\omega) = R(j\omega) + jX(j\omega)$ is the ratio of the Fourier transforms of the voltage and current where R is the resistance, X the reactance, $j = \sqrt{-1}$ and $\omega = 2\pi f$ where f is frequency. The admittance $Y(j\omega) = G(j\omega) + jB(j\omega)$ where G is the conductance and B the susceptance. For a constant capacitance C the admittance is $j\omega C$. We may define the admittance of an unknown two-terminal network as $Y_\mu = j\omega C^*$ where $C^* = C' - jC''$. C^* is referred to as the complex capacitance. We have $Y_\mu = j\omega C^* = \omega C'' + j\omega C'$. What is usually meant by the "capacitance" of the membrane is C', which one hopes is obtained, after correction for any frequency-independent series resistances, by dividing the susceptance by ω. The conductance, or sometimes the conductivity, is the real part of the admittance after correction. This will be equal to $\omega C''$. In general, C' and C'' will be functions of frequency. If we know that our system has a passive, frequency-independent conductance in parallel, we would like to subtract this from the conductance before dividing by ω to obtain C''. When these operations have been performed correctly, we can try to relate the macroscopic complex capacitance to some microscopic model.

A graph of C'' versus C' is often referred to as a Cole-Cole plot,[14] and nothing is implied regarding the physical nature of a network by its use.

Leakage

"Leakage" is a term that was introduced by Hodgkin and Huxley to account for the current carried by ions other than sodium and potassium, as described by them in their empirical equations.[15] It is small in good axons, its origin is obscure, it is not linear,[16] and it would be unprofitable to dwell on it here.

Hodgkin-Huxley Model Admittance

Although much can be learned from certain inadequacies of the model developed by Hodgkin and Huxley,[15] it describes very well the time and voltage-dependent ionic conductances for the membrane of the squid giant axon under normal conditions. From the Laplace transforms of the linearized equations[17] the generalized admittance, where p is the complex transform variable, is $\bar{A}(p) = g_\infty + pC + g_m/(1 + p\tau_m) + g_h/(1 + p\tau_h) + g_n/(1 + p\tau_n)$. If we replace p with $j\omega$ we have $\bar{A}(j\omega) = Y(j\omega)$, which is the ordinary admittance obtained with use of small signals.

The infinite frequency conductance g_∞ is the sum of that for sodium, potassium, and leakage, and for most purposes can be taken to be independent of potential, as it is in the model. As noted above, the leakage is not quite independent of potential and neither is the sodium ion contribution. We take this conductance to be the reciprocal of the slope of the instantaneous current voltage curve where "instantaneous" refers to times long compared to ion redistribution times and short compared to the time-dependent conductance changes.

Because of the values of the H-H model parameters, the admittance measured at sufficiently high frequencies—above 5 kHz, say—will provide information only about $g_\infty + j\omega C$. The far-reaching results of Curtis and Cole in 1938 (see Ref. 1) of measurements of the 20 kHz impedance during the course of an action potential were that the conductance fell by a factor of about 40 and the capacitance changed by less than 2%. At those frequencies the capacitance is not a pure static capacitance as in the model, of which more later, but there has been no indication of any marked time or voltage-dependence. Each of the other terms can be written in the form $g/(1 + pLg)$, which can be represented as a conductance g in series with an inductance L, with an admittance of $g/(1 + j\omega Lg)$. The time constant is always positive, but the individual L's and g's can be positive or negative depending on the value of the potential of which they are functions. It is the negative values for L_m and g_m in the appropriate potential range which provides the instability resulting in action-potential production. These admittance terms are important at frequencies below a few kHz.

It is important to emphasize that during an action potential or a voltage clamp step the parameters that determine the potential-dependence of the g's and τ's are changing with time. The admittance under discussion is defined only for constant g's and τ's and can be measured only in a steady state condition. The time necessary to reach a steady state depends on the τ's which also determine the frequency range of interest. Thus these admittances cannot be measured during an action potential or during the early part of a potential step. They have been measured in the steady state.[1,18,31] By suddenly switching from the zero current to voltage clamp mode, the infinite frequency conductance was measured during the course of the action potential by Rojas *et al.*[19] There has been general agreement between these experimental results and the predictions of the H-H model, which is satisfactory but hardly surprising.

Gating Current Contribution

As mentioned above, both the potential and the time-dependence of the ionic conductances responsible for the excitability of excitable membranes are intimately related to the movement of membrane-confined charges or dipoles, very much as predicted by Hodgkin and Huxley.[15] Much of the ionic current can be eliminated with the use of blocking agents such as tetrodotoxin (sodium channels) or tetraethyl ammonium ions (outward potassium current) or by substitution of the ions with ones that do not pass through the channels. Using internally perfused squid axons and removing the linear components of membrane current with the use of electronic averaging techniques, a current with many of the expected properties was reported by Armstrong and Bezanilla and by Keynes and Rojas in 1973 (see Armstrong and Bezanilla[20] for references). A number of investigators are actively working in this area at the present time. Dr. Meves (this volume) will be discussing gating (also called asymmetry or displacement) currents and Dr. Takashima (this volume), related impedance measurements, so I shall make only a few comments.

What is usually measured is the time course of the nonlinear components of the current during a step of potential where an attempt is made to compensate for series-resistance using electronic feedback.

As far as I know, the question of whether or not an initial delay is present is still unanswered. To the extent that the gating current decline can be represented by a single exponential, the time constant and the integrated charge are sufficient to predict the contribution to the measured impedance. Dr. Bezanilla and I did this several years ago, using data which he and Dr. Armstrong had obtained and published, with the result that the predicted extra capacitance, i.e., the difference between the zero and infinite frequency real part of the complex capacitance, had a maximum in the region of small negative potentials of about 0.3 $\mu F/cm^2$, falling to small values at plus or minus 50–80 mV and characteristic frequencies somewhat less than a kHz for small potentials and somewhat more than 2.5 kHz for large. Armstrong and Bezanilla[20] also computed the zero frequency capacitance directly, without using averaging techniques, from the time course of the integral of the membrane current during a potential step, in 10 mV intervals, and found that it increased from about 1.0 $\mu F/cm^2$ at plus or minus 50 mV to about 1.35 $\mu F/cm^2$ at about -10 mV. They compared these results to the curve of total gating charge movement *in the same axon* and found that the maximum slope of this curve gave a differential capacitance of 0.3 $\mu F/cm^2$ at -30 mV, quite compatible with Dr. Takashima's[21] measurements of admittance at 1.0 kHz.

They have found that the gating current is at least the sum of two exponentials,[20] and a longer time constant component would, of course, introduce another dispersion at a lower frequency.

An important point for measurements of impedance is that the inactivation of the sodium conductance (h in the H-H model) affects the magnitude of the gating current.[20]

Intermediate Frequency Impedance

By "intermediate," I mean high enough (a few kHz) to be out of the range of the Hodgkin-Huxley and gating current reactances and less than a few hundred kHz. Geometrical and other complexities make it difficult to say very much in detail about the membrane contribution to the dispersions observed in impedance measurements of any preparation other than the squid giant axon (cf. Ref. 18).

Early work[1] led to the conclusion that the impedance could be represented as a constant phase angle element (of about 80°) in series with a resistance. A consistent result was obtained by Hodgkin *et al.*[13] in the transient analysis of their voltage clamp results and by Takashima and Schwan,[18] (using alternating currents). Matsumoto *et al.*[22] have suggested a two time constant equivalent circuit which has been criticized on technical grounds by Takashima and Schwan.[18] Everyone seems to agree that there is very little in the way of potential-dependence in this frequency region.

The measurements I did with Knox Chandler[23] indeed confirmed, using internal and external electrodes with the squid axon, that this membrane may be represented as a constant phase angle impedance plus series-resistance and thus can be considered as such for the equivalent circuit in voltage clamping or computation of the spread of current in a cable.[24,25] This does not mean that, in fact, that is what it is physically.

In trying to deduce something about the physical nature of the membrane, other possible equivalent circuits can be investigated. Polarization capacitance resulting from restricted diffusion appeared to be unlikely because one would expect a significant potential-dependence. By taking the series resistance and parallel conductance to be adjustable parameters, it was possible[26] to achieve a reasonably good fit for the impedance data with the use of an equivalent circuit containing a static capacitance in parallel with a single time constant dielectric with an infinite frequency capacitance of 0.6 $\mu F/cm^2$ and 0.8 $\mu F/cm^2$ at zero frequency. The fit to the transformed data in terms of the complex capacitance was not quite as good, requiring a moderate distribution of time constants centered around 5.7 microseconds. Analyzed in this way the positive temperature coefficient of 1.0%/°C (and see Palti and Adelman[27]) resided in the nonlossy part.

Beyond this it is pure speculation, but it would seem reasonable that some 20% of the membrane is composed of protein with an infinite frequency permittivity of about 3. If so, then the 80% of the membrane being a nonlossy lipid bilayer (cf. White, this volume), of specific capacitance about 0.6 $\mu F/cm^2$, would contribute 0.48 $\mu F/cm^2$ and the protein, referred to *its* surface area, would have an infinite frequency capacitance of 0.6 $\mu F/cm^2$ and a zero frequency capacitance of 1.0 $\mu F/cm^2$. I do not have any idea at this time of how to estimate the thickness of the membrane in the region of a penetrating protein—somewhere between 20 and 200 Angstroms, I suppose, depending on the protein.

Our data covered the range of 10–70 kHz. For the squid giant axon the Schwann cell layer will interfere above 100 kHz.[12] It is now possible to eliminate most of the ionic current contribution at low frequencies, as has been done for gating current measurements. In order to distinguish the low-frequency impedance contributed by the nongating current but still lossy dielectric of the membrane, it would be good to see the results of impedance measurements made after the additional elimination of the gating current with internal zinc ions.[28,29]

With good impedance data over a wide frequency range, it might be possible to think about the use of dipole correlation functions[30] in addition to distributions of relaxation times in attempts to relate the data to physical models.

Conclusion

From a functional point of view the impedance in many cases is quite well understood. For the squid axon a resistance in series with a constant phase angle element is adequate for frequencies above a few kHz and within the band width of most measuring devices. At lower frequencies the admittance as described by the linearized Hodgkin-Huxley model is predictable to the extent that the model is adequate. The displacement currents related to gating of ionic channels are small under most experimental conditions and will contribute significantly to the linearized admittance with dispersions where the dominant frequencies are potential-dependent and below 3 kHz. The gating currents represent a saturable dielectric with maximum differential capacitance in the region of -10 to -30 mV.

From the structural point of view there are still many unresolved problems. Above 100 kHz the external Schwann cell layer dominates the squid axon impedance results. Below a few kHz the Hodgkin-Huxley model admittances obscure the impedance contributed by the membrane as such. If these are eliminated, the gating current contributions to the admittance remain. Work on gating currents has sparked new interest in low-frequency impedance studies, as evidenced by the results of Armstrong and Bezanilla using steps, Takashima with bridge techniques, and Poussart *et al.* with the application of noise and correlation procedures.

Acknowledgments

I should like to thank Dr. Francisco Bezanilla, Dr. Gerald Ehrenstein, and Dr. Kenneth S. Cole for many helpful discussions. I have benefited enormously from the deep understanding of these subjects of Dr. Harold Lecar.

References

1. Cole, K. S. 1968. Membranes, Ions and Impulses. Univ. California Press. Berkeley, Calif.
2. Taylor, R. E. 1963. Cable theory. *In* Physical Techniques in Biological Research. W. L. Nastuck, Ed. Vol **6:** 219–262. Academic Press. New York, N.Y.
3. Hodgkin, A. L. & W. A. H. Rushton. 1946. The electrical constants of a crustacean nerve fibre. Proc. R. Soc. (Lond.) [Biol.] **133:** 444–479.
4. Katz, B. 1966. Nerve, muscle, and synapse. McGraw-Hill. New York, N.Y.
5. Cole, K. S. 1947. Four Lectures on Biophysics. Institute of Biophysics. Univ. of Brazil. Rio de Janeiro, Brazil.
6. Cole, 1949. Some physical aspects of bioelectric phenomena. Proc. Nat. Acad. Sci. USA **35:** 558–566.
7. Ehrenstein, G. & H. Lecar. 1972. The mechanism of signal transmission in nerve axons. Ann. Rev. Biophy. Bioeng. **1:** 347–368.
8. White, S. This volume.
9. Fricke, H. 1923. The electric capacity of cell suspensions. Physic. Rev. **21:** 708–709.
10. Gorter, E. & F. Grendel. 1925. On bimolecular layers of lipoids on the chromocytes of the blood. J. Exp. Med. **41:** 439–443.
11. Davson, H. & J. F. Danielli. 1952. The permeability of natural membranes. 2nd edit. Cambridge Univ. Press. London, England.

12. Cole, K. S. 1976. Electrical properties of the squid axon sheath. Biophys. J. **16:** 137–142.
13. Hodgkin, A. L., A. F. Huxley & B. Katz. 1952. Measurement of current-voltage relations in the membrane of the giant axon of *Loligo.* J. Physiol. Lond. **116:** 424–448.
14. Cole, K. S. & R. H. Cole. 1941. Dispersion and absorption in dielectrics. I. Alternating current characteristics. J. Chem. Phys. **9:** 341–351.
15. Hodgkin, A. L. & A. F. Huxley. 1952. A quantitative description of membrane current and its application to conduction and excitation in nerve. J. Physiol. **117:** 500–544.
16. Adelman, W. J. & R. E. Taylor. 1961. Leakage current rectification in the squid giant axon. Nature **190:** 883–885.
17. Chandler, W. K., R. FitzHugh & K. S. Cole. 1962. Theoretical stability properties of a space-clamped axon. Biophys. J. **2:** 106–127.
18. Takashima, S. & H. P. Schwan. 1974. Passive electrical properties of squid axon membrane. J. Membrane Biol. **17:** 51–68.
19. Rojas, E., F. Bezanilla & R. E. Taylor. 1970. Demonstration of sodium and potassium conductance changes during a nerve action potential. Nature **225:** 747–748.
20. Armstrong, C. M. & F. Bezanilla. 1975. Currents associated with the ionic gating structures in nerve membrane. Ann. N.Y. Acad. Sci. **264:** 265–277.
21. Takashima, S. 1976. Membrane capacity of squid giant axon during hyper- and depolarizations. J. Membrane Biol. **27:** 21–39.
22. Matsumoto, N., I. Inoue & U. Kishimoto. 1970. The electrical impedance of the squid axon membrane measured between internal and external electrodes. Jap. J. Physiol. **20:** 516–526.
23. Taylor, R. E. & W. K. Chandler. 1962. Effect of temperature on squid axon membrane capacity. Biophy. Soc. (abstr.) TD1.
24. FitzHugh, R. & K. S. Cole. 1973. Voltage and current clamp transients with membrane dielectric loss. Biophys. J. **13:** 1125–1140.
25. FitzHugh, R. & K. S. Cole. 1974. Dielectric loss transients. Biophys. J. **14:** 625–626.
26. Taylor, R. E. 1965. Impedance of the squid axon membrane. J. Cell. Comp. Physiol. **66** (Part II): 21–25.
27. Palti, Y. & W. J. Adelman, Jr. 1969. Measurement of axonal membrane conductance and capacity by means of a varying potential control voltage clamp. J. Membrane Biol. **1:** 431–458.
28. Begenisich, T. & C. Lynch. 1974. Effects of internal divalent cations on voltage-clamped squid axons. J. Gen. Physiol. **63:** 675–689.
29. Bezanilla, F. & C. M. Armstrong. 1974. Gating currents of the sodium channels: three ways to block them. Science **183:** 753–754.
30. Williams, G. 1972. The use of the dipole correlation function in dielectric relaxation. Chem. Rev. **72:** 55–69.
31. Guttman, R. & L. Feldman. 1975. White noise measurement of squid axon membrane impedance. Biochem. Biophys. Res. Comm. **67:** 427–432.
32. Taylor, R. E. 1974. Excitable membranes. Ann. Rev. Phys. Chem. **25:** 387–405.

INVESTIGATION OF VOLTAGE-DEPENDENT MEMBRANE CAPACITY OF SQUID GIANT AXONS

Shiro Takashima* and R. Yantorno†

Department of Bioengineering D2
University of Pennsylvania
Philadelphia, Pennsylvania 19174

Introduction

Ever since Cole and Curtis[1,2] and Cole and Baker[3] observed that membrane capacity of nerve axons is a passive quantity and does not change during activity, it is generally agreed that nerve excitation is a process that involves very little molecular motion.‡ On the other hand, Hodgkin and Huxley[4] postulated that the opening and closing of ionic channels may be a motion of dipole molecules in response to the change in electrical fields in the membrane. Also, Chandler *et al.*,[5] using the linearized Hodgkin-Huxley Equation, discussed the presence of frequency-dependent capacitances due to sodium and potassium currents.

These two sets of information, i.e., (1) early observation of invariance of membrane capacitance during depolarizations and (2) the prediction of displacement currents due to gating particles or to the reactance component associated with time-dependent ionic currents, are not necessarily compatible with each other and indicate some missing information that is still to be uncovered. Schwan[6] pointed out that the conductance increase observed by Cole and Curtis[1,2] during the action potential must be paired with an increase in membrane capacitance. However, his claim was not substantiated by experiments. Recent observations of gating currents[7-13] added a new dimension to this problem and suggested the presence of capacitive currents associated with the opening and closing of ionic gates. Furthermore, Schneider and Chandler[14] and Adrian and Almers,[15,16] using time-domain techniques, observed that membrane capacitance of skeletal muscles is voltage dependent and increases with depolarization. Although these new developments indicate that capacitance of excitable membranes is voltage dependent and possibly related to the gating mechanism, there has been no renewed effort on frequency-domain measurements of membrane capacitance during activity. Takashima,[17,18] using the frequency domain measurement, recently reported that membrane capacitance of squid axons is indeed dependent on the membrane potential and changed by as much as 50% between −160 mV and +40 mV. He points out (see also Kishimoto[19]) that the change in membrane capacitance during depolarization was not uncovered because of the masking of possible changes due to the presence of series resistances. The correction for them was therefore a

*Recipient of ONR-N-00014-76-C-642.

†Recipient of Predoctoral Training Grant NIH GM-527875.

‡ It should be emphasized that measurements by Cole *et al.* were limited to 20 kHz where only the limiting conductance change can be observed.

crucial step in observing the decrease and/or increase in membrane capacitance during hyper- and depolarizations.

This report is an extension of previous ones and encompasses results of further investigations on the voltage dependence of membrane capacitance in the presence and absence of pharmacological agents that block sodium and potassium channels selectively. In addition to these measurements (during externally applied pulses), we will report the result of measurements of capacitance during the action potential. This is the condition under which Cole and Curtis[1,2] observed the constancy of membrane capacitance, originally using external electrodes. It is important and opportune to reinvestigate the behavior of membrane capacitance under this condition with, however, a better electrode arrangement; i.e., an internal electrode. It is generally agreed that the use of internal axial electrodes is more straightforward and less ambiguous than external electrode arrangements for the measurement of membrane capacitance and conductance

Experiments

All experiments were performed at the Marine Biological Laboratory in Woods Hole, Mass., with giant axons of squid *Loligo pealeii* with circulating seawater at a temperature of about 10°C. Internal electrodes are Pt-Ir wire with a diameter of 75 microns, and the length of the electrodes is 25–27 mm. These long electrodes are chosen in an attempt to minimize the effect of fringe capacitance at the tip of the axial electrode. Electrodes are coated with platinum black in order to reduce the effect of electrode polarization. Ground electrodes are platinum plates, also coated with Pt-black, which are placed on both sides of axons and pressed to the axon wall as closely as possible. We observed that the size and position of ground electrodes have considerable effects on the value of series resistance. In order to decrease the value of series resistances, therefore, ground electrodes were designed to be as wide as possible.

The values of series resistances are determined for each axon by first transforming measured capacitance and conductance into reactance and resistance and plotting them in the complex plane. The intersection of impedance locus (for $f = \infty$) with the resistance axis gives the value of series resistance. In our electrode arrangement, the value of R_s is usually 6–10 ohms/cm^2. The presence of series resistance which is due to seawater, Schwann cell layers, and axoplasm causes a substantial error in the value of membrane capacitance at high frequencies. In particular, for depolarized membranes, the error due to R_s is serious even at low frequencies, as analyzed later. As has been discussed, the membrane immersed in an electrolyte solution can be represented by an equivalent circuit with a resistance (R_s) in series with the membrane admittance (C and G). Under these circumstances, the measured capacitance is given by the following equation:[20]

$$C_p = \frac{CR^2/(R + R_s)^2}{1 + \left(\dfrac{R_s R C\omega}{R + R_s}\right)^2} \tag{1}$$

where $R = 1/G$ and ω is angular frequency $(2\pi f)$. At sufficiently low frequencies, this equation reduces to a simple form, since $\omega \rightarrow 0$, namely,

$$C_p = \frac{C}{(1 + R_s/R)^2}. \tag{2}$$

This equation clearly shows that measured capacitances are always smaller than true membrane capacitance for all nonzero values of series resistance. For passive membranes, i.e., hyperpolarized or at rest, the value of R is much greater than that of series resistances (membrane resistivity (R is about 1 Kohm/cm^2, whereas R_s is 6–10 ohms/cm^2). Therefore, the difference between measured capacitances and true membrane capacitances at low frequencies is very small. For depolarized membranes, however, the membrane resistivity decreases substantially and the ratio R_s/R approaches a value of 1. Under these circumstances, measured capacitances are considerably smaller than the true membrane capacitance. The ratio R_s/R increases progressively with the extent of depolarization. Therefore, the possible increase in the capacitance of depolarized membranes can be compensated by the increase in the ratio R_s/R, and the net change could be very small. For example, if the observed capacitance and resistance during the action potential is 1 μF/cm^2 and 40 ohms/cm^2, Equation 2 gives the value of true membrane capacity to be 1.56 μF/cm^2 if we assume the value of R_s to be 10 ohms/cm^2. The repeated corrections of measured capacitances and conductances at each frequency are cumbersome and tedius. This difficulty is circumvented easily by use of a small computer, PDP-9, and the whole procedure can be completed within a few minutes. Particularly, the choice of values for R_s is not necessarily straightforward and the use of a computer enables us to investigate the extent of errors introduced by the choice of wrong values of series resistances. According to our calculations, the error in the value of corrected membrane capacitance due to a choice of wrong R_s values (within ± 0.7 ohms/cm^2) is relatively small. Additionally, this method for the determination of values for R_s's is based on an assumption that the series resistance is independent of frequency. According to Cole,[21,22] the resistivity of axoplasm is purely ohmic and no frequency dependence was detected. Furthermore, the impedance of Schwann cells has an anomalous dispersion in the frequency range above 5 MHz. We therefore concluded that the assumption that the series resistance is frequency independent in the range of our measurements is well justified.

The long internal electrode was inserted from one end of axons almost all the way to the other end, the tip of the electrode nearly reaching the air-gap. Thus, the stray field from the tip of the electrode extends over this portion where the axon may not be viable. This arrangement was purposely selected so that the tip effect or stray capacitance could be minimized. Membrane capacitance (μF/cm^2) measured with internal electrodes having different lengths ranging from 5 mm to 26 mm are shown in FIGURE 1. As one can see, membrane capacitance measured with a 5-mm electrode (for example) is larger than those measured with longer electrodes. We interpret these higher values as arising from the fringe effect, and only the interpolated value is considered free of the tip capacitance. Likewise, the phase angles measured with electrodes with varying lengths are shown in FIGURE 2. Obviously, the phase angle also depends on the length of the internal electrode approaching a value between 80° and 83° as the length increases. Therefore, the errors introduced in the value of membrane capacity and phase angle by the use of short electrodes are clearly demonstrated by these discussions. It is important to note that the phase angle approaches a value not far from 90° as the contribution of fringe effects decreases. In practice, the corrections for the tip capacitance were performed by repetition of the measurements of the capacitance and conductance of resting as well as depolarized membranes, using an electrode with its tip exposed only for 1 mm. The values obtained with this electrode were subtracted from those obtained with a long electrode. Because of this correction, the membrane capacity became slightly smaller than the value

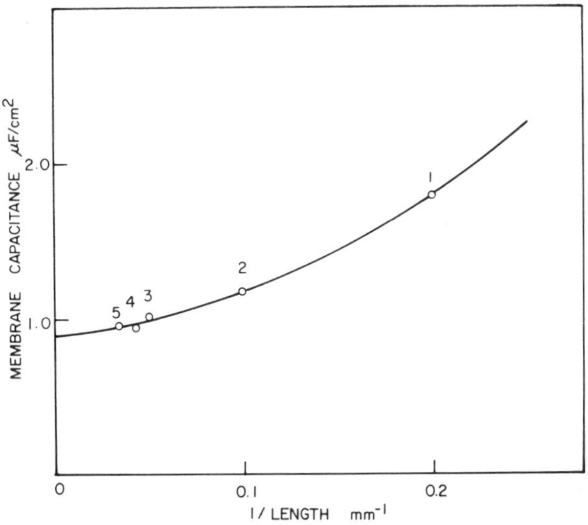

FIGURE 1. Membrane capacitances measured with internal electrodes with varying lengths. The abscissa is the reciprocal of the length of electrodes. Three points (1, 2, 3) are measured with one axon and points (4 and 5) with other axons.

reported by Cole and Curtis originally (1 $\mu F/cm^2$ vs. 0.85–0.9 $\mu F/cm^2$). The increase in the tip capacitance during depolarization was found to be very small, perhaps because of the position of the tip of the axial electrode.

Capacitance and conductance were measured with a Wayne-Kerr wide-band admittance bridge B-221, which covers a frequency range between 100 Hz and 30 kHz. A PAR-124 Lock-in amplifier was used as an oscillator and detector. The variable Q's of the amplifier was set at 2–5 in order to minimize the artifacts caused by external pulses without loss of sensitivity. The neutral lines of the isolation transformer is connected with a large capacitor of 10 μF and external pulses are applied across the capacitor.[23] With the use of a glass microelectrode, it was confirmed that the potential applied across the capacitor is reproduced across the membrane. External pulses with a duration of 20–30 msec are used for hyper- and depolarizations of the membrane. Likewise, a space-clamped action

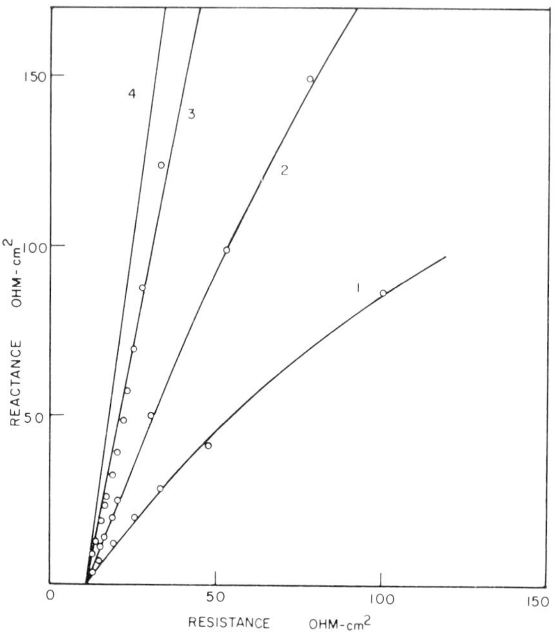

FIGURE 2. Phase angles measured with internal electrodes with varying lengths. Curves 1, 2, and 3 are obtained with 1.3, 5, and 26 mm electrodes, and phase angles are 50°, 67°, and 78°. Curve 4 is obtained after the correction described in the text (electrode length 26 mm and 1.3 mm). Phase angle is 83°.

potential can be triggered by applying a short pulse across the capacitor. It was learned that large polarizing pulses can be applied by using this technique uniformly over the entire axon without noticeable deterioration of nerves. The membrane potential was varied from −160 mV to +40 mV.

The voltage of AC signals for the measurement of admittance is kept minimum at the level of 1–2 mV ($p - p$) across the membrane. The admittance of the membrane, particularly when depolarized, is sensitive to the field strength, and it is crucial to use small AC signals for the bridge measurement. For measurements during polarizing pulses or during the action potential, the bridge was first balanced with the resting membrane. Application of polarizing pulses causes a change in the admittance of that membrane, and this, in turn, unbalances the bridge. The AC currents that flow through the bridge arms are monitored and displayed on an oscilloscope. Secondly, the bridge is rebalanced to the polarized membrane and the values of capacitance and conductance are obtained. For externally applied pulses, the bridge is rebalanced after the steady state is reached. For the measurement during the action potential, the bridge is rebalanced at the peak of the impulse. (The maximum admittance change may not be identical with the peak of the action potential).

Sodium current is blocked by use of TTX in the bathing solution at the concentration of 3 × 10^{-7} mol, and potassium current is eliminated or greatly reduced by replacing potassium ions with cesium fluoride by internal perfusion.[24] To block both sodium and potassium currents, TTX in ASW combined with internal Cs is used. Voltage clamp experiments indicate that there is a minute outward current. This may be due to the fact that a small amount of potassium fluoride is added in the Cs internal solution in order to stabilize the membrane. The resting membrane potential with the Cs internal solution was found to be −30 mV instead of the usual value, −55 to −60 mV.

Results

Externally Applied Polarization

The membrane capacitance and conductance of resting, hyperpolarized, and depolarized membranes were investigated during externally applied pulses under normal conditions. Membrane capacitances under these conditions are shown in Figure 3. As shown, membrane capacitance at rest is slightly frequency dependent even after corrections for series resistance and fringe effect. The magnitude and frequency dependence of membrane capacitance diminishes with hyperpolariza-

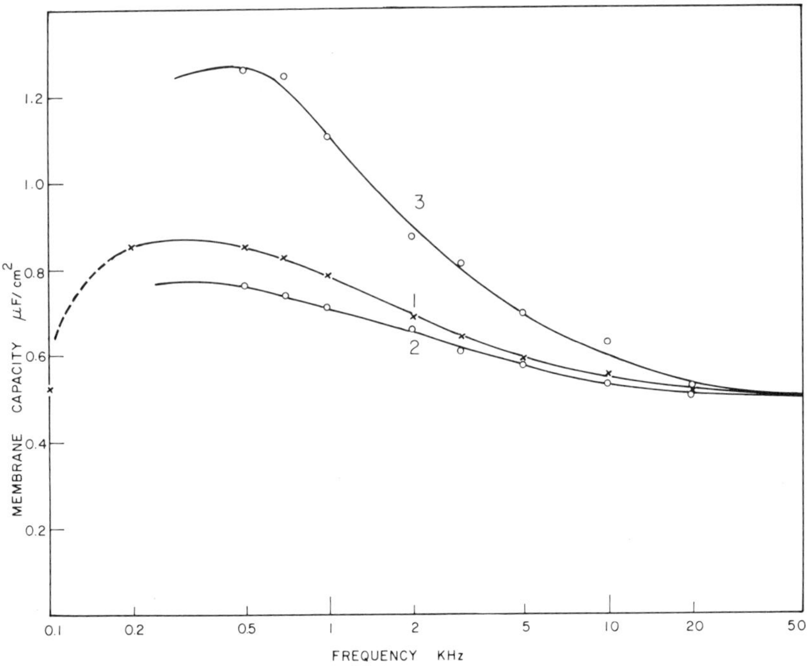

Figure 3. Membrane capacities of squid giant axons at rest (curve 1), and during hyperpolarizing (curve 2) and depolarizing pulses (curve 3). Curve 2 and 3 were obtained at membrane potentials −140 and +10mV. NSW at 10°.

tions to some extent. However, even with large hyperpolarizations, the frequency dependence cannot be abolished, indicating a certain degree of freedom of rotation of polar components in the passive membrane. If the membrane is depolarized, large increase in capacitance as well as conductance are observed. The high-frequency limiting value of the capacitance remains constant, however, and thus the frequency dependence becomes much more pronounced. The low-frequency static capacitances are determined for a wide range of membrane potential between −160 mV and +40 mV, and the results are shown in FIGURE 4 along with membrane conductances at the same membrane potentials. First of all, membrane conductance undergoes an enormous increase, and if the value at

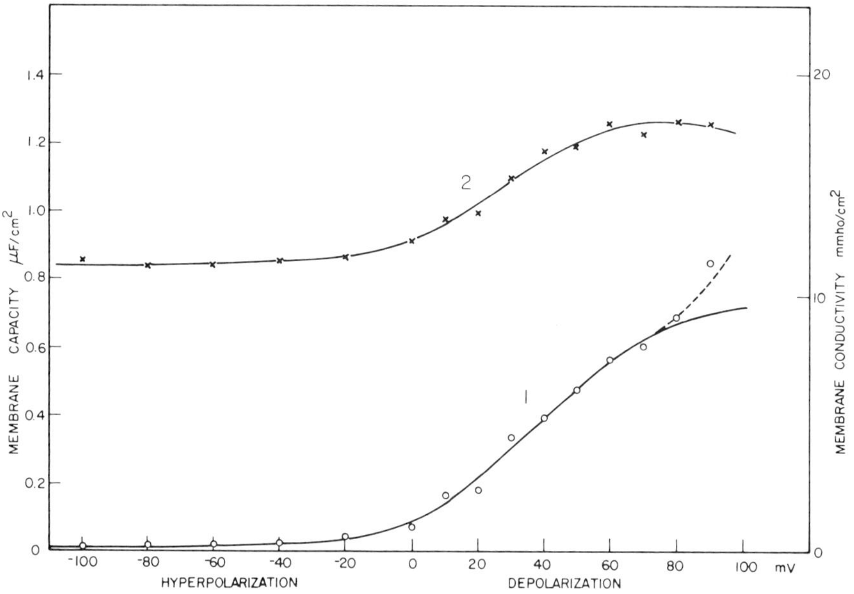

FIGURE 4. Voltage dependence of membrane conductance (curve 1) and capacitance (curve 2). Capacitances and conductances are determined at 500 Hz. Intact axons in NSW. Resting potential about −60mV.

−160 mV is compared with the values at +30 or +40 mV, the increase is almost 100-fold. This result is obviously in qualitative agreement with those obtained by Cole and Baker.[3] Secondly, the capacitance increases from 0.8 $\mu F/cm^2$ to 1.3 $\mu F/cm^2$ and reaches a saturation at the membrane potential around +20 to +30 mV. This result could be compared with those obtained by Meves,[24] indicating that the charge movement during depolarizations measured by gating current methods also reaches a saturation at the same membrane potential.

The question whether the voltage-dependent membrane capacitance is related in any way to ionic channels or currents can be answered by measuring the membrane capacitance blocking sodium and/or potassium channels, using pharmacological agents. First, the sodium channel is selectively blocked by TTX in ASW and the voltage dependence of membrane capacitance was reinvestigated.

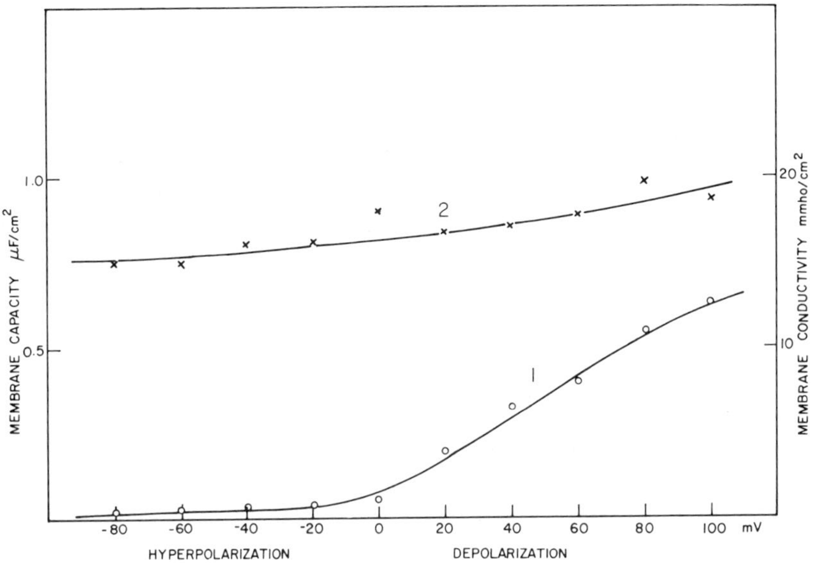

FIGURE 5. Voltage dependence of membrane conductance (curve 1) and capacitance (curve 2) in the presence of 3×10^{-7} mol TTX in ASW at 10°.

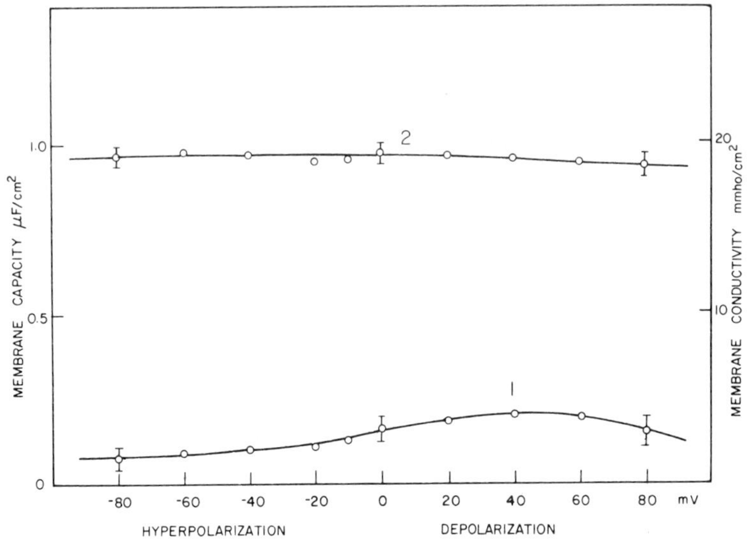

FIGURE 6. Voltage dependence of membrane conductance (curve 1) and capacitance (curve 2) in the absence of ionic currents. Axons are internally perfused with CsF internal mixture. 3×10^{-7} TTX in ASW. Vertical bars are standard deviations. Resting potential −30 mV. No correction for fringe effects.

The membrane capacitances with external TTX are illustrated in FIGURE 5. As shown, the voltage dependence of the capacitance is nearly abolished by TTX, whereas conductance remains essentially the same. This indicates that voltage-dependent capacitance originates from the sodium channel, but conductance has a different origin. Second, both sodium and potassium channels are blocked by TTX and Cs. Membrane capacitance and conductance measured under these conditions are shown in FIGURE 6. Membrane capacitance is nearly independent of membrane potential, but the conductance is slightly voltage dependent. This small voltage dependence, however, may be due to the presence of a small outward current detected by the voltage clamp experiment. Nevertheless, blocking potassium channels results in greatly suppressed voltage dependence of conductances. From these results we can conclude that the change in membrane capacitance and conductance with membrane potential is a phenomenon closely related to the gating of ionic channels. More specifically, the capacitance change is likely to stem from sodium channels, while conductance changes are due mainly to potassium channels. Finally, the change of time constants with depolarization is tabulated in TABLE 1.

TABLE 1

TIME CONSTANT OF SODIUM ACTIVATION MEASURED BY MEMBRANE CAPACITY METHOD AND CALCULATED BY HODGKIN-HUXLEY EQUATION

Depolarization	0.	20.	40	60	80	100	mV
Time Constant (Obs)	0.052	0.054	0.073	0.076	0.074	0.051	msec
Time Constant (Cal)	—	0.479	0.357	0.111	0.075	0.057	msec

Capacitance Change During the Action Potential

So far, the changes in membrane capacitance and conductance during externally applied pulses have been discussed. The logical extension of these experiments is to investigate the change in capacitance during the action potential. Unlike externally applied pulses (20–30 msec), the action potential lasts only for 1.5–2 msec, and this fact imposes a severe restriction on the transient measurement at low frequencies. The lowest frequency Cole and Curtis used was 10 kHz, and many measurements were made at 20 kHz. We extended the measurement down to 2 kHz by careful designs such as shielding of the experimental setup by use of a well-grounded cage to eliminate low-frequency noises and the use of very small AC signals for the measurement of capacitance and conductance. FIGURE 7 shows the admittance change during the action potential and rebalancing of the bridge at the peak of the action potential. First of all, *without correction* for series resistance, membrane capacitance was found to be constant or slightly smaller than the resting value, as reported by Cole and Curtis.[1,2] As pointed out, however, the correction for series resistance must be applied to measured capacitances in order to obtain real values of membrane capacitance. The consequence of this correction has been clearly demonstrated in the previous section of this report.

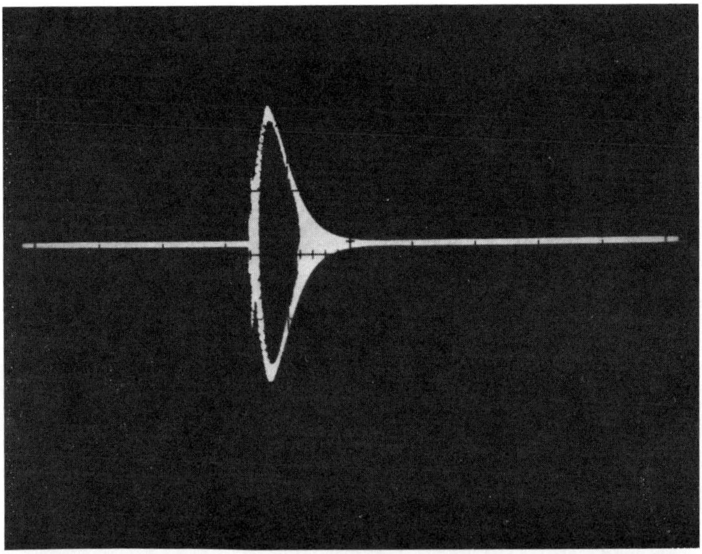

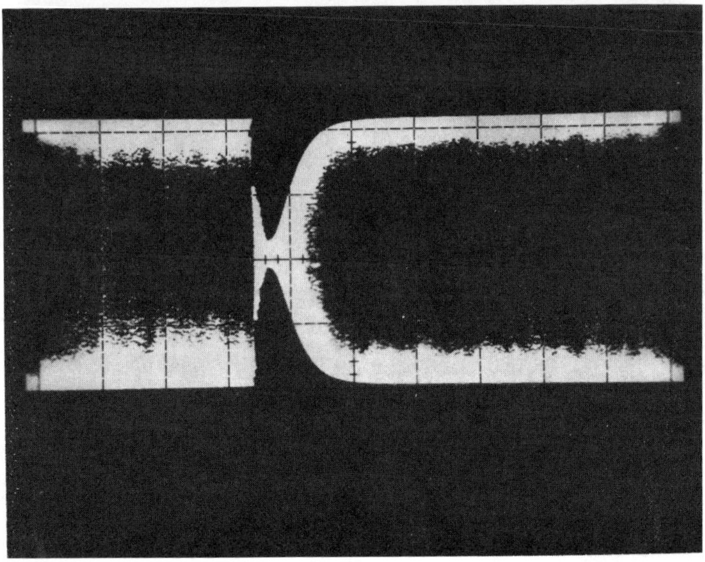

FIGURE 7a (*top*): Unbalancing of admittance bridge during the action potential. Frequency at 5 kHz.

FIGURE 7b (*bottom*): Rebalancing of the bridge at the peak of the admittance change (the peak may differ slightly from that of the action potential).

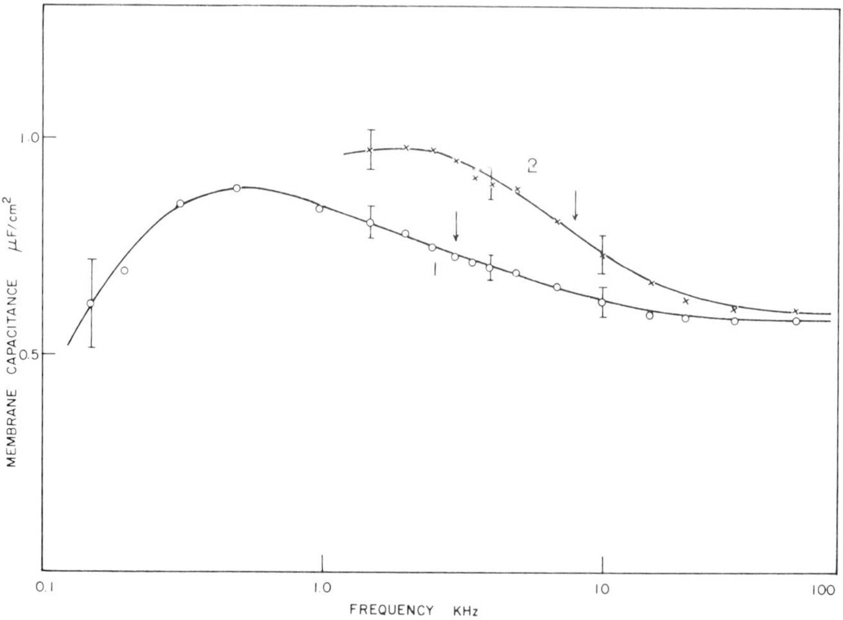

FIGURE 8. Membrane capacitances of resting membrane (curve 1) and during the action potential (curve 2) at various frequencies. NSW at 10°. Arrows indicate the relaxation frequencies, and time constants are calculated by $\tau = \frac{1}{2}\pi f_r$.

FIGURE 8 shows the capacitances after corrections for series resistances and fringe effects for resting membrane and at the peak of action potentials. This figure reveals two interesting features. Membrane capacitance increases during the action potential, but the extent of the increase is relatively small (only from 0.9 $\mu F/cm^2$ to 1 $\mu F/cm^2$). The time constant, τ, ($\tau = \frac{1}{2}\pi f_r$, where f_r is relaxation frequency), on the other hand, indicates a large change from 0.53 msec (f_r = 3 kHz) to 0.199 msec (f_r = 8 kHz). Using the conductivity data, we can calculate the imaginary part (C″)§ of complex permittivity, namely:

$$C'' = (G - G_0)/\omega \qquad (3)$$

where G and G_0 are conductances at given and zero frequencies and ω is angular frequency. C'' is normally interpreted as the measure of absorption of electromagnetic energy by dipoles. The maximum value of C'' is, theoretically, located

§The quantities C' and C'' used by us are related to the more common notations ϵ' and ϵ'' by

$$C' = \epsilon'\epsilon_r/d \quad \text{and} \quad C'' = \epsilon''\epsilon_r/d$$

where d is membrane thickness and ϵ_r is the permittivity of free space, i.e., $8.3 \times 10^{-14}\ \mu F/cm^2$.

at the relaxation frequency and its magnitude C''_m is stipulated by the following equation:[25,26]

$$C''_m \leq \Delta C/2 \quad \text{with} \quad \Delta C = C_0 - C_\infty \tag{4}$$

where C_0 and C_∞ are low- and high-frequency-limiting values of capacitance. According to FIGURE 7, the value of ΔC for resting membranes is 0.3 $\mu F/cm^2$ and the observed value of C''_m is 0.14 $\mu F/cm^2$ from FIGURE 9. Therefore, the inequality stipulated by (**4**) is satisfied and the results are internally consistent. However, for active membranes during the action potential, the value of C''_m is as high as 0.46 $\mu F/cm^2$) whereas the value of ΔC is only 0.39 $\mu F/cm^2$. These two values are therefore mutually inconsistent. According to the value of C''_m, the capacitance increment ΔC should be about 0.9 $\mu F/cm^2$ instead of 0.39 $\mu F/cm^2$. Therefore, either the capacitance increase during the action potential should be much higher or the conductance increase from which C'' is calculated should be smaller than the observed.

DISCUSSION

In this research we investigated the change in membrane capacitance during externally applied pulses and during the action potential. In both cases we found that membrane capacitance and time constant depended on membrane potential. These results seem to agree with the gating current measurements with squid

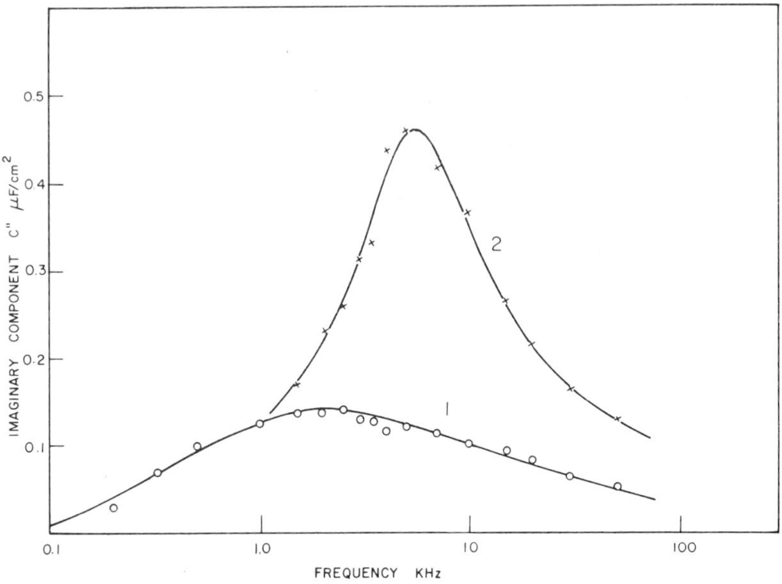

FIGURE 9. Imaginary part (C'') of complex permittivity for resting membrane (curve 1) and during action potential (curve 2). These are calculated from conductance data using (**3**).

axons[7-12] and frog nodes.[13] We found, however, that the voltage dependence of membrane capacitance and conductance was profoundly affected by blocking sodium and potassium channels, whereas the time-domain analysis uncovered the voltage-dependent capacitance or capacitive current under similar conditions. Although our results indicate that capacitance changes of polarized membranes are closely related to the gating of channels, these observations demonstrate a curious difference between those obtained by time-domain analyses and our frequency-domain technique. Fishman *et al.*[27] have been performing admittance measurements with a pseudorandom noise technique in the absence of ionic currents. They report that a small decrease in membrane capacitance was observed with depolarizations, rather than an increase. This result is consistent with ours and seems to reinforce our view that there are missing links between time-domain analyses and frequency domain techniques.

Membrane capacitance as well as time constants were found to change markedly during externally applied hyper- and depolarizing pulses. However, during the action potential, only time constants underwent a considerable change, while the increase in membrane capacitance is nominal. With externally applied pulses, the capacitance was measured after the steady state had been reached, and therefore the time delay between the onset of the pulse and the determination of capacitance was of the order of 3–4 msec. On the other hand, it takes less than 1 msec to reach the peak of the action potential, and this difference may be used to account for the different results obtained in these two sets of experiments. This disparity, however, may be connected to the internal inconsistency pointed out previously; that is, the magnitude of C'' does not satisfy the condition set forth by (**4**) and thus violates one of the fundamental rules of the dielectric theory. If the value of C'' is used (instead of observed membrane capacitance, i.e., ΔC), the increment of membrane capacitance during the action potential should be about 0.9 μF/cm^2 (total capacity, 1.5 μF/cm^2 rather than 1 μF/cm^2). Although this value is not quite comparable to that observed with externally applied pulses (see FIGURE 2), the two sets of results become more consistent with each other. Therefore, the problem reduces, once again, to a question: Why does the membrane violate the rule of inequality (Equation 4)?

The voltage-dependent capacitance indicates the presence of dipolar species or ionic processes that give rise to dipole moments. According to calculations by Levitan and Palti[28] and by Takashima *et al.*,[17] the dipole moment change of nearly 1000 D.U. (10^{-18} esu cgs) is predicted with a depolarization of 100 mV. The capacitance change observed by us is related to the dipole moment change by the well-known equation:[29]

$$\Delta\mu^2 = \frac{9000\,\mathrm{kTW}}{4\,NhC\rho}\,\Delta\epsilon \qquad (\mathbf{5})$$

where N is Avogadro's number, W is molecular weight, and h is an empirical parameter which has a numerical value 5.8 for amino acids or small proteins. C_ρ is a concentration of active proteins in the membrane in gram/liter.[29] Since N, k, and T are known and $\Delta\mu$ and $\Delta\epsilon$ are obtained by calculation and by measurement, (**5**) reduces to a simple form when all these numbers are substituted.

$$W/C_\rho = 3.9 \times 10^4. \qquad (\mathbf{6})$$

Unfortunately, the quantity C_p is not easy to assess because of the uncertainty regarding the number of sodium channels in axon membranes. Therefore, the calculated W is subject to large errors and the result will be only qualitative.

As mentioned, the results we obtained indicate the presence of dipolar species in the membrane or ionic processes that produce dipole moments. The second possibility has been thoroughly treated by Chandler *et al.*[5] and recently by Fishman,[30] using the linearized Hodgkin-Huxley Equation

$$Y(im) = i\omega\left\{C + \frac{\tau_m g_m}{1 + (\omega\tau_m)^2} + \frac{\tau_h g_h}{1 + (\omega\tau_h)^2} + \frac{\tau_n g_n}{1 + (\omega\tau_n)^2}\right\} \quad (7)$$

where all the parameters conform to those used by Hodgkin and Huxley.[4] This equation suggests that the time-dependent ionic currents give rise to frequency-dependent capacitances. Of these, the terms due to potassium '*n*' and sodium '*h*' give rise to negative values and only the sodium term '*m*' gives rise to a positive number. This reinforces the notion that the capacitance change may be due to sodium channels. However, since this equation is derived for small perturbations

TABLE 2

RADII OF GATING PARTICLES AT REST (R_r) AND AT THE PEAK OF ACTION POTENTIAL (R_a) CALCULATED FROM THE RELAXATION FREQUENCY AND EQUATION 8*

Viscosity (poise)	R_r (A)	R_a (A)	V_r ($A^3 \times 10^5$)	V_a ($A^3 \times 10^5$)	V_r/V_a
5.0	33	23	15.02	5.08	~3

* V_r and V_a are calculated with use of R_r and R_a.

by truncating nonlinear terms in the Taylor expansion, it is questionable to use this equation for large depolarizations.

Since the evidence for or against existing hypotheses is scanty, options for ionic processes or for gating particles are still wide open. We pointed out that ionic processes can give rise to voltage-dependent capacitances as a consequence of the linearized Hodgkin-Huxley Equation. However, it was found by us that the concept of gating particles responsible for the voltage-dependent capacitance led us to a notion that is not inconsistent with the Hodkin-Huxley formalism. From the change in time constant during the action potential (see FIGURE 8), we can calculate the change in effective radii of active elements or gating particles at rest and during activity, using the Debye-Stokes Equation.[26]

$$\tau = 4\pi a^3 \eta / kT \quad (8)$$

where a is the radius of spherical particle, η is the viscosity of the medium and T is temperature. Substituting the values of time constants that we observed, i.e., 5.3 msec and 1.99 msec at rest and during the action potential, and also assuming the viscosity of the membrane to be about 5 poises[31] and constant during activity, we can calculate values for R_r (radius at rest) and R_a (radius during action potential) and consequently their volumes. These values are tabulated in TABLE 2. The ratio of V_r (volume at rest) and V_a (volume during action potential) is found to be 2.98 and thus excitation of the membrane produces smaller particles whose volumes are about one third of the one at rest. Conversely, this can be interpreted as an indication that the particle in the channel consists of three subunits and that upon excitation, the elements dissociate into smaller particles. Although the number of subunits we found was three and is a remarkable coincidence with the Hodgkin-Huxley "m^3" formalism, it is still uncertain whether this should be taken seriously or should be considered fortuitous.

SUMMARY

Membrane capacitance and conductance of squid axons were investigated at the resting state and during hyper- and depolarizations using externally applied pulses. It was observed that membrane capacitance is voltage dependent and increases from 0.84 μF/cm^2 to 1.27 μF/cm^2 between -160mV and $+30$mV. Furthermore, measurements of capacitance in the absence of ionic currents indicate that the voltage dependence stems from sodium channels and has a different origin from that of voltage-dependent conductance.

Secondly, membrane capacitance and conductance were reinvestigated during the action potential. Conductance was found to change drastically, but the increase in membrane capacitance was small (0.9 μF/cm^2–1.0 μF/cm^2), in agreement with those originally obtained by Cole and Curtis. However, the relaxation frequency of capacitance dispersion shifted substantially from 3 kHz to 8 kHz, indicating the change in time constant from 0.53 msec to 0.19 msec. This change indicates either some rearrangements of gating particles in channels or those due to the reactance components associated with ionic currents as predicted by the linearized Hodgkin-Huxley Equation.

ACKNOWLEDGMENTS

The authors are indebted to Dr. K.S. Cole for his guidance, stimulating discussions and encouragement. They are also grateful to Dr. H.P. Schwan for his continuing interest in this work. Mr. Roger Novack participated in this work as part of the requirements for a course entitled Excitable Membrane Training Program at the Marine Biological Laboratory, Woods Hole, Mass.

REFERENCES

1. COLE, K. S. & H. J. CURTIS. 1938. Nature **142:** 209.
2. COLE, K. S. & H. J. CURTIS. 1939. J. Gen. Physiol. **22:** 5, 649.
3. COLE, K. S. & R. F. BAKER. 1941. J. Gen. Physiol. **24:** 535.
4. HODGKIN, A. L. & A. F. HUXLEY. 1952. J. Physiol. **117:** 500.
5. CHANDLER, W. K., R. FITZHUGH & K. S. COLE. 1962. Biophys. J. **2:** 105.
6. SCHWAN, H. P. 1965. Int. Biophys. Congr. (abstr.) Paper Sy-IA. Vienna, Austria.
7. KEYNES, R. D. & E. ROJAS. 1973. J. Physiol. **233:** 28.
8. KEYNES, R. D. & E. ROJAS. 1976. J. Physiol. **255:** 157.

9. Armstrong, C. M. & F. Bezanilla. 1973. Nature **242:** 459.
10. Bezanilla, F. & C. M. Armstrong. 1974. Science **183:** 753.
11. Armstrong, C. M. & F. Bezanilla. 1974. J. Gen. Physiol. **63:** 533.
12. Meves, H. 1974. J. Physiol. **243:** 847.
13. Neumke, B., W. Nonner & R. Stampfli. 1976. Pflugers Archiv. **363:** 193.
14. Schneider, M. F. & W. K. Chandler. 1976. J. Gen. Physiol. **67:** 125.
15. Adrian, R. H. & W. Almers. 1976. J. Physiol. **254:** 317.
16. Almers, W., R. H. Adrian & S. R. Levinson. 1975. Ann. N. Y. Acad. Sci. **264:** 278.
17. Takashima, S. 1976. J. Membrane Biol. **27:** 21.
18. Takashima, S., R. Yantoro & R. Novack. 1977. Biochim. Biophys. Acta. In press.
19. Kishimoto, U. 1972. Adv. Biophys. Vol. **3:** 199.
20. Takashima, S. & H. P. Schwan. 1974. J. Membrane Biol. **17:** 51.
21. Cole, K. S. 1975. J. Gen. Physiol. **66:** 133.
22. Cole, K. S. 1976. Biophys. J. **16:** 137.
23. Block, M. & E. F. Haynes. 1970. Trans. Farad. Soc. **574:** 66, 2512.
24. Gilbert, D. 1971. *In* Biophysics and Physiology of Excitable Membranes. W.J. Adelman, Ed.: Chap. 14, p. 264. Van Nostrand-Reinhold. New York, N.Y.
25. Cole, K. S. & R. H. Cole. 1941. J. Chem. Phys. **9:** 341.
26. Debye, P. 1929. Polar Molecules. Dover. New York, N.Y.
27. Fishman, H. M., L. E. Moore & D. Poussart. 1977, Abstract, Ann. Mt. Biophys. Soc. (abstr) New Orleans, La.
28. Levitan, E. & Y. Palti. 1975. Biophys. J. **15:** 239.
29. Oncley, J. L. 1943. Proteins, Amino Acids and Peptides. E.J. Cohn and J.T. Edsall, Eds.: Chap. 22. Reinhold Pub. Co. New York, N.Y.
30. Fishman, H. M. 1975. First Symposium on Testing and Identification of Non-Linear Systems, Calif. Inst. Technology. Pasadena, Calif.
31. Edidin, M. 1974. Ann. Rev. Biophys. Bioeng. Vol. **3:** 179.

ACTIVATION, INACTIVATION, AND CHEMICAL BLOCKAGE OF THE GATING CURRENT IN SQUID GIANT AXONS

Hans Meves

Laboratory of the Marine Biological Association
Plymouth, England

In the past twenty years the interest of those who study excitable membranes was focused on the Na inward current and the K outward current, their dependence on voltage and time, and their sensitivity to various drugs. During the last four years the situation has changed; the main interest is now on intramembrane charge transfer, i.e., movement of electrically charged organic molecules (either single charges or dipoles or multipoles) inside the membrane. Currents due to orientational changes of molecules inside the membrane were first recorded in frog muscle fibers;[1] they probably play a role in excitation-contraction coupling. Similar but faster currents have been found in nerve fibers of squid[2-15] and frog.[16-19] It is generally believed that the intramembrane movement of charged molecules which is seen in nerve fibers is responsible for the increase in Na permeability—the opening of the gates in the Na channels; the currents are therefore called gating currents.

The object of this article is to summarize what is known about the presumed connection between intramembrane charge movement and changes in Na permeability. Most people agree now that the connection is complicated and that the precise relation between intramembrane charge movement and permeability change has not yet been established. This cautious attitude expresses itself in the statements[11,5] that gating current and function of the Na gates are "in some way related," "but regarding the precise manner of their interrelationship there are perhaps more questions than answers at present."

As a basis for the discussion, the following three equations, taken from Hodgkin and Huxley,[20] are needed. The first describes the total current through the nerve membrane

$$I = I_C + I_{Na} + I_K + I_{leak} \quad (1)$$

as the sum of the capacitative current I_C and the ionic currents I_{Na}, I_K, and I_{leak}. The capacitative current consists of two parts, a large early part (representing the charging or discharging of the membrane capacity through the resistance in series with the membrane) and a small later part (thought to arise from the slow movement of charged molecules inside the membrane). The second equation gives the Na current

$$I_{Na} = g_{Na}(V - V_{Na}) \quad (2)$$

as the product of Na conductance g_{Na} and driving force $(V - V_{Na})$. The third equation (which is the most important in the present context) describes the Na conductance

$$g_{Na} = \bar{g}_{Na} m^3 h \quad (3)$$

as the product of a constant $\bar{g}_{Na}$ and the two variables m^3 and h; the latter depend on voltage and time (time constants τ_m and τ_h) and represent activation and inactivation of the Na conductance. The essential question is: Does the voltage and time dependence of the intramembrane charge movement directly or indirectly reflect the known voltage and time dependence of m and h?

FIGURE 1 shows the basic experimental finding in its simplest form. The membrane current produced by a +140 mV pulse (left) and a −140 mV pulse (right) is recorded from a perfused squid axon in which the ionic currents I_{Na} and I_K are blocked by appropriate poisons (tetrodotoxin, tetraethylammonium chloride) so that only the capacitative current I_C and the leakage current I_{leak} remain. Most

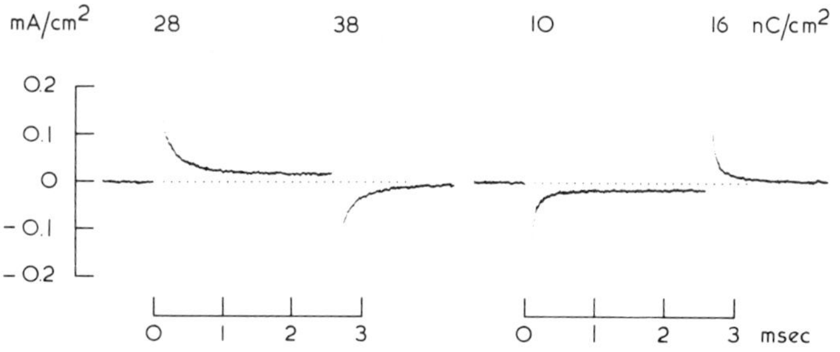

FIGURE 1. Membrane currents produced by a positive and a negative voltage clamp pulse of exactly equal size. Left: +140 mV pulse; right: −140 mV pulse; pulses superimposed on a holding potential of −111 mV. External solution: artificial sea water, Na totally replaced by Tris. Internal solution: 275 mM RbF + sucrose. 2 μM tetrodotoxin in external solution, 50 mM tetraethylammonium chloride in internal solution. Resting potential −71 mV. Temperature 2.5°C. Figures give areas under the current transients at the beginning and end of the pulses; areas were determined by tracing the enlarged record on graph paper and weighing the cut-out areas (omitting the first 125 μsec).

of the large early capacitative current is off screen, but the late capacitative current is clearly visible; it decays slowly to a small sustained current (leakage current) which is outward for the +140 mV pulse and inward for the −140 mV pulse. Comparison of the two records shows immediately that the transient current at the beginning and end of the pulse is larger in the left record (+140 mV pulse) than in the right record (−140 mV pulse). This is confirmed by measuring the area (in nC/cm^2) under the late capacitative current, starting at an arbitrarily chosen time of 125 μsec. The areas (indicated by the figures above the records) are larger for the left than for the right record, the difference being 18 nC/cm^2 at the beginning of the pulse and 22 nC/cm^2 at the end of the pulse. In both records the area at the beginning of the pulse is smaller than that at the end, probably because the sustained current (which had to be subtracted from the current transient at the beginning) was overestimated (see Ref. 21). The difference ($I_+ - I_-$) or ($I_{depol} - I_{hyperpol}$) is the nonlinear or asymmetrical part of the late capacitative current and is called gating current, asymmetry current, or difference current.

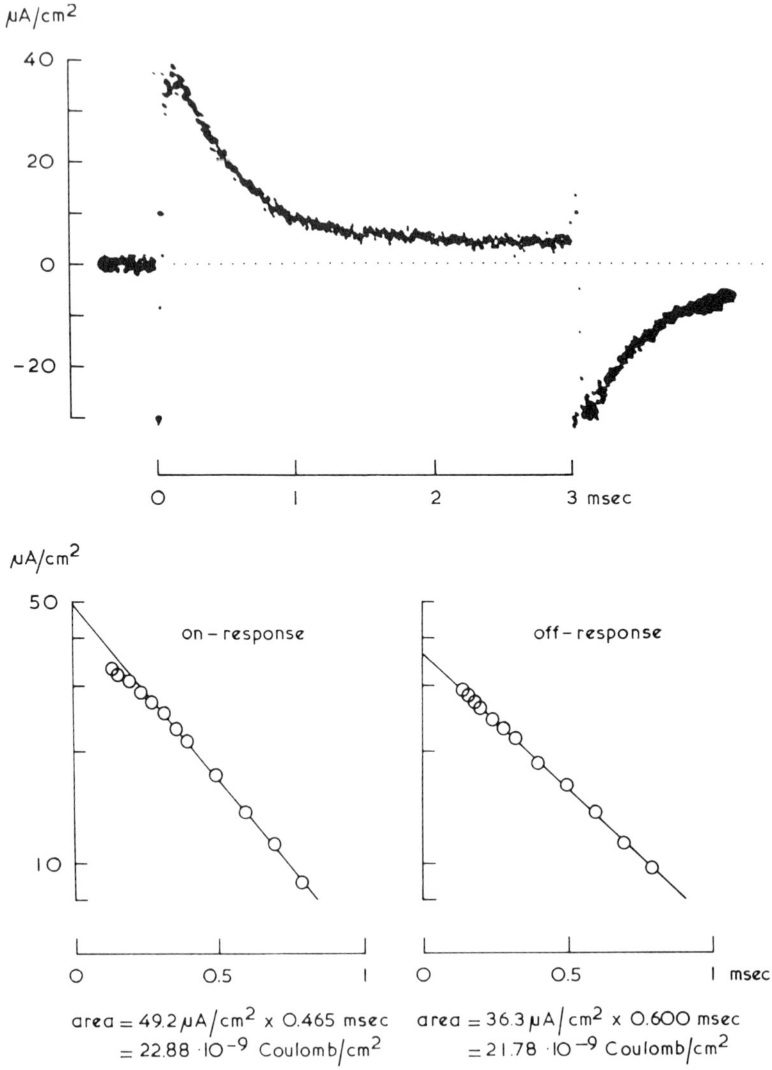

FIGURE 2. Gating current obtained by averaging the membrane currents associated with 32 positive and 32 negative pulses of 110 mV amplitude. The positive and negative pulses are applied alternately (frequency 1.5/sec) and superimposed on a holding potential of −88 mV. External and internal solution as in FIGURE 1. Resting potential −68 mV. Temperature 1°C. First 30 μsec blanked out to avoid overloading of averager by the large early capacitative current. In the lower half the transient outward current at the beginning of the pulses (on-response) and the inward current at the end of the pulses (off-response) are plotted on a logarithmic scale against time. (From Meves.[11] By permission of the Journal of Physiology.)

The method of determining the area under the late capacitative current from single-pulse records is laborious. It is more convenient to use a signal averager or computer. The fiber is subjected to an equal number of exactly equal positive and negative clamp pulses (superimposed on a potential of −80 to −100 mV) and the currents associated with these pulses are algebraically summed by the computer. The output signal is the gating current. An example is shown in FIGURE 2. The signal consists of a transient outward current at the beginning of the pulses (on-response) and a transient inward current at the end of the pulses (off-response). The on-response decays to a small sustained outward current which represents the asymmetrical part of the leakage current (leakage rectification). In the lower part of FIGURE 2 the on-response (after subtracting the sustained current) and the off-response are plotted on a logarithmic scale against time. Both responses follow an exponential time course, apart from a deviation at the beginning of the on-response (see CHEMICAL BLOCKAGE section). The areas under the on- and off-response were obtained from the extrapolated current at zero time and the time constant and represent the amount of charge movement during the nonlinear component of the late capacitative current. The figures given in FIGURE 2 show that the charge movement during the on-response is almost the same as that during the off-response. The equality suggests that the charges are an integral part of the membrane. A violation of the equality rule will be mentioned later.

Gating current records from squid giant axons were first published by Armstrong and Bezanilla[2] and later by Keynes and Rojas[8] and Meves.[11] Similar records were obtained in frog myelinated nerve fibers,[16–19] *Myxicola* axons,[22] and *Aplysia* nerve cells.[23] In all the early studies the method of equal and opposite pulses was used (upper pulse program in FIGURE 3); it consists of an equal number of exactly equal positive and negative pulses, superimposed on a holding potential of −80 to −100 mV. Later, it was replaced by the divided pulse pro-

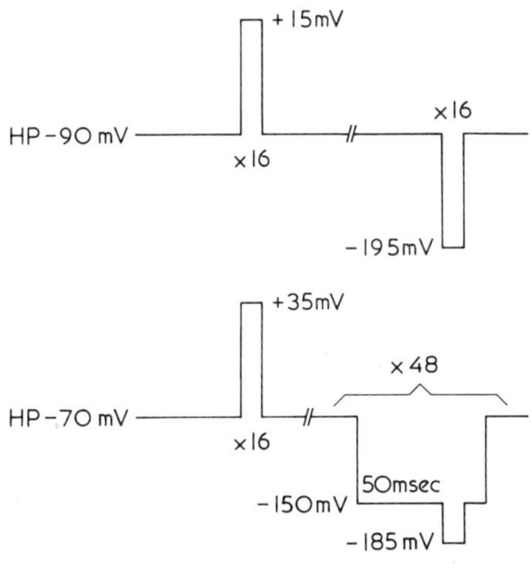

FIGURE 3. Pulse programs for the measurement of gating currents. *Top:* method of equal and opposite pulses. *Bottom:* divided pulse procedure. See text for details.

cedure of Armstrong and Bezanilla.[3] The lower pulse program in FIGURE 3 shows the version used by Meves and Vogel:[13-15] n positive pulses of amplitude P (superimposed on a holding potential of -70 mV) and $3 \times n$ negative pulses of amplitude $P/3$ are applied alternately; the negative pulses are superimposed on a potential of -150 mV; this ensures that the currents produced by the negative pulses contain very little late capacitative current (see FIGURE 6B in Ref. 12).

ACTIVATION

The area under the on- and off-response of the gating current increases with increasing pulse height. This is illustrated in FIGURE 4, which shows the asym-

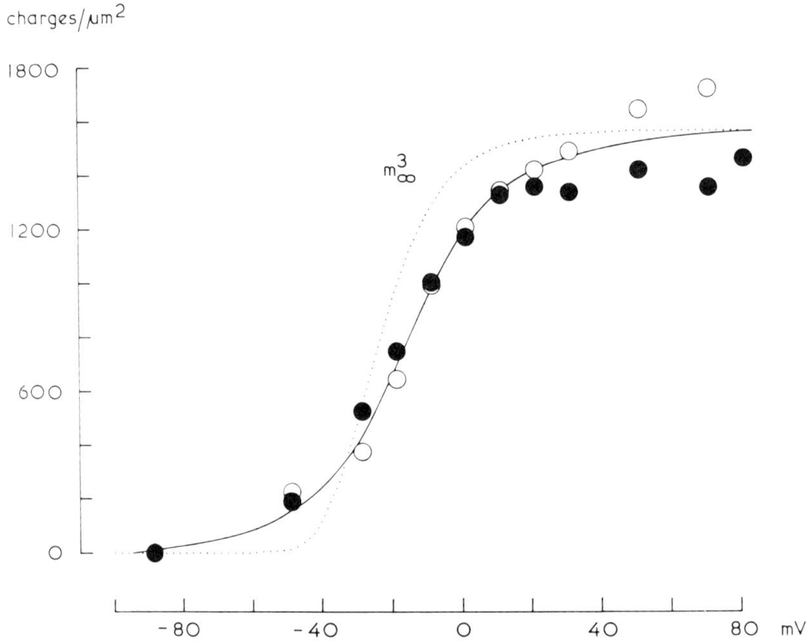

FIGURE 4. Charge movement during the on-response (○) and during the off-response (●) plotted against membrane potential during the positive pulses. Same experiment as in FIGURE 2. The charge movement was determined from semilogarithmic plots of the on- and off-responses (see FIGURE 2) and is given in electronic charges/μm^2, Q_{max} = 1600 electronic charges/μm^2 = 26 nC/cm^2. The continuous curve was calculated from the Langevin-Debye Equation

$$y = \coth x - \frac{1}{x}$$

where $x = \mu \Delta V/kTl$ and $\Delta V/l = (V - V_{half})/l$ with μ = 840 Debye units, l = 7 nm and $V_{half} = -16$ mV; a very similar curve was obtained from the flip-flop equation

$$y = \tanh x$$

with μ = 270 Debye units. The dotted curve represents $m_\infty{}^3$ and was calculated from Hodgkin and Huxley,[20] assuming a resting potential of -62 mV in their experiments. (From Meves.[11] By permission of the Journal of Physiology.)

metrical charge movement plotted against the membrane potential during the positive pulses. The measurements were made with the method of equal and opposite pulses at a holding potential of −88 mV. The on-response Q_{on} (empty circles) and the off-response Q_{off} (filled circles) increase along a sigmoid curve as the pulse potential becomes more positive. At positive potentials the off-responses are consistently somewhat smaller than the on-responses, whereas at negative potentials no systematic deviation is seen. The averages from on- and off-response are used to draw the continuous curve. Similar sigmoid Q vs. V curves for the squid axon have been published by Keynes and Rojas[8,9] and Armstrong and Bezanilla.[5] Q vs. V curves have also been measured in nodes of Ranvier[16,17] and in *Myxicola* axons.[22]

The sigmoid Q vs. V curve is important for the description and interpretation of the asymmetrical charge movement. The curve is characterized by three parameters: its maximum slope $(dQ/dV)_{max}$, the maximum charge displacement Q_{max} obtained with very strong pulses, the potential V_{half} at which $Q = 0.5\ Q_{max}$. Differentiation of the sigmoid Q vs. V curve leads to a bell-shaped curve that represents $C_1 = dQ/dV$ as a function of potential. As the potential is shifted from the holding potential to more positive values, the potential-dependent capacity C_1 rises, reaches a maximum at V_{half} and decreases again (see also Ref. 24 and note added in proof).

A sigmoid Q vs. V curve is expected if the membrane contains permanent dipoles or charged particles that change their position inside the membrane under the influence of the electric field. The orientation of permanent dipoles in an electric field is described by the Langevin-Debye Equation (see legend for FIGURE 4), which applies to the case that the angle between dipole axis and field can have any possible value. A similar equation (see also legend for FIGURE 4) is valid when only two positions, parallel or antiparallel to the field, are possible (flip-flop). The continuous curve in FIGURE 4 has been calculated from the Langevin-Debye equation with a dipole moment $\mu = 840$ Debye units; a similar curve was obtained for the flip-flop case with $\mu = 270$ Debye units.

The dipole moments mentioned above are within the range of dipole moments reported for proteins,[25] but there is no proof that permanent dipoles of this size are really present in the membrane. The sigmoid Q vs. V curve could also be explained by the assumption that the dipoles are not permanently present but are induced by the potential change (possibly from a limited store) through a fast chemical reaction.[26]

In FIGURE 4 the m^3 vs. V relation (calculated from Hodgkin and Huxley[20]) is plotted as a dotted curve for comparison with the Q vs. V curve; the latter is clearly much less steep, whereas V_{half} is similar for both curves (−23.5 and −16 mV, respectively). The slope ratio is between 2 and 3 in all preparations studied (squid axon,[8,9] node of Ranvier,[16,17] *Myxicola* axon[22]). Keynes and Rojas[8] tentatively explained the difference in slope by assuming that each Na channel incorporates three gating particles and that there is cooperativity between the three particles; their explanation is based on the flip-flop model. An alternative explanation starts from the assumption of permanent dipoles which can take up any position in the field; most transitions are not reflected in a change of g_{Na}, but there is one critical movement that leads to the opening of the gate and occurs at a critical potential which may be somewhat different for individual Na channels.

The hypothesis that there are movements between different "open" positions and between different "closed" positions (or between "closed" and "fully closed," see Armstrong and Bezanilla[3]) accounts for charge movements in those potential regions where m^3 does not change at all. A multistate hypothesis of this kind, "postulating multiple transitions of each charged particle before *and* after the one which opens the channel," has been discussed by Neumcke *et al.*[17]

Although there is general agreement about the shape and position of the sigmoid Q vs. V curve, opinions are divided with regard to the exact time course of

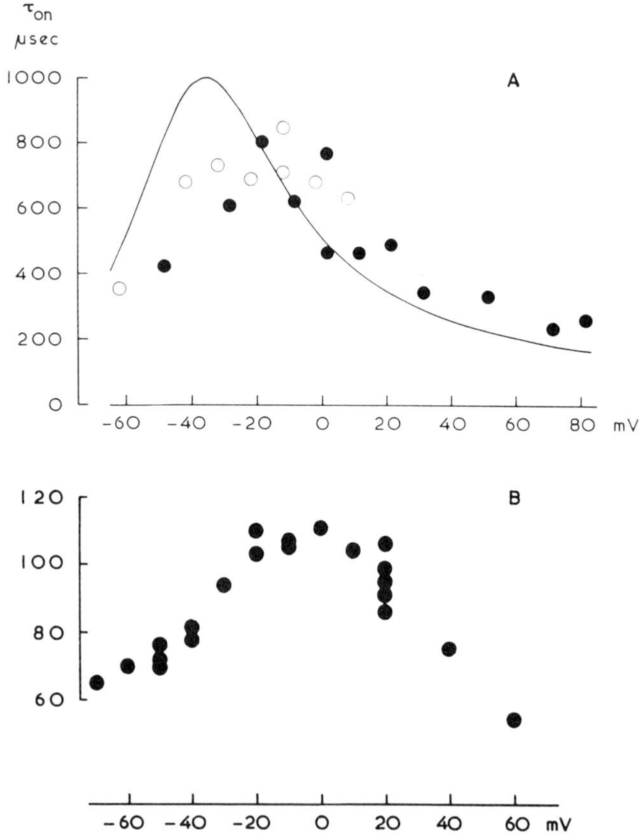

FIGURE 5. Time constant of on-response, τ_{on}, plotted against membrane potential during positive pulses. A: from Meves[11] (by permission of the Journal of Physiology), using method of equal and opposite pulses. Same experiment as in FIGURES 2 and 4, temperature 1°C. • = measurements at a holding potential of −88 mV (as in FIGURES 2 and 4); ○ = measurements at a holding potential of −102 mV. The continuous curve gives the time constant τ_m, calculated from the equations of Hodgkin and Huxley[20] for 0°C. B: from Bezanilla and Armstrong[6] (by permission of the Royal Society), using divided pulse procedure; the measuring points represent the time constant of the fast exponential component. Ordinate scale ten times expanded. External solution: Na-free Tris sea water with tetrodotoxin. Holding potential −80 mV. Temperature 8°C.

the gating current. The essential question is whether the on- and off-response can be fitted by a single exponential or consists of two or more components, corresponding possibly to two or more populations of charges. In the early experiments (which were done with the method of equal and opposite pulses) the time course of the on-response was found to follow a single exponential, both in squid axons[3,8,9,11] and in nodes of Ranvier.[16] An example has already been given in FIGURE 2 which shows that on- and off-response can be fitted by straight lines in a semilogarithmic plot. The more recent experiments (done with the divided pulse procedure) gave conflicting results: The on-response followed a single exponential in the experiments of Meves and Vogel[14,15] (see measuring points ○ in FIGURE 7), whereas Armstrong and Bezanilla[5] regularly observed two components; J. E. Kimura (unpublished) finds one component in some fibers, two or more components in other fibers. The only possible conclusion at present is that the time course depends on the experimental conditions.

Numerical values for the time constant τ_{on} of the on-response are given in FIGURE 5; they are plotted against the membrane potential during the positive pulses. Part A is from my experiments[11] with equal and opposite pulses, giving on-responses with a simple exponential time course; part B is from measurements of Bezanilla and Armstrong[6] with the divided pulse procedure and shows only the time constant of their fast component. Both sets of measuring points suggest a bell-shaped relation between τ_{on} and V with a maximum at a potential close to V_{half}. The time constants in B are smaller than those in A because they refer only to a fast component and because the temperature is higher.

To compare the time constant of the on-response (τ_{on}) and the time constant of the sodium activation variable m(τ_m) the τ_m vs. V curve (calculated from Hodgkin and Huxley[20]) has been plotted in FIGURE 5A. The two time constants are similar (but not identical) in their numerical values and in their voltage dependence. Almost perfect agreement has been reported by Keynes and Rojas.[9] However, Neumcke *et al.*[17] found on the node of Ranvier $\tau_m > \tau_{on}$ for small pulses and $\tau_m < \tau_{on}$ for large pulses, i.e., the same deviation as in FIGURE 5A. The discrepancy $\tau_m > \tau_{on}$ at small pulses could simply indicate an additional reaction intermediate between the charge movement and the change in m; the reversed inequality at large pulses could mean that the channel opens before the charge movement has come to an end.

INACTIVATION

In discussing inactivation of gating current and Na current it should be remembered that in addition to the usual relatively fast inactivation process h (see Equation 3) there is also a more slowly developing and more slowly removed inactivation.[27-29] The molecular mechanism underlying slow Na inactivation may be different from that of the h process.[30]

The intramembrane charge movement in muscle fibers can be inactivated by prolonged lowering of the membrane potential.[31] Inactivation of the gating current in squid axons was first demonstrated by Bezanilla and Armstrong.[4] The potential was held at +56 mV for two minutes (temperature 2°C) and then re-

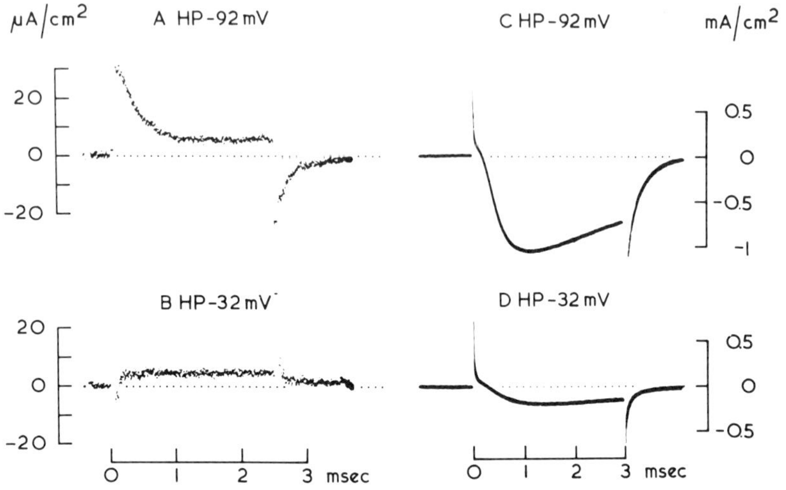

FIGURE 6. Suppression of gating current and Na current after holding the potential at −32 mV for 3 minutes. Holding potential indicated above each record, holding time 3 minutes. Records from two different fibers, showing gating current in left column (records A–B) and Na current in right column (records C–D). A–B: gating current obtained by averaging the membrane currents associated with 32 positive and 32 negative pulses of 90 mV. A = pulses superimposed on −92 mV; B = potential held at −32 mV, but shifted to −92 mV 440 msec before each positive or negative pulse; solutions as in FIGURE 1. C–D: Na current in Na sea water, elicited by a single test pulse. C = +90 mV test pulse superimposed on −92 mV; D = potential held at −32 mV, but shifted to −92 mV 440 msec before the +90 mV test pulse. Temperature 3.5°C in A–B, 2°C in C–D. (From Meves and Vogel.[15] By permission of the Journal of Physiology.)

turned to its normal value; gating current and Na current had completely disappeared and recovered very slowly with a half-time of about 50 seconds. As illustrated in FIGURE 6, the same effect can be produced by holding the potential at −32 mV for three minutes (temperature 3.5°C). Record A shows the normal gating current recorded with the method of equal and opposite pulses at a holding potential of −92 mV. In record B the potential was held at −32 mV for three minutes but shifted back to −92 a short time (440 msec) before each positive and negative pulse; the on- and off-response are completely absent; only the sustained outward current (leakage rectification) is visible. The Na current recorded under the same conditions from another fiber is also markedly reduced in size (compare record D with control record C). Similar results were obtained with the divided pulse procedure.[15] The observations suggest that the mobile charges in the nerve membrane are immobilized or neutralized by a severe and long-lasting depolarization. The agreement between recovery of Na current and recovery of gating current is quite good (see Ref. 15); i.e., *slow* Na inactivation is directly reflected in the gating current measurements.

The relation between gating current and *fast* inactivation (which is characterized by the variable h and operates during a normal action potential) turned out to be more complicated. It was studied with use of *short* conditioning depolarizations. The pulse program was the divided pulse procedure (see inset in FIGURE 7); short prepulses (10–20 msec duration) preceded the depolarizing test pulses and were separated from them by a recovery interval of a few msec duration. Plots A and B show the on-response of the gating current on a semilogarithmic scale. The prepulse was +80 mV, 10 msec in A (filled circles) and +40 mV, 20 msec in B (filled circles); the recovery interval was 0.94 msec in A and 0.95 msec in B. The on-response measured without prepulses is represented by the empty circles. The areas under the conditioned on-responses in A and B are, respectively, 67 and 62% of the control; lengthening the recovery interval to 5 msec increased the areas to 81–84%. Similar measurements have been reported by Armstrong and Bezanilla.[5]

The experiment in FIGURE 7 indicates that a short-lasting depolarization blocks part of the charge movement, but blockage is much smaller than with a long-lasting depolarization. For comparison, reduction of the Na current by a +40 or +80 mV prepulse of 10–20 msec duration was studied (FIGURE 8). With a

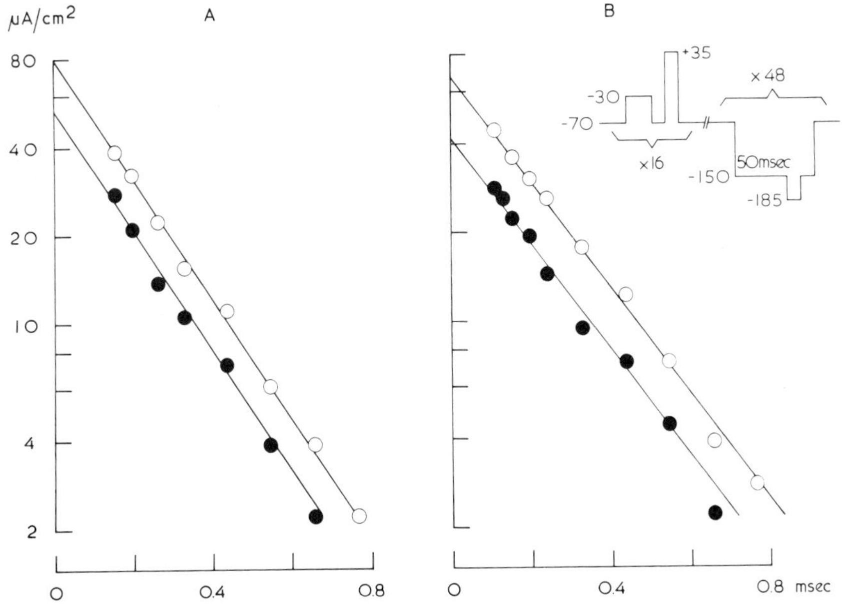

FIGURE 7. Effect of a short positive prepulse on the on-response of the gating current. Semilogarithmic plot of on-responses as in FIGURE 2; sustained outward current subtracted. Divided pulse procedure (see inset): 16 pulses + 105 mV (superimposed on a holding potential of −70 mV) and 48 pulses − 35 mV (superimposed on a potential of −150 mV); positive prepulse (followed by short recovery interval) precedes each positive test pulse. A: ○ = control record without prepulse; ● = +80 mV, 10 msec prepulse, recovery interval 0.94 msec. B: ○ = control record without prepulse; ● = +40 mV, 20 msec prepulse, recovery interval 0.95 msec. Solutions as in FIGURE 1. Temperature 2.7°C. Straight lines drawn with τ = 213 μsec (A) and τ = 244 μsec (B). (From Meves and Vogel.[14] By permission of the Journal of Physiology.)

recovery interval of about 5 msec the peak Na current was only a third of its normal size (record A: reduction to 37% for recovery interval 4.6 msec; record D: reduction to 37% of control for recovery interval 4.5 msec). Even with 11.5 msec recovery interval (record B), the Na current was only 60% of the control value; with 1.25 msec (record C) it was reduced to 25%. Thus, reduction of the Na current is more pronounced than reduction of the gating current.

If a 10–20 msec depolarization leads to partial blockage of the charge movement, one should expect that the off-response of the gating current (which is about equal in size to the on-response for short test pulses, see FIGURE 2) becomes smaller for 10–20 msec long test pulses. FIGURE 9 shows that this is indeed the case. The size of the off-response can be determined in various ways: by measuring simply the amplitude at a given time after pulse end, by determining the area under the total off-response (Q_{off}) or the area under its early part or by measuring the ratio Q_{off}/Q_{on}. Whatever parameter one chooses, the result is the same, a decrease to about 50% for long pulses (10–20 msec duration) compared with short pulses (1 msec duration). This is shown in FIGURE 9 for the area under the early part of the off-response (○) and for the ratio Q_{off}/Q_{on} (●). An exponential curve with a time constant of 7 msec has been drawn through the measuring points ○. These observations are similar to those of Bezanilla and Armstrong,[7] who reported a decrease of Q_{off}/Q_{on} to $\frac{1}{3}$ at long pulse durations. The phenomenon could also be demonstrated by using only positive pulses. The currents at the end of 32 10 msec pulses +90 mV were subtracted from the currents at the end of 32 1 msec pulses +90 mV. The difference was displayed (inset of FIGURE 9: ↓ marks pulse end). A clear difference (indicating $I_{off\,1\,msec} > I_{off\,10\,msec}$) is seen during the

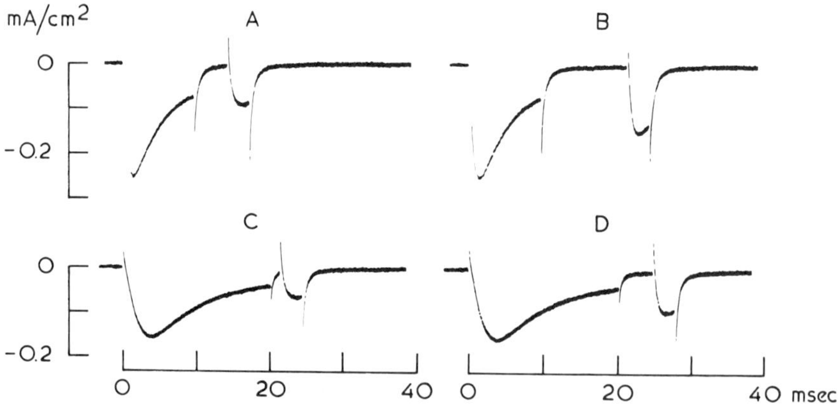

FIGURE 8. Effect of a short positive prepulse on the Na current. Records show current during prepulse and current during test pulse. Test pulse +80 mV, superimposed on a holding potential of −70 mV. Prepulse: A = +80 mV, 10 msec, recovery interval 4.6 msec; B = same but recovery interval 11.5 msec; C = +40 mV, 20 msec, recovery interval 1.25 msec; D = same but recovery interval 4.5 msec. External solution: artificial sea water, Na concentration reduced to 118 mM by replacing $\frac{3}{4}$ of the Na with Tris. Internal solution as in FIGURE 1. Temperature 1.5°C. (From Meves and Vogel.[14] By permission of the Journal of Physiology.)

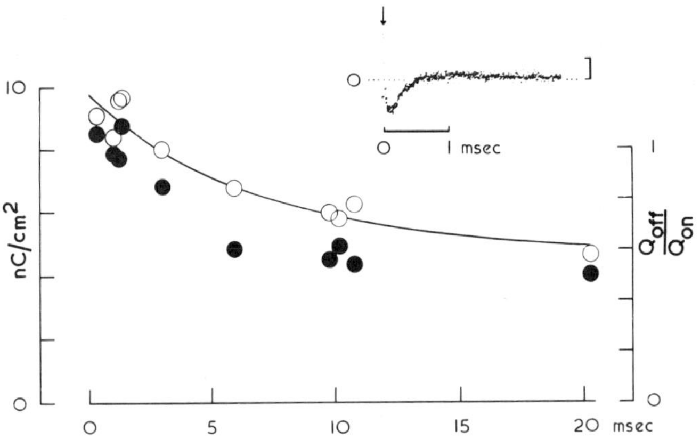

FIGURE 9. Effect of pulse duration (abscissa) on early part of off-response (○, left ordinate) and on Q_{off}/Q_{on} ratio (•, right ordinate). Area under early part of off-response (67–558 μsec after pulse end) determined by weighing. Q_{off}/Q_{on} ratio calculated from total areas of off- and on-responses (omitting the first 67 μsec). Divided pulse procedure: 16 pulses +90 mV (superimposed on a holding potential of −70 mV) and 48 pulses −30 mV (superimposed on a potential of −150 mV). Inset: see text, vertical bar 10 μA/cm^2. Solutions as in FIGURE 1. Temperature 8.5°C. (From Meves and Vogel.[13] By permission of the Journal of Physiology.)

first 0.5 msec after the end of the pulses; at later times the signal slightly overshoots the zero line. Again, the results were compared with measurements of the Na current under similar conditions. The Na current during a 20 msec pulse of +90 mV decreased to 10–26% of its peak value with a time constant of 1.35 msec; i.e., the decrease was stronger and faster than that of the off-response.

Similarly, the reduction of the asymmetrical charge movement by ultraviolet light is much smaller than the reduction of the Na current; this was shown by Fox *et al.*[18] on the node of Ranvier. A difference of this kind is expected if each Na channel contains two or three gating particles and blockage of one by inactivation, ultraviolet light, or any other agent is sufficient for blockage of the whole channel (see Ref. 18). A related explanation would be that Na conductance g_{Na} is proportional to the cube of the charge movement ($g_{Na} \sim Q^3$, resembling $g_{Na} \sim m^3$ in (**3**)); in this case a small reduction of gating current (e.g., to 80%) would be associated with a much larger reduction of g_{Na} (e.g. to 51%) (see Ref. 14). This assumption could also explain why the time constant for the decrease of the off-response in FIGURE 9 is so much larger than the time constant of Na inactivation.

CHEMICAL BLOCKAGE

Blockage of the charge movement inside the nerve membrane cannot be achieved only by lowering the membrane potential for a longer or shorter period; it can also be achieved by chemical means. Gating current and Na current can be blocked in perfused squid axons by intracellular application of 30 mM glutaraldehyde;[11] this is probably due to the cross-linking effect of glutaraldehyde on the membrane proteins. Keynes and Rojas[8] have reported a substantial reduction of the charge movement by a high external concentration of procaine (1%). Simultaneously, the time constants τ_{on} and τ_{off} fell significantly, in contrast to the general observation that local anesthetics delay the rise of the Na conductance. A similar discrepancy has been noticed in nodes of Ranvier; procaine delays the turning-on of the Na current but does not affect τ_{on} and τ_{off} (see Ref. 19). It is noteworthy that a drastic change of the external pH (between pH 5 and 8) does *not* affect the charge movement in the squid axon membrane.[8]

The most carefully investigated blocking agent for the gating current is intracellular Zn in a concentration of 3.3–10 mM, which also blocks the Na current.[4,12] It not only abolishes the nonlinear part of the late capacitative current (gating current) but also substantially reduces the linear part of the late capacitative current.

Varying the holding potential in normal and Zn-treated squid axons revealed an interesting phenomenon, illustrated by FIGURES 10 and 11. In normal axons, the asymmetrical charge movement (recorded with the method of equal and opposite pulses) reverses its sign when the holding potential is shifted from about

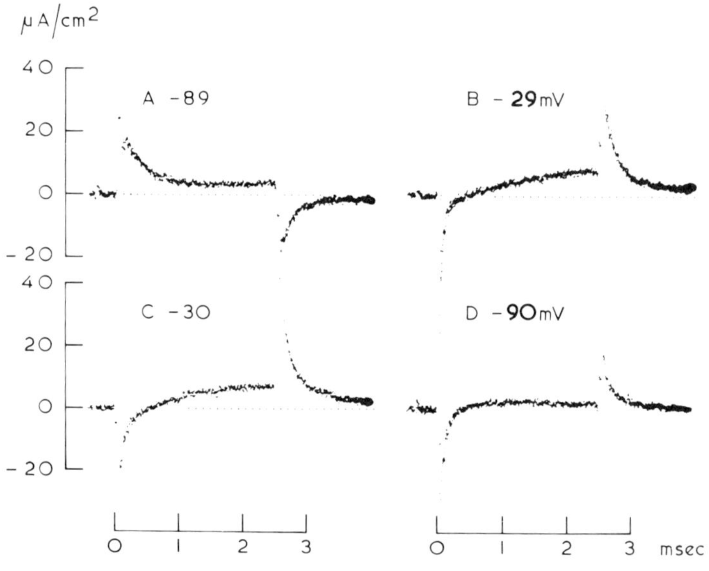

FIGURE 10. Reversal of the asymmetrical charge movement at a holding potential of −30 mV. Records obtained by averaging the membrane currents associated with 32 positive and 32 negative pulses of 90 mV. The potential upon which the pulses are superimposed is indicated above each record. A: membrane held at −89 mV for three minutes. B: same, but +60 mV, 44 msec prepulse preceding each depolarizing and each hyperpolarizing pulse and outlasting each pulse by a few msec. C: membrane held at −30 mV for three minutes. D: same, but −60 mV, 44 msec prepulse preceding each depolarizing and each hyerpolarizing pulse and outlasting each pulse by a few msec. Solutions as in FIGURE 1. Temperature 4°C.

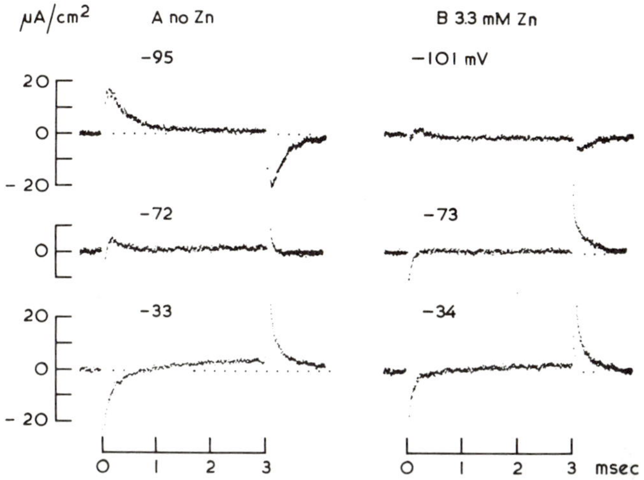

FIGURE 11. Effect of internal Zn on the asymmetrical charge movement recorded at different holding potentials. Records obtained by averaging the membrane currents associated with 32 positive and 32 negative pulses of 90 mV. Holding potential indicated above each record; holding time three minutes. Solutions as in FIGURE 1; in B the internal solution contained 3.3 mM $ZnCl_2$. Temperature 4°C. (From Meves.[12] By permission of the Journal of Physiology.)

−90 to about −30 mV; i.e., the on-response becomes inward and the off-response outward.[3,10,11] This is shown by records A and C in FIGURE 10, which were obtained after three minutes holding at −89 and −30 mV, respectively. To achieve the reversal, a very short prepulse of a few msec duration from −89 to −29 mV was sufficient (see record B with a 44 msec prepulse); however, restoration of the normal response by a 44 msec prepulse from −30 to −90 mV was not possible (see record D, compare also record B in FIGURE 6). FIGURE 11 compares the asymmetrical charge movement (recorded with the method of equal and opposite pulses at different holding potentials) in the absence (A) and in the presence (B) of 3.3 mM Zn. It is clear that Zn suppresses the "normal" asymmetrical charge movement (recorded at a holding potential of −95 or −101 mV) completely, but does not affect the reversed signal seen at a holding potential of −33 or −34 mV. The reason for the Zn-resistant reversed asymmetry current is obscure (see discussion in Ref. 12), but it seems simplest to treat it as a separate phenomenon which has nothing to do with the normal gating current. The transient inward current visible at the beginning of the pulse at a holding potential of −73 mV in FIGURE 11B is probably responsible for the initial deviation of the on-response from a strictly exponential time course in FIGURE 2 (lower part).

CONCLUSION

The results can be summarized with the statement that the relation between gating current and activation or inactivation of the Na conductance is complex;

its understanding requires hypothetical assumptions and certainly more experimental work. However, the evidence at present available supports the view that the two phenomena are in some way related. This evidence can be summarized in the following four points: (1) the simultaneous blockage of gating current and Na current by Zn, glutaraldehyde, and ultraviolet light; (2) the voltage-dependence of the asymmetrical charge movement which turns on steeply at a potential close to the threshold of the action potential; (3) the time course of the asymmetrical charge movement (characterized by the time constants τ_{on} and τ_{off}) which is similar (but not identical) to the time course of the Na activation variable m or Na conductance g_{Na}; and (4) the total or partial inactivation of the charge movement during a long or short depolarization which qualitatively (but not quantitatively) resembles the inactivation of the Na conductance.

[**Note Added in Proof:** Several further publications about gating current in squid giant axons, nodes of Ranvier, and molluscan neurons have recently appeared or are currently in press. Armstrong & Bezanilla (J. Gen. Physiol., in press) have studied inactivation of the gating current: the decrease of the off-response with increasing pulse duration is faster and more pronounced than in the experiments of Meves & Vogel[13,14] (see Figure 9); consequently, the agreement with the time course of Na inactivation is better. Kimura & Meves (J. Physiol., in press) measured gating currents at different temperatures; increasing the temperature substantially increased the area under the on-and-off response (Q_{on}, Q_{off}) and decreased the time constants τ_{on} and τ_{off}. Short communications concern the effect of the local anesthetic QX-314 (Almers & Cahalan, 1977, Biophys. J. **17:** 205a) and of the insecticide DDT (Dubois & Bergman, 1977, Nature **266:** 741) on gating currents; Kostyuk, Krishtal & Pidoplichko (1977, Nature **267:** 70) describe gating currents in *Helix pomatia* neurons. Takashima & Yantorno (this volume) have again measured the membrane capacity C of the squid axon in an admittance bridge; the previously reported increase of C upon depolarization (Takashima[24]) was found to disappear after application of tetrodotoxin, an observation difficult to reconcile with the generally accepted view that tetrodotoxin has no effect on the intramembrane charge movement. Admittance measurements by Fishman, Moore & Poussart (1977, Biophys. J. **17:** 11a) showed a decrease (rather than an increase) of C upon depolarization. Field-induced chemical transitions in macromolecules and their possible relevance for the excitation process have been discussed by Schwarz (this volume). Blatt (1977, Biophys. J. **18:** 43–52) points out that electrostriction leads to an asymmetry current opposite in direction to the normal gating current.]

References

1. Schneider, M. F. & W. K. Chandler. 1973. Voltage dependent charge movement in skeletal muscle: a possible step in excitation-contraction coupling. Nature **242:** 244–246.
2. Armstrong, C. M. & F. Bezanilla. 1973. Currents related to movement of the gating particles of the sodium channels. Nature **242:** 459–461.
3. Armstrong, C. M. & F. Bezanilla. 1974. Charge movement associated with the

opening and closing of the activation gates of the Na channels. J. Gen. Physiol. **63:** 533–552.
4. Bezanilla, F. & C. M. Armstrong. 1974. Gating currents of the sodium channels: Three ways to block them. Science **183:** 753–754.
5. Armstrong, C. M. & F. Bezanilla. 1975. Currents associated with the ionic gating structures in nerve membrane. Ann. N.Y. Acad. Sci. **264:** 265–277.
6. Bezanilla, F. & C. M. Armstrong. 1975. Kinetic properties and inactivation of the gating currents of sodium channels in squid axon. Philos. Trans. R. Soc. Lond. [Biol Sci.] **270:** 449–458.
7. Bezanilla, F. & C. M. Armstrong. 1975. Inactivation of gating charge movement. Biophys. J. **15:** 163a.
8. Keynes, R. D. & E. Rojas. 1974. Kinetics and steady-state properties of the charged system controlling sodium conductance in the squid giant axon. J. Physiol. **239:** 393–434.
9. Keynes, R. D. & E. Rojas. 1976. The temporal and steady-state relationships between activation of the sodium conductance and movement of the gating particles in the squid giant axon. J. Physiol. **255:** 157–189.
10. Keynes, R. D., E. Rojas & B. Rudy. 1974. Demonstration of a first-order voltage-dependent transition of the sodium activation gates. J. Physiol. **239:** 100–101 P.
11. Meves, H. 1974. The effect of holding potential on the asymmetry currents in squid giant axons. J. Physiol. **243:** 847–867.
12. Meves, H. 1976. The effect of zinc on the late displacement current in squid giant axons. J. Physiol. **254:** 787–801.
13. Meves, H. & W. Vogel. 1976. Effect of pulse duration on the off-response of the 'gating current.' J. Physiol. **263:** 125–126 P.
14. Meves, H. & W. Vogel. 1977. Inactivation of the asymmetrical displacement current in giant axons of *Loligo forbesi*. J. Physiol. **267:** 377–393.
15. Meves, H. & W. Vogel. 1977. Slow recovery of sodium current and gating current from inactivation. J. Physiol. **267:** 395–410.
16. Nonner, W., E. Rojas & R. Stämpfli. 1975. Displacement currents in the node of Ranvier. Voltage and time dependence. Pflügers Arch. Ges. Physiol. **354:** 1–18.
17. Neumcke, B., W. Nonner & R. Stämpfli. 1976. Asymmetrical displacement current and its relation with the activation of sodium current in the membrane of frog myelinated nerve. Pflügers Arch. Ges. Physiol. **363:** 193–203.
18. Fox, J. M., B. Neumcke, W. Nonner & R. Stämpfli. 1976. Block of gating currents by ultraviolet radiation in the membrane of myelinated nerve. Pflügers Arch. Ges. Physiol. **364:** 143–145.
19. Kniffki, K.-D., E. Koppenhöfer & W. Vogel. 1976. Effects of procaine on gating and sodium currents at the nodal membrane. Pflügers Arch. Ges. Physiol. **365:** R33.
20. Hodgkin, A. L. & A. F. Huxley. 1952. A quantitative description of membrane current and its application to conduction and excitation in nerve. J. Physiol. **117:** 500–544.
21. Cole, K. S. & R. FitzHugh. 1971. Dielectric loss and voltage clamp transients. Biophys. Soc. 15th Ann. Meet. Abstr. TPM-F4.
22. Rudy, B. 1976. Sodium gating currents in *Myxicola* giant axons. Proc. R. Soc. Lond. [Biol.] **193:** 469–475.
23. Adams, D. J. & P. W. Gage. 1976. Gating currents associated with sodium and calcium currents in an *Aplysia* neuron. Science **192:** 783–784.
24. Takashima, S. 1976. Membrane capacity of squid giant axon during hyper- and depolarizations. J. Membrane Biol. **27:** 21–39.
25. Cohn, E. J. & J. T. Edsall. 1943. Proteins, Amino Acids and Peptides. Reinhold Publishing Co. New York, N.Y.
26. Schwarz, G. 1972. Dielectric relaxation of biopolymers in solution. Adv. Mol. Relaxation Processes **3:** 281–295.
27. Narahashi, T. 1964. Restoration of action potential by anodal polarization in lobster giant axons. J. Cell. Comp. Physiol. **64:** 73–96.
28. Chandler, W. K. & H. Meves. 1970. Slow changes in membrane permeability and

long-lasting action potentials in axons perfused with fluoride solutions. J. Physiol. **211:** 707–728.

29. Schauf, C. L., T. L. Pencek & F. A. Davis. 1976. Slow sodium inactivation in *Myxicola* axons. Evidence for a second inactive state. Biophys. J. **16:** 771–778.
30. Rudy, B. 1977. Slow inactivation of the sodium conductance in squid giant axons. Pronase resistance. J. Physiol. In press.
31. Chandler, W. K., R. F. Rakowski & M. F. Schneider. 1976. Effects of glycerol treatment and maintained depolarization on charge movement in skeletal muscle. J. Physiol. **254:** 285–316.

DISCUSSION

Robert S. Eisenberg, *Moderator*

Department of Physiology
Rush Medical College
Chicago, Illinois 60612

DR. S. SANFORD (*Eastman Kodak Co., Rochester, N.Y.*): Dr. Taylor, what is the physical interpretation of the presence of an inductive element in the equivalent circuit of a membrane? Can one get a similar result by assuming negative conductances and rectification?

DR. TAYLOR: It is precisely the same as the inductance and the capacitance that one can measure in a tungsten or carbon filament. The point is that it is nonlinear, and the nonlinearity is associated with a time delay. One can get very large inductive or capacitive reactance in this way.

DR. FISHMAN: Dr. Meves, you summarized the evidence for calling the difference current a gating current. However, Dr. Takashima showed that when he suppressed all ion conductances, virtually no change in capacitance was observed with a change in potential. We are in agreement with him, in the sense that we have done the same measurements in using pseudorandom noise. I wonder how this squares with the idea of gating currents, whether or not they actually totally inactivate. If they do not totally inactivate, one would expect to see some increase in capacitance with depolarization. Neither of us sees it, and in fact our measurements indicate a decrease in capacitance with depolarization in steady state. Would you comment on that?

DR. MEVES: It is a bit difficult to explain the difference. Takashima's previous paper indicated that the change in capacitance with depolarizations seemed to agree with the general idea of a charge movement in the membrane. Gating current measurements are extremely difficult because of the small magnitude (microamps per cm^2). I also feel that capacitance measurements are extremely difficult, as we have seen that the results depend entirely on the correction for series resistance. So I would not jump to any conclusion.

DR. TAKASHIMA: Actually, there are quite a few similarities between these, such as the saturation phenomenon and the values of time constants. However, when it comes to the effect of blocking ionic currents, the difference becomes pronounced. I applied a 20 millisecond pulse, and the measurement of capacitance begins around 10 milliseconds after the onset of the pulse. Therefore, it is possible that the inactivation process may have been completed during that time span.

DR. FISHMAN: I would comment that I expect you would have seen some component of capacitance increase in that time. Our measurements are completely steady states, so that inactivation is a possible explanation, although it would have to be total inactivation. Dr. Takashima does see a capacitance increase in the membrane with depolarization, if the conductances are not suppressed. It can, however, be shown that they are not essentially associated with the dielectric properties of the membrane but merely reactive components that are associated with time-variant conductances.

DR. TAYLOR: It seems to me that there would be a lot of virtue in using either noise and correlation techniques or impedance bridges to measure the capacitance associated with gating, because one could get a better signal-to-noise ratio.

DR. YANTORNO (*University of Pennsylvania, Philadelphia, Pa.*): It deals with sodium inactivation with glutaraldehyde. I cannot remember whether you actually put glutaraldehyde inside the axon as well as outside. The question is, if it was added outside, how do you explain the fact that the inactivation is an internally controlled mechanism? Could you discuss whether you can totally obliterate the activation process and look at the inactivation itself by that method?

DR. MEVES: The glutaraldehyde was applied internally. It was suggested to us that glutaraldehyde might cross-link proteins and block this effect. As to your second question, that you affect the inactivation gate only from inside, this is perhaps debatable. One of the most prominent agents to prolong an activation process is scorpion venom, which acts only from the outside.

DR. YANTORNO: I do not know what happens to the scorpion venom in terms of its movement across the membrane. It would be important to note whether it can move in and out.

UNINDENTIFIED SPEAKER: Dr. Meves, my question is related to the negative conductance of a current voltage curve and its interpretation in terms of molecular mechanism, which can be described as gating mechanism. Is it not possible that the gate might be analogous to a receptor protein at the postsynaptic junction which flips from one state to another as one changes the voltage across the membrane? The flip is accompanied by a large change in the sodium conductance associated with that channel.

DR. MEVES: If one looks at channels in terms of a molecular model, it is a charged molecule which changes its position in the membrane according to the electric field across the membrane, and so opens and closes the channel. If one uses an equivalent circuit, we come up with the circuit Dr. Taylor has shown. If one wants to make an electronic model, he incorporates a little capacitor parallel to the normal capacitance of the membrane. They are all, however, to some extent only an approximation.

DR. R. EISENBERG: One could summarize by saying that although we have many models, I do not believe there is an explicit model that has both gating currents and conductance as an output. It would be very hard, if we do not have an explicit model, to know what is going on inside, because you would be observing two outputs that were not in a series—causal relationship. I (and perhaps the audience) would look forward to an explicit model that has both conductance and gating currents as outputs.

DR. K. S. COLE (*NIH, Bethesda, Md.*): My first comment is to emphasize even more strongly than Takashima and Yantorno that the data and analyses of Cole and Curtis and Cole and Baker were restricted to frequencies of 10 KHz and above. Consequently, they can be expected to show only about the infinite frequency conductance changes from Hodgkin-Huxley (Cole,*†) as are confirmed†

*K. S. Cole. 1955. *In* Electrochemistry in Biology and Medicine. T. Shedlovsky, Ed. John Wiley and Sons. New York, N.Y.

†K. S. Cole. 1968. Membranes, Ions and Impulses. Part III: 303; Part I: 87. University of California Press. Berkeley, Calif.

from 1941 data and this paper. It is interesting to note that Hodgkin and I‡ measured the strictly DC or zero frequency membrane resistivity, thus bracketing the ionic conductances. Over the years I have been getting more frustrated and more beguiled by the varieties of cell membrane phase angles: why many membranes usually show $\phi < 90°$ and others always give $\phi = 90°$, while a few may produce either behavior in different experiments. I have come to suspect both inadequate electrode arrangements and protein surface conductances, and I hope to test electrode geometry on squid axons this coming summer. But Takashima and Yantorno have neatly and elegantly shown a geometry effect in their Figure 2 and used it to predict a squid phase angle approaching 90°. Their experiment can be crudely represented by parallel components of a 90° capacity over the electrode length, x, and semiinfinite cables at each end with 45° angles. The impedance Z is then:

$$1/Z = Aj\omega x + B(j\omega)^{1/2}$$

which approximates $1/Z \sim C(j\omega)^{\alpha}$ to within about 10% for $1/2 \leq \alpha \leq 1$ as $0 \leq x \leq \infty$. Even if a more detailed analysis improves this primitive attempt, there will still be some of the dozen or more examples that seem to require a surface conductance such as Fricke and Parker§ found for gelatine. I congratulate the authors on this opening wedge, as well as for their confirmation of work that Curtis and Baker and I¶‖ did so long ago, and for some other material I do not yet understand.

‡K. S. Cole & A. L. Hodgkin. 1939. J. Gen. Physiol. **22:** 671.
§H. Fricke & E. Parker. 1940. J. Phys. Chem. **44:** 716.
¶K. Cole & H. J. Curtis. 1939. J. Gen. Physiol. **22:** 649.
‖ K. S. Cole & R. F. Baker. 1941. J. Gen. Physiol. **24:** 771.

MEASUREMENT, MODELING, AND ANALYSIS OF THE LINEAR ELECTRICAL PROPERTIES OF CELLS*

Robert S. Eisenberg and R. T. Mathias

Department of Physiology
Rush Medical College
Chicago, Illinois 60612

J. S. Rae

Departments of Ophthalmology and of Physiology and Biophysics
University of Texas Medical Branch
Galveston, Texas 77550

Most cells and tissues use the flow of electric current to help perform their natural functions. The widespread use of electricity in the function of cells is natural from the physical point of view, since so many of the substances transported in the body are charged and since electrical signals have properties suitable for information-transduction, transfer, and processing. The widespread use of electricity is also natural from an evolutionary point of view. Cells, in most cases, must have a resting potential across their surface membrane if they are to exist at constant volume; the resting potential thus was probably present in the earliest cells in biological history. It seems that evolution is a highly conservative process; that is, if a structure or property is present, it will be modified and used in diverse ways in evolutionary development. Thus, one expects what one finds; most cells and tissues use their resting potential and associated electrical mechanisms to help perform their natural function.

It is interesting to note that electrical processes seem to be used even where they need not be used. Most systems that transport nonelectrolytes appear to involve the obligatory transport of ions as well, even though physically there is no need to transport charge when uncharged solutes are being moved across membranes. One can rationalize this fact by imagining that the active transport of ions was developed early in biological history to help maintain the volume of cells in the presence of small but persistent leakage of "impermeant" ions. Then the mechanism for nonelectrolyte transport developed as a modification of an ion-transport system rather than as an entirely new, electrically silent mechanism.

The study of the mechanisms that underlie the electrical activity of cells is of considerable biological interest. It is sometimes possible to study such mechanisms by the simple observation of natural electrical activity and by observing the change in natural electrical activity produced by ionic or experimental intervention. But it is usually necessary to study such mechanisms by direct electrical intervention; that is, by applying a perturbing current and measuring the responding voltage change. This paper is concerned with the study of the response of cells and tissues to the application of current and illustrates with two examples some of

*This work was supported by NIH grants HL 20230 to R.S.E. and EY 01207 to J.S.R., and by NSF grant MBS 75-5500 to R.S.E.

the information that can be determined from the measurement of the voltage response to applied current.

Our particular interest here will be the construction and test of electrical models of two preparations: the skeletal muscle fibers of frogs and the lens of the eye. These models predict the voltage response to applied current and include as the components of the model the electrical and morphological properties of the cell and its component structures and organelles. The need for an explicit electrical model may not be apparent but is, in our opinion, a general requirement for the analysis of the electrical properties of any preparation. We therefore will discuss the role of electrical models before we discuss the specific cases just mentioned.

The first role of the electrical model is to enumerate and describe the pathways for current flow. Current applied to a preparation flows in many directions. It leaves the current-passing electrode (in our cases a microelectrode inserted into the cell) and flows in the cytoplasm to a membrane, crosses the membrane into an extracellular space, and then flows in the extracellular solution to a collecting electrode, an indifferent electrode in the external bathing solution. The flow of current in the extracellular space may be rather complicated, since in many preparations some of the extracellular space is a specialized compartment within the cell. The flow of current in the intracellular saline solution is also important. In cylindrical cells like muscle fibers and nerve axons, most of the intracellular current flows longitudinally, the flow of intracellular current in other directions, both away from the microelectrode source and toward the outer membrane, producing significant effects only under extreme conditions. The longitudinal flow of current is significant because a longitudinal pathway of sufficient length (of a length constant, namely millimeters) has a resistance equal to the impedance of a similar length of outer membrane. Thus, the potential drop down the length of a cylindrical cell must be analyzed, as well as the potential drops across the outer membrane. In spherical cells there is no longitudinal pathway, and so the important intracellular potential drops occur in a different manner. As current leaves the microelectrode source in the region near the microelectrode, the great density of the lines of current flow causes significant potential drops. These three-dimensional effects, as they are called, since they involve the flow of current in all directions, are large in large spherical preparations, like the lens of the eye, for example.

Another role of an electrical model is to provide the relationship between the observed electrical properties and the properties of the individual structures and organelles that make up a cell. The measured response to applied current depends on all the pathways for current flow, many of which are not involved in the biological process of interest. Thus, a model is needed to determine the properties of the individual cell components from the observed properties of the whole preparation.

A model should include all the important characteristics of the preparation and should depend explicitly on the properties of the various structures that make up the cell (the size and shape of these structures can be measured by morphometric techniques), and it should depend on the electrical properties of individual membranes and compartments of intra- and extracellular saline solution. It is usually difficult to determine the individual electrical properties of cell components *a priori*. For example, if one is studying muscle fibers, it is not clear how one should describe the nonlinear properties of the surface and tubular membranes, since (for the most part) measurements are available only of the nonlinear properties of the combined surface membrane and tubular system. If, however, attention is restricted to strictly *linear* properties, it is much easier to specify the electrical properties of individual membranes, since it is widely believed that the linear electrical properties of individual membranes can be represented, to a reasonable approximation, as a resistor and capacitor in parallel. The procedure we take in constructing an electrical model is then to consider only the linear electrical properties of the cell and represent each membrane as a resistor and capacitor in parallel, the unknown value of the resistor and capacitor depending on the unknown specific value of the membrane resistance and capacitance and on the known amount of membrane area. We represent the properties of the cytoplasm and extracellular solutions as resistances, with areas, volumes and shapes determined morphologically.

A quantitative description of the model—an equation—is derived with use of whatever method seems expedient. In the simplest cases a lumped equivalent circuit can be directly written; in other cases, partial differential equations that describe the continuous variation of potential are written (with boundary conditions to describe the membrane); in still other cases, difference equations are written to describe the variation in potential in a discrete number of locations; and finally, it is occasionally necessary to combine all three types of analysis. The equations are then solved; that is to say, an approximate solution is sought that has *physical meaning* and a known mathematical error. The requirement for physical meaning is critically important since the qualitative properties of the mathematical solutions are usually a particularly useful description of the physiological properties of the preparation. Furthermore, the physical meaning of the solution allows speculative extension of the mathematical approximation to cases not strictly within the domain of the analysis. For example, one often guesses that nonlinear effects will be confined to terms in the mathematical solution that depend on membrane properties and will not effect terms that depend only on intra- or extracellular resistances.

By following the procedure just described, it is possible to construct an electrical model of a cell or tissue and be in a position to test the model and interpret the electrical properties of the preparation. In this way a qualitative analysis of mechanism is possible, and indeed a reasonably precise quantitative description can be made.

Frog Skeletal Muscle Fibers

The electrical properties of the muscle fibers of the frog have received a great deal of attention over the years because of the intrinsic interest in skeletal muscle and the important role of electrical properties in the initiation of contraction. Interest is also aroused because skeletal muscle fibers probably represent the least complicated structure of excitable membranes except for nerve axons.† It has been widely felt that the techniques and forms of analysis necessary to unravel the electrical properties of skeletal muscle would be a guide to the methods necessary to unravel the electrical properties of more complicated tissues of great biological and clinical importance, like smooth muscle, cardiac muscle, and the various epithelia.

The overall model of the electrical properties of frog muscle utilizes one-dimensional cable theory (see Jack *et al.*[18] for an excellent review) to describe the longitudinal spread of current and three-dimensional cable theory (see Peskoff *et al.*[29] for the most recent work and Eisenberg & Johnson[10] for an early review) to describe the flow of current near the microelectrode source (important at short times or high frequencies). One-dimensional cable theory includes the longitudinal and shunt (i.e., transverse) pathways for current flow. Since the longitudinal pathways are essentially resistive,[22,23] we confine our attention here to the shunt pathways, the pathways by which current can leave the sarcoplasm in a single sarcomere. We are mostly interested in modeling the linear properties and so will not discuss the recent work on asymmetrical capacity currents. We must emphasize, however, the importance of that work in the understanding of the nonlinear processes in skeletal muscle (Chandler *et al.*[5] describe and analyze the asymmetry currents and give references to other work on the subject).

There are three systems of membranes that one might suspect would contribute importantly to the electrical properties of a sarcomere of skeletal muscle: the surface membrane, the tubular system, and the sarcoplasmic reticulum. The surface membrane itself comprises a small part of the total membrane within frog muscle fibers (some 1–2% for a fiber of typical diameter) but obviously is the location of substantial current flows. The surface membrane, however, is not a simple structure, since it has infoldings called caveolae[8] whose function is not known. Indeed, it is not clear that all the caveolae behave electrically as part of the surface membrane. If many caveolae serve as openings of the tubular system,[35] they would modify the current flowing into the tubular system and so would act more as extensions of the *t*-system than as components of the surface membrane. Unfortunately, until the anatomical relationship of the opening of the tubular system and the caveolae is settled, and until the function of the caveolae is known, it will be difficult to evaluate their contribution to the overall electrical properties of the muscle fiber. At present we model the surface membrane as a simple resistance and capacitance in parallel, assuming that the only effect of the caveolae is to modify the amount of membrane that produces these electrical properties.

The electrical properties conferred on the fiber by the sarcoplasmic reticulum are customarily ignored in models, since it is widely believed that no current can flow across the tubular membrane into this compartment. Certainly there is no

†The structure of the excitable membranes of skeletal muscle is not all that simple, of course.[27,21] But the membranes are highly organized in a reasonably periodic arrangement with characteristics suitable for observation in the electron microscope (except at the openings of the *t*-system to the extracellular space). Thus, the complications of the structure are rather well known; indeed, the qualitative and quantitative organization of muscle membranes is known at least as well as that of the much simpler, but less studied axon membranes!

convincing evidence of substantial current flow in that pathway, and the structure of the junction between the sarcoplasmic reticulum and the tubular membrane (reviewed in Franzini-Armstrong[13]) is not reminscent of junctions that do electrically couple adjoining cells. We follow the customary procedure and ignore the contribution of the sarcoplasmic reticulum to the overall electrical properties of the fiber, but we are painfully aware that some forms of coupling between the sarcoplasmic reticulum and the tubular system might influence the electrical properties of the fiber in a sufficiently subtle manner to have avoided detection but in a sufficiently substantial manner to have important effects.

The main determinant of the linear electrical properties of skeletal muscle fibers is the tubular system, since the great majority of the membrane which delimits the cell from the bathing solution is in the *t*-system. The tubular system of muscle fibers is a branching network of tiny tubules that arise as invaginations of the surface membrane and trap a component of extracellular space within the fiber. The tubules contain solution that appears to exchange readily with the extracellular bathing solution (see reviews[7,24,9]). Because the cross section of the tubules is small, and yet they have considerable length and membrane surface, there can be significant radial potential gradients in the lumen of the tubules. It is the description of this potential drop—and concomitantly of the entire pattern of current flow and potential distribution within the *t*-system—that we will analyze here.

The first description of the electrical properties of the *t*-system[12] represented the tubular network either as lumped elements—a resistance in series with the tubular membrane—or as a distributed network, a pair of discs of membrane enclosing a small extracellular space. The former description seemed sufficient to fit the experimental results, but the data did not rule out the disc or distributed representation, and a foreshortened theoretical analysis of the disc model was presented. A variety of subsequent experimental results[15,3,31,16,17,4,33,34] showed, however, that there were significant radial potential drops within the *t*-system and ruled out a lumped representation.

The central problem in modeling the *t*-system is then the correct representation of the branching nature of the network and radial potential drop produced by the network. This problem is sufficiently difficult that some[14,26] have chosen preparations with unbranched tubules, and it is probably not coincidental that the clearest results come from such preparations. Others[12] have ignored the branching or have analyzed the effect of branching in a few special networks with morphological properties distinct from those of frog muscle.[2,31] We will discuss here models that seek to include explicitly the structure of the *t*-system as reported by morphologists.[27,21]

Mathias[19] numerically constructed explicit exact solutions for two-dimensional branching networks of tubules of various structures. These exact solutions (and unpublished solutions for a hexagonal *t*-system in a hexagonal fiber; Mathias, personal communication) and the methods of deriving and computing them are keystones to future analysis, since they are the only results that one can be certain are correct, at least in a mathematical sense. All other analyses contain some form of irrational approximations; that is, approximations that do not permit complete computation of their own mathematical error. However, the exact solutions are so cumbersome as to preclude qualitative understanding. Furthermore, they would be difficult to compute and fit to a substantial amount of experimental data. The fundamental reason that the exact solutions are cumbersome is that they use a two-dimensional description of the structure of the *t*-system; they describe each node of the tubular system with two spatial coordinates. For that reason, Mathias *et al.*[20] constructed a model of the *t*-system that requires only one spatial coordinate to describe each node. They classified the nodes of the *t*-system into concentric shells by a geometric construction that would tend to minimize the circumferential variation of nodal potential within any shell. Mathias *et al.* then presented a rather complex statistical derivation of a finite difference-differential equation to describe the average potential in each shell. When the shell width is small compared to the tubular length constant, the difference equation reduces to a differential equation that is of the same form as the disk model and could be easily and exactly solved. Also, it is fortunate that a quite simple and precise approximation to the solution of the difference equation for the admittance of the *t*-system could be found, this solution being valid over the entire range of length constants. Nonetheless, the complexity of their analysis is rather forbidding.

Here we shall present a physical analysis of the structural model of Mathias *et al.* We do not seek as complete an approximation as they have presented, but rather derive and solve an equation that is quite accurate within most of the physiological range, to something like 1000 Hz under normal conditions. The reader is referred to the original paper[20] for the more precise result and for a discussion of the applicability of the present more limited approximation.

Consider a random or deterministic network of tubules, each tubular branch being of length L, having a resistance R_b between nodes due to the lumen of one branch, having N_B branches per node (note that the number of *distinct* branches per node is $N_B/2$, since each branch spans two nodes), with circumferential spacing δ_N between nodes, and with a total length of tubule per unit cross-sectional area of fiber of L_T/A_F. Such a network enclosed in a finite boundary can usually be decomposed into shells within which tubular branches can be defined as radial (that is, crossing a boundary of a shell) or circumferential (lying entirely within a shell). The spacing between *radial* branches is called δ_B. The potential drop ΔU in a shell of thickness Δr when there is radial current i_r flowing in the lumen of the tubules is

$$\Delta U = \frac{(\text{radial current}) \times (\text{resistance of one tubular branch})}{(\text{no. of radial tubular branches})} \tag{1}$$

$$\Delta U = \frac{i_r R_b}{2\pi r/\delta_B} \tag{2}$$

$$\frac{\Delta U}{\Delta r} = \frac{i_r R_b}{(2\pi r/\delta_B)\cdot \Delta r} \rightarrow \frac{dU}{dr} \text{ as } \Delta r \rightarrow 0. \tag{3}$$

The effective radial resistance is therefore $R_b(\delta_B/\Delta r)$, so that the factor $\delta_B/\Delta r$ is a network parameter that depends on the manner in which the nodes of the network are connected and also on the degree of tautness in the network. For example, a network with a small spacing between radial branches will have—*all other morphometric parameters being constant*—a lower value of effective radial resistance than a network with a larger spacing between radial branches, since the former network will have more parallel pathways in which radial current can flow. On the other hand, a network with a small value of Δr will have a larger effective radial resistance than one with a larger value—*all other morphometric parameters being constant*—because the total amount of tubular system in the two networks is the same, but in the former network the radial tubules that connect shells must twist and bow. Thus, the effective path length and radial resistance is greater in the network with the smaller shell thickness, all other things being equal.

In general, then, it is necessary only to determine the shell thickness from the measured morphometric parameters of the network. This can be done by equating two expressions for the total number of tubules (both radial and circumferential) in a shell

$$\text{(no. of tubules in shell)} = \text{(no. of tubules of length } L \text{ per unit area)} \times \text{(area of one shell)} \quad (4)$$

$$\text{(no. of tubules in shell)} = \text{(no. of distinct tubules per node)} \times \text{(no. of nodes per shell)} \quad (5)$$

$$[(L_T/A_F)(1/L)][2\pi r \Delta r] = (N_B/2)(2\pi r/\delta_N) \quad (6)$$

or

$$\Delta r = L(N_B/2)/(\delta_N L_T/A_F). \quad (7)$$

In a general expression it is preferable to write R_b in terms of resistivity of the lumen of the tubules R_L and the morphometric parameters of the network.‡ Then,

$$R_b = LR_L/A_T = LR_L \frac{L_T/A_F}{S_T/A_F} \frac{S_T}{V_F} \frac{V_F}{V_T}. \quad (8)$$

It should be noted that the expression is more useful than it is awkward, since all the morphological parameters on the right-hand side have been experimentally measured.

To complete the derivation we write Kirchoff's Current Law

$$\frac{di_r}{dr} \simeq I_w \cdot 2\pi r = U(G_w + j\omega C_w)(S_T/A_F)(2\pi r) \quad (9)$$

where I_w is the density of current crossing the tubular wall, G_w and C_w are the specific conductance and capacitance of the tubular wall, and S_T/A_F is the surface area of tubular wall per cross-sectional area of fiber.

Equations 3 and 9—using (7) and (8) if desired—can be combined to give a differential equation for the potential within the tubular system. This resulting equation is of the same form (but with different parameter values) as that used by virtually all workers in the field.[2,31,16,17,25,32,6]

‡Where A_T is the cross-sectional area of the tubule, S_T/A_F is the surface of tubular wall per unit cross-sectional area of fiber, S_T/V_F is the surface of tubular wall per unit volume of fiber, and V_F/V_T is the volume of fiber per volume of tubule.

$$\frac{d^2U}{dr^2} + \frac{1}{r}\frac{dU}{dr} - \Gamma_m{}^2 U = 0. \tag{10}$$

The propagation constant of the tubular system Γ_m is a generalization of the reciprocal of the usual tubular length constant, more easily applied to sinusoidal and transient problems. The propagation constant of the tubular mesh is related to the propagation constant of a single tubule Γ

$$\Gamma_m = \left(\frac{\delta_B}{\delta_N}\frac{N_B}{2}\right)^{1/2} \Gamma(L/\Delta r) \tag{11}$$

and the tubular propagation constant can be written in terms of the specific properties of the tubular wall and lumen

$$\Gamma^2 = R_L(S_T/V_T)(G_w + j\omega C_w). \tag{12}$$

It is also useful to write (from (**3**) and (**9**)) the propagation constant of the tubular network in terms of the properties of an effective radial resistivity of the lumen and the effective wall conductivity and capacitance per unit volume of fiber:

$$\Gamma_m{}^2 = \left(R_b \sigma \frac{\delta_B}{\Delta r}\right)\left(\frac{S_T}{V_F}[G_w + j\omega C_w]\right)$$

$$\Gamma_m{}^2 = (\text{effective radial resistivity in units of ohm-cm}) \times (\text{wall admittance per unit volume of fiber in mho/cm}^3) \tag{13}$$

where σ is the sarcomere spacing and $\sigma S_T/V_F = S_T/A_F$.

The previous differential equation (10) can be trivially solved and has been applied to a wide range of situations involving the tubular system. Indeed, it serves as an adequate electrical model of the tubular system. The substantive difference between this model and earlier results is the explicit dependence of the propagation factor on the measured morphometric parameters of the tubular system. No tortuosity factors computed from deterministic models of the *t*-system appear in our result.

The question then arises, how is the model checked? What experimental interventions can be made that can be compared with the predictions of the model? First, one can determine if the model fits the impedance data measured from muscle fibers. Second, one can determine if the model describes the variation of impedance with the diameter of the fiber. Third, one can determine if the values of the parameters of the model agree with independent evidence. Fourth, one can determine whether a nonlinear extension of the model allows prediction of the shape and conduction velocity of the action potential. Fifth, one can determine whether the parameter values of the model change as they should with the change in length of the fiber. Sixth, one can determine whether the parameter values of the fiber change as they should with changes in the resistivity of the external bathing solution.

In this context it is important to point out that one essential check has already been performed. Exact solutions of networks of branching tubules similar to the *t*-system have been simulated with digital computers, and the present model has been shown to fit the properties of the exact solutions closely. The experimental checks are not really necessary to determine whether the present model describes a branching network of tubules; rather, the experimental checks are important to see if the assumptions from which the model is constructed are reasonably correct: to see if the circuit elements in the model are reasonably apt descriptions of the structural components of muscle, to see if the division into longitudinal and shunt pathways is correct, even in the presence of a spiral *t*-system,[28] and, most importantly, to check if the model is reasonably complete (to check if any significant pathways of current flow have been omitted).

We shall refrain from elaborate discussion of each of the experimental checks, since the data is incomplete in most cases. Mathias *et al.*[20] show that the model well describes experimental data taken in a variety of conditions in which the membrane is reasonably linear. The values of the components of the model and the variation of the values, however, often do not agree with preconceptions. Some of the problem is undoubtedly caused by experimental error, particularly in the morphometric parameters which are subject to change by fixation and other preparatory procedures. However, these problems may also derive from some incompleteness of the model, the result of some structure or process not presently included. To cite a few obvious possibilities, it may be important to include explicitly the properties of the sarcoplasmic reticulum, the junction between the sarcoplasmic reticulum and the tubular system, the caveolae, the openings of the *t*-system, the asymmetrical capacity currents, and perhaps even the longitudinal components of the *t*-system.

As our knowledge of the structure of muscle increases, and as exciting new techniques such as the optical recording of membrane potentials become available, we anticipate that a much wider range of precise experimental data will become available to describe the electrical behavior and properties of muscle. We hope that theoretical models will develop hand-in-hand with the experimental developments so that the electrical behavior and properties can be properly analyzed. Perhaps in that way all the components of current flow important in the processes of excitation-contraction coupling can eventually be both measured and analyzed.

Crystalline Lens of the Eye

The lens of the eye is an interesting organ because of its clinical and biological properties: clinically it is the source of many impairments of vision, cataracts unfortunately being quite a common occurrence. Biologically, the question of how the lens is able to perform the normal metabolic maintenance functions of any tissue and still preserve its transparency is of considerable interest. The electrical properties of the lens are of interest for at least two reasons: first, the electrical coupling between cells may well be importantly involved in the processes that maintain the transparency of the lens; second, the transport processes in the surface of the lens, which clearly are important to the metabolism of the tissue, involve the flow of current. An extensive critical review of the electrical properties of the lens develops these themes in some detail (Rae[30]) and should be consulted.

The models of the electrical properties of the lens are much less complete than those of skeletal muscle primarily for historical reasons; the preparation has been studied by only a few workers for a rather short time. Thus, the role of the model presented here, approximating the lens as a giant spherical cell, is analagous to the role of cable theory in the historical development of an electrical model of muscle fibers; it allows the gross analysis of the voltage response to current flow into a membrane component and an internal component. The model ignores the complex structure of the extracellular space within the lens, however, and it also ignores the specialized layer of cuboidal epithelial cells on the anterior side of the lens. Surely, the model will need to be extended, and revised, to deal with these important structural complexities. But just as in skeletal muscle, the original oversimplified model of one-dimensional cable theory has served as the foundation for later complexities, so in the lens of the eye the spherical cell model may still be of use as newer, more precise models are developed.

We crudely represent the entire lens as a spherical cell, or more precisely, as a finite cell, one with dimensions of the same approximate size in all directions. Eisenberg and Rae[11] have applied the three-dimensional theory of a number of authors (see Ref. 11 for references that describe the assumptions and analysis of the theory) to the lens and have shown that the overall potential recorded by a microelectrode in response to a step of current is given by

$$V = \frac{IR_m}{4\pi a^2}\left(1 - e^{-t/R_mC_m}\right) + I \cdot F(R_i, a, r_1, R_1, \theta) + I \cdot G(R_B, a) \quad \textbf{(14)}$$

where

$$\begin{aligned} F(\cdots) = & \left(\frac{R_i}{4\pi a}\ \frac{r_1^2 + R_1^2}{a^2} - \frac{2r_1R_1}{a^2}\cos\theta\right)^{-1/2} \\ & + \frac{R_i}{4\pi a}\left\{\left(1 + \frac{r_1^2R_1^2}{a^4} - \frac{2r_1R_1}{a^2}\cos\theta\right)^{-1/2} - 2 + \ln 2\right\} \\ & - \frac{R_i}{4\pi a}\ln\left\{1 - \frac{r_1R_1}{a^2}\cos\theta + \left(1 + \frac{r_1^2R_1^2}{a^4} - \frac{2r_1R_1}{a^2}\cos\theta\right)^{1/2}\right\} \end{aligned} \quad \textbf{(15)}$$

$$G(\cdot\cdot) = \frac{R_B}{4\pi a} \quad \textbf{(16)}$$

and where V is the potential measured, I the current applied, a the radius of the lens, t the time after onset of current, C_m (in F/cm^2) the "specific" membrane capacitance, R_m (in ohm-cm^2) the "specific membrane" resistance, R_i (ohm-cm) the "resistivity" of the lens interior, R_1 the radial location of the current electrode measured from the center of the lens (as is r_1 the radial location of the voltage electrode), θ the angular separation of the electrode tips, and R_B (ohm-cm) the resistivity of the extracellular bathing solution.

The important characteristic of this equation is the separation of the response into three components: one, which depends on membrane properties, is independent of spatial location and has a slow time course governed by the time constant R_mC_m; the second, which depends on internal resistance, is independent of membrane properties, is highly localized, and has so fast a time course that it appears to be established instantaneously; the third is a component that arises in the extracellular bathing solution and is small and of little interest.

Experimental results show the qualitative behavior predicted by the equation: the potential can be separated into two components either by analyzing the variation of potential with position of the microelectrode or by analyzing the time course of the potential. The components then have the properties predicted by the model; namely, the spatially uniform component of potential is established slowly, while the local component of potential is established essentially instantaneously.

The structural complexity of the lens, not included in this model, precludes too much further analysis. Although the local potential can be analyzed to give a figure for the effective internal resistivity of the lens, the figure is undoubtedly a composite of the properties of the coupling between cells, the properties of the cytoplasm, and perhaps the properties of the trapped extracellular space within the lens. It is better to have such a composite figure than none at all, but it would be better yet to have a more complete theory to allow the separation of the effective internal resistance into its components. Analysis of the slow, spatially uniform component of potential is even more complicated: experimentally, it is found to have a time course quite distinct from that described by the model (Equations 14–16), and the parameter values necessary to crudely fit the model to the data are wildly different from the properties of an isolated membrane. Pursuing the analogy with frog muscle, we suspect that the cause of the discrepancy is the failure of the model to include the properties of the trapped extracellular space within the lens. The membranes lining this internal extracellular space would certainly be expected to contribute to the slow component of potential, to modify its time course, and to modify enormously the apparent parameter values. Thus, we look forward to the development of a model that will explicitly include current flow in these trapped extracellular spaces, and that will deal perhaps with the epithelial coating of cells on the anterior surface. Such a model might even help resolve the essential paradox of the electrical properties of the lens—its extremely long time constant.

It seems then that at least in the case of these two preparations—the lens of the eye and skeletal muscle fibers of the frog—the construction of linear electrical models has fulfilled many of the criteria we described earlier. The models are necessarily incomplete, both for theoretical and experimental reasons, but the general success and utility of the approach in the understanding of the electrical function of the tissues seem clear. Perhaps similar success can be expected in the application of this approach to other preparations.

References

1. Adrian, R. H. & W. Almers. 1974. Membrane capacity measurements on frog skeletal muscle in media of low ion content. With appendix by R. H. Adrian, W. Almers and W. K. Chandler. J. Physiol. **237:** 573.
2. Adrian, R. H., W. K. Chandler & A. L. Hodgkin. 1969. The kinetics of mechanical activation in frog muscle. J. Physiol. (Lond.) **204:** 207–230.
3. Adrian, R., L. L. Costantin & L. D. Peachey. 1969. Radial spread of contraction in frog muscle fibers. J. Physiol. (Lond.) **204:** 231–257.
4. Barry, P. H. & R. H. Adrian. 1973. Slow conductance changes due to potassium depletion in the transverse tubules of frog muscle fibers during hyperpolarizing pulses. J. Membrane Biol. **14:** 243–292.
5. Chandler, W. K., R. F. Rakowski & M. F. Schneider. 1976. A nonlinear voltage dependent charge movement in frog skeletal muscle. J. Physiol. **254:** 245–283.

6. Chandler, W. K. & M. F. Schneider. 1976. Time course of potential spread along a skeletal muscle fiber under voltage clamp. J. Gen. Physiol. **67:** 165–184.
7. Costantin, L. L. 1975. Contractile activation in skeletal muscle. Prog. Biophys. Molec. Biol. **29:** 197–224.
8. Dulhunty, A. F. & C. Franzini-Armstrong. 1975. The relative contribution of the folds and caveolae to the surface membrane of frog skeletal muscle fibers at different sarcomere length. J. Physiol. (Lond.) **250:** 513–539.
9. Ebashi, S. 1976. Excitation contraction coupling. Ann. Rev. Physiol. **38:** 293–314.
10. Eisenberg, R. S. & E. A. Johnson. 1970. Three dimensional electric field problems in physiology. Prog. Biophys. Molec. Biol. **20:** 1–65.
11. Eisenberg, R. S. & J. L. Rae. 1976. Current-voltage relationships in the crystalline lens. J. Physiol. **262:** 285–300.
12. Falk, G. & P. Fatt. 1964. Linear electrical properties of striated muscle fibers observed with intracellular electrodes. Proc. R. Soc. Lond. [Biol.] **160:** 69–123.
13. Franzini-Armstrong, C. 1975. Membrane particles and transmission at the triad. Fed. Proc. **34:** 1382–1389.
14. Gilai, A. 1976. Electromechanical coupling in tubular muscle fibers. II. Resistance and capacitance of one transverse tubule. J. Gen. Physiol. **67:** 343–367.
15. Gonzalez-Serratos, H. 1966. Inward spread of contraction during a twitch. J. Physiol. **185:** 20–21.
16. Hodgkin, A. L. & S. Nakajima. 1972. The effect of diameter on the electrical constants of frog skeletal muscle fibers. J. Physiol. (Lond.) **221:** 105–120.
17. Hodgkin, A. L. & S. Nakajima. 1972. Analysis of the membrane capacity in frog muscle. J. Physiol. (Lond.) **221:** 121–136.
18. Jack, J. J. B., D. Noble & R. W. Tsien. 1975. Electric current flow in excitable cells. Oxford: Clarendon Press.
19. Mathias, R. T. 1975. A study of the electrical properties of the transverse tubular system in skeletal muscle. Ph.D. Dissertation. Univ. California. Los Angeles, Calif.
20. Mathias, R. T., R. S. Eisenberg & R. Valdiosera. 1977. Electrical properties of frog skeletal muscle fibers interpreted with a mesh model of the tubular system. Biophys. J. **17:** 57–93.
21. Mobley, B. A. & B. Eisenberg. 1975. Sizes of components in frog skeletal muscle measured by methods of stereology. J. Gen. Physiol. **66:** 31–45.
22. Mobley, B. A., J. Leung & R. S. Eisenberg. 1974. Longitudinal impedance of skinned frog muscle fibers. J. Gen. Physiol. **63:** 625–637.
23. Mobley, B. A., J. Leung & R. S. Eisenberg. 1975. Longitudinal impedance of single frog muscle fibers. J. Gen. Physiol. **65:** 97–113.
24. Nakajima, S. & J. Bastian. 1976. Membrane properties of the transverse tubular system in amphibian skeletal muscle. *In* Electrobiology of Nerve, Synapses, and Muscle. Raven Press. New York, N.Y.
25. Nakajima, S., Y. Nakajima & J. Bastian. 1975. Effects of sudden changes in external sodium concentrations on twitch tension in isolated muscle fibers. J. Gen. Physiol. **65:** 459–482.
26. Nicolaysen, K. 1976. The spread of the action potential through the *t*-system in hagfish twitch muscle fibers. Acta Physiol. Scand. **96:** 29–49.
27. Peachey, L. D. 1965. The sarcoplasmic reticulum and transverse tubules of the frog's sartorius. J. Cell Biol. **25** (3) pt. 2: 209–231.
28. Peachey, L. D. & B. Eisenberg. 1975. The *t*-system and striations of frog skeletal muscle are spiral. Biophys. J. **15:** 253a.
29. Peskoff, A., R. S. Eisenberg & J. D. Cole. 1976. Matched asymptotic expansions of the Green's function for the electric potential in an infinite cylindrical cell. SIAM J. Appl. Math. **30:** 222–239.
30. Rae, J. L. 1977. The electrophysiology of the crystalline lens in current topics in Eye Research. Vol. 1 J. Zadunaisky and H. Davson, Eds. Academic Press. New York, N.Y.
31. Schneider, M. F. 1970. Linear electrical properties of the transverse tubules and surface membrane of skeletal muscle fibers. J. Gen. Physiol. **56:** 640–671.

32. SCHNEIDER, M. F. & W. K. CHANDLER. 1976. Effects of membrane potential on the capacitance of skeletal muscle fibers. J. Gen. Physiol. **67:** 125–163.
33. VALDIOSERA, R., C. CLAUSEN & R. S. EISENBERG. 1974. Measurement of the impedance of frog skeletal muscle fibers. J. Gen. Physiol. **14:** 295–315.
34. VALDIOSERA, R., C. CLAUSEN & R. S. EISENBERG. 1974. Circuit models of the passive electrical properties of frog skeletal muscle fibers. J. Gen. Physiol. **63:** 432–459.
35. ZAMPIGHI, E. G., J. VERGARA & R. RAMON. 1975. The connection between the *t*-tubules and the plasma membrane in frog skeletal muscle. J. Cell Biol. **64:** 734–740.

ION MOVEMENTS AND KINETICS IN SQUID AXON I. COMPLEX ADMITTANCE*

D. Poussart

Department of Electrical Engineering
Université Laval
Québec, Canada G1K 7P4

L. E. Moore and H. M. Fishman

Department of Physiology and Biophysics
University of Texas Medical Branch
Galveston, Texas 77550

Introduction

Small signal impedance measurements on squid axon provided some of the first clues regarding the time-dependent, nonlinear character of ionic conduction. In spite of the far-reaching implications of the early data, complex impedance has been rather neglected as a useful probe over these years since Cole's original contributions.[1] In large part, this certainly resulted from the sustained success of the voltage-clamp concept, which, in concert with the use of large-step changes of membrane potential, has yielded much of today's description of the mechanisms of ionic conduction in excitable membrane. Another reason, perhaps significant as well, relates to the relative efficiency of step-clamp experiments. Extensive sets of data can be gathered over a short experimental time, and wide variations of experimental conditions can be studied on a single preparation with good expectation of adequate control data. By contrast, techniques in common use for impedance/admittance measurements have changed little since Cole's original work. Typically, they make rather inefficient use of experimental time.

In addition to convenience there are many instances where it is necessary to minimize observation time. This is the case for systems that are not strictly time-invariant, and whose kinetics one would like to follow as time evolves. It is also for systems which, although they can be considered as relatively time-invariant, cannot have some of their important parameters modified for prolonged periods. The squid axon under axial wire conditions is a good case in point. Although this preparation can yield useful data over periods of several hours, it still undergoes gradual changes of significance as time proceeds, and control data must be obtained frequently. Furthermore, as a result of internal poisoning following the application of prolonged currents, even of moderate magnitude, membrane potential can be displaced from rest for only brief periods.[2] Consequently, the available impedance/admittance data has been generally limited to rest or near rest potential levels. Outside of that range, it has been necessary to rely upon data

*This work was supported in part by Canadian National Research Council grant A5274 (DP) and National Institute of Health, United States Public Health Service, research grants NS11764 (HMF) and NS13520 (LEM).

extracted indirectly from linearized step-clamp records obtained for large, brief changes of membrane potential.

Several communications at the Conference on which this volume is based underline the importance of precise and detailed characterization of impedance/admittance of excitable and "passive" membrane and tissues. Foremost is the observation that such data is a direct reflection of the kinetics of ionic conduction. This is also the case for the spontaneous, microscopic fluctuations of electrical quantities which, together with others, we have been studying for a number of years. And it is no coincidence that our own involvement with admittance was motivated by membrane noise measurements, nor is it that the technique about to be described also exploits noise as a useful quantity. Some time ago we reported spectral features of potassium-conduction noise, in squid axon, which suggested that the overall kinetics were not first-order.[4] These data predicted a particular feature of the low frequency impedance function which we confirmed in subsequent measurements.[2] The lengthy (minutes) periods of data acquisition required by the frequency range of interest—1 to 30 Hz—underscored the relative inefficiency of the conventional techniques that were being used and prompted the method described below. This method typically offers a 100-fold reduction in the observation time needed.

Step-clamp and impedance/admittance measurements share generally similar requirements as to the preparation and its instrumentation interface. The central objective expressed by Marmont nearly thirty years ago has not changed: "What one would really like to do is to take a known area of the axon membrane and be able to measure directly a current of uniform density through the membrane as well as the potential across it."[5]

The present communication focuses primarily on these aspects which are specifically related to the impedance/admittance function. The required observation time is considered a key parameter. Conventional techniques and some practical aspects of their application are first reviewed briefly. The relative merits of various test stimuli are discussed and the unique qualities of pseudorandom signals outlined. The implementation of an instrument designed for transfer-function analysis in real time and the procedure developed for measurements on the squid axon are described. Illustrative measurements obtained for different experimental conditions are then presented and compared with predictions issued from the linearized Hodgkin-Huxley model.

Common Methods for Impedance/Admittance Measurements

For "small" perturbations, the excitable membrane appears to behave linearly. Efficient experimental methods for systematic studies of nonlinearities, in the form of Wiener kernels, have recently generated considerable interest.[6] These methods, which would allow precise qualification of the term "small," have not yet been applied to squid axon. However, the first relevant data,[7,8] and our own (see below) indeed suggest that in squid axon the linear component of the total response substantially dominates if the dynamic excursions of membrane potentials are limited to a few millivolts.

The measurement objective, then, consists of evaluating the complex impedance or admittance function (i.e., under current or voltage clamp conditions, respectively), $H(f)$, expressed, say, as magnitude $\mathrm{Mag}_H(f)$ and phase $\mathrm{Arg}_H(f)$. It is assumed that data is to be obtained with a frequency resolution Δf, and over a specified range $K\Delta f$, where K is some constant.

Fundamentally, two factors play a significant role in the required observation time T. First, from basic properties of time-frequency domain analysis, a resolution Δf in the frequency domain implies a minimum $T \approx 1/\Delta f$. Second, if the measurement is conducted in noisy conditions, averaging results from successive, independent observations may be required to reduce the statistical uncertainty within acceptable limits.

The most common technique is based on the sinusoïdal response. An excitation at frequency f_0 is applied, and once the response is in steady state, the quantities $\mathrm{Mag}_H(f_0)$ and $\mathrm{Arg}_H(f_0)$ are evaluated. The two-terminal AC bridge has frequently been used for that purpose. Four-terminal measurements are preferable, in principle, since the electrode characteristics are removed. Designs of four-terminal bridges have been published,[9] but apparently not been applied to squid axon. Other evaluation techniques include Lissajous patterns, zero crossing, and coherent (phase-lock) detection. References 2, 10, and 11 are representative recent examples and additional references and further technical discussions can be found in 12. With coherent detection, accurate data can be extracted even in cases of poor signal/noise ratio. These techniques, however, are extremely time-consuming, since the procedure must be repeated sequentially with frequency changed in discrete steps or continuously swept (linearly or logarithmically) with time. A lower bound for the total measurement period is $K/\Delta f$, since, without *a priori* knowledge of $H(f)$, the settling time allocated to each data point is of the order of $1/\Delta f$. For the present application, several minutes have been required, typically, for magnitude and phase evaluation.

The function $H(f)$ may be directly obtained from the Fourier transform of the response $h(t)$ to an impulse perturbation. Since we shall be dealing with discrete time and frequency series, it will be convenient to express H as a function of the index, k, of its components in the Fourier expansion. In principle, the impulse method provides a minimum observation time, since $h(t)$, which is transient, has a duration of significance of $1/\Delta f$. If the time required for computing the Fourier transform is shorter than $1/\Delta f$, the method operates in real-time. However, it has more conceptual than practical interest, and two difficulties are likely. The very high peak-to-average power ratio of the stimulus may drive the system into nonlinear response where $h(t)$ loses its simple meaning. The large dynamic range of the response also requires special care for avoiding significant computation errors in the Fourier transform, with corresponding costs in time or in hardware complexity. These problems may be alleviated somewhat by processing step rather than impulse responses.[24]

The key property of the impulse—simultaneous and equal energy perturbation at "all" frequencies—can also be obtained by using random noise as a test stimulus. Typically, the random perturbation consists of a Gaussian process with white magnitude spectrum, i.e., statistically uniform power density over the frequency range of interest. An estimate of the magnitude of the transfer function can be simply extracted, within a known constant, from the power-density autospectrum of the response. This quantity is obtained directly from a set of filters or as the result of a Fourier transform computation. Spectrum analyzers that can operate in real-time—i.e., with speed sufficient to process incoming data without gaps—are readily available for that purpose. Since the method is direct, computation errors can be economically avoided and a dynamic range in excess of 60 decibels (dB) is typically achieved. We have found this method convenient to use because it essentially involves the same instrumentation already required for studies of the spontaneous membrane noise.[2,13] In order to obtain magnitude and phase data, it is necessary to derive the cross-spectrum of the system, a computation which involves both the stimulus and the response signals. Classically, this quantity has been obtained through a single Fourier transform from the cross-correlation function. Since the transform is computed on a product quantity, overall dynamic range is restricted accordingly. The technique has been applied successfully to squid axon,[2,8,14] but in our experience, the limited dynamic range, typically 30 dB with available hard-wired processors, has been found to be a serious limitation. However, the cross-spectrum may be computed directly, without the intermediate step of the cross-correlation function, by appropriate combination of individual Fourier transforms of the stimulus and the response. This approach is now preferable with the availability of fast Fourier processors.

These noise methods have several advantages. Since the perturbation signal has a wideband spectrum, data for all frequencies of interest is acquired simultaneously. Since the perturbation has a low peak-to-average power ratio, nonlinear response can be more easily avoided. In addition, the coherence function which can be derived with the required quantities provides an indication of significant nonlinear components and/or inadequate signal/noise ratio. However, the specified spectral characteristics of the perturbation hold only in the mean; i.e., any particular sample of white noise over a finite time is not strictly white. Typically, it is necessary to average a large number of individual measurements in order to reduce statistical uncertainties within acceptable limits. In our experience in measuring admittance of squid axon in this manner, the required observation time, nearly a minute for a 400-point function (2.5–1000 Hz, 128 averaged responses), is still prohibitively long for many experimental conditions of interest.

Measurement in Response to Pseudorandom Perturbations

It is apparent that an appropriate test signal for efficient impedance/admittance measurement would include the following properties:

1. low peak-to-average power ratio in the time domain in order to make membrane nonlinearities and computation errors less likely;
2. uniform power-density distribution in the frequency domain, over the range of interest;
3. nonrandomness, so that averaging, if required, relates only to unavoidable noise in measurement, and not to the statistical characteristics of the test signal.

A particular class of signals with these and other useful properties is known as linear pseudorandom binary sequences[15] (PRBS). PRBS signals have been widely used for system identification. However, there have been relatively few applications to biological systems.[16] In loose terms, the main feature of such sequences is that over a finite period T, their first- and second-order statistics are essentially similar to those of a purely random telegraph wave (two-state with successive transitions being uncorrelated) observed for an asympotically long time. Unlike the discrete impulse, the PRBS or signals derived thereoff have a low peak-to-average power ratio, and the perturbation energy is spread out uniformly in time. A PRBS is periodic, with period T, and can be expanded in complex Fourier series expressed, say, as magnitude and phase. As a result of its detailed time structure, a PRBS stimulus has a magnitude function, $\mathrm{Mag}_S(k)$, with the same distribution—with $\sin^2 x/x^2$ shape—as that of the impulse. With appropriate low-pass filtering, a close approximation to a true, calibrated, white spectrum can be obtained. Consequently, a single Fourier computation of the system response $\mathrm{Mag}_R(k)$ can yield the desired quantity $\mathrm{Mag}_H(k)$ within a known constant.

To exploit these properties of the PRBS fully, however, two restrictions must be resolved.

1. The above properties of a PRBS hold only if the relevant quantities are evaluated over precisely one period T of the sequence. The simplest PRBS generator—a digital shift register with linear feedback[18]—yields periods that consist of $N = 2^P - 1$ pulses, p some integer. However, computations such as implemented by the fast Fourier algorithm are fastest and/or require less complex hardware when operating on data arrays of 2^P samples.

2. The unknown phase function $\mathrm{Arg}_H(k)$ corresponds to the difference between the response phase $\mathrm{Arg}_R(k)$ and the stimulus phase $\mathrm{Arg}_S(k)$. For a PRBS, the phase function takes widely scattered values, and so does that of the response. Precise computation of the individual phase data must be implemented to maintain overall accuracy.

FIGURE 1 shows the block diagram of an instrument that embodies these considerations. We briefly describe its implementation below. Additional technical details are available elsewhere.[17] The next section describes its application to admittance measurements in squid axon.

Basically, the instrument consists of a PRBS generator and a fast processor. Its prototype utilizes a combination of special-purpose circuits and commercial real-time spectral analyzer (Schlumberger/EMR Model 1510). These units are supervised by control and synchronization circuits. Extensive facilities for convenient operation in a real-time environment are included.

The PRBS generator, with $N = 1023$, uses the common technique of a digital shift register with appropriate feedback.[15,18] The register is driven from a clock with frequency f_{PRBS}, which determines the time (and frequency) scale of the sequence. The PRBS is processed by a preconditioning analog filter available for prewhitening purposes (see DISCUSSION) and thence applied as a perturbation stimulus to the unknown system (in the present case squid axon membrane and current/voltage clamp interface).

The system response to the filtered PRBS is processed for discrete Fourier transform (DFT) computation. This processing includes postconditioning filters

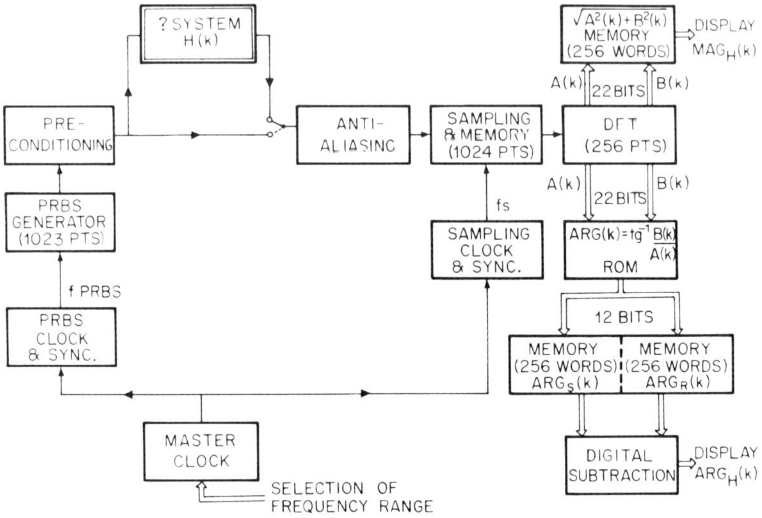

FIGURE 1. Block diagram of analyzer developed for rapid evaluation of the complex transfer function $H(k)$ of an unknown system. The magnitude $\text{Mag}_H(k)$ and phase $\text{Arg}_H(k)$ functions are computed from the discrete Fourier transform (DFT) of the system response to a pseudo-random perturbation (PRBS).

(6th order Tchebycheff) for antialiasing, analog-digital conversion at sampling frequency f_S, storage of a 1024-point data array in an input memory, and computation of 256 pairs of Fourier components.

Restriction 1) regarding the compatibility between linear PRBS and efficient DFT computation is solved by a form of time compression-expansion technique. The PRBS period, which consists of 1023 f_{PRBS} clock pulses, and the sampling window, which consists of 1024 f_S samples, are precisely matched in duration and occur in exact coincidence. This is achieved by frequency and phase-lock circuits that derive f_S and f_{PRBS} from a single master clock. Phase jitter is prevented by periodically resetting the state of the PRBS shift register and the initial address of the input-sampling memory to preassigned values. Since synchronization is validated at the beginning of each measurement frame, the image of the filtered PRBS, hence the basic reference phase spectrum of the stimulus, $\text{Arg}_S(k)$ is fixed and independent of the actual time scale. Frequency range is determined by the frequency of the master clock. Its selection also assigns the appropriate cut-off frequency to the programmable antialiasing filter.

Effectively, with this procedure, the values of the filtered PRBS are redefined over a 1024-point baseline with the antialiasing filter acting as an interpolation unit. Once the 1024 data samples have been accumulated, the DFT takes place. Within 90 milliseconds, it is completed and the 256 pairs of real and imaginary Fourier components are transformed into magnitude and phase functions. $\text{Mag}(k)$ is computed in floating-point format within the EMR 1510 and stored in a memory bank of 256 words. Simultaneously, each pair of real and imaginary Fourier components is fed, in succession, to a digital divider and read-only-memory programmed for the $\tan^{-1}$ function. A series of 256 phase data points (0–360°, 12-bit resolution) is computed. Two additional memory banks, each with 256 words, are available for phase storage. Depending upon whether the current measurement has been computed for a stimulus or for a response, the data points are stored in the $\mathrm{Arg}_S(k)$ or in the $\mathrm{Arg}_R(k)$ memory.

Once the memory banks have been filled with magnitude and phase data, the system switches to read-out mode. The memory addresses are driven from a single bus, with repetitive scanning of the 256 discrete frequency lines. The contents of the phase memories feed a digital substractor that computes the desired quantity $\mathrm{Arg}_R(k) - \mathrm{Arg}_S(k)$. Following digital-analog conversion, scaling, and buffering, $\mathrm{Mag}_H(k)$ and $\mathrm{Arg}_H(k)$ are available for display and plotting. Additionally, a set of numeric displays driven from a movable cursor provide direct reading of frequency (Hz), magnitude (dB to known reference level, ± 0.1 dB) and phase (0–360°, ± 0.2°) of any feature of interest.

The control provides for seven analysis bands that range from 0.1–25.6 Hz to 10–2560 Hz, with corresponding measurement periods ranging from 10 msec to 100 msec. In operation, the stimulus mode is initially selected. This automatically connects the input of the DFT to the preconditioned PRBS. Within 100 msec after the end of the next full sequence, the reference phase $\mathrm{Arg}_S(k)$ is computed and stored. Next, the response mode is selected. This now connects the DFT input to the output of the unknown system. Again at the end of the next full sequence, the response phase $\mathrm{Arg}_R(k)$ is computed and stored. Immediately thereafter, the complete $\mathrm{Mag}_H(k)$ and $\mathrm{Arg}_H(k)$ are displayed and available for

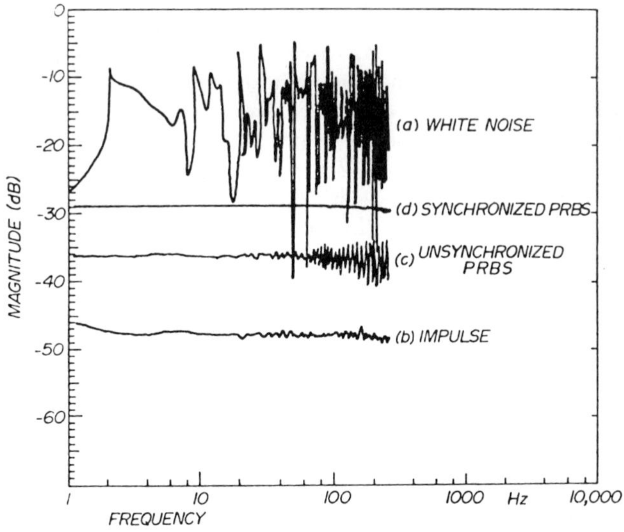

FIGURE 2. Magnitude spectra of various wide-band stimulus signals. The synchronized PRBS used in the present instrument has $\sin^2 x/x^2$ shape which, over range used, is essentially white with gradual (0.9 dB maximum) droop at the high-frequency limit. (By permission of IEEE.[17])

simultaneous plotting on XY recorders. The control also provides for additional modes. In continuous mode, the data displays are automatically updated at the end of each PRBS period, for instance ten times/second in the 10–2560 Hz band. In this way it is possible to follow the kinetics of a system that is slowly evolving in time. In the transient mode, a single response is captured and repetively displayed until a manual or external command requests another measurement.

Prior to its application to admittance measurements in membrane, the present approach has been evaluated extensively on systems of known characteristics. FIGURE 2 illustrates the key features of PRBS and shows single-magnitude functions of different wide-band test signals discussed previously: (a) sample of white, random noise; (b) impulse; (c) PRBS without 1023–1024 interpolation; and (d) synchronized PRBS. All signals have been processed with full utilization of the dynamic range of the instrument but replotted at different levels for clarity. The large statistical fluctuations of random noise, which would require averaging, are clearly visible. The significant computation noise of the impulse response illustrates the large dynamic range required for Fourier processing of an impulse or correlation function. The need for precise PRBS can also be appreciated. In this example, a normal 1023-bit sequence augmented by a single bit has been loaded in the DFT unit. Even such a minute mismatch introduces significant deviations that are especially prevalent at high frequencies. Finally, the figure shows that DFT-compatible PRBS generated in the present instrument provides a near-perfect magnitude baseline, which is white and smooth, except, of course, for the predictable droop associated with the $\sin^2 x/x^2$ character of the magnitude spectrum of the discrete time series. Note that (a) and (b) would improve with averaging, (c) would not, and (d) does not require it.

The basic accuracy of the present method is essentially determined by those of sampling and DFT computation. The EMR 1510 utilizes a 9-bit analog-digital conversion, which, together with dithering techniques, provides a usable dynamic range of 60 dB. In conditions of a favorable signal/noise ratio, the magnitude noise is less than 0.2 dB and the phase noise is less than 0.5°. This accuracy in absolute phase, although obtained as a difference, results from the digital computation, which nearly eliminates systematic errors, and also, paradoxically, from using a stimulus-phase spectrum with widely scattered values. The scattering actually increases the usable accuracy, since the various computation errors do not tend to be correlated over successive values of k, thus producing more randomly distributed errors than for a stimulus with white-phase spectrum.

MATERIALS AND METHODS

The method described above was applied to admittance measurements on axons from the squid *Loligo pealei*. Animals were received live at the Marine Biological Laboratory, Woods Hole, Mass., where the present experiments were conducted.

The general experimental procedure and details, including mounting of preparation, chamber design, perfusion technique, electrodes, and voltage-clamp electronics were as described previously for axial wire conditions.[2,13]

Artificial sea water (ASW) was used as the external solution, 423 mM NaCl, 9 mM KCl, 9.3 mM $CaCl_2$, 23 mM $MgCl_2$, 25.5 mM $MgSO_4$, 2.15 mM $NaHCO_3$, pH 7.6 at 22°C. The standard internal perfusate was .5 M KF and 5 mM Tris. HCl, pH 7.5 at 22°C. A high-cesium perfusate consisted of 290 mM CsF, 5 mM Tris. HCl, and .4 M sucrose. Cooling of the external, gravity-fed solution was provided by an ice bath. Flow rate within the chamber and preparation temperature could be finely adjusted by means of a needle valve at the inlet and adjustable outlet level.

An important objective of the present experiments is to explore the membrane admittance function over a wide range of membrane potentials. Two experimental constraints must be satisfied.

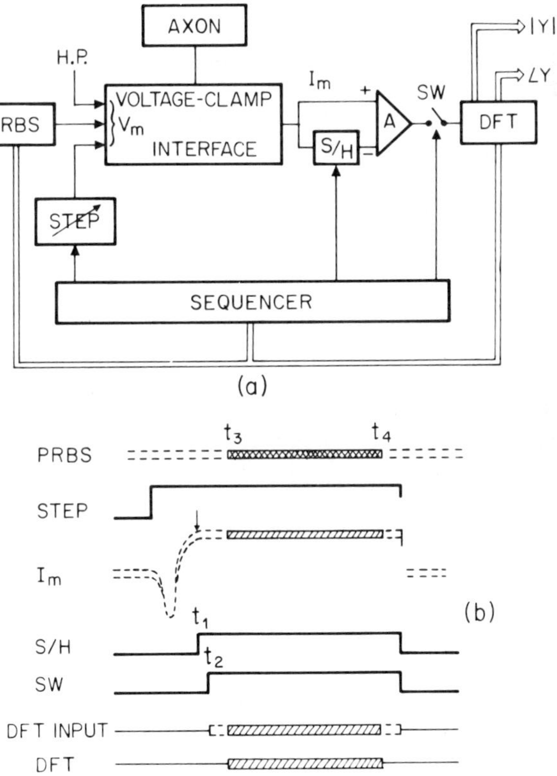

FIGURE 3. Simplified block diagram of implementation and timing sequence of PRBS method applied to admittance measurement on squid axon. Sequencer ensures that measurement is performed during steady-state response of membrane and provides protection against switching transients. Sample-and-hold (*S/H*) and differential amplifier (*A*) generates automatic base-line compensation and maintains wide dynamic range of the analysis. Total time required for full admittance measurement is comparable to that required for conventional step-clamp data.

1. Under axial wire conditions, the preparation does not survive well to the prolonged passage of a current of magnitude such as required for substantial displacement from rest potential. Consequently, it is imperative to synchronize the measurement properly with the potential profile imposed upon the membrane.

2. Since the admittance is to be derived from a small-signal response, the "mean" polarizing currents, which are much larger (up to 100 times in the present case) must be adequately removed prior to processing in order to preserve dynamic range.

Provisions to satisfy these requirements are included in the system whose simplified block diagram is shown in FIGURE 3. The figure also illustrates the timing sequence used for a typical admittance measurement. The sequence is determined by a circuit that operates under automatic or manual control.

Initially, the membrane potential V_m is clamped to the holding value selected (H.P.) on which a small (e.g. 2 mV peak-peak) PRBS is superimposed, without gaps, between successive PRBS periods. When a measurement is to be made, the sequencer issues the request for a step pulse of chosen polarity and amplitude. The request occurs at a time, t_3, which anticipates the next complete PRBS period. The value of the lead time, typically 100–200 msec, is chosen so that the total membrane current response I_m will be well in "steady-state" when the forthcoming period occurs. In the meantime, an analog switch (SW) maintains the input of the DFT front-end to ground potential to prevent transient overloading. Current I_m undergoes its usual time course, with a small pseudorandom component which carries the admittance information and which must be extracted. At time t_1, also selected so that the general response is in "steady-state" (e.g. 80 msec), I_m is sampled and its value held in memory in a sample-and-hold (S/H) module. Sampling is preceded by low-pass filtering (dual RC network, time constant = 680 μsec) for rejection of the pseudorandom spikes that reflect dominance of membrane capacity at high frequency. In conjunction with the nearly zero mean property of linear PRBS signals, this filtering insures that the DC value in memory is well centered within the dynamic excursions of the response. A differential amplifier (A) subtract this DC value from I_m. Immediately after sampling transients have subsided, at time t_2, the analog switch SW closes and the pseudorandom response component, which is now superimposed on a minimal pedestal, is applied to the DFT input. This automatic restoration of DC baseline allows full utilization of the dynamic range of the DFT processor. At time t_3, precisely when the new PRBS period is beginning, the synchronization circuits contained within the instrument initiates sampling and storage in the input memory. Immediately after completion of the period (which lasts from 100 msec to 10 sec, depending on the selected frequency range), at time t_4, the system is reset to rest condition. Within 100 msec, the magnitude and phase of the admittance corresponding to membrane potential of (H.P. + step) have been computed and are displayed and available for plotting. This paradigm ensures that the duration of the large currents applied to the preparation is nearly as short as the frequency resolution of the recorded admittance data allows.

In a typical experiment, the reference phase $\text{Arg}_S(k)$ is first computed. Next, the frequency range is selected and the above procedure is applied repeatedly to obtain a set of admittance data as a function of membrane potential. This might require a few minutes, which are mostly spent for plotting the data. The sequence may then be repeated for other analysis bands to obtain overlapping data over a wider frequency range. Frequently, this data is complemented by conventional step-clamp measurements, which are recorded photographically. The operating condition, such as temperature or solutions, is then modified and the entire sequence repeated. At the conclusion of the experiment, a 1 K resistor is substituted for the preparation. The admittance, measured for identical system settings, provides absolute calibration and correction for the response of the voltage-clamp and amplifier circuits.

Provisions for impedance measurements were also included. The general operating procedure was identical. A composite signal, representing holding current, step current, and PRBS perturbation was applied to the preparation as I_m through a controlled current source and axial wire. The response component contained in V_m was extracted and processed as described above.

Occasionally, in conditions of limited signal/noise ratio, the magnitude functions from several (e.g. 2–32) measurements obtained in immediate succession were digitally averaged within the EMR 1510. Since this unit computes a normalized, true average of accumulated data, averaging could be stopped at any time, without loss of absolute calibration, as soon as a satisfactory accuracy was obtained. No means for averaging successive phase functions were provided for these initial experiments.

Illustrative Results and Discussion

The following data are intended to illustrate the potential of the technique described above. As a result of its speed, it has yielded the first complex admittance/impedance measurements of squid axon obtained over a wide range of experimental parameters. These data are qualitatively compared to predictions derived from the linearized Hodgkin-Huxley equations. We also further comment on the method as well as on relevant experimental questions related to axon membrane measurements.

Linearity

Earlier data[2,7] has suggested that the squid axon membrane behaves linearly as long as perturbations of membrane potential are limited to a few mV. A quantitative study of nonlinearities would have required complex techniques, such as those derived from Wiener theory for system identification, which were beyond the scope of these first experiments. It is of interest, however, to note that the method readily provides a relevant indication of the presence of nonlinearities.

When deriving transfer function estimates through auto/cross-spectral measurements in response to random noise, a useful indicator of confidence in the observed data is the coherence function.[19] This quantity essentially represents, as a function of frequency, the fraction of the system output that is causally related to system input. For a noiseless, linear, time-invariant system, the coherence function is unity. Output components that arise from other sources or processes,

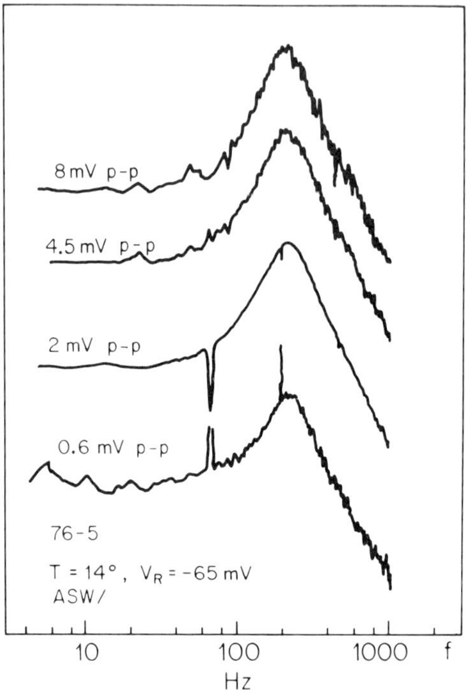

FIGURE 4. Magnitude of impedance function observed for different peak-to-peak amplitudes of perturbation of membrane potential. Lower curve corresponds to measurement with significant background noise. For 2 mV peak-peak level, signal-to-noise ratio is much improved (large local peaks corresponds to power-line leakage). With larger stimulation level, data becomes "noisy" because of the presence of significant nonlinear components.

such as noise, nonlinearities, additional inputs, and so on, lead to a coherence function that is less than unity. Simple inspection of the magnitude and phase estimates from the present method provides somewhat analogous information. As a result of its synchronization, the magnitude spectrum of the PRBS is white and smooth (see FIGURE 2). Well-behaved, real systems, such as nerve membrane, can be described by differential equations of limited order which, when linearized, yield transfer functions that have a limited number of poles and/or zeros in the complex plane and "smooth" magnitude and phase spectrums. It follows that in actual measurements, randomlike deviations of the response spectra from their general outline can be ascribed to error such as computation or system noise, nonstationnarity or nonlinearities. Similar comments apply to phase.

A typical pattern observed for membrane impedance is shown in FIGURE 4. For a small stimulus, which results in approximately 0.6 mV peak-peak excursions of membrane potential, the measured response is noisy, presumably because of significant membrane, electrode, and amplifier noise and drift. With the stimulus intensity increased for a 1–2 mV response, the improved signal/noise ratio yields a virtually smooth response. With further increase, e.g. for a 5–10 mV response, the magnitude spectrum again becomes "noisy," except that the deviations are now systematic and reproducible. Without doubt, these reflect nonlinear components that have become detectable for these perturbation levels. A similar pattern occurs for phase as well as for admittance measurements. When plotted with constant system gain, the data obtained for different levels largely superimpose within their erratic deviations, although close examination suggest slight, systematic differences, especially visible in the phase function. The two regions of nonlinear and noisy response occurring on either side of a linear, noise-free region were not always as clearly separated as on the figure. Especially for guarded measurements, which are intrinsically noisier, the linear region often occurred for relatively poor, though usable, signal/noise.

The nonlinear response observed in this manner was not systematically investigated in these early experiments. The data in FIGURE 4 is compatible with previous observations. They also indicate that even without further computation, the method provides an immediate, on-line indication of noisy or nonlinear response. In conjunction with particular assumptions on the internal structure of the membrane considered as a system to be identified, it is also plausible that meaningful information might also be extracted from the deviations observed in a response that clearly includes nonlinear components.

Admittance Features in ASW

FIGURE 5 shows the general features of complex admittance typical of normal conditions near rest-potential. Preparation 76–50 is bathed in ASW and unper-

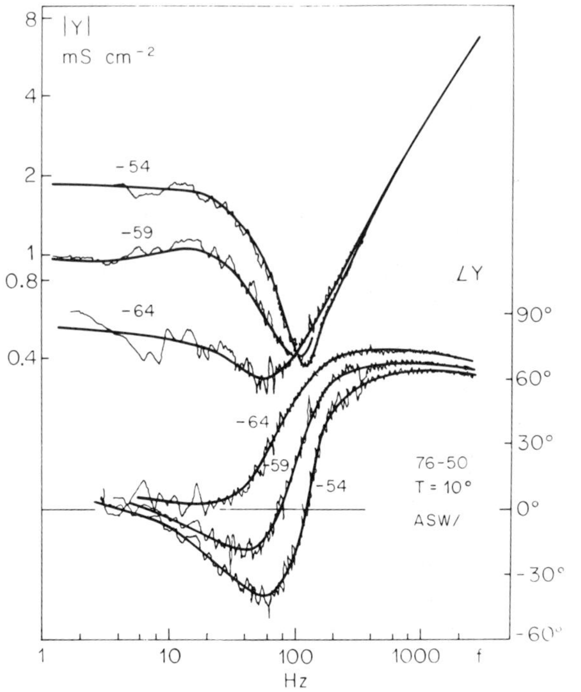

FIGURE 5. Unretouched and smoothed data for magnitude and phase of admittance in normal conditions and near rest potential. The magnitude in these and following data is not corrected for slight droop at high-frequency edge. Data obtained over two bands of analysis. Each pair of curves corresponds to 1.1 sec of data.

fused. The current electrodes consist of three *Pt-Pt* plates, each approximately 4 mm long and 3 mm wide. The two outermost plates are grounded and act as guards. The central plate drives the current amplifier. The measurements extend from 1 to 2560 Hz, using two overlapping bands of analysis. Each pair of magnitude and phase function corresponds to 1.1 sec. of data. In order to appreciate the actual measurement, unretouched plots are shown together with visually smoothed curves.

In the 1–30 Hz region, we note that the magnitude may not converge monotonically to its low frequency asymptote. Rather, for a certain range of membrane potential (e.g. see curve for −59 mV) it undergoes a local resonance around 10 Hz. Accordingly, the phase does not converge to zero but actually crosses the zero-degree axis. The zero-crossing frequency, around 4–8 Hz, decreases with depolarization. Below 1 Hz the behavior of admittance is still uncertain, since an unusually stable preparation is mandatory and only a few measurements, down to 0.1 Hz, have been attempted so far. These measurements, obtained by a different technique, are in agreement with those of a previous report,[2] which had also indicated that this low-frequency feature was related to potassium ion conduction.

The main feature in the admittance occurs in the 20–200 Hz region. For hyperpolarized membrane potentials, such as −70 mV (not shown on the figure), the response is generally similar to that of a simple RC circuit and the phase remains positive for all frequencies. With depolarization, the antiresonance appears with a definite region of negative phase for frequencies below the resonance. The envelope of the magnitude curves for different potential clearly intercepts the asymptotic characteristic observed for hyperpolarization. For this limited range of potentials the magnitude and curves for different potentials merge beyond 200 Hz and overlap within the accuracy of the plot. By contrast, the phase plots show strong potential dependence and can be readily distinguished.

In the 200–2560 Hz region, the magnitude plots have reached their asymptotic behavior. The slight droop, on these log-log coordinates, mostly reflect the $\sin^2 x/x^2$ magnitude spectrum (see FIGURE 2), which was not removed. The phase develops toward its asympotic value, which is clearly less than 90°, in confirmation of the frequency-dependent nature of membrane capacity. Finally we note a significant phase roll-off that suggests the presence of a series-resistance component.

FIGURE 6 shows similar data recorded over a single frequency band, from 4 to 1024 Hz. Each set of curves is computed from 250 msec of data. The current is recorded from a single (unguarded) *Pt-Pt* electrode. Over this wider range of membrane potential we note that the sharpness of resonance is quite potential-sensitive. The figure also illustrates a typical observation in the high-frequency region. As the membrane is depolarized beyond −40 mV, the magnitude curves, which had merged and overlapped, start to move upwards with a corresponding rapid change in the phase. With further depolarization, the resonance frequency continues to increase, the curves become broader, and progressively get closer to each other for a given potential increment.

Such data has been recorded for membrane potential up to +15 and occasionally +30 mV without sign of abnormal axon deterioration and with excellent controls. For very large depolarizations, the base line current corresponding to the mean, steady-state potassium conduction may become two orders of magni-

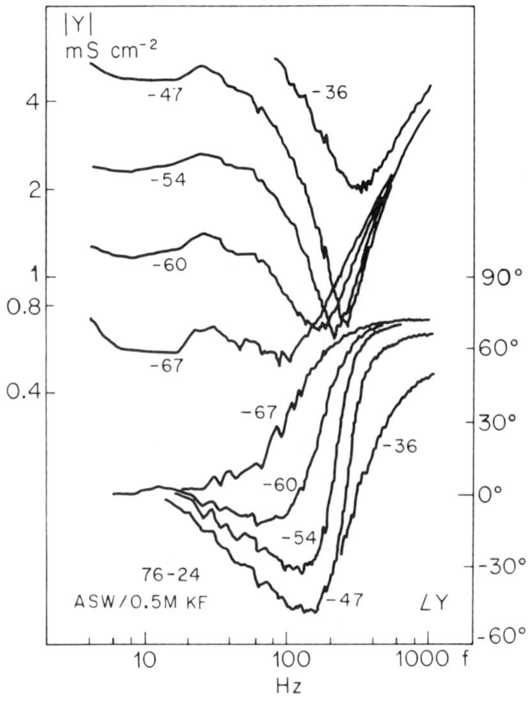

FIGURE 6. Unretouched set of admittance data in normal conditions and over appreciable range of membrane potentials. Each pair of curves corresponds to 250 msec of data.

tude larger than the pseudorandom response. In such conditions, the automatic baseline compensation becomes quite critical.

Admittance Features in TTX

FIGURE 7 shows some of the complex admittance data (smoothed) obtained on preparation 76-50 (described previously) bathed in 10^{-6} M TTX ASW. The magnitude and phase plots have been obtained in the sequence −59, −49, −44, etc., to +16mV, and the data at −54 constitutes the control. Quite clearly, the preparation has suffered little if any deterioration from the measurement. Under these conditions of suppressed sodium conduction, the behavior of the antiresonance feature is considerably modified. The sharpness of the resonance is substantially decreased. The envelope of the magnitude function does not cross the asymptotic boundary defined by rest and hyperpolarized (not shown) levels. Finally, the magnitude curves do not merge as early as in normal ASW.

Admittance Features in High Internal Cesium

In a continuing study of membrane noise[3] we have been looking for conditions that suppress potassium conduction. Noise data has shown that some agents that

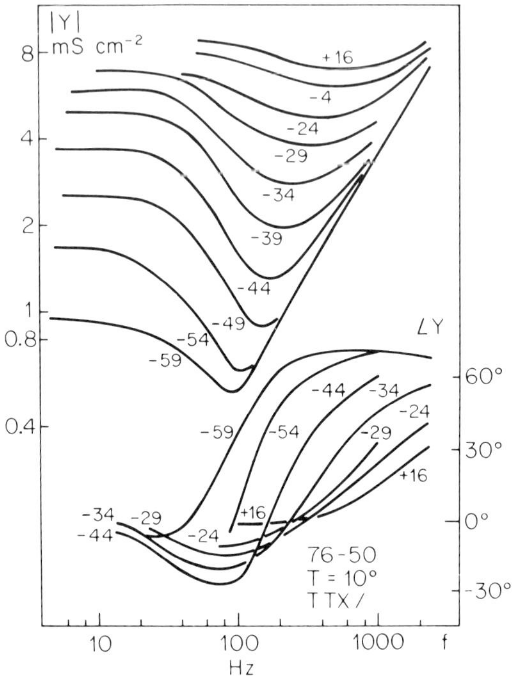

FIGURE 7. Smoothed admittance data for axon in TTX ASW. Acquisition time is 250 msec for lower curves and 100 msec for upper three curves at large depolarization. Speed of present method allowed wide range of potentials to be explored without abnormal axon deterioration.

block potassium current on a macroscopic scale, such as TEA, induce new, additional fluctuation components as a result of the dynamics of their interaction with membrane. In an earlier report, a perfusion solution of 50 mM Cs and 50 mM *K* was shown to substantially reduce the potassium noise, which is normally overwhelming, to the point where sodium noise became marginally detectable. This condition did not produce induced *K* noise. In the most recent experiments, a .29M CsF perfusate was found to decrease *K* noise further and allow Na noise to be readily observed. Step-clamp data showed sodium currents that appeared normal except for partial suppression or slowing down of inactivation.

The Cs perfusion produces a membrane with unusual electrical characteristics that reflect in the admittance data shown in FIGURE 8. Preparation 76-47 was initially perfused with normal 0.5 KF solution, with corresponding rest admittance shown in dotted lines. The antiresonance is particularly sharp, a suggestion of an axon with low leakage. Upon Cs perfusion, the membrane virtually becomes an open circuit at low frequency at the holding potential of −62 mV. The magnitude is somewhat analogous to that of a pure capacitance although the phase, which increases toward and beyond 180° rather than decreases at low frequency, shows that a very low but negative conductance is present across the capacitance. With depolarization, the conductance remains negative and increases in magnitude, reaching a maximum around −27 mV. With further depolarization, to −17 and −7 mV, the net negative conductance decreases. Under these conditions, we predicted that this nonresonanting membrane could be made to resonate with a purely resistive shunt. This was indeed verified. In order to bypass the unpredictable characteristics of the axial wire, the shunt was synthetized through the addition of an in-phase PRBS component to the input of the current amplifier. The resulting antiresonance was extremely sharp, with measured rejection limited by noise, and at a frequency in agreement with the 180° point on the phase spectrum of the unshunted axon. Other axons, in similar 290 mM Cs, showed rather similar antiresonance without an artificial shunt, probably as a result of a larger leakage. In such cases, the resting admittance converged toward zero phase. With depolarization, the magnitude first decreased, as a result of the negative parallel conductance, reached a minimum of essentially zero conductance, then increased (with a phase of 180° or more), and then decreased, as described before. A rather similar pattern had been typical of the 50 mMK 50 mM Cs condition used previously. It is likely that in that case the residual *K* conduction acted as an effective "shunt" (with characteristics more complex than a pure conductance), which in turn produced a sharp antiresonance.

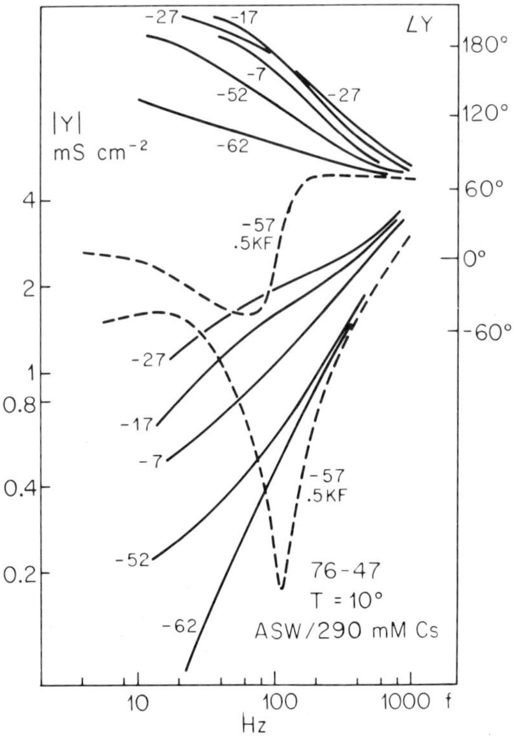

FIGURE 8. Internal perfusion with 290 mM CsF produces substantial *K* blockage and near-normal Na system. Corresponding admittance has negative real part. Note phase behavior at low frequencies.

Admittance Function from the Hodgkin-Huxley Model

FIGURES 9, 10, and 11 show admittance functions derived from the linearized Hodgkin-Huxley model. Standard HH values are used throughout. The computations also include the effects of a series resistance.[20] In FIGURE 9, the full model is used, in simulation of normal conditions of potassium and sodium conduction. In FIGURE 10, $\bar{g}_{Na} = 0$, in simulation of data obtained in TTX. In FIGURE 11, $\bar{g}_K = 0$, in simulation of potassium conduction blocked by high internal cesium perfusion. Although the experimental data shown in FIGURE 5, 6, 7, and 8 correspond conceptually to these conditions, caution must be exercised in any comparison. At this moment, the theoretical curves are only presented to suggest that the experimental admittance data provide valid information on ionic movement and kinetics. A quantitative comparison cannot yet be made because:

1. There is no assurance that the parameters of the HH model are adequate for *Loligo pealei*, used in Woods Hole. In a previous report[2] we indicated some of the discrepancies, but a systematic study will be required.

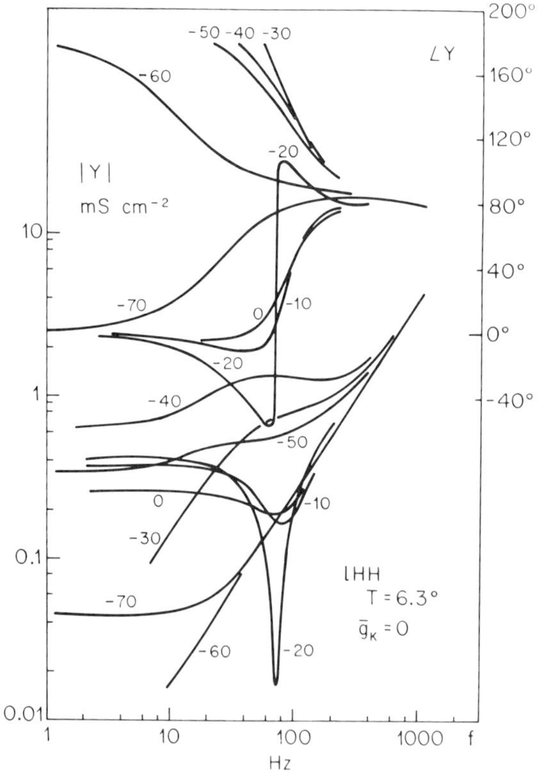

FIGURE 9. Admittance computed from linearized Hodkin-Huxley with normal conductances. The model includes an additional series resistance of 7 ohm · cm^2.

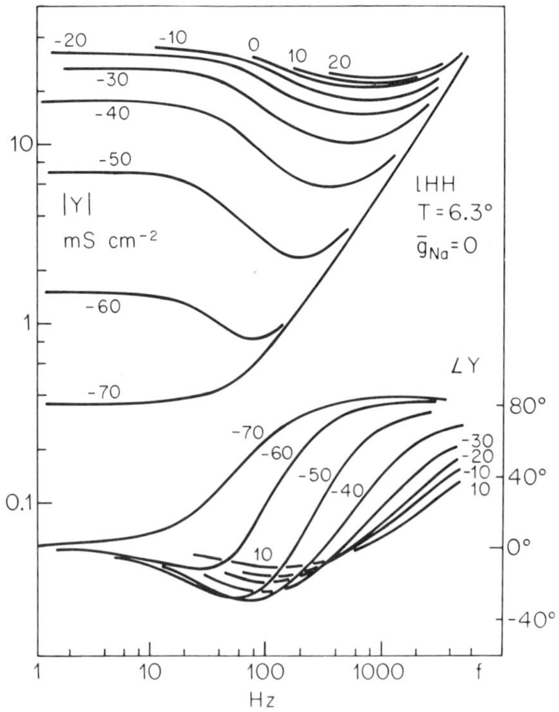

FIGURE 10. Admittance computed for linearized Hodkin-Huxley model with suppressed sodium conductance.

2. The original 6.3°C temperature was used in the computations. The effect of temperature on rate parameters reflects in complex changes in total admittance.

3. The computation assumes a constant membrane capacity.

4. Detailed curve fitting and model analysis of the admittance data now available will be required and are in progress.

Nevertheless, there is a striking qualitative similarity between the data and the predictions of the HH model. There are also noticable differences. In ASW (FIGURES 5, 6, & 9), the antiresonance is sharper for the model and the envelope of the peak of the magnitude crosses the asymptotic boundary for much less depolarized values. The high-frequency tails of the magnitude merge at higher frequencies, and a feature present in the model around −30 mV in the form of humps in the magnitude and phase curves are not present in the experimental data.

The qualitative agreement between model and experiment is improved in the case of suppressed Na conduction (FIGURES 7 & 10). In both cases, the envelopes of the magnitude curves lie on the same side of the boundary defined by hyperpolarized levels. The phase functions also behave similarly, including the range of membrane potential over which the phase curves cross one another. This is significant, since by its very nature phase is generally a sensitive indicator and precursor of transfer function features. As discussed previously, however, the disagreement noted at very low frequencies requires correction of the model.

For suppressed, *K* conductance, the comparison between model and experiment (FIGURES 8 & 11) is more hazardous. The Na system, with its *m* and *h* processes, is more complex than the *K* system of the previous case. The interplay of a conductance with a negative real part, leakage, and possible residual *K* process is subtle. Before extensive analysis is completed, we may note only that some of

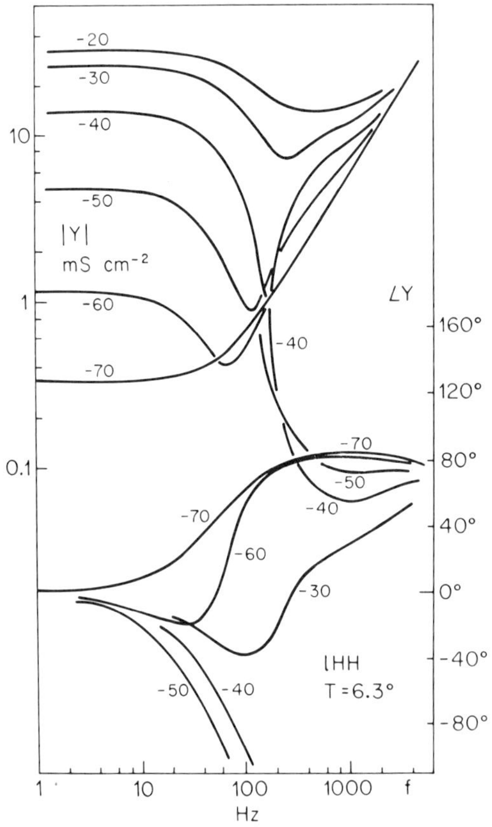

FIGURE 11. Admittance computed for linearized Hodkin-Huxley model with suppressed potassium conductance.

the key features, such as low-frequency phase of 180° or more, successive increase and decrease of magnitude, and general behavior of high frequency asymptotes, are present both in the data and in the linearized HH model.

Assessment of the Method

Several characteristics of the present method have been mentioned in the preceding text. We now briefly summarize its main features.

A key aspect is speed, both in data acquisition and in data processing. A basic notion of signal analysis states that an instrumental resolution Δf requires a minimum observation time $T \approx 1/\Delta f$. A repetitive waveform with period of 1 second, for instance, yields Fourier components spaced 1 Hz apart. The present method operates near this fundamental limit. Strictly speaking, it requires that the response achieve "steady-state," hence have lasted for a duration which is a few times the inverse band-width of the narrowest spectral feature of significance. For a well-behaved system such as the squid axon under voltage-clamp, which has relatively broad features, the response reaches "steady-state" rapidly. Experience confirmed that a minimal delay such as that generated by the sequence of FIGURE 3 is adequate. Consequently, admittance measurements can be performed on the same time scale as conventional step-clamp experiments and with comparable effects on the preparation. The frequency resolution of the present measurements may appear excessive for resolving these broad features mentioned above. However, as for membrane-noise analysis,[21] such resolution is convenient. It allows erroneous data, such as that occurring at power-line frequency and the harmonics thereof, to be easily edited prior to a systematic procedure of curve-fitting.

The interpolation used to synchronize the PRBS also allows maximum processing speed, since the DFT is computed over a data array of 2^p points. An alternate solution was used by Husimi and Wada, who programmed a general purpose computer for a 1023-point transform.[22] However, special purpose hardware, available for DFT computation, is most invariably designed to process arrays of 2^p points. Such units typically operate with a speed advantage of tenfold or more over software implementations. Although not a strict requisite, real-time processing is desirable, for instance, in conditions when an unfavorable signal/noise ratio requires contiguous measurements to be averaged.

We have mentioned before that this method fully exploited the dynamic range of the DFT processing because of the low peak-to-average power ratio of the stimulus. In our experience, the measurement accuracy was, in most instances, limited by external constraints such as residual noise in preparation, in electrodes, and in interface electronics (which were not especially optimized for the low source impedance typical of axial wire measurements). For the lowest frequencies, drift in the preparation and imperfections in the sample/hold circuit were significant. Some of these aspects can be improved in the future.

A rather unique convenience of the method is that: 1) there is no need for a pair of antialiasing filters, and 2) the phase of any preconditioning filter is automatically canceled. Most other methods involve computations on simultaneous observations of both stimulus and response signals. For example, a digital cross-correlation measurement requires that both signals be band-limited to less than half the sampling frequency in order to avoid aliasing errors. These filters must be carefully matched for phase. In a real-time environment, this is inconvenient, since band-switching of analysis requires a time-consuming matching and/or an extremely accurate (and expensive) set of filters. In the present method, the stimulus and response are not combined at the same physical time. Rather, they appear to be so because of the fixed synchronization between the PRBS period and the sampling window. Consequently, only a single filter is required. If its characteristics are stable in time (a condition easy to achieve to the needed accuracy), its

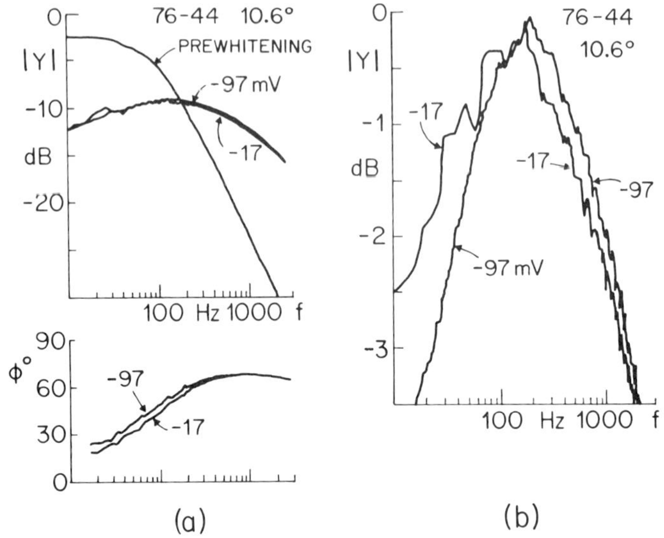

FIGURE 12. Illustration of prewhitening capabilities of present method. (a) Magnitude and phase of axon following TTX application and 290 mM CsF perfusion. Preconditioning filter shapes stimulus spectrum for more uniform response, resulting in increase in accuracy and dynamic range of analysis. (b) Similar magnitude data plotted at larger magnification, showing that differences as small as 0.1 dB (1.2%) are resolved.

arbitrary phase automatically self-cancels. As a result, prewhitening—i.e., the preshaping of the stimulus magnitude spectrum such that the response spectrum is approximately uniform—becomes a practical, on-line operation even when experimentation time is at premium. Prewhitening extends the usable dynamic range and reduces computation errors.

In squid axon, there are several instances where such prewhitening is indicated. An example is shown in FIGURE 12, obtained while studying changes in admittance that occur under "gating" current conditions. For this measurement, ionic current has been largely suppressed by the combined treatment of external TTX and internal perfusion with 290 mM Cs. The membrane is now dominated by its capacitative behavior. Because of the large response at high frequency and finite dynamic range of instrument, difficulties are encountered in the low-level, low-frequency region. To improve resolution, a tunable analog filter is then introduced, and as shown in FIGURE 12(a), its response is simply and rapidly adjusted to approximate the inverse of the admittance magnitude being measured. A new reference phase is computed and stored as $\mathrm{Arg}_S(k)$. The response magnitude is now fairly uniform, as indicated for −17 and −97 mV. The corresponding phase curves are read-out without additional correction and with much-improved low-frequency accuracy. Since the magnitude is flatter, it can be viewed at high gain, as shown on FIGURE 12(b). The small difference is now well resolved, within the 0.1–0.2 dB accuracy mentioned earlier. Its correction factor is simply given by the filter response plotted previously. These data are discussed elsewhere[23] and presented only to show how easily prewhitening can be applied.

The method also provides for rapid change of frequency range of analysis. Because the PRBS is time-locked to the DFT sampling, its power spectral density is automatically related to the inverse of the effective resolution band-width of analysis. No magnitude recalibration is required when switching bands and plots overlap directly (except for the predictable $\sin^2 x/x^2$ droop). Furthermore, since the image of the PRBS is always the same in the input memory, the reference phase $\mathrm{Arg}_S(k)$ is fixed and needs to be computed only once. This property holds as long as the antialiasing filter has a phase function that is invariant for the various bands when plotted on a normalized frequency axis (i.e., as a function of k). This was indeed the case, within 2°, for the present instrument. Phase data from different bands did not need relative correction and also produced overlapping plots. This facility for essentially instantaneous band-switching was enhanced by a circuit that automatically offset the logarithmic frequency axis of displays and plotters with the band selected.

A Significant Experimental Problem Remains

The present method appears to provide solutions to the speed, convenience, and efficiency required for extensive admittance measurements on the squid axon. A serious problem remains, however, regarding the definition of the preparation itself. In a companion communication,[3] we discuss at some length the relative merits of various spatial configurations for "large" and "small" area measurements. The questions associated with "edge effects," "shunt," "space clamp," "guards," although well understood under theoretical assumptions, still remain quite difficult to assess on any given preparation. For instance, in principle, guarded measurements are regarded as preferable when dealing with large areas. However, as we mentioned previously,[13] the additional source shunt arising from the guard system further degrades the noise performance of instrumentation amplifiers. Even for admittance measurements, with signals much larger than spontaneous membrane noise, the degradation was quite noticeable. Furthermore,

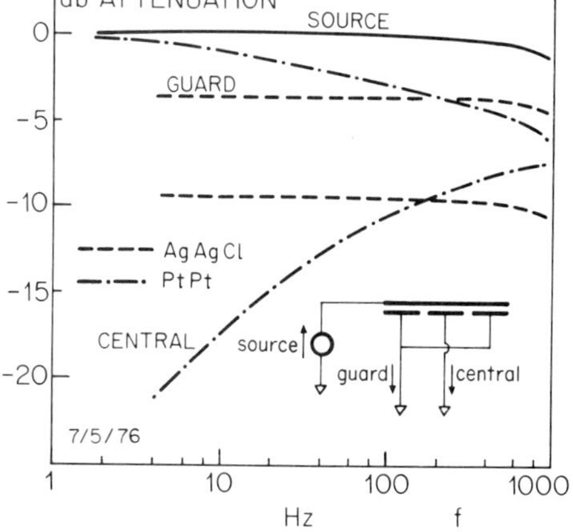

FIGURE 13. In principle, guarded current measurements are preferable for large-area recording. However, the figure illustrates a potential problem arising from mismatch in frequency-dependence of impedance of guard and central electrodes (size no. 30 wire, uninsulated sections 4 mm long), especially for *Pt-Pt* material. This behavior, if overlooked, may lead to significant errors, since effective area corresponding to measured current is frequency-dependent.

routine checks revealed possible aberrations of guard systems that can be easily overlooked. FIGURE 13 shows magnitude plots of the current flowing into the guard and into the central active electrode. The current is supplied through the *Pt-Pt* axial wire in the absence of a preparation. The data suggest that for the *Pt-Pt* electrodes, the effective membrane area "seen" by the active electrode is frequency-dependent to a considerable degree. Admittedly, these electrodes are of small dimensions, and it is known that uniform plating of *Pt-Pt* electrodes is difficult to control. These data are presented to indicate that although efficient data processing techniques are available, the search for "a known area of the axon membrane" and "a current of uniform density" wished by Marmont still remains a challenge.

These difficulties aside, the task of precise curve-fitting, modeling, and parameter extraction from the present admittance data is considerable. However, the preliminary comparisons with the general prediction of the linearized HH model are encouraging. They give strong support to the validity of this general method. They also suggest that an extensive description of membrane kinetics can be extracted from such measurements. These provide a complementary probe to spontaneous membrane noise studies and other approaches.

ACKNOWLEDGMENTS

We wish to thank Prof. U. Ganguly for his assistance in developing the present analyzer, Dr. R. Grisell for supplying the linearized model data, and Dr. R. Guttman and L. Feldman for helpful discussions and support during experiments.

REFERENCES

1. COLE, K. S. 1968. Membranes, Ions and Impulses. Univ. California Press. Berkeley, Calif.
2. FISHMAN, H. M., D. POUSSART, L. E. MOORE & E. SIEBENGA. 1977. *K* conduction description from the low frequency impedance and admittance of squid axon. J. Memb. Biol. **32:** 255–290.
3. FISHMAN, H. M., L. E. MOORE & D. POUSSART. This monograph.
4. FISHMAN, H. M., L. E. MOORE & D. J. M. POUSSART. 1975. Pottasium-ion conduction noise in squid axon membrane. J. Memb. Biol. **24:** 305–328.
5. MARMONT, G. 1949. Studies on the axon membrane. I. A new method. J. Cell. Comp. Physiol. **34:** 351–382.
6. MARMARELIS, P. & K. NAKA. 1974. Identification of multi-input biological systems. IEEE. Trans. Biomed. Eng. **21:** 88–101.
7. GUTTMAN, R., L. FELDMAN & H. LECAR. 1974. Squid axon membrane response to white noise stimulation. Biophys. J. **14:** 941–955.
8. GUTTMAN, R. & L. FELDMAN. 1975. White noise measurement of squid axon membrane impedance. Biochem. Biophys. Res. Com. **67:** 427–432.
9. BERBERIAN, J. G. & R. H. COLE. 1969. Low-frequency bridge for guarded 3-terminal and 4-terminal measurement of admittance. Rev. Sci. Instrum. **40:** 811–817.
10. CONTI, F. 1970. Nerve membrane electrical characteristics near the resting state. Biophysica **6:** 257–270.
11. TAKASHIMA, S. 1976. Membrane capacity of squid axon during hyper and depolarizations. J. Memb. Biol. **27:** 21–39.
12. VALDIOSERA, R., C. CLAUSEN & R. S. EISENBERG. 1974. Measurement of the impedance of frog skeletal muscle fibers. Biophys. J. **14:** 295–315.

13. FISHMAN, H. M., D. POUSSART & L. E. MOORE. 1976. Noise measurements in squid axon membrane. J. Memb. Biol. **24:** 281–304.
14. FISHMAN, H. M. 1975. Rapid complex impedance measurements of squid axon membrane via input-output cross-correlation function. Proc. 1st Symp. Testing and Identification of Nonlinear Systems. G. D. McCann and P. Z. Marmarelis, Eds.: 257. Calif. Inst. Technol. Pasadena, Calif.
15. GOLOMB, S. W. 1967. Shift Register Sequences. Holden-Day. San Francisco, Calif.
16. O'LEARY, D. P. & V. HONRUBIA. 1975. On-line identification of sensory systems using pseudo-random binary noise perturbations. Biophys. J. **15:** 505–532.
17. POUSSART, D. & U. S. GANGULY. 1977. Rapid measurement of system kinetics—an instrument for real-time transfer function analysis. Proc. IEEE. **65:** 741–747.
18. DAVIES, W. D. T. 1970. System Identification for Self-Adaptative Control. Wiley-Intersciences. London, England.
19. BENDAT, J. S. & A. G. PIERSOL. 1971. Random Data: Analysis and Measurement Procedure. Wiley-Intersciences. New York, N.Y.
20. BINSTOCK, L., W. J. ADELMAN, JR., J. P. SENFT & H. LECAR. 1975. Determination of the resistance in series with the membranes of giant axons. J. Memb. Biol. **21:** 25.
21. POUSSART, D. 1977. Comments on the analysis of membrane noise by the integral spectrum procedure. Biophys. J. **18:** 134–137.
22. HUSIMI, Y. & A. WADA. 1976. Time-domain measurement of dielectric dispersion as a response to pseudo-random noise. Rev. Sci. Instrum. **47:** 213–219.
23. FISHMAN, H. M., L. E. MOORE & D. POUSSART. 1977. Asymmetry currents and admittance in squid axons. Biophys. J. In press.
24. SCHEIDER, W. 1977. This annal.

DISCUSSION

A. A. Verveen, *Moderator*

Department of Physiology
University of Leiden
Leiden, The Netherlands

DR. SCHEIDER: As to the comparison on the advantages and disadvantages of the different types of stimulation, we worked on the method of applying the step function. A comparison with impulse is very unfair to the single-shock approach, because it is different from step-function approach. The question is, have you considered making the comparison with a step function in which you put in the same amount of energy and therefore approximately the same amount of interference with the condition of the membrane as you do in your system and as you would in the impulse?

DR. POUSSART: No, we have not used the step response for the membrane. The PRBS is so easy to use and with this type of stimulus you are perturbing the system on a continuous mode. The step fits your purpose nicely. In case of the membrane, we probably have to deal with a system that is more readily nonlinear, and the range over which we can work between an adequate signal-to-noise ratio of the measurement and nonlinearities is not very wide. So these are all interlaced conditions that have to be appreciated for the squid axon preparation. This is why we use a PRBS. Also, the PRBS has a zero mean property such as your mean property, which is natural polarization.

DR. R. EISENBERG: It seems to me that there are three aspects of step functions. First of all, if you look at responses to step functions and try to curve-fit them in the time domain, you have essentially a hopeless problem in trying to identify the system. If you take the Fourier transform, on the other hand, then you have an input whose waiting function is $1/s$ so you have almost all your energy at low frequencies, and therefore from just the step functions alone, you are not going to see the equivalent of high-frequency behavior. However, it has been found by Adrian *et al.* and Schneider *et al.* that possibly for a wide variety of circuits, there is circuit invariance, that is combination of parameters that can be determined by the area underneath the response to a step-function voltage or current.

DR. PILLER (*Columbia University, New York, N.Y.*): If you derive the property of a membrane from the first principle, and you can do this by including every function and every sequence of steps that you wish, from aqueous diffusion to phase transfer, you can in fact derive unique sets of solutions that can be represented by equivalent electric circuits whose topology is determined only by the physics of the situation. I would like to put something else forward, and that is the use of step function. If you treat them not by FFT but program your computer to do Laplace transform it turns out to be an interesting approach by which you can readily obtain measurements from one hertz to a hundred megahertz.

DR. R. EISENBERG: My point is not that you cannot take the Laplace transform but that if you take a model and try to curve-fit it in the frequency domain, it will not work. If you try to curve in the frequency domain you have the problem that most of your energy is 1/s.

DR. NICHOLSON (*New York University, New York, N.Y.*): Dr. Eisenberg, with regard to the muscle model, to what extent is it necessary to measure the individual topology in geometry of every muscle? Do you think it will be possible to derive a general topology for muscles, and your expressions would have much better generality?

DR. R. EISENBERG: The answer to the question is that we do not have any need to look at individual fibers. There is very little variance in the average morphometric properties from fiber to fiber. The problem is that the network within a given fiber in a given Z-band is quite random. That cannot be described as hexagons, or something of that sort. But the average morphometric properties are really not very dispersed from fiber to fiber. However, one has to be very careful, because what I have just said is fine if you really know what is there. Every physiologist who has looked at skeletal muscle has been convinced that what he sees is a set of sarcomeres which are stacked disks on top of each other. However, van Leeuwenhoek in 1706 clearly showed that the sarcomeres were, rather, a spiral. If you interpret the morphometric parameters with the wrong image in mind, you can have big trouble. On the other hand, if you have the structure approximately right you need remarkably little data, and you can deal with average parameters.

NOISE ANALYSIS AND CHANNEL PROPERTIES

Yi-der Chen

Laboratory of Molecular Biology
National Institute of Arthritis, Metabolism and Digestive Diseases
National Institutes of Health
Bethesda, Maryland 20014

INTRODUCTION

In a recent paper[1] Fishman and his colleagues reported an observation of "peaking" phenomenon in the current noise power spectrum of squid axon membranes associated with potassium conduction; that is, the noise power spectrum at the transition region is "sharper" than a simple relaxation process would predict (see FIGURE 1). This suggests that the usual Hodgkin-Huxley formalism may have to be modified if the "peaking" phenomenon is truly derived from the open-close kinetics of potassium channels. In this communication we first present a brief derivation of the noise power spectrum of a general channel kinetic model (for a complete treatment, see Reference 2) and then discuss some simple models that may produce the "peaking" phenomenon.

THE NOISE POWER SPECTRUM OF A GENERAL MEMBRANE MODEL

Throughout this paper we consider only K^+ channels. Similar treatment can be applied to Na^+ or leakage channels.

The Basic Current-Voltage Relation

Assume a (K^+) channel can exist in z channel states. Let $\bar{g}_i$ represent the unit conductance of a channel in state i and $N_i(t)$ be the number of channels in state i in the membrane at time t. Then, the total ionic current passing through the channels at constant (clamped) voltage $V°$ can be expressed as

$$I(t) = \sum_{i=1}^{z} \bar{g}_i(V° - V_K)N_i(t) \tag{1}$$

where V_K is the equilibrium potential of K^+, which is determined by the concentrations of K^+ across the membrane. The $N_i(t)$ in Equation 1 obeys the usual chemical rate equations:

$$\frac{dN_i}{dt} = F_i(N_1, N_2, \ldots N_z; \alpha_1, \alpha_2, \ldots)$$

$$(i = 1, 2, \ldots, z) \tag{2}$$

where the α's represent the rate constants of the transition reactions and F_i can be

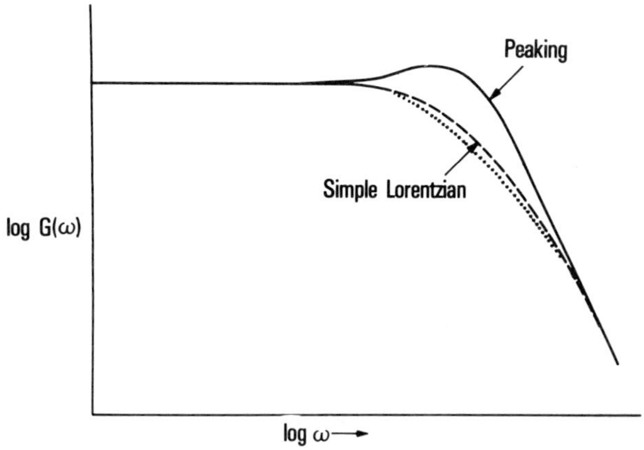

FIGURE 1. Schematic noise power spectra with and without "peaking." The dashed curve is obtained by fitting the low- and high-frequency asymptote of the experimental spectrum with a simple one-term Lorentzian, $K[1 + (f/f_c)^2]^{-1}$. A spectrum (solid curve) is defined to have a "peaking" if it is "sharper" than the asymptotic Lorentzian curve in the transition region. For equilibrium kinetic models whose $G(\omega)$ contains only the ℓ-summation in (**18**), the spectrum is always less sharp than the asymptotic Lorentzian (dotted curve).

explicitly expressed in terms of the N's and α's, using "mass action" law of chemical reactions. In general, F_i is always linear in α_i, but may be nonlinear in N_i. Some of the α's are voltage-dependent. The functional form of F_i is determined by the particular "model" of consideration.

The Phenomenological Relaxation Matrix

The noise power spectrum of membrane current fluctuations can be calculated from the time correlation function $C(t)$ according to the Wiener-Khintchine Theorem[2] as

$$G_I(\omega) = 4\int_0^\infty C(t)\cos\omega t\,dt. \qquad (\mathbf{3})$$

The time correlation function is defined as

$$C(t) = \langle \delta I(t)\,\delta I(0) \rangle \qquad (\mathbf{4})$$

where $\delta I(t)$ represents the deviation of membrane current from the steady-state value and $\langle\ \rangle$ means ensemble average.

From (**1**) and (**2**), we have

$$\delta I(t) = \sum_{i=1}^{z} \bar{g}_i(V^\circ - V_K)\,\delta N_i(t) \qquad (\mathbf{5})$$

$$\frac{d\delta N_i}{dt} = \sum_{j=1}^{z} \bar{F}_{ij}\,\delta N_j$$

$$(i = 1, 2, \ldots, z) \tag{6}$$

where

$$\bar{F}_{ij} = \left(\frac{\partial F_i}{\partial N_j}\right)_s. \tag{7}$$

The subscript s in $(\partial \bar{F}_i/\partial N_j)_s$ denotes the steady-state value at $V°$. The matrix F is usually referred to as the phenomenological relaxation matrix of the kinetic system.

One must note here that $\delta N_i(t)$ represents a deviation (fluctuation) of $N_i(t)$ from the steady-state value N_i^s. Thus, not all of the $\delta N_i(t)$ in (**5**) and (**6**) are linearly independent, if the system has conservation relations.[2,3] In general, the number of independent N_i is equal to the number of species (z) minus the number of conservation relations. Let δN_n ($n = 1, 2, \ldots, x$) refer to the set of independent species. Then the remaining z-x (dependent) species obey the relation:

$$\delta N_m = \sum_{n=1}^{x} \lambda_{mn}\,\delta N_n$$

$$m = x+1,\ x+2,\ \ldots,\ z, \tag{8}$$

where the λ_{mn} are constants determined by the system's conservation relations.[2,3]

After substituting (**8**), Equations 5 and 6 become

$$\delta I(t) = \sum_{i=1}^{x} g_i(V° - V_K)\,\delta N_i(t) \tag{9}$$

where

$$g_i = \bar{g}_i + \sum_{m=x+1}^{x} \lambda_{mi}\,\bar{g}_m, \tag{10}$$

and

$$\frac{d\delta N_i}{dt} = \sum_{j=1}^{z} M_{ij}\,\delta N_j.$$

$$(i = 1, 2, \ldots, x) \tag{11}$$

where

$$M_{ij} = \left(\frac{\partial F_i}{\partial N_j}\right)_s + \sum_{m=x+1}^{z} \lambda_{mj}\left(\frac{\partial F_i}{\partial N_m}\right)_s. \tag{12}$$

The matrix **M** with its elements expressed by (**12**) is called the "irreducible" phenomenological relaxation matrix of the kinetic system. The eigenvalues of **M** are always nonsingular.

The Noise Power Spectrum

The solution of (**11**) can be expressed in terms of eigenvalues and eigenfunctions of **M** as

$$\delta N_i(t) = \sum_{j=1}^{x} \phi_{ij}(t)\, \delta N_j(0). \tag{13}$$

where $\delta N_j(0)$ is the deviation of N_i at $t = 0$. The $\phi_{ij}(t)$ can be expressed formally as[3]

$$\phi_{ij}(t) = \sum_{\ell} r_{i\ell} h_{\ell j} e^{\xi_\ell t} + \sum_{n} (r_{in} \sin \theta_n t + w_{in} \cos \theta_n t) h_{nj} e^{\xi_n t}$$
$$+ \sum_{m} (r_{im} \cos \theta_m t + w_{im} \sin \theta_m t) h_{mj} e^{\xi_m t}. \tag{14}$$

where the $r_{i\ell}$, $h_{\ell j}$, ξ_ℓ, w_{in}, and so on, are related to the eigenvalues and eigenfunctions of matrix **M**, as discussed in Reference 3. The summation in ℓ refers to the real eigenvalues, and summations in n and m refer to the conjugate complex eigenvalues.

With (**9**) and (**13**), (**4**) can be shown to be

$$C(t) = \sum_{\ell} \tilde{r}_\ell \tilde{h}_\ell e^{\xi_\ell t} + \sum_{n} (\tilde{r}_n \sin \theta_n t + \tilde{w}_n \cos \theta_n t) \tilde{h}_n e^{\xi_n t}$$
$$+ \sum_{m} (\tilde{r}_m \cos \theta_m t + \tilde{w}_m \sin \theta_m t) \tilde{h}_m e^{\xi_m t} [V^\circ - V_k]^2 \tag{15}$$

where

$$\tilde{r}_k = \sum_{i=1} g_i r_{ik}$$
$$\tilde{h}_k = \sum_{i=1}^{x} \sum_{j=1}^{x} g_i h_{kj} \sigma_{ij}$$
$$\tilde{w}_k = \sum_{i=1} g_i w_{ik}$$
$$k = \ell, n, m, \tag{16}$$

and σ_{ij} represents the (i, j) element of the variance matrix of the system:

$$\sigma_{ij} \equiv \langle \delta N_i(0) \delta N_j(0) \rangle. \tag{17}$$

Substituting (**15**) into (**3**), we have the general expression for the noise power spectrum:

$$\frac{G(\omega)}{4} = \left\{ \sum_{\ell} \frac{-\tilde{r}_\ell \tilde{h}_\ell \xi_\ell}{\xi_\ell^2 + \omega^2} + \sum_n \frac{\tilde{h}_n}{2} \left[\frac{(\theta_n - \omega)\tilde{r}_n - \xi_n \tilde{w}_n}{\xi_n^2 + (\theta_n - \omega)^2} + \frac{(\theta_n + \omega)\tilde{r}_n - \xi_n \tilde{w}_n}{\xi_n^2 + (\theta_n + \omega)^2} \right] \right.$$

$$\left. + \sum_m \frac{\tilde{h}_m}{2} \left[\frac{(\theta_m - \omega)\tilde{w}_m - \xi_m \tilde{r}_m}{\xi_m^2 + (\theta_m - \omega)^2} + \frac{(\theta_m + \omega)\tilde{w}_m - \xi_m \tilde{r}_m}{\xi_m^2 + (\theta_m + \omega)^2} \right] \right\} [V^\circ - V_k]^2. \quad \textbf{(18)}$$

Peaking Phenomena

In general, the noise power spectrum of (**18**) poses an asymptotic feature for all models: constant at low frequencies and proportional to f^{-2} at high frequencies where f = frequency. Depending on whether the last two summations are present or not, the spectrum in the transition region differ sharply from one model to the other. If the matrix **M** of the model does not have complex eigenvalues, the last two summations drop out. In this case, $G(\omega)$ contains only terms proportional to $(\xi_\ell^2 + \omega^2)^{-1}$ and is a typical Lorentzian curve centered at zero frequency. If matrix **M** has complex eigenvalues, then $G(\omega)$ contains nonzero Lorentzians ($\theta_n \pm \omega$, etc.) and may have a "sharper" transition region (FIGURE 1).

As discussed by Fishman *et al.*,[1] it is always possible to obtain a one-term Lorentzian curve based on the high- and low-frequency asymptote of the experimental spectra. If the spectrum in the transition region is "sharper" than its one-term Lorentzian, the spectrum is referred to as having a "peaking." It is easy to show that the spectrum containing more than one ℓ-term (the term in the ℓ-summation of (**18**)) without n- and m-term is always less sharp (roll-off) than its one-term Lorentzian (see FIGURE 13 of Reference 1). In order to have "peaking," the $G(\omega)$ in (**18**) must contain *n*- and *m*-terms.

It is well known in relaxation theory[4] that, for equilibrium systems, the relaxation matrix (**7**) can always be reduced to a real symmetrical matrix by a similar transformation. Since the eigenvalues of a real symmetric matrix are always real, matrix $\overline{\mathbf{F}}$ and therefore matrix **M** have only real eigenvalues. That is, the noise spectrum of an equilibrium system (linear or nonlinear) can never have "peaking." In order to have "peaking," the underlying kinetic reactions must be at a nonequilibrium steady state. One must note that not all nonequilibrium systems will give "peaking" spectra. However, if the noise spectrum shows "peaking," the system is definitely not at equilibrium.

Recently Fishman and his colleagues have observed the "peaking" phenomenon in some current and voltage noise power spectra of squid axon membranes. If these phenomena are really derived from the conductance fluctuations of potassium channels, then the usual Hodgkin-Huxley model[5] cannot be right, because it is a linear equilibrium model.[6]

Proposed Nonequilibrium Channel Model

Nonequilibrium cyclical biochemical reactions are very common in biological systems. Recently, a linear model with cyclic splitting of ATP (or GTP) and cyclic protein conformational transformation has been proposed to explain the mechanism of active transport, muscle contraction, protein synthesis, and so on.[7] However, most of the current nerve membrane models do not involve a cycling steady state. Here, we propose two simple cycling models that may be applicable to the nonequilibrium channel subunits.

In general, a thermodynamic system must be in contact with some driving reservoir or reservoirs in order to be at a nonequilibrium steady state. For example, a conductor must be in contact with a voltage difference in order to maintain a constant current flow. In biological systems, the driving free energy of a reservoir is usually provided by an enzymatic reaction (such as ATP → ADP + P_i) or by a concentration gradient (across a membrane). If only one driving force is involved in the system, the free energy dissipated by the system is used only to maintain the steady state. In case that more than one driving force is present, the system may show "energy transduction" phenomenon.[7] In FIGURE 2(a) and

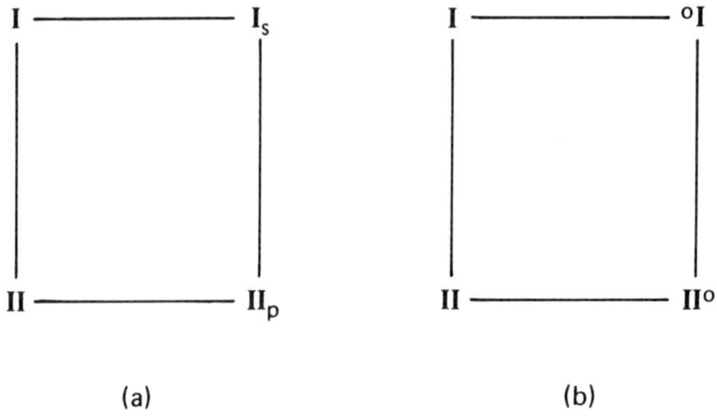

FIGURE 2. Two illustrative cycling steady-state mechanisms.

(b), two simple cycling mechanisms are proposed for channel subunits that are at a nonequilibrium steady state. A channel subunit is assumed to exist in two conformations, I (low-conductance state) and II (high-conductance state). In FIGURE 2(a), substrates (S) are assumed to bind only to I, and products (P) only to II. If the concentration of S is higher than the equilibrium value (with respect to P) in solution (where $S \rightarrow P$ reaction is negligible), the free energy difference of S and its equilibrium value drives the reactions in FIGURE 2 (a) cyclically: $I \rightarrow I_s \rightarrow II_p \rightarrow II \rightarrow I$. When S is in equilibrium with P in solution, the model in FIGURE 2 (a) reduces to the original Hodgkin-Huxley formalism. In FIGURE 2 (b), the concentrations of an ion (or a molecule) (denoted by a circle in the diagram) at the two sides of a membrane are assumed to be different. The ions at the left-hand of the membrane can bind only to a subunit in conformation I, and the ions at the right-hand side can bind only to II. If the concentration of ions at the left-hand side is higher than that at the right-hand side, the free energy difference of the two baths drives the reactions cyclically, as I → °I → II° → II → I. In model (a) a channel subunit acts as an enzyme; in model (b) a channel subunit serves as a translocator.

References

1. Fishman, H. M., L. E. Moore & D. J. M. Poussart. 1975. J. Membrane Biol. **24:** 305–328.
2. Chen, Y. 1977. Adv. Chem. Phys. In press.
3. Chen, Y. 1975. J. Theor. Biol. **55:** 229–243.
4. Schwarz, G. 1968. Rev. Mod. Phys. **40:** 206–218.
5. Hodgkin, A. L. & A. F. Huxley. 1952. J. Physiol. **117:** 500–544.
6. Hill, T. L. & Y. Chen. 1971. Proc. Nat. Acad. Sci. USA **68:** 2488–2492.
7. Hill, T. L. 1977. Free Energy Transduction in Biology. Academic Press. New York, N.Y.

SOME PROBLEMS IN MEMBRANE NOISE RESEARCH*

J. de Goede, M. W. Vonk, R. J. van den Berg,
H. van Rijn, and A. A. Verveen

Department of Physiology
The University of Leiden
Leiden, The Netherlands

Since Hodgkin and Huxley's description of the electrophysiological behavior of the giant axon of the squid in 1952,[1] the total current I_m flowing through excitable membranes is usually written as

$$I_m = I_D + I_K + I_{Na} + I_L \tag{1.1}$$

where I_D is the displacement current, I_K and I_{Na} are the potassium and sodium current, and I_L is an unspecific ohmic ionic current, the leakage current. The ionic conductances g_K, g_{Na} and g_L are defined by

$$I_{ion} = g_{ion}(E - E_{ion})$$

where E_{ion} are the ionic equilibrium or reversal potentials. The g_K and g_{Na} are membrane voltage- and time-dependent and given by

$$g_K = \bar{g}_K n^4 \tag{1.2a}$$

$$g_{Na} = \bar{g}_{Na} m^3 h \tag{1.2b}$$

with maximal conductances $\bar{g}_{ion}$. In the frog node of Ranvier, permeabilities must be used instead of conductances (cf. Cole[2]). The parameters $i = n$, m, or h are assumed to obey first-order differential equations

$$\frac{di}{dt} = \alpha_i(1 - i) - \beta_i i, \tag{1.3}$$

in which the forward rate constants α_i and the backward rate constants β_i are taken as functions of the momentary membrane potential. The parameters are obtained from experiments in which the membrane voltage is controlled and the currents are measured after steps in voltage (voltage-clamp experiments) and subsequently analyzed. From the possible transport mechanisms reluctantly suggested by Hodgkin and Huxley (carriers and pores), the conduction through pores or channels is considered preferable. Charged particles (dipoles) are displaced by electrical field changes across the membrane, and open the channels. They are denoted by n, m, and h. m^3h, for instance, is interpreted as the probability that the three activating m-particles and the one inactivating h-particle are all in the right position to open a channel for sodium ions. The kinetics are taken to be independent. A very slow inactivation process for the potassium system with time constants of seconds has been incorporated[3,4] in the Hodgkin-Huxley (H.H.)

*This work was supported by a grant from the Netherlands Organization for the Advancement of Pure Research (Z.W.O.).

model by the postulation of an independent process k. Slow processes in the sodium system are dealt with in a similar way.[5] One of the asymmetrical displacement (gating) currents predicted by Hodgkin and Huxley has been measured recently. (*cf.* References 6 and 7.) It exhibits kinetics similar to those of the activation process of the sodium system and is, therefore, at least partly attributed to the m-process.

Electrical Membrane Noises

After the initiation of fluctuation analysis of electrical membrane noise by Verveen and Derksen in the 1960s, it became clear that one of the main features of biomembranes is the generation of many different kinds of noise in excess of the fundamental thermal noise component,[8] each with its own spectral structure and its own dependence on ionic species or channel system, on membrane potential, and on current intensity. To unravel these components is one of the main problems in fluctuation analysis. Judicious experimentation almost always reduces the measurable components to at least a combination of one over f ($1/f$) noise and relaxation noise:

$$S_I(f) = \frac{a}{f^{\alpha}} + \frac{b}{1 + (2\pi f\tau)^2} \qquad \textbf{(2.1)}$$

with kinetic parameter τ, constants a and b, and the $1/f$ slope constant α, which is *about* one. The slope constant α cannot be *taken* to be one, and this creates two problems (H.v.R., unpublished work). The problem how to compare the $1/f$ noise arises when the values of α are different from one and from each other. An experimental problem follows from the fact that in fits of (**2.1**) to the spectral data, the obtained value of τ appears to be quite sensitive to the value of α. Therefore, it seems certain that α must be measured accurately. This is difficult to realize in the low frequency-restricted pulse noise experiments.

Up to three noise components have been detected, generated by the potassium system: a $1/f$ component and two types of relaxation noises, of which one qualitatively follows H.H. n-kinetics. In the frog node of Ranvier this component is found in pulse noise experiments only,[9,10] in which the inactivation of this system is still small or absent.

The second relaxation noise component found in frog nodes[10,11] is present in the stationary state in which the potassium system is partially inactivated. This one does not follow H.H. kinetics. This interesting problem is substantiated by the finding that the corresponding $1/f$ components behave differently as a function of I_K (R. v.d. B., unpublished).

"Peaking" of these relaxation noise components[12] may complicate the issues. Preliminary calculations (J. d. G. and M. W. V., this paper) indicate that the periaxonal space may generate both this behavior and another relaxation noise component. Its effect on the measured voltage clamp parameters is perhaps even more important.

For the sodium channel system, one $1/f$ component,[13] and two relaxation noise components associated respectively with H.H. h-kinetics[13,14] and m-kinetics,[15-17] have been found. Since both gating currents and noise, attributed to the m-gate mechanism, have been measured,[17] a simple noise model obtained from the continuous state model has been developed for further tests and discrimination between models (J. d. G. and M. W. V.).

For the leakage system a $1/f$ component has been found,[13] but burst noise, present upon hyperpolarization, further complicates the picture.[8] Metabolic pump activity may be another source of noise,[18] while the generation of chemical relaxation noises due to ionic-channel-pharmacon interaction[12,19] must be kept in mind with respect to experiments in which membrane-active pharmaca are present or applied.

Formal Theory

The microscopic equations of motion of matter contain, in principle, all the information about a system, but they cannot be solved in practice. One therefore usually starts with the macroscopic equation of motion and introduces the fluctuations as small perturbations. There are two main approaches, the master equation approach and the Langevin approach.

In the master equation approach it is assumed that the probability $P(q, t)$ (the probability that the macroscopic variable has the value q) obeys the master equation:

$$\frac{d}{dt} P(q,t) = \sum_{q'} \{W(q \mid q')P(q',t) - W(q' \mid q)P(q,t)\} \tag{3.1}$$

when q is a discrete variable. $W(q \mid q')$ represents the transition probability per unit time. Equation 3.1 has the appearance of a loss-gain equation.

In the Langevin approach, a fluctuating "force term" $l(t)$ is added to the macroscopic equation. For the macroscopic variable q, obeying a linear equation, we have

$$\frac{d}{dt} q(t) = -\lambda q + l(t) \tag{3.2}$$

in which the Langevin term $l(t)$ is assumed to obey certain statistical properties, namely (in the time domain),

$$< l(t) > = 0 \tag{3.3a}$$

and

$$< l(t)l(t') > = \Gamma\delta(t - t') \tag{3.3b}$$

where $< \ldots >$ denotes an ensemble average. For a system in equilibrium Γ can be found with the "fluctuation-dissipation theorem," usually introduced by the statement that the average relaxation toward equilibrium and the autocorrelation function of the fluctuations around equilibrium obey the same (linear) transport equation. We refer to the article by van Kampen[20] and the references cited therein for a detailed discussion of these methods.

The development of noise models from the H.H. equations is mainly due to Stevens[21] and to Hill and Chen.[22,23] For channel-system noise-models, derived with the master equation approach, see the presentation by Chen (this volume).

Another class of models, *continuous state models,* are obtained with the Langevin approach[20] or, directly, with the use of the fluctuation-dissipation theorem for which we first have to linearize the H.H. equations. Generalization to a nonlinear Langevin approach is dubious.[21] This gives the following expression for the power spectrum of the potassium system:

$$S_{I_K}(f) = 64 \frac{I_{K,\infty}{}^2 \tau_n}{1 + (2\pi f \tau_n)^2} \cdot \frac{<\delta n^2>}{n_\infty{}^2} \tag{3.4}$$

and similar equations for the sodium system, where

$$\tau_n = \frac{1}{\alpha_n + \beta_n} \tag{3.5}$$

and

$$n_\infty = \frac{\alpha_n}{\alpha_n + \beta_n}. \tag{3.6}$$

The condition $(<\delta n^2> | n^2) \ll 1$ serves as a consistency check. The Langevin approach need not be restricted to membrane models with *independent* channels with conductances $\gamma_K n^4$. Our use of the linearized H.H. equations is purely phenomenological. Note that the interpretation of the H.H. equations in a number of N-independent channels with conductances $\gamma_K n^4$ leads to[21,24]

$$S_{I_K}(f) = \frac{64}{N} \frac{I_{K,\infty}{}^2 \tau_n}{1 + (2\pi f \tau_n)^2} \cdot \frac{<\delta n^2>}{n_\infty{}^2}. \tag{3.7}$$

It may be argued that the formal difficulty that the variable i (n, m, h, or k) lies between one and zero has no consequence for the derived spectrum in the Langevin approach. For a discussion, we refer to Gichman and Skorochod.[25]

Conductance Fluctuations in the Sodium System

In the sodium system relaxation noise of the h-gates was observed by Anderson and Cahalan[14] and by van den Berg *et al.*,[13] together with a $1/f$ component.[13,14]

Upon blocking of the potassium system of the node, the noise was measured in the stationary state as voltage noise. Although not explicitly mentioned in their paper, the impedance needed to convert to a current spectrum was measured by van den Berg *et al.*,[13] and found to be frequency-independent in the frequency range 10–10^3 s^{-1} at the membrane potentials from -70 to -20 mV in the true steady state (v. d. B., unpublished). The Lorentzian had a maximum at -40 mV. Upon the addition of TTX or after the replacement of sodium by choline, this component disappeared together with an appreciable amount of $1/f$ noise. With the binary state channel model and the standard node parameters as given in Hille's thesis, the corner frequency was found[13] within 30% of that of the h-process. After correction for possible rundown and for a slow process with a time constant of 10 s, a single-channel conductance γ_{Na} (at zero voltage) of about 2–5 10^{-12}S was obtained. It followed from their work that it is advisable to avoid the slow sodium inactivation for the investigation of the m-process, as done by Conti *et al.*,[17] who measured the m-component in pulse-noise investigations. They analyzed their current-noise spectra in terms of an independent sum of $1/f$ noise and the relaxation component. Near-simultaneous measurement of the H.H. sodium system parameters for the same nodes were used. Application of the binary state model resulted in a single-channel conductance of 7.9 ± 0.9 10^{-12}S. The γ_{Na} turned out to be rather insensitive to errors in m_∞ and to changes in its power, which may be understood from the work of Chen. The small discrepancy between the results of van den Berg *et al.*[13] and Conti *et al.*[17] may be due to the uncertainties involved in the assumptions of the former authors or they may be real and related to changes in the sodium system with time after the step in voltage, such as has been found for the potassium system. In a subsequent paper, Conti *et al.*[26] showed that modification of the h-parameter with venoms had no major effect on γ_{Na} and on the m-process, from which results they argued that the h-gating mechanism is an all-or-nothing mechanism and can be functionally removed.

The fact that an asymmetrical displacement current has been measured and related to the m-gate mechanism gives one the opportunity to connect this with the noise measurements.

Gating Currents and Noise

The displacement current I_D (**1.1**) consists of a symmetrical capacitive current $I_C = C_m\, dt/dt$ and an asymmetrical component I_A:

$$I_D = I_C + I_A. \qquad \textbf{(5.1)}$$

The ionic gates $i = n, m, h, k, \ldots$ open and close the channels in response to changes in the electric field across the membrane. Asymmetrical displacement currents I_i are therefore expected to exist upon movement of the gating particles with changes of the electrical field. Since "The particle idea should probably be regarded as just an electrically equivalent representation of protein conformational changes involving movement of charged or dipolar residues," (Hille[27]) nonspecific or not yet specified asymmetrical currents I_N not associated with these ionic gating processes cannot be ruled out beforehand, i.e.

$$I_A = \sum_i I_i + I_N. \qquad \textbf{(5.2)}$$

The symmetric component can be experimentally removed by subtracting and averaging hyper- and depolarizing voltage clamp pulses of identical sizes. In this way asymmetrical displacement currents I_A during and after these steps have been reported recently to exist in squid[6] as well as in the node of Ranvier[7] (with blocked sodium and potassium currents). The current found possesses many of the characteristics of the m-process, and the first experiments seem to show a direct relationship between the gating currents and the sodium-ion currents. The work of Neumcke *et al.*[28] indicates that at least in the node, I_A may have some connection with the sodium gating mechanism, but it cannot be considered as directly correlated with the H.H. m-process. The fundamental problem at present is thus to find suitable criteria—if any—to separate I_A into true gating current (and to divide this into its components I_i), and a part relating to other conformational displacements in the membrane. The following approach (J. d. G. and M. W. V.) to connect an asymmetrical displacement current with the corresponding activation conductance mechanism is given as a possible contribution to this problem and applied to the sodium system.

The power spectral density for m-process current fluctuations under voltage clamp, assuming m^3h conditions and calculated with the continuous state model, is given by

$$S_{I_{\mathrm{Na}}}(f) = \frac{<\delta m^2>}{m_\infty^2} \frac{36\tau_m}{1 + (2\pi f \tau_m)^2} I_{\mathrm{Na},\infty}^2 \qquad (5.3)$$

(cf. (**3.4**)). We now assume that the charge displacement Q due to true gating of the activation ($m-$) process is

$$Q(t) = Q_m m(t) \qquad (5.4)$$

where Q_m is the maximal displacable charge. We may therefore write

$$<\delta Q^2> = Q_m^2 <\delta m^2>. \qquad (5.5)$$

We next assume that the formula for the fluctuating charge on a capacitor from equilibrium statistical mechanics may be applied,[29] *viz*

$$<\delta m^2> = \frac{\mathrm{k}T}{Q_m} \frac{\partial m_\infty}{\partial E}, \qquad (5.6)$$

which gives the current spectral density

$$S_{I_{\mathrm{Na}}}(f) = \frac{36\tau_m}{1 + (2\pi f \tau_m)^2} I_{\mathrm{Na},\infty}^2 \frac{\mathrm{k}T}{Q_m m_\infty^2} \frac{\partial m_\infty}{\delta E}. \qquad (5.7)$$

The low-frequency spectral limit is then

$$S_{I_{\mathrm{Na}}}(0) = \frac{36\tau_m I_{\mathrm{Na},\infty}^2}{m_\infty^2 Q_m} \frac{\delta m_\infty}{\delta E}. \qquad (5.8)$$

With this noise model Q_m can be calculated from the low-frequency limits of experimentally determined m-process current spectra at several levels of membrane potential. For the preliminary calculations we use the voltage clamp and noise data given by Conti *et al.*[17] From the data we find Q_m values of about 7.5 10^{-15}C at the membrane potentials of −62, −54, and −46 mV. For −38 and −30 mV Q_m is much smaller (1.8 10^{-15} and 0.2 10^{-15}C, respectively). With the use of τ_m from their noise data, which are in fair agreement with the macroscopically determined time constants τ_m, and a current calculated from their Figure 5, we find Q_m = 2.0 10^{-15}C for the three mentioned membrane potentials, and again, much smaller values at −38 and −30 mV.

With the expectation that the maximally displacable charge Q_m should be constant, both this continuous state model and the two-state channel model predict a decrease of the low-frequency spectral limit of the m-process at membrane potentials above about −40 mV. Since its decrease for the two-state channel model is slower than for the continuous state model, experimental differentiation between these models should be possible. The experiments of Conti *et al.*[17] do not show a tendency to decrease, however, and an extension of the noise measurements to larger depolarizations is needed to differentiate between the two classes of models, assuming that one of them is consistent with experiments.

The disturbing feature of the continuous state model is not so much the fact that Q_m is small (about 5% of the measured maximal asymmetrical charge displacement of 140 10^{-15}C[7]), but the fact that it decreases at larger depolarizations. It might be speculated that the amount of the gating charge that can participate in the fluctuations is not a function of *m* only but depends on *h* for long depolarizations. It would be interesting to test the above-described continuous state model for the squid axon.

Periaxonal Space

The influence on the noise of the perinodal space in frog (periaxonal space in squid) has not been investigated in detail, but it has been mentioned by Fishman *et al.*[30] for squid as an unlikely source for peaking of the relaxation noise component. When a node[31] or a squid axon[32] is repolarized after the break of a depolarizing pulse, a "tail current" is observed which is carried by potassium ions accumulated in the periaxonal space outside the membrane during depolarization. For squid this effect has been treated by Adelman and FitzHugh,[33] who incorporated an additional variable of state K_s into the H.H. equations:

$$I_K = \bar{g}_K n^4 (E - E_K) \quad \textbf{(6.1)}$$

$$\frac{dn}{dt} = -\frac{1}{\tau_n} n + \alpha_n \quad \textbf{(6.2)}$$

$$\frac{dK_s}{dt} = \frac{1}{\Theta}\left\{\frac{I_K}{F} - P_K{}^s (K_s - K_0)\right\} \quad \textbf{(6.3)}$$

and E_K is now given by

$$E_K = \frac{RT}{F} \ln \frac{K_s}{K_i} \quad \textbf{(6.4)}$$

where Θ is the radial thickness of the phenomenological periaxonal space. $P_K{}^s$ is the potassium permeability of the diffusion barrier between the periaxonal space and the external bulk solution. Electromotive forces across the barrier were and will be neglected. The sodium system is supposed to be blocked.

We will use the Langevin approach to discuss the effect of the periaxonal or perinodal space on the potassium relaxation noise, in order to connect the potassium noise with an equivalent circuit augmented with appropriate white noise sources, which is related to the method used by Fishman, *et al.*[30] for the potassium noise in squid.

Upon linearization of Equations 6.1, 6.3, and 6.5 around *E* in voltage clamp conditions and the introduction of Langevin terms l_n in **(6.2)** and l_s in **(6.3)** we find after Fourier transformation ($\omega = 2\pi f$)(neglecting slow k-process inactivation):

$$S_{I_K}(f) = \frac{16\,\bar{g}_K{}^2 n_\infty{}^6 (E - E_{K,\infty})^2 S_n{}^1}{\left(\omega^2 + \dfrac{1}{\tau_n{}^2}\right)\left(1 + \dfrac{\lambda_s{}^2 + 2\lambda_s/\tau_s}{\omega^2 + 1/\tau_s{}^2}\right)} + \frac{\theta^2 F^2 \lambda_s{}^2 S_s{}^1}{(\omega^2 + 1/\tau_s{}^2 + \lambda_s{}^2 + 2\lambda_s/\tau_s)} \tag{6.5}$$

in which $S_n{}^1$ and $S_s{}^1$ are the spectral densities of l_n and l_s respectively, and assumed to be frequency-independent. Here

$$\lambda_s = \frac{\bar{g}_K\, n_\infty{}^4\, RT}{\theta K_{s,\infty} F^2} \tag{6.6}$$

$$\tau_s = \theta / P_K{}^s. \tag{6.7}$$

The Langevin terms l_n and l_s are assumed to be uncorrelated, which condition can be relaxed if necessary. It is possible to incorporate the net potassium current from the periaxonal space to the external solution across the external barrier,[32] into the previously mentioned noise equations by redefining θ and $P_K{}^s$ (see Equations 3 through 7 in Adelman *et al.*[32]).

The first term in (**6.5**) has a maximum for

$$\omega^2 = -\frac{1}{\tau_s{}^2} \pm \sqrt{\beta\left(\frac{1}{\tau_n{}^2} - \frac{1}{\tau_s{}^2}\right)} \tag{6.8}$$

where

$$\beta = \lambda_s{}^2 + 2\lambda_s/\tau_s. \tag{6.9}$$

We see that the n-process noise may have a maximum for real ω, different from zero, depending on the values of τ_s and β. It also follows that another relaxation noise component may exist: the second term in (**6.5**).

The equivalent electrical circuit, derived in the standard way from the fully modified H.H. equations under the assumptions that the sodium system is blocked and that the reversal potential of the leakage E_L is unmodified, is given in FIGURE 1.

The specification of the circuit parameters in terms of the parameters of the fully modified H.H. equations is straightforward and will not be given here. Much more work has to be done in order to find out whether or not this circuit can satisfactorily describe the impedances of frog node and squid axon satisfactorily and whether peaking can be explained in this way (cf. (**6.8**)). These points are quite important for the evaluation of microscopic models, for peaking is also possible in kinetic systems in which the principle of detailed balance is not obeyed, as has been worked out by Chen.[23] Using the numerical values as given by Adelman *et al.*[32] for squid axon at 6°C, we find peaking for the potassium noise in the voltage range from −30 mV checked to +20 mV. At 0 mV the order of magnitude of the maximum (which lies at about 21 Hz) over the value at $\omega = 0$ is a factor of about 1.5. The effect of the periaxonal space on peaking thus cannot be excluded at present. This cannot be evaluated for the node as yet because of a lack of numerical data.

Perhaps more important is the effect of these spaces on the voltage clamp parameters, where the modified τ_n are about a factor of 2 larger than the unmodified ones in the range −40 to 0 mV in squid. This may have some bearing on Fishman's findings[34] that the corner frequency of the potassium relaxation noise is in better agreement with the continuous state model. Although not stated explicitly, it seems that Conti *et al.*[17] have corrected their voltage clamp data used in their noise calculations for this effect (cf. Conti & Wanke,[24] p. 480). It would be very interesting to investigate the squid potassium system in pulse noise experiments in this respect, and in the node of Ranvier, whether or not the perinodal space can explain the discrepancies between the corner frequencies of the noise and those ob-

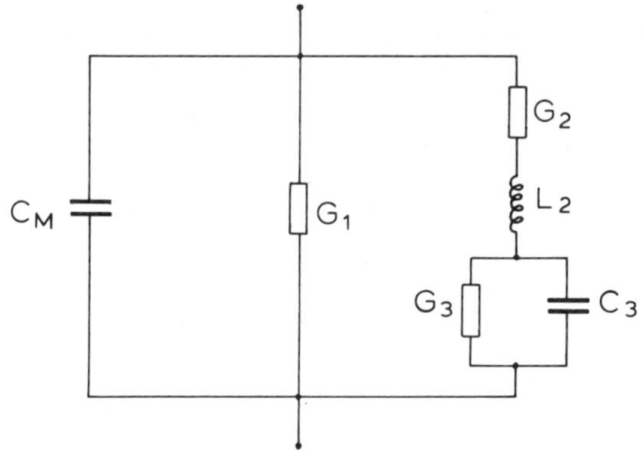

FIGURE 1. Potassium system: equivalent circuit after inclusion of potassium accumulation (small signal impedance). Conductances G, capacitances C, and self-inductance L.

tained from the voltage clamp data found by van den Berg *et al.*,[13] using the multistate channel model and the continuous state model. Calculations of the maximal potassium channel conductance γ_K from the noise variances seem to indicate that a two-state model is consistent with the experimental data. For a more detailed discussion we refer to Chen[23] and his contribution in this volume.

In summarizing the main problems of present-day membrane noise research, challenges are offered both by the many different noise components existing in these structures and by the models developed and yet to be developed to describe these processes. Careful and judicious experimentation into the limits of accuracy obtainable for these fragile biological systems will be necessary to unravel these components and to differentiate between different hypotheses. We expect that fluctuation analysis will thereby be of considerable aid in the elucidation of the structure and function of biomembranes and other biological systems.

REFERENCES

1. HODGKIN, A. L. & A. F. HUXLEY. 1952. J. Physiol. **117:** 500–544.
2. COLE, K. S. 1968. Membranes, Ions and Impulses. Univ. California Press. Berkeley, Calif.
3. SCHWARTZ, J. R. & W. VOGEL. 1971. Pflügers Arch. **330:** 61–73.
4. EHRENSTEIN, G. & D. L. GILBERT. 1966. Biophys. J. **6:** 553–564.
5. FOX, J. M. 1976 Biochim. Biophys. Acta **426:** 232–244.
6. BEZANILLA, F. & C. M. ARMSTRONG. 1976. *In* Cold Spring Harbor Symposia on Quantitative Biology **40:** 297–304.
7. ROJAS. E. 1976. *In* Cold Spring Harbor Symposia on Quantitative Biology **40:** 305–320.
8. VERVEEN, A. A. & L. J. DEFELICE. 1974. Prog. Biophys. Molec. Biol. **28:** 189–265.
9. BEGENISICH, T. & C. F. STEVENS. 1975. Biophys. J. **15:** 843–846.
10. VAN DEN BERG, R. J., E. SIEBENGA & G. DE BRUIN. 1977. Nature **265:** 177–179.
11. SIEBENGA, E., A. MEYER & A. A. VERVEEN. 1973. Pflügers Arch. **341:** 87–96.
12. FISHMAN, H., L. E. MOORE & D. J. M. POUSSART. 1975. J. Membr. Biol. **24:** 305–328.
13. VAN DEN BERG, R. J., J. DE GOEDE & A. A. VERVEEN. 1975. Pflügers Arch. **360:** 17–23
14. ANDERSON, C. R. & M. CAHALAN. 1974. Proc. Int. Union Physiol. Sci. **11:** 8.
15. CONTI, F., L. J. DEFELICE & E. WANKE. 1975. J. Physiol. **248:** 45–82.
16. FISHMAN, H. M., L. E. MOORE & D. J. M. POUSSART. 1976. Biol. Bull. **151:** 408–409.
17. CONTI, F., B. HILLE, B. NEUMCKE, W. NONNER & R. STÄMPFLI. 1976. J. Physiol. **262:** 699–727.
18. SEGAL, J. R. 1972. Biophys. J. **12:** 1371–1390.
19. ULBRICHT, W. 1976. *In* Phénomènes Électriques au Niveau des Membranes Biologiques. Société de Chimie Physique. 29 Réunion Internationale. Orsay, France.
20. VAN KAMPEN, N. G. 1976. Physica **83B:** 1–5.
21. STEVENS, C. F. 1972. Biophys. J. **12:** 1028–1047.
22. HILL, T. L. & Y-D. CHEN. 1972. Biophys. J. **12:** 948–959.
23. CHEN, Y-D. 1977. *In* Advances of Chemical Physics. S. A. Rice, Ed. In press.
24. CONTI, F. & WANKE. 1975. Quart. Rev. Biophys. **8:** 451–506.
25. GICHMAN, I. I. & A. W. SKOROCHOD. 1971. Stochastische Differentialgleichungen, Akademie-Verlag. Berlin, Germany.
26. CONTI, F., B. HILLE, B. NEUMCKE, W. NONNER & R. STÄMPFLI. 1976. J. Physiol. **262:** 729–742.
27. HILLE. B. 1976. Ann. Rev. Physiol. **37:** 139–152.
28. NEUMCKE, B., W. NONNER & R. STÄMPFLI. 1976. Pflügers Arch. **363:** 193–203.
29. VAN VLIET, K. M. & J. R. FASSET. 1965. *In* Fluctuation Phenomena in Solids. R. E. Burgess, Ed.: Chap. 7. Academic Press. New York, N.Y.
30. FISHMAN, H., L. E. MOORE, D. J. M. POUSSART & E. SIEBENGA. 1977. J. Membr. Biol. **32:** 255–290.
31. DUBOIS, M. & C. BERGMAN. 1975. Pflügers Arch. **358:** 111–124.
32. ADELMAN, Jr., W. J., Y. PALTI & J. P. SENFT. 1973. J. Membr. Biol. **13:** 387–410.
33. ADELMAN, Jr., W. J. & R. FITZHUGH. 1975. Fed. Proc. **34:** 1322–1329.
34. FISHMAN, H. 1973. Proc. Nat. Acad. Sci. USA **70:** 876–879.

ION MOVEMENTS AND KINETICS IN SQUID AXON
II. SPONTANEOUS ELECTRICAL FLUCTUATIONS*

H. M. Fishman and L. E. Moore

Department of Physiology and Biophysics
University of Texas Medical Branch
Galveston, Texas 77550

D. Poussart

Department of Electrical Engineering
Université Laval
Quebec, Canada G1K 7P4

INTRODUCTION

Until very recently, fluctuation phenomena or noisy processes were viewed as undesirable nuisances that set the limit on the minimum signal strength required for intelligible communication or the sensitivity of measurements. This notion, which is correct and prevalent, ignored the application of noise concepts and measurements as a useful tool to acquire information about systems and processes. With respect to biological phenomena, there are essentially three categories in which stochastic methods are of importance: 1) measurements of spontaneous fluctuations, which arise from statistical variations in a physical parameter from its mean value, and which constitute a probe into microscopic phenomena; 2) measurements in which random noise is applied to a system as a small-signal perturbation in order to obtain a linear description of an input-output relationship; and, 3) measurements in which random noise is applied to a nonlinear system in order to ascertain and characterize the degree of nonlinearity. The focus of this presentation is on measurements in category 1. Examples of measurements in category 2 are found in this conference[1,2] and measurements in category 3 have been reported elsewhere.[3] In this paper we shall attempt to provide an account and to indicate the status of various noise measurements in the squid axon based upon our own involvement in this work.

The application of fluctuation phenomena to conduction processes in biological membranes has grown rapidly since the early stage of development about a decade ago. Measurements of noise in the node of Ranvier of single frog fibers by Derksen and Verveen[4,5] and in an isolated area of lobster axon by Poussart[6,7] showed a spectral form which was f^{-1} with an intensity that related to the driving force for potassium ions. From a substantial existing physical literature[8,9] on noise associated with conduction processes in solid state materials, f^{-1} was not the type of noise expected from the ion conduction "channels" implicit in the Hodgkin and Huxley (HH) description of current flow in a squid axon membrane. In fact, Hill

*This work was supported in part by National Institutes of Health, United States Public Health Service research grants NS 11764 (HMF) and NS 13520 (LEM), and by Canadian National Research Council grant A 5274 (DP).

and Chen[10] showed that the HH conduction scheme did not give a f^{-1} power spectrum.

Noise measurements in the classical squid axon preparation began in 1971. Using a new method[11,12] for isolating small areas of squid axon, which produced low extraneous noise, Fishman[13] confirmed the presence of f^{-1} noise, but also observed a new noise component that had the apparent form of a relaxation process; namely, $[1 + (f/f_c)^2]^{-1}$, in power spectra of fluctuations. Apparently, at the same time a relaxation component was observed in the node.[15,16] The new component was clearly established as arising from *K* conduction kinetic phenomena.[14,17-19] The observation of the relaxation-like component as well as the lack of a f^{-1} component in noise from "channel formers" in lipid bilayers[47-49] leave little doubt that the f^{-1} noise arises from the interaction of ions with conducting ion channels in a restricted diffusion sense,[9,30,50,51] rather than from the lifetime of *K* channel conduction states.

Noise measurements from isolated large areas[20-22] of squid axon essentially confirmed the above results. However, some details of the measurements from large areas differed from those found in small-area measurements. These aspects will be dealt with later in this presentation.

The concept that measurements of spontaneous fluctuations constitute a probe of ion conduction processes on a microscopic level received strong support when measurements of noise induced in *K* channels by the interaction of specific agents was reported.[18] The assumption by investigators[19,21-24] that macroscopic blockers of *K* current (such as tetraethylammonium ion) also suppressed *K* channel noise was shown to be false. The measurement of induced noise clearly indicates that macroscopic concepts must be applied with caution to noise measurements. These measurements also suggest that the inducer molecules constitute new microscopic probes of membrane conduction.

In contrast to *K* channel noise, which is quite large and dominates, Na channel noise is difficult to measure under normal ionic conditions in the squid axon. DeFelice *et al.*[21] and Conti *et al.*[22] claim to have measured it under normal conditions, but our own extensive search[17,18] for it and positive identification[29] only under special ionic conditions leaves considerable doubt about their interpretations.

As a not too distant goal for fluctuation measurements, it should be possible to discriminate between some types of models of conduction. In this paper, we begin to compare some simple models with accumulated data. The form of power spectra and their dependence on parameters such as voltage and temperature are used to compare kinetic schemes of the Hodgkin-Huxley type. Also, the variance of the noise, together with mean conduction parameters, affords an opportunity to distinguish between two versus more than two conduction-state models.

Noise Measurements from Small Versus Large Isolated Areas

The methods and technical details of recording and analyzing membrane noise have been discussed in numerous articles[5,7,12,17,20-26] and will not be repeated here. One aspect of noise methods over which there has been some controversy is the selection and isolation of the area of membrane from which fluctuations are recorded. The following discussion is intended as a nonrigorous treatment of the salient points concerning this issue.

The prime motivation for the development of a method of isolating and recording from a small area (patch) of squid axon was the desirability of obtaining low extraneous noise performance from the instrumentation[7,17,25] without resorting to special techniques such as cross correlation.[26] In addition to this consideration, there are a number of other important aspects that can now be discussed in the light of existing published works. In the only comparison[25] between small (10^{-4}–10^{-5} cm^2) and large (0.1–0.3 cm^2) area measurements, it was shown that the low source impedance inherent in large-area measurements can produce emphasis of extraneous instrumentation noise as well as electrode noise. The noise figure of instrumentation used on large areas was measured to be 10–12 dB as compared to 0.1–0.3 dB in patch instrumentation over the frequency range of 10–1000 Hz. Despite simple[17,25] or more complicated[7,21] calculations of signal-to-noise (S/N) ratio, which indicate that the best S/N performance is obtained with present-day (field effect transistor) amplifiers when measuring from areas of 10^{-3}–10^{-4} cm^2, the proponents of large-area measurements[20-22] dismiss this point as unimportant on two bases. The first is that S/N performance of instrumentation can be improved in large areas by paralleling input stages.[21] However, in practice this is rather futile. For example, most FET devices have their best noise figure (< 1 dB) when driven from 1 MΩ sources.[17,25,26] A reduction in this optimum source resistance (for low noise operation) by paralleling n identical input stages goes as n^{-1}. Since typical source resistances of large-area measurements range from 2–15 kΩ,[20-22] approximately 100 paralleled stages would, in principle, be necessary to change the optimum source resistance from 1 MΩ to 10 kΩ. Secondly, since excess noise is so large in squid axons, it is implied that measurements can be made despite poor noise performance from instrumentation. There are many instances where this may be valid; however, there are also cases (see Na channel noise) where differences between spectra are the only means of extracting noise components, and instrumentation noise could be a crucial factor in accurately describing these components.

In addition to the above consideration, there are several other important aspects that bear on the selection of membrane area. A widely accepted method for "space," voltage clamping of squid axons, uses an internal axial current electrode[25] and appears to be the major motivation for large-area noise measurements. However, axial wire measurements from air-gap isolated 2-cm lengths of axon[20-22] violate the conditions for collection of current from an area of membrane that is under uniform potential control. The mere presence of a current electrode inside the length of an axon does not guarantee an isopotential condition over the entire axon length. Figure 10 of Cole and Moore[27] indicates that the potential variation in the axoplasm can be as much as 5–10 mV over a 0.46-cm length even with a low effective resistance current electrode. In fact, Cole and Moore state (p. 147) that "... the membrane current measurements should be made both as close to the potential electrodes and over as small an area as possible." Ineffective control of potential during current noise measurements could produce a dispersion in noise components identified in power spectra. Such effects could lead to ambiguous conclusions about the applicability of models which are compared with voltage dependent parameters obtained from spectra. Furthermore, changes in potential uniformity, after blocking specific ion conduction, could produce artifactual noise components in difference spectra.

In contrast to large-area measurements, patch measurements[12] of current are confined to the smallest areas ever isolated and voltage clamped in a squid axon. Furthermore, isopotential (< 1 mV) conditions were demonstrated in the axoplasm during large-step voltage clamps, and capacitance measurements indicated collection of current from an area that was only 25% greater than the patch pipette.

Secondly, a large membrane area has associated with it a large effective† membrane capacitance. The effective membrane capacitance becomes a very important factor in accurately observing membrane current noise in the upper region (200–1000 Hz) of the measurements band without complicated corrections for instrumentation noise.[7,25] Since the intrinsic voltage noise from input instrumentation produces a frequency-dependent current noise through effective membrane capacitance, current noise from large membrane areas must be corrected for instrumentation noise at frequencies above 200 Hz (see Ref. 25, Figure 8A, and Ref. 22, Figure 3). This correction could be crucial for positive identification of a noise component in the difference between two noise spectra that are obtained under different conditions of effective membrane capacitance. By contrast, the patch capacitance (100 pF) is so small that amplifier noise is not significant in spectra over a 1 kHz band.[25] Thus, patch current noise measurements yield uncontaminated power spectra over a wider band than in large-area measurements.

Thirdly, it has been noted[28] that in axial current electrode experiments, depolarizations of more than 15–20 mV for extended periods (1 min) produce axon degradation. This effect severely limits the use of this method in the investigation of potential change on membrane noise (see *K* channel noise section). This restriction does not occur in patch measurements.[18]

Finally, there are some unique aspects of the patch method that circumvent various disadvantages inherent in large-area measurements. One misconception[22–24] about the patch method is that a finite isolation of the membrane area with sucrose solution produces poor performance or unreliable measurements. The shunt of a patch has been shown[12,17,25] to present no insoluble problems. For example, in step clamp measurements an enhanced leakage is observed that can be corrected to give the usual ion-conduction parameters. In noise measurements (FIGURE 1) the isolation is sufficient to assure that noise from the axon membrane outside the patch (which may not be under good potential control) has a negligible effect on patch noise.[25] In addition, the finite isolation actually has some advantages. It has been shown[12] that the patch impedance at rest is between one and two orders larger than the isolation. Consequently, during potential changes, although the patch impedance can change considerably, the shunt dominates the total impedance and presents a nearly constant source impedance to the subsequent instrumentation. Thus the excellent *S*/*N* performance of the instrumenta-

†The effective membrane capacitance includes not only the contribution from the dielectric portion of the membrane, but also a contribution from the time-varying behavior of ion conductances.[28] The latter contribution can be as much as 30% of the total capacitance and is potential-dependent.[28,44,45] Thus, significant changes in effective membrane capacitance occur when voltage-dependent ion conduction is blocked, which complicates instrumentation noise corrections.

tion is maintained over a wide potential range by the dominance of the shunt, and thermal noise is nearly constant. In large-area measurements, both the S/N performance and thermal noise change as the impedance of the large-area changes.[21] A dominant constant shunt thus offers advantages in taking the difference between power spectra obtained at different potentials or ion-conduction conditions. Spectra can be subtracted directly without determination of the impedance at each potential and without correction of each spectrum for different S/N conditions or thermal noise. This point is particularly important with respect to Na conduction noise and is dealt with later in this presentation. Thus, in summary of the preceding discussion, the patch technique employs a finite isolation that is adequate to accomplish the measurement goals of recording from a very small membrane area which is under uniform potential control. It also achieves additional benefits in noise performance from instrumentation and in the manipulation of spectral data that are obtained under a wide range of conditions.

Criteria for Identification of Ion Channel Noise

Three criteria form a basis for positive identification of spectral features that relate to the noise produced by fluctuations in conduction states of ion channels.

1. It is necessary that spectral features be eliminated by conditions that suppress a specific ion-current flow *without inducing substantial noise.*
2. It is necessary that a spectral feature conform to criterion 1 after elimination of active transport.
3. Assuming the fluctuation dissipation theorem, it is necessary that kinetic parameters, which are obtained from noise spectra, relate to those determined from macroscopic measurements.

These criteria have been used in our measurements,[14,17,18] whereas others[21-23] have relied primarily on the effects of macroscopic blockers of specific ion currents as a basis for interpretation. The reliance on macroscopic blocking agents has been shown[18] to lead to equivocal microscopic interpretations, particularly with respect to measurements of the Na conduction system. These problems will be discussed in subsequent sections.

K Channel Noise

Under normal‡ ionic conditions in a squid axon, power spectra of steady-state current fluctuations, measured in small patches held at fixed potentials, show a dominant noise component (Figure 1). This observation is independent of the state or condition of the axon membrane outside the patch, since the lower set of spectra in Figure 1 were obtained from the same patch with the rest of the axon depolarized in 0.5 M KCl solution. At rest potential (0 mV), a f^{-1} type component is evident. With increasing depolarization, another component (hump), which ap-

‡Normal ionic conditions: externally, sea water or artificial sea water (430 mM NaCl, 10 mM KCl, 10 mM $CaCl_2$, 50 mM $MgCl_2$, and 5 mM Tris HCl); internally, nonperfused or perfused with 0.5 M KF and 5 mM Tris HCl.

pears to be Lorentzian in form (but is actually more complicated), becomes dominant. In axons in which the apparent Lorentzian component is not present[18] (inexcitable or leaky axons), the exponent on noise of the f^{-1} type is variable between 0.5 and 1.5. This variation in the exponent is similar to that observed in power spectra of noise from various artificial porous membranes[30] and is indicative of a restricted diffusion mechanism[9,50,51] involving interaction between moving

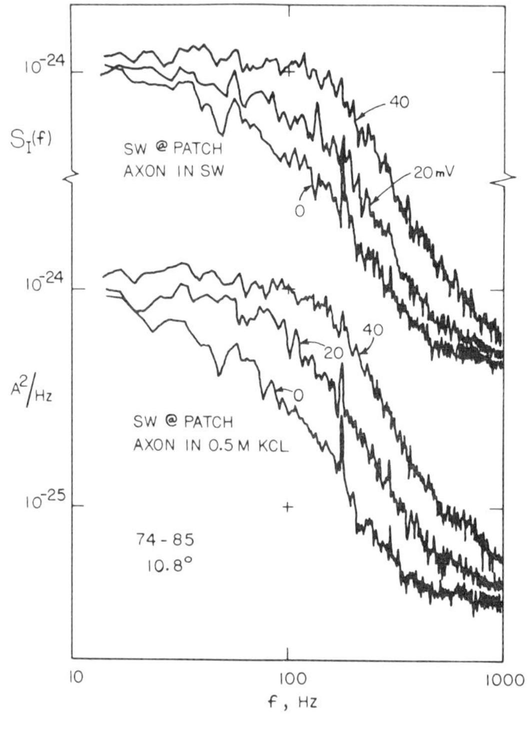

FIGURE 1. Power-density spectra of current noise, $S_I(f)$, in an isolated small patch ($\sim 10^{-4}$ cm^2) of squid axon under normal ionic conditions (SW externally unperfused internally) at the designated depolarized potentials away from rest. Note the sharp corner region of the noise component at depolarized potentials that dominates over a f^{-1} type component, obvious at rest potential (0 mV). Upper set of spectra, SW external to both patch and axon outside of the patch. Lower set of spectra, same patch but after replacement of the SW bathing the axon outside of the patch with .5 M KCl (axon exclusive of patch is depolarized and inexcitable). The similarity between the two sets of spectra indicates that the patch isolation is sufficient to make the noise contribution from the axon outside the patch negligible. (By permission of the publishers of J. Membr. Biol.[25])

ions and conducting channels. Since a f^{-1} function is only an approximation of the actual spectral form, the simple assumption of f^{-1} to facilitate its removal, which would permit accurate characterization of other components, is problematical because of the variability of the exponent. Nevertheless, other components and their spectral form are measurable at depolarized potentials where they dominate the spectrum as in FIGURE 1.

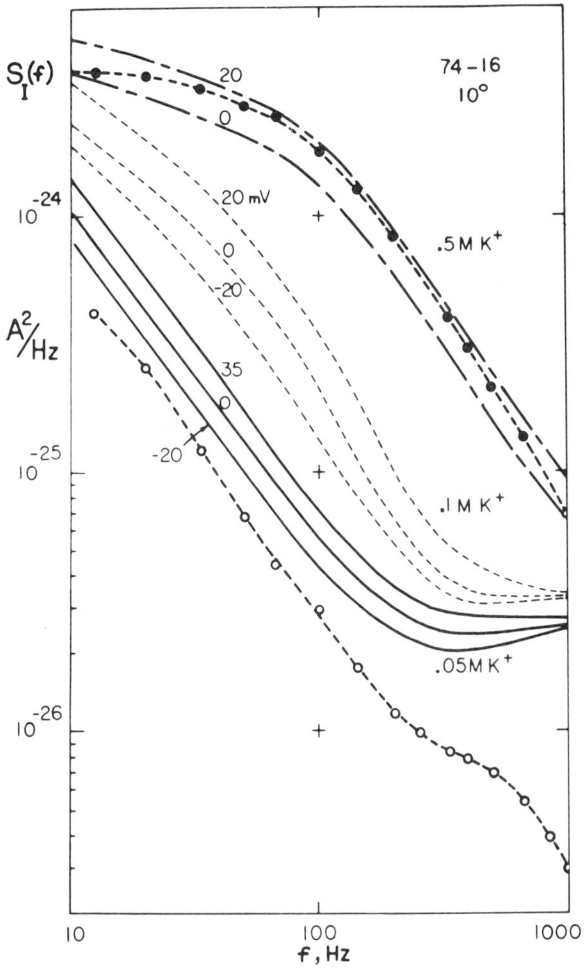

FIGURE 2. Power spectra of current noise from the same patch of an axon internally perfused with three concentrations of K^+ (dilution of KF perfusate with isosmotic sucrose solution) in the perfusate. Spectra of the non f^{-1} type were not observed at concentrations of .05 M or below. The filled circles represent a difference spectrum between .5 M K_i at V = 20 mV and .05 M K_i at V = 35 mV. "Humps" in spectra become apparent at .1 M and obvious at control concentration (.5 M). This axon gave patch spikes of 90 mV or more at all concentrations of K^+, and thus Na^+ conduction was operative. The open circles represent a difference spectrum in .05 M K_i between V = 35 mV and V = −20 mV. The appearance of a hump, in addition to f^{-1} type noise, suggests a Na noise component that was not evident in 6 axons with TTX (1 μM) applied to the patch. In this and subsequent figures, smooth curves are drawn through spectra. (By permission of J. Membr. Biol.[18])

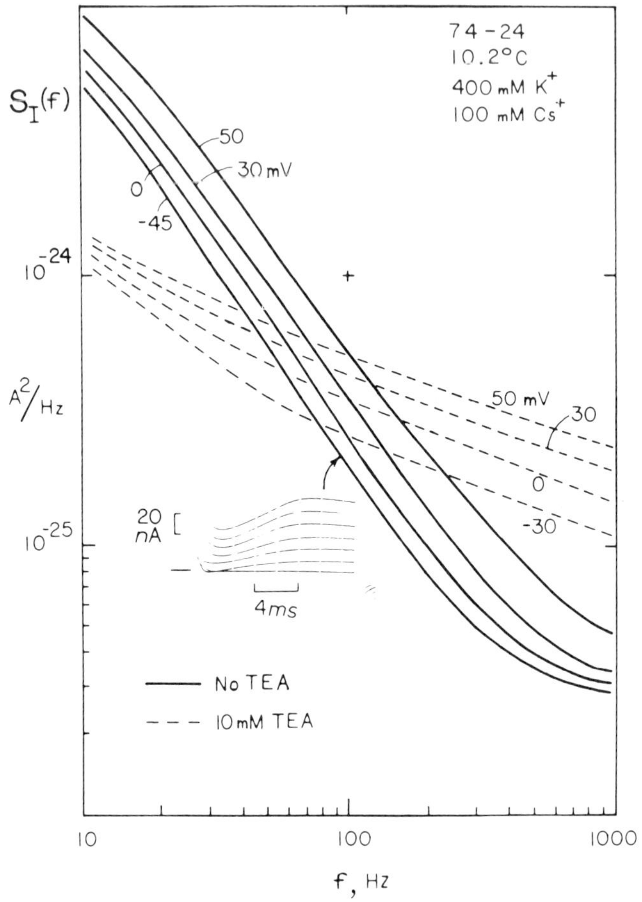

FIGURE 3. Power spectra of current noise from a patch of axon internally perfused with Cs added to the perfusate. Steady-state (K) currents (inset), for step voltage clamps of 50–170 mV in increments of 20 mV, are about half of normal. The normal relaxation-like component is not evident in these f^{-1} type-dominated spectra under a condition of reduced K current. Dashed curves:spectra after perfusion with the same Cs perfusate with 10 mM TEA added. Note increase in high-frequency noise induced by TEA at all potentials. (By permission of J. Membr. Biol.[18])

The non f^{-1} component has been clearly established according to the previously stated criteria as relating to *K* channel conduction. An effective demonstration of the compliance with criterion 1 is illustrated in FIGURE 2, where the *K* concentration in an internally perfused§ axon was changed from 50 mM to 0.1 M and then to 0.5 M in the same axon. The power spectra with 50 mM K_i, where *K* current in the patch is quite low, show only f^{-1} type spectra. With increasing *K* con-

§Spectra in perfused (0.5 M KF) or unperfused axons are indistinguishable.

centrations and current, an additional component becomes measurable and dominant at 0.5 M. Further evidence is provided from agents that are known to reduce K current in the membrane. For example, FIGURE 3 (solid curves) shows additional evidence of the absence of non f^{-1} components in an axon internally perfused with Cs, which reduced K current in the presence of 0.4 M K_i. K currents measured in step-clamp records (inset, FIGURE 3) were about half of their values without Cs. In axons internally perfused with tetraethylammonium (TEA) ion added in concentrations of 10 mM or greater (FIGURE 4), K current is drastically reduced (see inset of FIGURE 6) and the spectral hump at low frequencies is again eliminated. The effect of TEA injected into axons is similar.[21,22] Thus, whenever K current is reduced significantly in an axon, the usual dominant spectral feature disappears or is altered. In contrast to the drastic effect of reducing K conduction in an axon, blockage of Na conduction by application of tetrodotoxin (TTX, 1 μM) externally has no obvious effect on the spectral features observed under normal ionic conditions.[14,17,18] An exception to this latter statement has been reported in large-area noise measurements[21-23] and will be discussed in the section on Na noise.

Criterion 2 is satisfied[14,17,18] by the fact that the anion fluoride, which is known to block active transport,[31,32] was used to obtain the data in FIGURES 2 and 4.

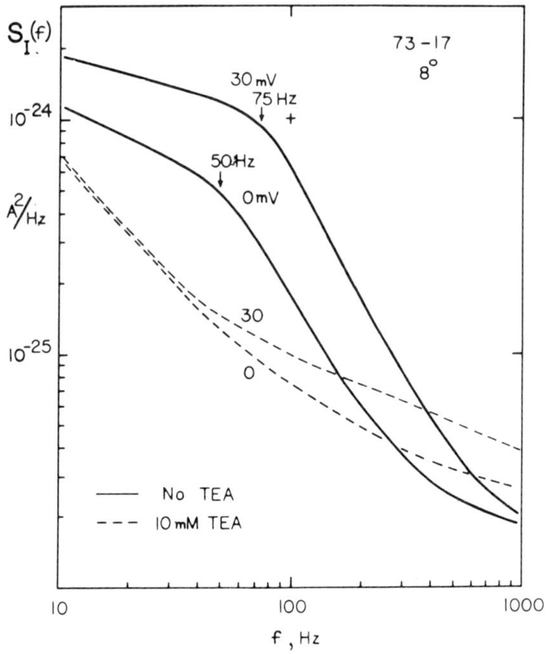

FIGURE 4. Power spectra of current noise from a patch of internally perfused axon (solid curves are controls). Dashed curves are spectra after 10 mM TEA, which blocks K current, is added to perfusate. Note disappearance of low-frequency relaxation-like spectra with TEA present, and the increase in noise power (crossing of dashed and solid curves) at high frequencies. (By permission of J. Membr. Biol.[18])

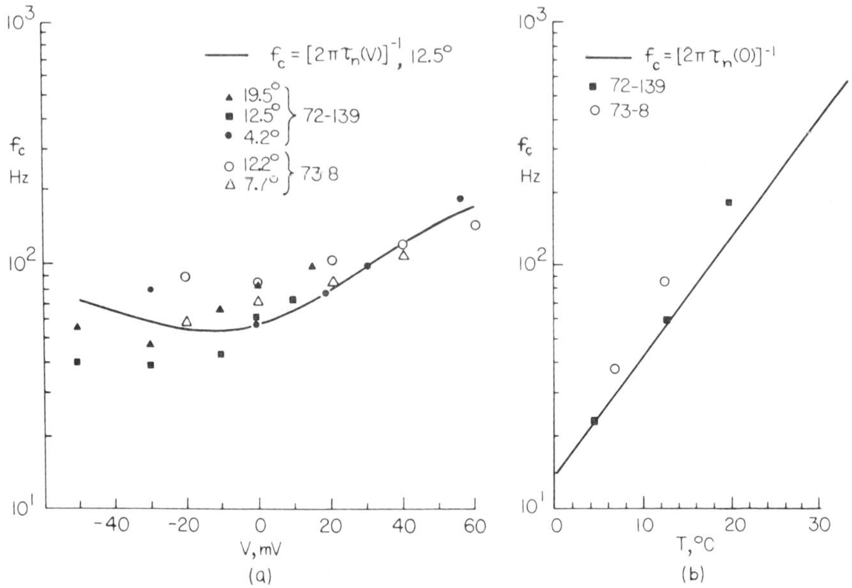

FIGURE 5. Comparison of corner frequency data from power spectra with changes in voltage (a) and temperature at rest potential (b) with the relation (curve) $f_c = (2\pi\tau_n)^{-1}$, where $\tau_n(V)$ is the HH relaxation time for K conductance and a Q_{10} of 3 is assumed. In (a) the curve is drawn for 12.5°C and data have been scaled to this temperature. (By permission of J. Membr. Biol.[18])

With respect to criterion 3, the apparent corner frequency f_c, (intersection of low- and high-frequency asymptotes of spectra) of humps in spectra for wide variations in potential (110 mV range) and temperature (5–25°C) relates to the Hodgkin-Huxley relaxation time, τ_n, for *K* conduction according to $f_c = (2\pi\tau_n)^{-1}$, as shown in FIGURE 5. These data are also in agreement with an assumed Q_{10} of 3. A more accurate comparison awaits determination of τ's for Woods Hole squid axons from small-signal measurements of complex admittance.[2,28] (Such a comparison is necessary because the exact relationship between macroscopic measurements and the corner frequencies of noise measurements is model dependent.) Conti *et al.*[22] confirmed the large temperature dependence ($Q_{10} = 3.3$) of f_c, and for a restricted potential range observed corner frequencies somewhat greater than $(2\pi\tau_n)^{-1}$. However, their data (see Ref. 22, Figure 7), with respect to potential, were confined to a limited range (20 mV), presumably because of axon degradation effects mentioned earlier, which occur during prolonged current flow in an internal axial electrode. Another complication (see the section on noise measurements from small versus large isolated areas) in the interpretation of large-area measurements[22] of f_c variation with potential may be the nonuniform potential conditions in the long lengths (2 cm) of axon used in these experiments. Finally, after consideration of the majority of observations in the squid axon, all the previously stated criteria have been met and therefore establish that the humped feature in spectra, obtained under normal conditions, is *K* channel conduction noise.

INDUCED *K* CHANNEL NOISE

Quaternary ammonium compounds such as TEA are known to block *K* current outflow when applied internally in squid axon. When TEA was used to test the dependence of the relaxation-like component on *K* current flow (FIGURE 4), TEA eliminated the hump at low frequencies, but the noise power at high frequencies actually increased above the intensities before application of TEA.[14,17,18] Furthermore, upon application of large voltages (50 mV or more), a high-frequency noise component was present (FIGURE 6). The high-frequency noise was evident only when TEA was in the internal perfusate, and there was sufficient $[K_i]$ present for the observation of the normal *K* spectrum before addition of TEA (compare solid and dashed curves in FIGURE 3). These data suggest that TEA induces noise in *K* channels by virtue of its binding to sites that control *K* conduction. This noise was explored further through synthesis of triethyldecylammonium (TEDA) by substitution of decane for one of the TEA ethyl groups. Internal TEDA in squid axons

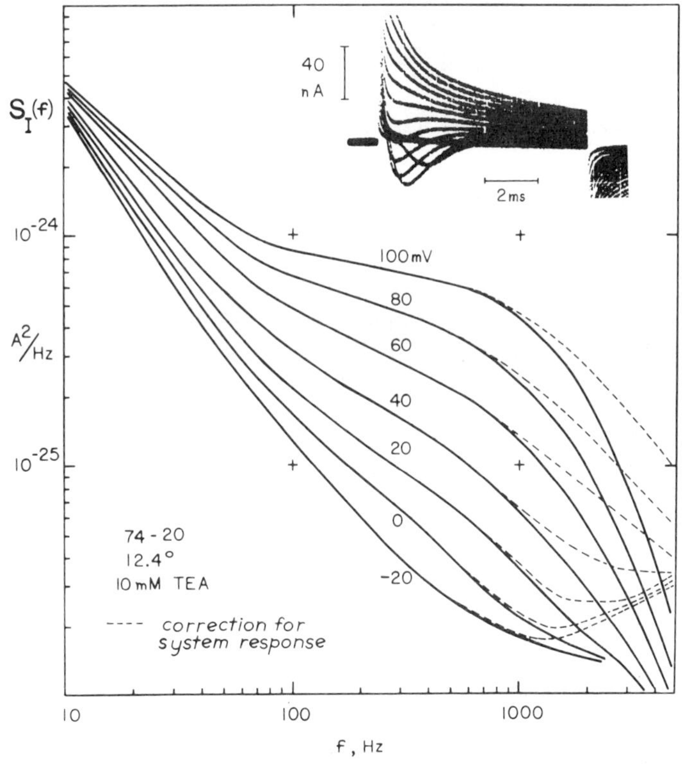

FIGURE 6. Power spectra of current noise from a patch of axon internally perfused with 10 mM TEA added to perfusate. Note the appearance of a new hump (at higher frequencies than occurs without TEA, compare FIGURE 4) with increasing depolarization. (By permission of J. Membr. Biol.[18])

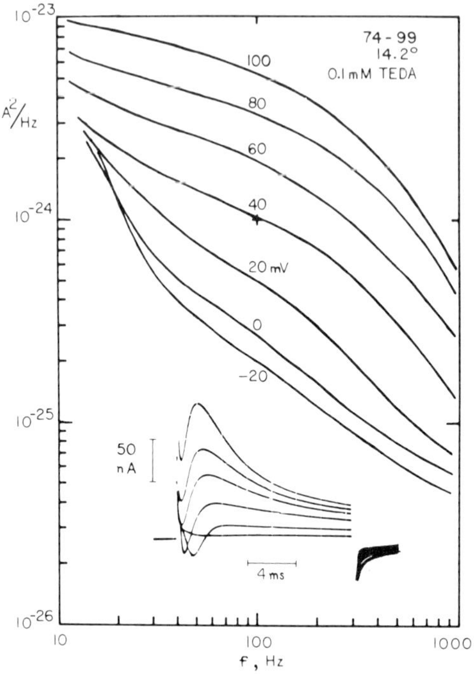

FIGURE 7. Power spectra of current noise from a patch of axon internally perfused with .1 M TEDA (Triethyldecylammonium) ion added to perfusate. Note spectral behavior with depolarization similar to TEA (FIGURE 6), but corner region is a decade lower, corresponding to slower kinetics of K current inactivation in step clamp currents (inset). (By permission of J. Membr. Biol.[18])

produced voltage clamp currents (insert FIGURE 7), which inactivated much more slowly than when TEA was used. The noise spectra (FIGURE 7) showed large-intensity noise, with depolarization, within the band of the instrumentation and indicated a relaxation component corresponding to the slower inactivation kinetics seen in the clamp current data. These results were reported previously.[18] In one further set of measurements, the presence of TEA inside the axon and TTX (1 μM) outside (see FIGURE 8) unequivocally demonstrates that the noise induced by TEA arises from the effects on *K* conduction[18] rather than on Na conduction.[22] The normal *K* conduction noise is quite large in relation to thermal noise, and TEA changes the character of this noise at low frequencies. Its effects at high frequencies in the presence of substantial internal [*K*] is to produce more noise than in the normal *K* condition (compare solid with dashed curve, FIGURE 8). Thus, TEA induces large-intensity noise in the frequency range where Na noise is likely to be observable (FIGURE 2) and thereby obscures rather than aids in measurement of Na noise.

This latter point is a critical confirmation of the earlier assertion that the use of ion-specific blocking agents to separate and study individual ionic processes, which has been effective on a macroscopic level, is not valid at the microscopic level, as assumed by some noise investigators,[21-24] without assurance that these agents do not induce or alter the noisy processes being studied. In fact, the noise spectra obtained by DeFelice *et al.*,[21] Conti *et al.*,[22] and Conti and Wanke[24] in TEA-injected axons agree with our axons, which were internally perfused with TEA. However, the interpretation of these investigators that the high-frequency component is due to Na conduction noise (under the assumption that TEA suppresses both *K* current *and* noise) is not substantiated by the data in FIGURE 8. It is also of interest that a report[19] of node noise with TEA applied externally indicated an increase in noise intensity in some experiments, but the increased noise was not attributed to an induced component.

A recent addition to the class of inducers of noise in *K* channels is 4-aminopyridine (4-AP). Reports[33,34] on the effect of 4-AP on *K* currents in step voltage clamp experiments have shown a complicated voltage, time, and frequency-dependent blockage. 4-AP was also shown[34] to be distinctly different in its action on *K* conduction from that of TEA in that 4-AP apparently binds to nonconducting channels and is released when channels conduct, whereas TEA binds only to conducting channels. FIGURE 9 shows that this ion also induces noise in *K* channels distinctly different from that induced by quaternary ammonium ions in that noise intensities are enhanced at low frequencies rather than at high frequencies. These ionic inducers of noise constitute microscopic probes of *K* channel conduction. Their distinguishing spectral properties provide additional information with which models of conduction may be compared. Our analysis of these induced noises will be the subject of a future publication.

NA CHANNEL NOISE

Because of the large, dominant *K* channel noise observed in the squid axon, it was apparent in early measurements[17] that noise from the Na conduction system would be difficult to measure under normal ionic conditions. Clearly, the most effective condition for observing Na conductance fluctuations is one in which *K*-related conductance fluctuations are suppressed. These points are illustrated in FIGURE 2, where the internal [*K*] was lowered to 50 mM, thereby suppressing both *K* current and channel noise, while the Na conduction system, as indicated by ex-

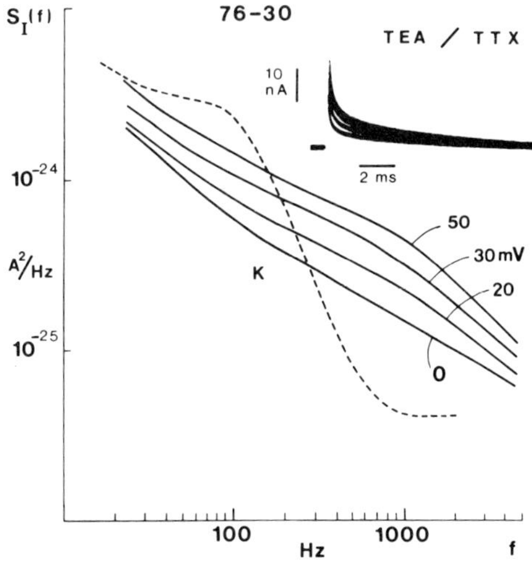

FIGURE 8. Power spectra of current noise from a patch of axon internally perfused with a perfusate containing 10 mM TEA and with TTX (1 μm) applied externally to the patch. Since TTX blocks Na conduction, the TEA-induced noise is due to its interaction with K conduction[18] rather than with Na conduction.[22] The dashed curve is a control spectrum at rest potential before TEA was added to the internal perfusate. Both the TEA-induced K channel noise and the residual noise are large enough to mask the apparent Na noise (FIGURE 2) under the condition of normal $[K_i]$.

cellent spikes and inward current under voltage clamp, was maintained. However, even under these conditions, there was no spectral feature that was obviously related to Na conduction. A spectrum obtained by taking the difference between spectra at depolarized and hyperpolarized potentials in 50 mM K_i (open circles, FIGURE 2) gave an apparent Na noise component at high frequencies that was an order or more lower in intensity than the normal K channel noise (compare filled and open circles in FIGURE 2). In six axons with TTX applied externally, the hump indicated in the open-circle curve of FIGURE 2 was not present.

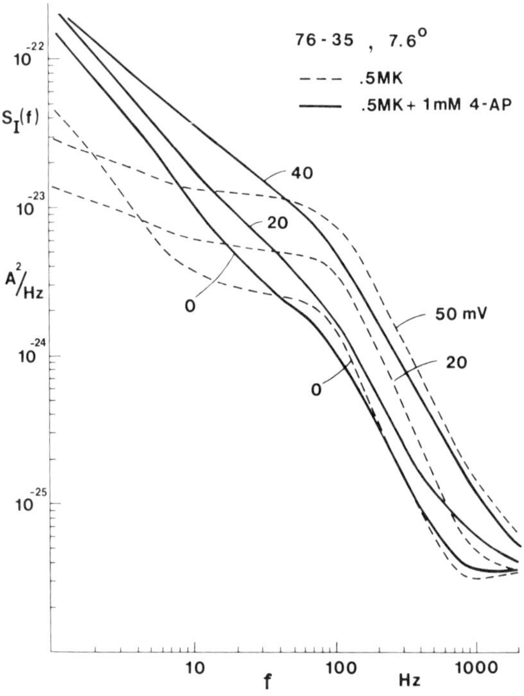

FIGURE 9. Power spectra of current noise in K channels in the presence of another inducer 4-aminopyridine (4-AP). Dashed curves, controls, and solid curves: spectra of noise from the same patch after addition of 4-AP to the perfusate. Note increased (induced) noise power at low frequencies in contrast to TEA (FIGURE 8) and TEDA (FIGURE 7), which induce a high-frequency component.

In contrast to the above observations, Na current noise was reported under normal ionic conditions using the blocking agents TTX and TEA.[21,22,24] These results have been criticized[18] with respect to the use of TEA to suppress K channel noise. FIGURE 8 in this presentation clearly indicates that measurements of Na noise that use TEA are obscured by the substantial noise that is induced in K channels in the frequency range where Na noise is expected to be evident. In addition, with normal $[K_i]$ both the residual noise at rest or hyperpolarized potentials and the TEA-induced noise completely mask Na noise (compare open circle curve of

FIGURE 2 with solid curves in FIGURE 8). The assumption of a Na noise component, as a significant portion of the spectral feature, measured in the presence of TEA, is therefore erroneous.

However, a TTX-sensitive spectral component was also reported[21,22,24] under normal conditions. There are several aspects that make the interpretation of this observation ambiguous. First there is the counterobservation that a TTX-sensitive spectral component is not measurable with normal $[K_i]$ in small-area measurements,[18] since it is completely masked by the K channel noise (FIGURE 2). Second, the assumed Na noise component occurs in the upper frequency region of the spectrum where large area measurements contain substantial amplifier noise due to the large effective membrane capacitance (see small- versus large-area section of this publication). As discussed earlier, an impedance change[2,28] (after TTX) in large-area measurements affects the S/N of instrumentation and possibly the thermal noise component in spectra. Corrections for both these effects could play an important role in identifying a noise component in the difference between two spectra obtained at different membrane impedances. Third, there is the question of constancy in the maintenance of uniform potential control over a large length of axon before and after application of TTX. Fourth, the component measured in difference spectra, before and after TTX, agrees with the component identified using TEA,[21,22,24] which is not a Na noise component. Finally, the TTX-sensitive difference component was adequately described by a single Lorentzian attributed to the m or activation process. This result is surprising, in view of the fact that the sodium system is known to be at least second order.

In contrast to all the above discussion on Na noise in the squid axon, an unambiguous TTX-sensitive sodium conductance noise has been observed[29] that is consistent with the criteria for identifying ion channel noise. After an extensive search for conditions in which all K-related conductance fluctuations as well as active transport are suppressed, an internal perfusion of either CsF or TMAF (tetramethyl ammonium fluoride) was found to be most effective. FIGURE 10 shows two sets of spectra in a Cs-perfused axon. The spectra in FIGURE 10 (a) at different potentials were measured from an axon with inward sodium transient currents that did not inactivate completely, leaving a maintained steady-state inward current (see inset of FIGURE 11). Generally, for these spectra, the intensity of the current noise increased to a maximum for a depolarization to about 0 mV from a holding level of -90 mV. Further depolarization led to a decrease in the intensity of the spectra. Thus the noise intensity parallels the behavior with potential of the transient and steady-state inward current. The spectra in FIGURE 10 (b) are from another patch in the same axon treated with TTX (1 μM) in the external solution to block the sodium inward currents. These spectra, in contrast with those in FIGURE 10 (a), decrease in intensity continuously with depolarization. Thus criteria 1 and 2 for identification of Na conductance noise are satisfied.

Because the Na noise is substantially above residual noise in only a limited frequency range at specific potentials, spectral details are more easily analyzed from difference spectra between a residual spectrum at hyperpolarized potential and each spectrum at a depolarized potential. Difference spectra are shown in FIGURE 11 (dashed curves) for an axon internally perfused with TMA. The difference spectra were calculated without any impedance correction, since the patch is shunted by a constant resistance (1 MΩ) at all membrane potentials. Thus,

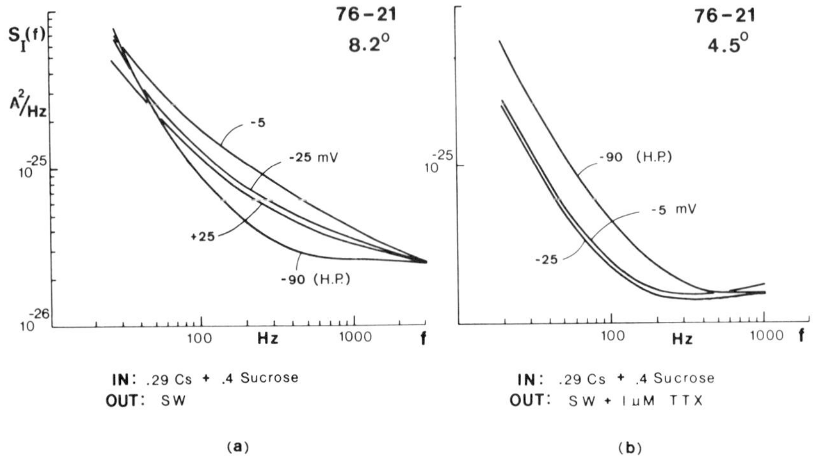

FIGURE 10. Identification of Na conduction noise in power spectra of noise from a patch of axon internally perfused with Cs (no K). (a): Residual spectrum at −90 mV with an increase in noise upon depolarization to −25 mV, which reaches a maximum at −5 mV and then diminishes upon further depolarization to +25 mV; thus the relationship of noise intensity to potential parallels Na current. (b): Another patch in the same axon with TTX applied externally (blocked Na conduction). Spectral behavior in (a) disappears with application of TTX. Potentials in this and subsequent figures are absolute.

despite patch membrane resistance changes of tenfold in FIGURE 13(b) (50–5 MΩ), the total patch impedance remains relatively constant at all membrane potentials. This conclusion is supported by a previous report of patch parameters[12] and by the data in FIGURES 10 and 11, where the spectra at different potentials merge beyond 2 kHz (at which frequency the thermal and instrumentation noise sources dominate). In addition, it has been assumed, in taking a difference spectrum, that changes in f^{-1} type noise intensity or form are not significant. At present, there is no effective way of evaluating the validity of this assumption.

The difference spectra in FIGURE 11 show two general features: (1) the spectra have broad corners which are not accurately fit by a single Lorentzian function, and (2) the corner appears to shift to a lower frequency on depolarization. These results contrast with the previous report of a single Lorentzian associated with the *m* process in squid axon[22,24] and a single Lorentzian associated with the h process in the node.[35] The data of FIGURE 11 are consistent with the hypothesis that the sodium-conductance spectra contain both *m* and *h* components at the initial depolarization and mainly the *h* component at higher depolarizations. Contribution from both *m* and *h* processes and interactions are expected from the power spectrum of Na conductance fluctuations expressed in the Hodgkin-Huxley formulation. Hence, the corner frequency appears to shift to low frequency in the overall spectra with depolarization. This behavior is probably a result of the incomplete inactivation of the sodium conductance seen in the inset of FIGURE 11. Thus it appears that criterion 3 is satisfied. Finally, it is unlikely that the observed Na channel noise is an induced noise, since both Cs and TMA give essentially the same results.

It also appears that the entire replacement of K_i by an impermeant cation achieves simultaneously two desirable conditions for observing Na noise: 1) It assures suppression of both K current and channel noise which otherwise would mask Na noise; 2) it produces a steady-state inward Na current which apparently does not occur or is negligible under normal ionic conditions. The basis for 2) is the relative magnitude of the steady-state inward current (see inset of FIGURE 11), which is about 25% of the peak inward current. This TTX-sensitive current is so large that, if it were present under normal conditions, it would be noticeable as a reduction in steady-state current after applying TTX to an axon. Since TTX is known not to produce any effect on the normal steady state clamp currents, the large steady-state inward current after replacement of K_i must be a direct result of the replacement. The substitution of an impermeant cation for K_i thus produces more Na noise than is measurable under normal conditions (compare open circle curve FIGURE 2 with FIGURE 11).

ION CHANNELS: CONDUCTING STATES AND KINETICS

Two important physical descriptors of conduction upon which models depend and which are most pertinent for noise data are the number of conduction states which an ion channel has and the channel kinetics that characterize the transitions

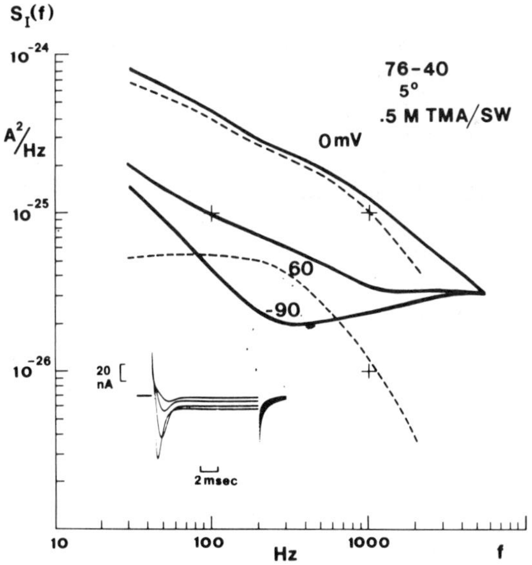

FIGURE 11. Na conduction noise in a patch of axon internally perfused with tetramethylammonium (TMA). Dashed curves are difference spectra obtained by subtraction of residual spectrum at −90 mV from spectrum at 0 and 30 mV. Inset, step clamp currents in the patch for potentials 60–90 mV in increments of 10 mV from a holding potential of −90 mV. Note that both inward peak and steady-state (Na) current occur at the same potential (0 mV) as the maximum intensity noise spectrum.

between all states, whether they are conducting or not. Aside from the obvious condition that the number of states in an applicable model of kinetics must be $\geq$ the number of conduction states, the two properties are, in general, not necessarily related. That is, a five-state kinetic scheme may have anywhere from two to five conduction states. In initial theoretical works,[36,37] computations of power spectra based upon the assumption of two conduction-state *K* channels and the HH formulation suggested[37] a means of testing the validity of two conduction state models that relied on the determination of kinetic parameters from noise spectra. The HH power spectrum consists of a sum of several relaxation terms, giving an overall spectrum that behaves asymptotically at low and high frequencies as a single relaxation process. The equivalent (single-Lorentzian) corner frequency-dependence of the noise data with potential and temperature change gave[14,18] $f_c = [2\pi\,\tau_n(v)]^{-1}$ (FIGURE 5), which was considerably lower than the equivalent f_c calculated from the HH power spectrum.

Thus it was suggested[14] that an HH *K* channel model, which consists of several two-state subunits and a channel with multiple states, was unlikely to have only two (open-closed) conduction states. An alternative description within the HH framework is the Hill and Chen[10] "κ" model, which also assumes several two-state subunits but a channel with *multiple* conduction states. The "κ" model fits the f_c data as well as the function $[2\pi\,\tau_n(v)]^{-1}$.

An uncertainty in the above method of evaluating two-conduction state models is that it relies on a comparison of corner frequency from noise spectral data with those calculated from power spectra that contain model-dependent macroscopic parameters, derived from data obtained under different measurement conditions (large signal step clamps). There is no reason to expect that kinetic parameters determined by two different types of measurements, on a highly nonlinear system, are the same. A much better method is to test conduction-state models on the basis of the variance of the total fluctuations and mean parameters, such as current and driving force, which are independent of kinetics.[37,38]

Another approach[16,22] is the assumption of a two-conduction state (open-closed) ion channel model, and the combined use of noise spectral parameters together with macroscopic parameters to calculate an open single-channel conductance. The ambiguity of this approach is exemplified by the fact that a misinterpretation of TEA-induced *K* channel noise for Na channel noise led to fortuitous estimates[22,24] of open channel conductance that happen to agree with values obtained by other methods.

Methods are now available for evaluating the validity of two-conduction state models which do not require determination of kinetic parameters.[37,38] An example of one method is presented in a later discussion of Na channel noise-conducting states. With respect to the kinetics of *K* channel conduction, it was suggested earlier that power spectra are not described by a single Lorentzian function.[18] In particular, the form of spectra in the corner frequency region are sharper than produced by a single relaxation process, although the low- and high-frequency behavior appears to be Lorentzian. Spectra in FIGURE 1 for depolarized potentials have sharper than Lorentzian corner regions, and FIGURE 12 shows a comparison of difference spectra, taken from the data in FIGURE 4, with a single Lorentzian fit to the low- and high-frequency asymptotes. Spectral peaking is a manifestation of a higher than first-order kinetic process. Because this result is a clear contradiction of the HH kinetics for *K* conduction and has profound implications on channel kinetics,[42] an alternative means of clarifying this issue was sought. The spectral data, together with simple assumptions about noise sources, were used to calculate the complex impedance of the squid axon. The calculations, using patch noise spectra, predicted a low-frequency feature (1–30 Hz), which was corroborated in impedance measurements,[12,28] using large axon areas and an internal axial current electrode.

The impedance data strongly support the patch noise measurements and interpretation of kinetics.[28] However, they are not sufficient to prove the point, since a low-frequency impedance feature does not necessarily yield peaking in a noise

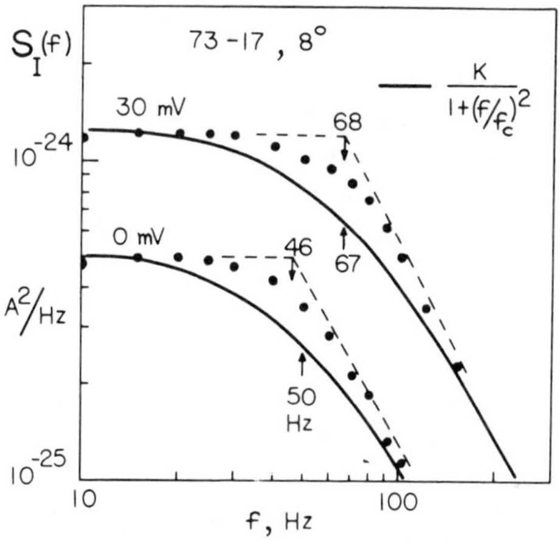

FIGURE 12. Difference spectra taken from the data in FIGURE 4 (filled circles) of K channel noise in the corner region compared with a single Lorentzian fitted to the low- and high-frequency asymptotes of each difference spectrum. This comparison as well as others that do not use TEA for difference spectra shows that noise spectra in the corner region are significantly sharper than a Lorentzian, which is indicative of a higher than first-order rate process.

power spectrum. A crucial matter in this respect is the question whether the phenomenon producing the peaking noise spectrum and impedance feature is in the axon membrane or arises from an extra membrane phenomenon such as *K* accumulation in the Schwann cell space. An extensive investigation[39] of accumulation models has not produced an adequate description of the impedance data, and thus the phenomenon presently appears to have a membrane origin. In any event, both the noise and impedance data indicate that overall *K* conduction kinetics in a squid axon are not adequately described by a single first-order rate process.[28]

With respect to kinetic schemes for *K* conduction in the squid axon, it appears that equilibrium rate processes of the HH type are inconsistent with our data. In particular, in the HH formulation a microscopic interpretation[10,37] of the *K* channel gating is that the n^4 gate consists of several subunits (the number is the power of n, which is 4 in the HH scheme). Each subunit exists in one of two possible states, as in (**1**).

$$\underset{[1-n]}{A} \underset{\beta}{\overset{\alpha}{\rightleftharpoons}} \underset{[n]}{B} \tag{1}$$

Thus, the *subunit kinetics* are described by four parallel independent "unimolecular reactions." A K channel can have five possible states such that the *channel kinetics* are described by the four consecutive reactions

$$[0] \underset{\beta}{\overset{4\alpha}{\rightleftharpoons}} [1] \underset{2\beta}{\overset{3\alpha}{\rightleftharpoons}} [2] \underset{3\beta}{\overset{2\alpha}{\rightleftharpoons}} [3] \underset{4\beta}{\overset{\alpha}{\rightleftharpoons}} [4] \tag{2}$$

with an arithmetic progression of rate constants, where the numbers in brackets correspond to the number of subunits in state B in (**1**). There are two types of possibilities with respect to channel conduction. First, transitions between state [0] and [3] are silent (nonconducting channel states), and the last transition from [3] to [4] yields an open channel in state [4]. Second, there is channel conduction in all states except [0]. The latter case corresponds to a multistate channel conductance within the HH framework. Although the second possibility is consistent with data suggesting more than two conduction states, both possibilities are inconsistent with peaking in noise power spectra.

In fact, if the peaking in noise spectra arises from K channel conduction fluctuations, this observation leads to very profound implications about the kinetics of K conduction. A peaking conduction noise spectrum excludes *all equilibrium reactions* (both linear and nonlinear, regardless of order of reaction) for which microscopic reversibility holds (i.e., most ordinary reactions). Thus Chen[40-42] has suggested a cyclic steady-state (nonequilibrium) scheme similar to an enzyme-substrate reaction to account for spectral peaking. This rate scheme does not obey microscopic reversibility, and produces noise spectra that conform to a single relaxation process at low and high frequency but can have peaking in the corner region. Alternatively, the kinetics may not reflect a chemically reactive scheme, but rather the movement of a charged group of a membrane-bound molecular constituent. The motion of such a harmonically bound particle yields a peaking power spectrum.[43] At this point in time, more experimental tests of the plausibility of such models are needed.

From data such as those presented in FIGURE 11 it is possible to comment on the number of conduction states and the kinetics of Na channels. To test for the applicability of two conduction state models a method[38,41] that does not depend on kinetic parameters (and therefore is independent of the type of measurement and formulation used to extract parameters for the calculation) was used to calculate an open-channel conductance, γ_{Na} over a wide potential range. In order that a two-state conductance be applicable, $\gamma_{Na}(V)$ must be a single-valued function.

Assuming that all "extraneous" components remain constant with potential change, the area under a difference spectrum in FIGURE 11 is the square of the variance of the sodium current fluctuation, $\Delta\sigma_I^2$. FIGURE 13(a) shows the de-

(a)

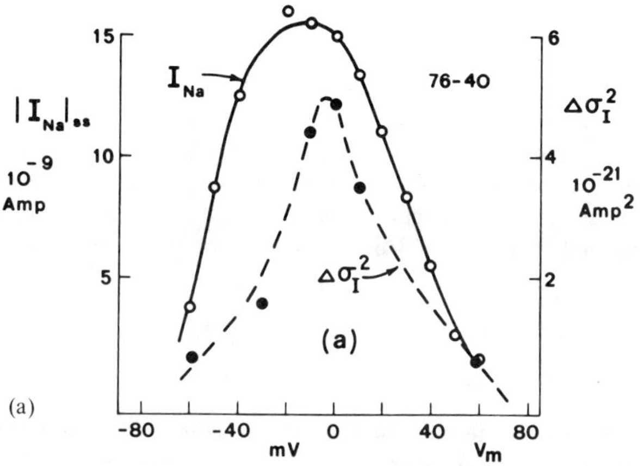

(b)

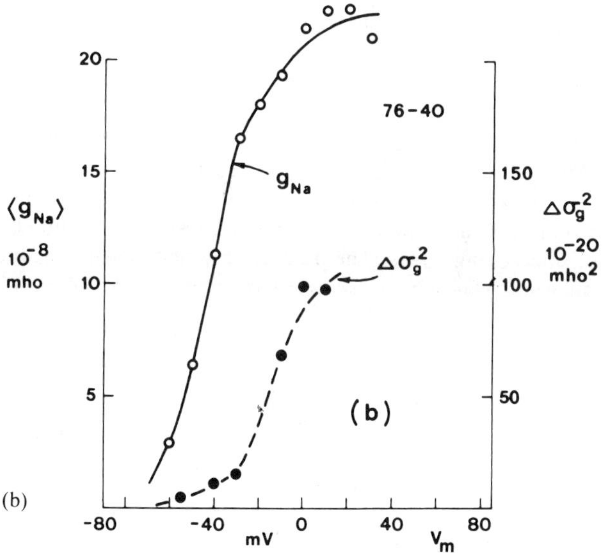

FIGURE 13. (a): Absolute value of Na current as a function of potential, $|I_{Na}|_{ss}$ measured in the steady state, and the difference in the variance of current noise $\Delta\sigma_I^{\ 2}$ at various depolarized potentials and that at hyperpolarized potential (−90 mV). (b): Mean Na conductance $<g_{Na}>$ and variance of conductance fluctuations $\Delta\sigma_g^{\ 2}$ with potential calculated from the data in (a). These data are for the same patch of axon as in FIGURE 11.

pendence of $\Delta\sigma_I^2$ and the leakage corrected steady-state inward current, $|I_{Na}|_{ss}$ (inset, clamp currents, FIGURE 11), on membrane potential. The $\Delta\sigma_I^2$ points were determined directly from the noise data (free of 60 Hz "pickup" and related components) by integrating the square of the current fluctuation for 15 seconds over a bandpass of 60 Hz–5 kHz (cascaded Butterworth low- and high-pass filters, 24 dB/octave) of depolarized and hyperpolarized patches to obtain σ_I^2 at each potential. $\Delta\sigma_I^2$ was then calculated by subtracting the residual σ_I^2 determined at hyperpolarized potential from that obtained at each depolarized potential. The mean sodium conductance, $\langle g_{Na}\rangle$, of the patch and the variance of sodium conductance fluctuations, $\sigma_{g_{Na}}^2$, as functions of membrane potential are plotted in FIGURE 13(b). They were obtained from the following relations:

$$\langle g_{Na}\rangle = I_{Na\,ss}/(V - V_r)$$

$$\Delta\sigma_{g_{Na}}^2 = \Delta\sigma_I^2/(V - V_r)^2$$

where V_r is the reversal potential for the leakage-corrected steady-state inward current. A simple two-state model predicts a maximum in $\Delta\sigma_{g_{Na}}^2$ when $\langle g_{Na}\rangle$ is half maximum. Thus the data of FIGURE 13(b) suggest that channel-conductance behavior is *not* two state. Furthermore, the test of two-state models outlined above is shown in FIGURE 14, where

$$\gamma_{Na} = \Delta\sigma_{g_{Na}}^2/\langle g_{Na}\rangle(1 - a)$$

$$\gamma'_{Na} = \Delta\sigma_{g_{Na}}^2/\langle g_{Na}\rangle$$

and a is the ratio, $g_{Na}/\bar{g}_{Na}$, of the sodium conductance to its maximum extrapolated value. The parameter γ_{Na} is the unit channel conductance expected from a generalized two-state model. These data clearly suggest that sodium channels have more than two conducting states. The parameter γ' is the unit conductance of a two state model when relatively few channels are open. For potentials close to rest,

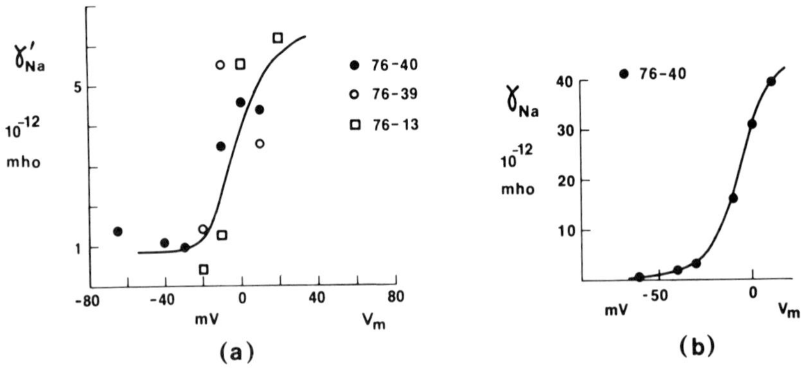

FIGURE 14. The Na channel conductance versus potential (see text for definition of γ'_{Na} and γ_{Na}). A generalized two-conduction state model should yield a single value of conductance with potential.

both γ_{Na} and γ'_{Na} appear relatively constant and less than 5 pmho. These data illustrate two points: (1) the importance of exploring a reasonable potential range before making a conclusion about the applicability of two state models, and (2) the assumption of a two-state conductance in cases where it is actually multistate can lead to estimates of maximum channel conductance that are in error by as much as an order of magnitude. The only other test of two state models[46] using the above method on node *K* channel noise data suggests two-state behavior over a limited potential range. However, Chen[38] later reevaluated these data and came to the opposite conclusion.

Finally, with respect to the kinetics of Na channels, difference spectra with potential change such as those in FIGURE 11 indicate the presence of complicated components in agreement with calculations of HH power spectra. Further comment on the kinetics and mechanism of Na channel conduction awaits detailed processing of spectral data and future experiments.

ACKNOWLEDGMENT

Dr. Ronald Grisell provided helpful assistance and discussion in various phases of this work. We thank Dr. Rita Guttman for reviewing the manuscript.

REFERENCES

1. HUSIMI, Y., H. NAKAMURA & A. WADA. 1977. Subsecond measurement of dielectric dispersion. This monograph.
2. POUSSART, D., L. E. MOORE & H. M. FISHMAN. 1977. Ion movements and kinetics in squid axon. I. Complex admittance. This monograph.
3. MARMARELIS, P. & K. NAKA. 1974. Identification of multi-input biological systems. IEEE Trans. Biomed. Eng. **21:** 88–101.
4. DERKSEN, H. E. & A. A. VERVEEN. 1966. Fluctuations of resting neural membrane potential. Science **151:** 1388–1389.
5. VERVEEN, A. A. & H. E. DERKSEN. 1968. Fluctuation phenomena in nerve membrane. Proc. IEEE **56:** 906–916.
6. POUSSART, D. J. M. 1969. Nerve membrane current noise: direct measurements under voltage clamp. Proc. Nat. Acad. Sci. USA **64:** 95–99.
7. POUSSART, D. J. M. 1971. Membrane current noise in lobster axon under voltage clamp. Biophys. J. **11:** 211–234.
8. KITTEL, C. 1958. Elementary Statistical Physics. John Wiley & Sons. New York, N.Y.
9. VAN VLIET, K. M. & J. R. FASSETT. 1965. Fluctuations due to electronic transitions and transport in solids. *In* Fluctuation Phenomena in Solids. R. E. Burgess, Ed.: **19:** 267–354. Academic Press. New York, N.Y.
10. HILL, T. L. & Y. CHEN. 1972. On the theory of ion transport across the nerve membrane. IV. Noise from the open-close kinetics of K^+ channels. Biophys. J. **12:** 948–959.
11. FISHMAN, H. M. 1973. Patch voltage clamp of excitable cell membranes without internal electrodes. J. Gen. Physiol. **61:** 267 (abstr.).
12. FISHMAN, H. M. 1975. Patch voltage clamp of squid axon membrane. J. Membr. Biol. **24:** 265–277.
13. FISHMAN, H. M. 1972. Excess noise from small patches of squid axon membrane. Biophys. J. **12:** 119 (abstr.).
14. FISHMAN, H. M. 1973. Relaxation spectra of potassium channel noise from squid axon membrane. Proc. Nat. Acad. Sci. USA **70:** 876–879.
15. SIEBENGA, E. & A. A. VERVEEN. 1971. The dependence of the 1/f noise intensity of

the node of Ranvier on membrane potential. Proc. First Europ. Biophys. Congress **5:** 219–223. Verlag Wiener Midizinischen Akademia. Vienna, Austria.

16. SIEBENGA, E., A. MEYER & A. A. VERVEEN. 1973. Membrane shot-noise in electrically depolarized nodes of Ranvier. Pflügers Arch. **341:** 87–96.
17. FISHMAN, H. M. 1975. Noise measurements in axon membranes. Fed. Proc. **34:** 1330–1337.
18. FISHMAN, H. M., L. E. MOORE & D. J. M. POUSSART. 1975. Potassium-ion conduction noise in squid axon membrane. J. Memb. Biol. **24:** 305–328.
19. SIEBENGA, E., J. DE GOEDE & A. A. VERVEEN. 1974. The influence of TTX, DNP and TEA on membrane flicker noise and shot effect noise of the frog node of Ranvier. Pflügers Arch. **351:** 25–34.
20. WANKE, E., L. J. DEFELICE & F. CONTI. 1974. Voltage noise and current noise in space clamped squid giant axon. Pflügers Arch. **347:** 63–74.
21. DEFELICE, L. J., E. WANKE & F. CONTI. 1975. Potassium and sodium current noise from squid axon membranes. Fed. Proc. **34:** 1338–1342.
22. CONTI, F., L. J. DEFELICE & E. WANKE. 1975. Potassium and sodium ion current noise in the membrane of the squid giant axon. J. Physiol. **248:** 45–82.
23. VERVEEN, A. A. & L. J. DEFELICE. 1974. Membrane noise. Prog. Biophys. Molec. Biol. **28:** 189–265.
24. CONTI, F. & E. WANKE. 1975. Channel noise in nerve membranes and lipid bilayers. Quart. Rev. Biophys. **8:** 451–506.
25. FISHMAN, H. M., D. J. M. POUSSART & L. E. MOORE. 1975. Noise measurements in squid axon membrane. J. Membr. Biol. **24:** 281–304.
26. POUSSART, D. J. M. 1973. Low-level average power measurements: noise figure improvements through parallel or series connection of noisy amplifiers. Rev. Sci. Instrum. **44:** 1049–1052.
27. COLE, K. S. & J. W. MOORE. 1960. Ionic current measurements in the squid giant axon membrane. J. Gen. Physiol. **44:** 123–167.
28. FISHMAN, H. M., D. J. M. POUSSART, L. E. MOORE & E. SIEBENGA. 1977. K^+ conduction description from the low frequency impedance and admittance of squid axon. J. Membr. Biol. **32:** 255–290.
29. FISHMAN, H. M., L. E. MOORE & D. J. M. POUSSART. 1976. Identification of Na conduction noise in squid axon. Biol. Bull. **151:** 408.
30. DORSET, D. L. & H. M. FISHMAN. 1975. Excess electrical noise during current flow through porous membranes separating ionic solutions. J. Membr. Biol. **21:** 291–309.
31. CANNESSA-FISHER, M., F. ZAMBRANO & E. ROJAS. 1968. The loss and recovery of the sodium pump in perfused giant axons. J. Gen. Physiol. **51:** 162S–171S.
32. BRINLEY, F. J. & L. J. MULLINS. 1967. Sodium extrusion by internally dialyzed squid axons. J. Gen. Physiol. **50:** 2303–2331.
33. MEVES, H. & Y. PICHON. 1975. Effects of 4-aminopyridine on the potassium current in internally perfused giant axons of the squid. J. Physiol. **251:** 60P–62P.
34. YEH, J. Z., G. S. OXFORD, C. H. WU & T. NARAHASHI. 1976. Dynamics of aminopyridine block of potassium channels in squid axon membrane. J. Gen. Physiol. **68:** 519–535.
35. VAN DER BERG, R. J., J. DE GOEDE & A. A. VERVEEN. 1975. Conductance fluctuations in Ranvier nodes. Pflügers Arch. **360:** 17–23.
36. STEVENS, C. F. 1972. Inferences about membrane properties from electrical noise measurements. Biophys. J. **12:** 1028–1047.
37. CHEN, Y. & T. L. HILL. 1973. Fluctuations and noise in kinetic systems. Application to K^+ channels in the squid axon. Biophys. J. **13:** 1276–1295.
38. CHEN, Y. 1976. Differentiation of channel models by noise analysis. Biophys. J. **16:** 965–971.
39. GRISELL, R. & H. M. FISHMAN. 1977. Analysis of models applicable to the low frequency admittance behavior of squid axon. In preparation.
40. CHEN, Y. 1975. Fluctuations and noise in genetic systems. III. Cycling steady-state models. J. Theor. Biol. **55:** 229–243.
41. CHEN, Y. 1977. Noise analysis of kinetic systems and its applications to membrane channels. Adv. in Chem. Phys. In press.

42. CHEN, Y. 1977. Noise analysis and channel properties of membranes. This monograph.
43. PAPOULIS, A. 1965. Probability, Random Variables and Stochastic Processes. McGraw Hill. New York, N.Y.
44. TAKASHIMA, S. 1976. Membrane capacity of squid axons during hyper and depolarizations. J. Membr. Biol. **27:** 21–39.
45. TAKASHIMA, S. 1977. Voltage-dependent membrane capacity of squid axon. This monograph.
46. BEGENISICH, T. & C. F. STEVENS. 1975. How many conductance states do potassium channels have? Biophys. J. **15:** 843–846.
47. SZABO, G. 1977. Electrical characteristics of direct and carrier mediated ion transport in lipid bilayer membranes. This monograph.
48. ZINGSHEIM, H. P. & E. NEHER. 1974. The equivalence of fluctuation analysis and chemical relaxation measurements: A kinetic study of ion pore formation in thin lipid membranes. Biophys. Chem. **2:** 197–207.
49. KOLB, H. A., P. LAUGER & E. BAMBERG. 1975. Correlation analysis of electrical noise in lipid bilayer membranes: kinetics of Gramicidin A channels. J. Membr. Biol. **20:** 133–154.
50. GREEN, M. E. 1976. Diffusion and 1/f noise. J. Membr. Biol. **28:** 181–186.
51. MIKULINSKY, M. A. & S. N. MIKULINSKY-FISHMAN. 1976. The theory of 1/f noise. Phys. Letters **58A:** 46–48.

DISCUSSION

Rita Guttman* and H. M. Fishman,† *Moderators*

**Department of Biology*
Brooklyn College
City University of New York
Brooklyn, New York 11210

†Department of Physiology and Biophysics
University of Texas Medical Branch
Galveston, Texas 77550

DR. ALAN FINKELSTEIN (*Albert Einstein Medical School, The Bronx, N.Y.*): I'd like to ask Dr. Fishman whether, on the basis of the noise analysis of the potassium system in normal sea water, you can come up with a conductance per potassium channel.

DR. FISHMAN: Yes. However, in light of our recent data and calculations on sodium conduction, which suggest that single-channel conductance varies with voltage, we want to go back and examine the validity of calculating a two-state conductance for potassium channels. We suggested in 1973 that potassium-channel conduction is also multistate, but this was based upon estimation of corner frequencies from spectra. We would like to compare the results obtained by methods that involve no kinetic parameters with those that rely on curve fits of spectra.

DR. FINKELSTEIN: It seems that single-channel conductances have been calculated at the neuromuscular junction on the basis of noise, they've been calculated for the potassium system for the node of Ranvier on the basis of noise, and just recently there's been a paper in the *Journal of Physiology* calculating the single-channel conductance for the sodium channel for the node of Ranvier on the basis of noise, and I just don't understand why it apparently has been done in those systems and yet there's so much difficulty doing it with the squid axon. Could you elaborate on this?

DR. FISHMAN: Yes. It appears that the assumption and calculation of a two-state channel conductance is valid at the neuromuscular junction, but this is certainly an unresolved issue with respect to axon membranes as indicated by our results on sodium conduction.

DR. FINKELSTEIN: Well that's okay, assuming you're dealing with a two-state system, what would you come up with?

DR. FISHMAN: If we assume that we're dealing with a two-state channel, a calculation from our published noise data gives a potassium-channel conductance on the order of 10^{-11} mhos.

DR. RONALD GRISELL (*University of Texas, Galveston, Texas*): Dr. Chen, you've prefaced your remarks with some very general statements about membrane modeling, and you have discussed linear, finite kinetic models. And you made use of an elementary result in differential equations; namely, that peaking would imply a nonequilibrium reaction. That statement is true only for linearizable, finite kinetic schemes. Moreover, this peaking theory is not structurally stable, in a sense. For instance, if second-order nonlinearities are admitted, I did some computer simulations with bimolecular reaction schemes that showed resonance due to second-order coupling between two-rate equations. That is to say, there was no resonance due to the linear terms that were in equilibrium, but the kinetic constants of the bimolecular interactions were opposite in two directions, and one got positive feedback and hence peaking due to nonlinearities alone.

Now, it's hard to do with a computer, but I think I can almost prove that one can get peaking that will override nonresonant linear terms. Your theory is interesting, but it does imply some difficult experimental aspects in excluding simple nonlinearities as sole causes of peaking. Can you make any generalization of linear to nonlinear? Suppose the linearization is in equilibrium, or linear terms are vanishingly small, and that the system is represented by coupled first-order differential equations with polynomial right-hand sides. Can you put conditions quadratic or higher-order coupling coefficients that will say the system is in nonequilibrium if there is peaking in the fluctuation spectrum? The theorems about diagonalization of matrices usually do not generalize to higher-order forms.

DR. CHEN: In any kind of nonlinear system you've got to have an energy source, or a cycling reaction. The linearization occurs in order to apply the microscopic fluctuation theory of chemical reactions.

DR. GRISELL: For equilibrium systems, you can make use of the Van Vliet and Fasset and, I think, Van Kampen results regarding nonlinear system noise spectra where some terms are due to nonlinearity.

DR. CHEN: As I showed in the noise-power spectrum, the $G(\omega)$, for example, corresponds to a second-order differential equation. In the admittance function there are RC branches that depend on a property of the relaxation matrix. If the relaxation matrix cannot be diagonalized using a similarity transformation, then a complex eigenvalue will result automatically.

DR. GRISELL: Yes, and your theorem must contain the hypothesis of linearizability.

DR. CHEN: I suppose general fluctuation theory depends on that, yes.

DR. GRISELL: And you also would exclude the diffusive systems, which are not finite schemes. One cannot have a finite number of time constants in a diffusive transient, for example.

DR. CHEN: It's a steady state—I mean, regarding fluctuations.

DR. GRISELL: Nonequilibrium diffusion can be steady state; e.g., the single-file diffusion of Hodgkin and Keynes.

DR. CHEN: Yes.

DR. GRISELL: Dr. Verveen, what would happen to the peak in the spectrum with membrane potential? And I'm thinking of that ω^2 in the equation. I believe it was something like $-1/\tau_s^2 \pm \sqrt{1/\tau_n^2 - 1/\tau_s^2}$.

DR. VERVEEN: I'll refer that to my colleague, deGoede, who actually did the computation.

DR. DEGOEDE: The situation is a bit awkward in the sense that the formula is very sensitive to the numbers you plug in, and we used values obtained from the data of Adelman and FitzHugh, but the general appearance is roughly that the peaking becomes stronger upon depolarization, and the frequency of resonance moves to higher frequencies.

DR. VERVEEN: But please note that you may also have peaking as an artifact when you subtract from one of the noises a slope of one, while the actual slope is 0.9, for instance.

DR. MICHAEL GREEN (*City College, CUNY, New York, N.Y.*): Dr. Chen, actually one part of my question is a small subset of what Dr. Grisell just asked, but also Schick and, earlier, Heyden and Lukes had done calculations of pulse-sequence noise, essentially depending on three parameters which may or may not couple one to another, the three parameters being the interval between pulses, the height of a pulse, and the length of the pulse. I was wondering whether in your calculations there is any way of taking into account the type or the time dependence within a pulse? I assume that there is not. I assume also that you can get from your calculation an average interval between pulses or distribution of intervals between pulses, which I think is essentially what you get.

I'm not sure, however, whether this applies to the length of an individual pulse, and I was wondering if you could comment on that?

DR. CHEN: I am not quite sure what you mean by the pulse.

DR. GREEN: During the time that the channel is open, one should observe essentially the pulse of a single channel.

DR. CHEN: In the calculation, I assumed that the pulse is square. Then when I say it's a constant conductance because of the separation of time domain, the transport of ions through open channels is of microsecond kinetics and the transition from one state to another is of millisecond kinetics. So there is essentially an assumption being made that the equation is possible.

DR. GREEN: Have you tried calculating it directly in the time domain to see whether you get anything approximating bursts or pulses by the kinetic scheme that you used?

DR. CHEN: No.

DR. BARRY SOKOL (*Emory University, Atlanta, Ga.*): I've got two questions for Dr. Verveen and Dr. Fishman. Actually, one question, in two parts. The first is: how do noise measurements differ in the three different preparations, basically? Two from the squid axon; that is, the patch clamp and the space clamp-air gap method, and then the node? And second, what type of artifacts might be involved in each of them, and could these explain the discrepancies between Dr. Fishman's work and Dr. De Felice and Dr. Conti's work?

DR. VERVEEN: That's quite a difficult question, I must say, because we have been talking about this over and over in our lab, and to give a list of the possible problems with regard to artifacts I think I had better refer to, let's say, Bendat and Piersol. It's difficult to know, but I think it's important also to realize that different species may also have different properties.

We very seldom have a node from a sensory fiber, and we never really did look at the properties of sensory fibers, for instance, but they show a completely different behavior with regard to burst noise. So that even in the same animal, with the same kind of nerve fiber, we infer that there are really big differences in behavior. That's about all I can say at present. I don't know whether squids from Naples and Woods Hole are different or not. I suppose that there is a difference.

DR. FISHMAN: With respect to squid axon measurements, we're in agreement on potassium noise in the sense that we observe humped components in noise spectra, and Conti *et al.* do, also. We also see peaking, which they don't see, but that's another matter. They see the corner frequency change with a Q_{10} of about three; we're in agreement there. Now we have a disagreement in terms of the potential dependence of the corner frequency, but there's a problem there in that their data is over a very limited potential range, something like 15 to 20 millivolts about rest, and so it's very difficult to compare potential dependence.

The major area of disagreement is with respect to sodium-conduction noise. As I discussed in my presentation, by far, the dominant source of noise in the squid axon under normal ionic conditions is from potassium conduction. We searched for three years for a sodium-noise component under a wide range of conditions, with limited success. We therefore concluded that a positive identification of sodium noise in the presence of the overhwelming potassium noise was marginal. We believe this conclusion to be substantiated by our recent identification of sodium noise only under conditions in which potassium noise is suppressed and an enhanced steady-state sodium current occurs. Furthermore, we have shown that TEA cannot be used to suppress potassium noise, since it induces large noise. Thus we interpret the Conti *et al.* noise measurements in the presence of TEA and with normal internal potassium concentration to be an induced noise in K channels, and *not* sodium noise. I have also presented other arguments which make their interpretation of spectra before and after TTX ambiguous.

Now, with respect to the techniques, there are many subtle complexities that have not really been adequately discussed previously. For example, the large areas they are claiming to voltage clamp are two centimeter lengths of squid axon. Cole and Moore showed in 1960 that a wire inside an axon does not assure isopotential conditions over a two-centimeter length. In fact, they showed that one can get 5 to 10 millivolt fluctuations over half a centimeter of length during a voltage clamp with the best available low impedance axial wire electrodes.

So it seems to me that in large-area measurements there's a clear problem with respect to uniformity of potential control over such large areas and the kind of interpretations one can make under these conditions.

Second, the large areas yield a large capacitance, and this presents problems. The effective capacitance of the preparation includes both the dielectric term as well as frequency-dependent conductances, as Dr. Poussart described in an earlier talk. The frequency-dependent terms, which are caused by the time-varying behavior of the ionic conduction mechanisms, can be about 30% of the total capacitance. In large-area measurements the intrinsic voltage noise of the clamp instrumentation drives a current through that effective capacitance and the resulting extraneous noise dominates a current noise measurement at 200 Hz and above. Consequently, if TTX is applied and a spectrum is obtained before and after, the effective capacitance is going to change. So the signal-to-noise ratio in large areas will change after application of a drug, and if the spectra are to be manipulated, then the different signal-to-noise ratio, before and after drug application, must be taken into account. Otherwise, it is possible to get a spectral component due to artifactual changes in signal-to-noise ratio of the instrumentation.

With the patch method we have about 100 pF of capacitance, and this problem is not significant until well past a kHz. Therefore, we can manipulate spectra out to a kHz or more without corrections for instrumentation noise. Thare are also some other advantages of the patch. Although the shunt through the sucrose solution is a low resistance relative to the patch, it's an adequate isolation for noise measurements, as we have shown here and previously. But, because of the low shunt, a constant-source resistance is presented to the instrumentation, so regardless of changes in patch impedance, the signal-to-noise ratio, as well as thermal noise, remains nearly constant. Consequently, in taking differences between spectra obtained under different conditions, the patch method required insignificant corrections.

However, I don't want to imply that the patch method is ideal. The patch has some handicaps as well; namely, the sucrose solution causes a hyperpolarization of the patch potential, and the viability of a patch is typically about one half hour, which is considerably less than under axial wire electrode conditions.

DR. O'KONSKI: Dr. Verveen, I'm interested in the values of the parameters that you calculate in the equation, which I believe you said de Goede and Vonk derived for the sodium-channel noise. You mentioned the number for Q_m—that's the maximal charge, and I didn't hear values at all.

DR. DE GOEDE: The Q_{max}, as measured in the node of Ranvier, is stated in the literature as 130 times 10^{-15} coulombs per node. That's Q_m, the maximal displaceable charge. And the τ_m is approximately as given by the Hodgkin-Huxley Equations and is a fraction of a millisecond.

CLOSING REMARKS

Rita Guttman

Department of Biology
Brooklyn College
City University of New York
Brooklyn, New York 11210

This conference, international in scope, has ranged over an enormous gamut of subjects, from the electrical properties of water, to those of biological and other polymers, to those of living membranes.

We have delved with mathematical, with physical, and with chemical probes into the basic nature of nonliving materials and into the basic nature of these same materials, organized in such a way that life has somehow been breathed into them.

We have heard reports about transport of ions through these living membranes and through man-made models of such membranes: the artificial membranes. We have wondered out loud about linear and nonlinear properties of membranes and about channels for the passage of ions through membranes. To our delight and satisfaction, we have been regaled with noise about noise, and the noise we have been hearing about noise seems not completely random.

The investigations reported here have been, on the whole, breathtaking in their daring, yet thoughtful and cautious in their execution. In all, this has been a successful coming together of co-workers and comrades in a common struggle that is eminently worthwhile.

Author Index

Bontemps, J., 170

Chen, Y., 382
Cole, K. S., 1
Cole, R. H., 59, 121

DeGoede, J., 389

Eisenberg, M., 281
Eisenberg, R. S., 339, 342
Emonds-Alt, X., 170
Essex, C. G., 142

Fishman, H. M., 355, 399, 424
Fredericq, E., 170

Gibbs, J. H., 10
Grant, E. H., 88, 142
Grinvald, A., 217
Guttman, R., 424, 429

Houssier, C., 170
Husimi, Y., 90

Kleinberg, M. E., 281
Kronick, P., 295

Mandel, M., 57, 74
Marchal, E., 123
Mathias, R. T., 342
Meves, H., 242, 322
Mills, G. L., 142
Minakata, A., 107
Moore, L. E., 355, 399

Nakamura, H., 90

O'Konski, C. T., 156, 159
Oncley, J. L., 214
Oosawa, F., 38, 88
Orttung, W. H., 22

Poussart, D., 355, 399
Pritchard, A. E., 159

Rae, J. S., 342

Scheider, W., 47
Scheraga, H. A., 2
Schwan, H. P., 198
Schwarz, G., 156, 190
Shaper, J., 281
Sheppard, R. J., 142
Slack, J., 142
South, G. P., 142
Stockmayer, W. H., 20
Symonds, M. S., 142
Szabo, G., 266

Takashima, S., 306
Taylor, R. E., 292, 298

Van den Berg, R. J., 389
van Rijn, H., 389
Verveen, A. A., 380, 389
Vonk, M. W., 389

Wada, A., 90, 214
Waggoner, A. S., 217
White, S. H., 243

Yantorno, R., 306

Subject Index

Absorption, dielectric
in bilayer lipid membranes, 246, 257–260, 292
in polypeptide analysis, 123–124, 135
temperature factor in, 140
Absorption, of dyes, *see* Dye absorption
Acetyl choline, 156
Acridine orange, 176–178
Actinomycin-D, 176–178
Action potential, cell membrane capacitance changes and, 314–317, 340
Acyl chains
in bilayer lipid membrane (BLM) structure, 246–247, 251–256, 260–263, 293
double-bond locations, 256, 263, 264
methyl terminal groups of, 252, 253, 255, 262
Admittance, *see* Impedance
Aggregation (Aggregate structure), *see also* Conformational changes; Disaggregation; Helix-coil transformations
antiparallel type, 131–132
in cellulose derivatives, 135
chain-length dependence in, 139–140
of copolymers, 131
electric field effects in, 131
flexibility in, 139
interaction energy and, 129–130
light-scattering measurement of, 127
molecular weight and, 123, 126–129
permittivity and, in PBLG, 104–105
of polyalkylisocyanates, 131–132
polymerization and, 137–140
in polymers, methods of study, 123
of polypeptides, 124–131
temperature factor in, 129, 131, 139
Alamethicin
as amphipathic molecule, 290
in channel formation, 282–289
conductive effects of, 275–277
structure of, 282–283
Alfalfa mosaic virus, 81
Alkane molecules, double bonds in, 247–263
4-Aminopyridine (4-AP), 411
Amphipathic molecules, 290
Amphotericin B
as channel former, 281
conductive noise of, 276
Ampullae of Lorenzini, 206
Amylose tricarbanilate (AC), conformational changes in, 135–140
Anhydroglucose rings, linkage of, 136
Annulus, of BLM, 256–257
Anisotropy
in dipole polarization, 64
in DNA polarizability, 172–175
in nucleosome helical structure, 185
in refractive indexes, 64–65
Antibiotics, membrane permeability effects of, 291
Aplysia nerve cells, 325
Aqueous solutions (Water solubility)
cluster model of, 3
hydrophobic effect in, 6–7, 13
of linear, charged macromolecules, 74–75, 81–86
of lipoproteins, 142–153
of nonpolar substances, 11–12
techniques in investigation of, 5–7
theoretical models of, 3
Ascitic cells, membrane potential of, 217
Association, *see also* Aggregation; Polymerization
of counterions and charged molecules, 74, 80–82
van der Touw-Mandel model of, 81–85
Asymmetrical current(s), *see* Grating current(s)
Attractive forces, in liquid state theory, 21
Axon, *see* Squid giant axon (SGA)

Bacteria, cell membrane potential in, 217, 219
Barker-Henderson approach, in simple liquids theory, 21
Bilayer lipid membrane(s) (BLM)
absorbtion in, 246, 257–260, 292
acyl chains in, *see* Acyl chains
alkane molecules in, 247–263
annulus of, 256–257
aqueous diffusion in, 266, 292
black, 281–292
bulk environment compared with, 258–273
see Capacitance, bilayer lipid membrane
characterization of, 281
cholesterol effects on, 272–274
conductive noise in, 274–277
diffusion in, 293
dipolar surface potentials and, 272–274
dye-absorbance changes in, 221, 222, 225
electrical properties of, 277–278
equilibrium in, 258, 260
in ethylene glycol, 295–296
fractional dye-absorption in, 222, 223

Bilayer lipid membrane(s) *(Continued)*
geometric capacitance of, 246
in glycerol, 292, 297
holding potential effects, 268–270
impedance response of, 269–270, 275
ion translocation in, 266
large-signal nonlinear response in, 266
large-signal pulse response in, 271–272
microlenses in, 257
molecular rigidity in, 293
noise in, 274–277
in nonaqueous solvents, 293–297
optical polarization in, 229
optical signals from, 235
permeability of, 281–282
resistor-capacitor series model of, 269, 272, 280
semiconductor analogy in, 293
small-signal electrical response in, 266, 275
as solutions, quasi two-dimensional, 264
squid axon compared with, 272
temperature and capacitance in, 257–260
thickness of, 246–247, 255–256, 264, 294
thinning of, 255–256
voltage dependence in, 267
Binary-state conduction model
experimental facilities of, 394–395
gating charge characteristics of, 394
ion translocation in, 415–420
potassium ion conduction in, 424
sodium system relaxation in, 392–393
Binding
of acetyl choline, 156
birefringence and, 178–179
in chromatin, 186
of counterions, 81
in DNA conformation, 159–168
of ethidium and DNA, 159–168
external, with DNA chain, 175–178
intercalative mode of, 166–167
nonintercalative mode of, 166
in nucleic acid intercalations, 175–182
viscosity and, 182
Biological membranes, characterization of, 281; *see also* Cell membranes
Biological systems, electrical field effects in, 198–211
Biomolecules
anisotropy of polarization in, 64–65
charged, in aqueous solution, 74–76
see Deoxyribonucleic acid (DNA)
dielectric effects in transitions of, 192–193
dielectric properties of, 74–77
dielectric relaxation in, 47–55
dye interactions with, *see* Dye(s)
free energy in structure of, 139
Kerr effect measurement in, 64–65
see Lipoprotein(s)
permeability and potentiality change in, 42
permittivity in, 76, 149–153
polarization and segmentation of, 173–174, 198
see Polypeptides
see Proteins; Protein molecule(s)
pulse response measurement in, 47, 49
state field fluctuation in, 40–41
Birefringence, *see also* Dichroism
decay signal for, 160
in DNA conformation, 170–186
in DNA dielectrical measurements, 88–89
in dye binding of polynucleotides, 178–179
elliptical polarization and, 64
fast polarization and, 174
helix-coil transition measurement by, 195–196
of histine-depleted chromatin, 215
relaxation time, 161–168
time dependent, 59, 64–67
Black lipid membrane(s), 281–290
preparation of, 285
Black-widow spider venom, in channel formation, 282
BLM, *see* Bilayer lipid membrane
Boiling, characteristics of, 13
Bonding, hydrophobic effect as, 6–7, 21
Bound ions, *see also* Counterion(s); Ion binding; Polyion(s)
longitudinal diffusion of, 111, 119
polarization relaxation of, 111–113
BPN, *see* Pseudorandom binary signal (PBS)
Brownian motion
in cell membranes, 38, 42
dielectric effects and, 89

Cable theory, of frog muscle properties, 345
Capacitance, *see* Capacitance, bilayer lipid membrane; Capacitance, cell membrane
Capacitance, bilayer lipid membrane
admittance response for, 270, 271, 292
measurement of, 69–70
specific geometric, 246, 257–260, 292
Capacitance, cell membrane
action potential and changes of, 314–318, 340
characterized, 299, 300, 306

equations for, 307, 308
frequency dependence of, 200, 311–312, 317–318
see Hodgkin-Huxley model
intramembrane-charge movement and, 339, 340
measured versus actual, 308
measurement of, 203, 204, 307–311, 318
in membrane-noise measurement, 402
polarization in, 311–314, 318
relaxation frequency, 317–318, 320
series resistance and, 302, 308, 314, 316
sodium channel blockage and, 312–314, 318
of squid axon, *see* Squid giant axon
voltage dependence of, 306, 307, 312–314, 318
Cartenoid shifts, 222
Cascade model, of vapor condensation, 15–16
Caveola(ae), 345, 350
Cavities, dipoles in, 32–36
Cell(s), biological
electrical activity in, evolutionary perspective of, 342
electrical model of, 342–344
eye lens as giant, 351
membranes of, *see* Cell membranes
modification of, dielectric, 208–210
optical signals from, 235
threshold-above-noise in, 210
Cell membrane(s)
admittance in, *see* Impedance, cell membrane
see Capacitance, cell membrane
see Channels, cell membrane
conductance in, *see* Conduction, cell membrane
depolarization of, 308
dielectric breakdown of, 210
dielectric properties of, 203–205, 298–304, 343–344
dye-absorbance changes in, 222
dye fluorescence in, 217, 219
dye movement in, 230–232
electrochromism in, 222
excitation of, 203–207, 298–304, 306–320
field fluctuation in, 38–46
impedance in, 204, 298–304; *see also* Impedance, cell membrane
lipid-to-protein ratio in, 300
see Noise, cell membrane
noise measurement in, 399–403
permeability changes in, 42
phase angles in, 341
see Potential, cell membrane
properties common to, 298
resistor-capacitator activity of, 307, 344
resistivity of, 308
shape of, and impedance, 204
Cell suspension, dyes in optical studies of, 219–220
Cellulose tricarbanylate, conformational changes in, 135–140
Cerebral responses, electrical currents in, 205
Cesium (Ce) ion
in cell membrane admittance, 370–371
potassium current blockage by, 407
Chain length, polymer, 139
Channel(s), membrane conductive
alamethicin in formation of, 283–288
amphipathic molecules in, 290
antibiotic mediation of, 281
in bilayer membranes, 266, 276–277, 281–282
channel molecules and, 42
conductive change in single, 284, 286
cylindrical barrel model of, 287–289
formation of, 281–289
see Gating current(s)
homoside hybrid, 288–289
H-H model of conduction by, *see* Hodgkin-Huxley model
ionic gates and currents in, 393
membrane capacitance and blockage of, 312–313, 318
monomer aggregation model of, 284–285, 289
pulse stimulation of, 323, 426
Chemical field effects
dipole moment in, 192–193
dipole polarization in, 194–196
electrical birefringence due to, 195–196
electrical relaxation due to, 194
Chloroform, in polymer-conformational changes, 134–135
Chlorophyl, fractional-absorbance change in, 223
Chloroplasts
cartenoid shifts in, 222
membrane potential in, 217
Cholesterol, in membrane changes, 272–273, 295–297
Chromatin
anisotropy in, 182, 183, 186
dye binding in, 176, 186
nucleosome substructure of, 183–186
properties of, electro-optical measurement, 171–175, 182–186
Chromophores, 222–223, 229–231
Clausius-Mossotti formulation, 62
Coloumbic interaction, of counterions, 113–114

Concentration-dependent dispersion curve, 75–79
Conductance change
in bilayer membranes, 295–297
in channel formation, 282–289
in membrane permeability, 281, 282
Conduction, cell membrane
in H-H model, 301
measurement of, in SGA, 309
noise-excitation measurement of, 399
nonequilibrium-cyclical model for, 386–387
in potassium ion channel blockage, 312–314, 318
in sodium ion channel blockage, 318
time constant in, 316–319
values for, 204, 312, 316, 320
voltage dependence of, 298–299, 312–314
Conductivity
in counterion fluctuation, 113, 119
Fourier transform in, 115
of low-density lipotropin (LDL), 142–144
in local field-effects model, 157–158
of outside medium, 116–120
of water, 201, 202
Conformational changes, *see also* Aggregation; Disaggregation; Helix coil transformations
in amylose tricarbanilate, 135–140
birefringence measurement of, 195
in cellulose derivatives, 135–140, 157
chain length and temperature in, 139
in DNA, 159–168, 175–186
electrical field effects in, 192–196
electro-optical techniques for, 182
of polypeptides, 129–134
solvent conductivity changes in, 133–134
transition temperature in, 129–134, 136–140, 157
Continuous-state conduction model
corner frequencies in, 397
gating charge in, 394
of noise, 392
potassium relaxation in, 397
Copolymers, aggregation of, 131
Corey-Pauling-Koltum models, of decanes, 243–245, 248, 253
Corner frequencies, in potassium conduction noise, 397, 408
Counterions, *see also* Polyions
binding of, 81
characterization of, 107
and charged macromolecules, 74, 80–81, 82
coloumbic interaction of, 113–114
in DNA dielectric effects, 80–81, 173
equilibrium distribution of, 82
fluctuations of, in polyelectrolytes, 70–73, 108–119
interaction energy between, 108–110, 113–114, 157
McTeague-Gibbs formulation on, 70–72
polarizability of bound, 107–119
in polyelectrolyte dipole effects, 74, 80–81
relaxation of, in lipoprotein dispersion, 142, 146–147
temperature coefficient, in mobility of, 157
van der Touw-Mandel model of, 81–82, 173
Curie point, in gas-liquid transition, 14
Cut-off radius, in ion translocation, 286–287
Cyanine dyes
adsorption changes of, in membranes, 221, 224–225
in cell membrane potential, 217–219
extinction change in, 237
fractional fluorescence changes, 234
as hydrophobic ions, 267

Debye factor
in DNA solutions, 72
in free and tissue water, 201
Debye-Falkenhagen effect, 81
Debye-Hückel potential, 110
Deoxyribonucleic acid (DNA)
binding modes, 166
birefringent relaxation of, 161–164
circular, 160, 162, 164
conformational transformations of, 159, 186
counterion relation in, 88–89
covalently closed, 160
dielectric properties of, 75–80, 88–89
dipole polarization analysis, 70–73
dye intercalation in, 175, 178, 179
electro-optical anisotropy of, 183
electro-optical measurements of, 159–161, 170–186
fluorescence, measurement of, 160–161, 166
frequency range of, 75–79
high molecular weight, 75
ionization in, 170
length of, 215–216
longitudinal dipole moment of, 172
nicked, 161–166
optical anisotropy on, 173–174, 182
polarizability of, 173, 174
relaxation time of, 215–216
segmental anisotropy of, 173–174

segmental orientation in, 171, 175
superhelical, 159–168
Dibutylproflavine, 176
Dichloracetic acid, 132–134
Dichroism, *see also* Birefringence
in DNA conformations, 170–186
in nucleic acid complexes, 175–186
potential change and, 242
Dielectric approach, 36
Dielectric constant
in bilayer lipid membranes, 294
in bound-counterion fluctuation, 112–113, 119
in chain dipole moment, 79
and concentration, in lipoproteins, 146
in dilute polyelectrolytes, 82
low-temperature values, 63
of membranes, 203
of muscle tissue, 199
of polar fluids, 59
polarization formula for, fl
of polyelectrolytes, 76–79, 107, 113, 198
of polyion lattice, 108
of rod-like molecule(s), 111
static, 76–79, 102, 108, 111, 113–114
temperature and, 71
values predicted for, 62–63
Dielectric decrement
characterization of, 149
of human LDL, 151–152
of low-density lipoproteins (LDL), 142–153
of water of hydration, 150–153
Dielectric dispersion
alpha dispersion, 200
beta dispersion, 200, 204, *see also* Maxwell-Wagner dispersion
in cell membranes, 199–200, 204
frequency levels of, 183, 199–200
gamma dispersion, 200
in low-density lipoproteins, 142–146
molecular-weight effects of, 83
outer-medium effects on, 116–119
in polyelectrolyte solutions, 107–120
subsecond measurement of, 90–105
two-level display of, 173
in van der Touw-Mandel model, 82
of water, 201–202
Dielectric measurement(s)
of DNA, 70–71, 75–80, 84–89
four-terminal null method, 68–70
frequency-domain methods, 90–91
low and high frequencies compared, 73
multiplex advantage in, 90
of polypeptide aggregations, 123–131
of synthetic, linear polyelectrolytes, 74–79
three-terminal null method, 69–70
of tissue properties, 198–200
Dielectric relaxation, *see* Relaxation, dielectric
Dielectric spheres, 26–32
Dielectric susceptibility, 23
Difference current, 339; *see also* Gating current
4,4-Dimethylaminonitrostilbene, fractional absorbance change in, 213
Dioxane, as solvent, 125–140
Dipicrylamine, as hydrophobic ion, 267
Dipole interaction, 59–63
Dipole moment(s)
absorption and, 140
in aggregation, of polypeptides, 123–132
cell membrane capacitance and, 318–319
counterion distribution and, 82
in counterion fluctuation, 110–111
in disaggregation, of polypeptides, 132–135
freely rotatable, 194
in gating currents, 327
in helix-coil stability, 132–134
longitidinal permanent, in DNA, 172
in molecular transition, 190–193
in polyelectrolytes, 79–82
in polymer aggregation, 123–124
of polypeptides, 123–124
vacuum, of PCBL solution, 103, 104
Dipole wave sums, 61
Disaggregation, of polypeptides, 129, 132–134
Displacement current, 393
Dual-state conduction model, *see* Binary-state conduction model
DNA, *see* Deoxyribonucleic acid
Dye(s)
absorbance of, *see* Dye absorption
aggregation of, *see* Dye aggregation
in cell membrane potential, 217
dipolar molecule, 230, 231
extinction changes, *see* Electrochromic changes
fluorescence molecule in, 233–234; *see also* Dye fluorescence
linear electrochromic changes in, 225, 229, 236, 237
metachromastic, 242
molecule movements of, *see* Dye molecule movements
optical change mechanisms in, 219, 220, 229–230, 236
Dye absorption
in bilayer lipid membranes, 221, 222, 225
changes in, 219–222

Dye absorption *(Continued)*
characterization of changes in, 221, 224, 231–232, 237
extinction ratio and, 237
maximizing of, 232–233
signal-to-noise ratio and, 236
spectra of, 225–229
Dye aggregation
absorption changes due to, 226
monomer-dimer mechanism of, 226–229
Dye fluorescence
in cell membranes, 217, 219
changes of, 217–218, 233–237
measurement of, 235–236
molecular orientation and, 229*n*
Dye-molecule movement
absorption change and, 231–232
characterization of, 220
membrane-potential change in, 230
microenvironment and, 225, 231
on-off movement in, 225–226
rotational, 220, 231–232
transitional, 220, 231

Elasticity, 19, 21
Electrical field effects
aggregation and, 131
biological aspects of, 38–46, 198–211, 342–344
in cell membranes, 205–206
cell membrane damage in, 208
cell modification by, 208–210
chemical transitions and, 190–193, 196
chromophore dipole orientation in, 222
electrochromic absorption and, 220–225
in fish membranes, 206–207
in microorganism swimming, 43–46
Electrical models, of cells and tissues, 342–344
Electricity, *see* Electrical field effects
Electrochromic changes
characterization of, 220–224
in dye molecules, 236–237
dye orientation and, 229–230
dye polarity change and, 225
in nerve fibers, 224–225
Electron pair interaction potential (EPEN), 3–8
Electro-optical techniques, in DNA studies, 159–161, 170, 175, 182
Electrosensitivity, in cell membranes, 203–207
Electrostatics, 22–23
Energy distribution functions, 5
Enthalpy
in bilayer membrane systems, 256–264
in chemical reaction systems, 190
in flicker model bonding, 21
in macromolecule aggregation, 129
Entropy
in bilayer membrane systems, 256–264
hydrophobic effect and, 11, 12, 18
negative, in hydrophobic effect, 21
thermal expansion coefficient and, 11–12, 19
EPEN, *see* Electron pair interaction potential
Erythrocytes, 217
Escherichia coli, 210
Ethidium bromide, 175–179
Ethidium ion, in DNA conformations, 159–168
Ethylacetate, 135–138
Ethylcellulose, 135
Ethylene glycol, 295
Euglena, 210
Excitability-inducing material (EIM), in channel formation, 282
Extremely-low frequency (ELF), 207
Eye lense(s), 350–352
as giant spherical cell model, 351
Fellgett advantage, 90
Finite element method, in molecular electrostatic problems, 22–36
Fish, electrosensitivity in, 206, 207
Flickering cluster model, of water, 20
Flicker noise ($1/f$)
as membrane-noise component, 390, 403–404, 407
occurence of, 399–400
in potassium ion conduction, 390, 406
in sodium ion conduction, 390, 392
Flotation rate, 143
Flory-Rehner model, 17–18
Fluctuation-dissipation theorem, 107
Fluorescence, *see* Dye fluorescence
Four-terminal null method, 68–70, 90
in impedance measurement, 357
Free energy
of alkanes, in bilayer membranes, 252, 253, 256–264
chain-length structure and, 139
computation problems of, 4–5
in DNA conformation, 168
in dye movements, 231
Free ions, *see* Counterions; Polyions
Freezing process, is cascade-ring formations, 16
Frequency-sweep four electrode method,
in permittivity analysis, 90, 94

Frog nerve tissue, *see* Node of Ranvier
Frog skeletal muscle, *see* Muscle fiber(s)
Fucus eggs, 210

Gas-liquid systems, 14
Gating current(s)
  asymmetrical and symmetrical components of, 293
  binary-state conduction model of, 394
  BLM ion translocation model, 272
  cell-membrane capacitance and, 317–318, 339
  cell-membrane impedance and, 299, 301–302
  chemical blockage of, 333–335
  depolarization inactivation of, 331
  dipole moments in, 327
  measurement of, 323–324, 326
  movement of, 326–329
  records of, for various species, 325
  in sodium-ion conduction, 335–336, 390
  time course for, 328–329
  ultraviolet-light inactivation, 333
  zinc-ion blockage of, 334
Gel, characterized, 14
Gibbs free-energy variation, in superhelical DNA, 168
Glassy state, characteristics of, 11
Glutaraldehyde, 334, 340
Glycerol, 292, 297
Glycerol monooleate (GMO)
  in bilayer membrane thickness, 246–249, 255, 293
  double-bond position in, 256, 262
Glycophorin, in bilayer membranes, 243
Gramacidin
  channel formation by, 282
  in bilayer membranes, 275–277

Hadamard transform-nuclear magnetic resonance spectroscopy, 90
Hamiltonian principle, 65
Hayakawa technique, 70
Heart muscle cells
  fractional fluorescence changes in, 234
  membrane potential of, 217
Helix-coil transformations, *see also* Conformational changes
  alpha-helix formation, 156
  birefringence in measurement of, 195, 196
  chain-length factors in, 139
  dipole moment in, 132–134, 193
  of DNA, 159–168
  hydrophobic bonding in, 6
  of PBLG, 132–134
  of PCBL, 100–104
  of PDLPA, 134–135
  of polypeptides, 132–134
  of PRLG, 104–105
Helix-coil transition(s), *see* Helix-coil transformation
Helix promatia, grating current in, 336
Hellman-Feynman theorem, 36
Hemocyanin, channel formation by, 282
H-H, see Hodgkin-Huxley model
High-frequency dispersion (HFD)
  ion fluctuation model explanation of, 119–120
  Maxwell-Wagner effect and, 119
Histone, 182, 186, 215
Hodgkin-Huxley (H-H) model
  binary-state ion conduction in, 416–418
  cell membrane capacitance and, 306, 310
  cell membrane impedance and, 300–301
  current flow, total, formulation, 398
  equations from, 322–323
  flicking noise spectrum in, 399
  gating current in, 390
  membrane noise models in, 390–392
  PBRS experimental findings compared with, 365, 372–378
  peaking phenomena, in K ion conduction and, 386
  potassium ion conduction and, 386, 389, 417–418
  relaxation time in K ion conduction, 408, 416
  sodium ion conduction and, 322–323, 329, 389
Human serum albumin, 47, 52–55
Hydrocarbons
  aqueous solution anomalies of, 12–13
  insolubility of, 19
  thermodynamic properties, in aqueous solution, 6; *see also* Hydrophobic effect
Hydrogen bonding
  breaking of, 140
  strength of, 18
Hydrophobic effect, 6–7, 10, 13, 17, 21
Hydrophobic ion(s)
  adsorption of, 267
  in BLM transport mechanisms, 266–267
  carrier-mediated transport of, 266
  current noise of, in BLMs, 274–277
  membrane pulse response and, 268–272
Hydrophobic proteins, 243, 244

Ice, dielectric properties of, 201

Impedance (Admittance)
in bilayer lipid membranes, 269–270, 275
in cell membranes, *see* Impedance, cell membrane
definition of, 300
intermediate frequency, characterized, 302–303
Impedance, cell membrane
cell shape and, 204
coherence function, 365–366
Cole-Cole equation in, 300
Hodgkin-Huxley model and derivation of, 300–301, 372–378
linearity of, 365–367
magnitude spectrum of, 366–369
measurement of, 310, 314, 355–358, 364–371, 377–378, 413
nonlinearity of, 365–367
PRBS technique in measurement of, 357–358, 374–377
of Schwann cell layer, 308
in squid giant axon, 298–304
test conditions and methods for measurement of, 358–359, 363–364
Impulse response function, 91
Infinite dilution, characterization of, 214
Interaction energy
bound-ion relaxation and, 111–113
of counterions, 108–110
of macromolecule aggregation, 129–130
Interactive volume, of bilayer lipid membranes, 247
Intramembrane charge movement
chemical blockage of, 333–335
inactivation of, 329–330, 339
time dependence in, 323, 328–329
Ion binding, *see also* Binding; Ion fluctuation
dielectric-constant values and, 107
and fluctuation, 108–109
Ion fluctuation
in molecular-field couplings, 41–43
outer-medium effects and, 116–120
in polyelectrolyte dispersions, 107–120
Ionic gates, 393; *see also* Channels(s); Gating currents
Ion-specific blocking, noise measurement effects of, 399–400, 410–411
Ion translocation (Ion conduction; Ion transfer)
by antibiotics, 281
in bilayer lipid membranes, 266–278
binary-state models of, 415–420
of blood, and tissue water, 202
cation selectivity in, 282
cell membrane admittance in, for SGA, 355–378
cholesterol and, 272
cut-off radius in, 286–287
derivation of expressions for, 277
dipolar-sufrace potential and, 272–274
evolutionary position of, 342
gating-current model of, 327–328; *see also* Gating current
see Hodgkin-Huxley model
kinetics of, 415–421
"K" model of, 416
measurement of, 355–358
mechanisms of, 266, 267, 281, 389; *see also* Conduction, cell membrane
noise identification in, 403–421; *see also* Noise, cell membrane
states of, 415–416
Ising model, 14–15, 20

Kerr constant, 64, 172
Kerr effect, *see also* Rotational diffusion
Benoit solution, 65–66
correlation function in, 67
DNA segmentation and, 174
measurement of, 64–67
polar liquid response in, 59
Kirkwood formulation, dipolar interactions and, 59–60
Kronig-Kramer concept, 179, 199

Langevin approach, in membrane noise theory, 391, 392, 395–396
Lattice model
characteristics of, 61–62
perpendicular diffusion and, 119–120
in rodlike polyelectrolytes, 108–115
Lattice-gas model, in gas-liquid transition, 14–15
LDL, *see* Low-density lipoproteins
Leakage current
in cell membranes, 300, 301, 323
in H-H model formulation, 289
as noise source, 391
Light transmittance, change of, 218, 235, 236
Linear response, analysis of, 90–94
Lipoproteins
aqueous-solution characteristics, 142–153
isolation of, 143
low density, *see* Low-density lipoproteins (LDL)
three-layer molecular model of, 148, 158
Liquid state, *see also* Aqueous solutions; Water
attractive forces theory, 21

Barker-Henderson approach, 21
gel model for, 14–15
lattice-gas model for, 14, 15
repulsive-force concept and, 21
ring and tree structure of, 16
simple-perturbation approach, 21
theoretical model parameters for, 13
thermal expansion coefficient of, 10
viscosity of, 11
Loligo pealeii (squid), 307, 362; *see also* Squid giant axon (SGA)
Low-density lipoprotein(s) (LDL), *see also* Lipoproteins
alpha dispersion in, 146
beta dispersion in, 147–148
characterization of, 143
dielectric increment for, 146
dispersion regions distinguished, 142, 145–146
gamma dispersion in, 145, 149, 150
from hyperbetalipoprotein (FH) sera, 150–151
molecular structure of, 147, 148
relaxation frequency in, 146
three-layer molecular model of, 148–149, 152–153, 158
Lutein, fractional-absorbance change in, 223

Macromolecules, *see* Biomolecules; Polypeptides; Protein molecules
McTague-Gibbs model, of counterion fluctuation, 70–72, 88
Magg (unit), characterization of, 127
Magnesium ion
in birefringence relaxation, in DNA, 162–166
DNA binding of, 166
in optical anisotrophy, 182
Magnetic field, as gas-liquid transition analogy, 14, 20
Manganese ions, in DNA-optical anisotropy, 182
Maxwell-Wagner effect
in counterion dispersion, 88
in high-frequency ion dispersion, 119, 120
in lipoprotein dispersion, 142, 148
in muscle tissue, 199–200
Melting
helix-coil transition analogy of, 16
properties of, 13
ring and tree cascade concept of, 15
Membranes, *see* Bilayer lipid membranes (BLM); Cell membranes; Muscle fibers
Merocyanine dyes, *see also* Merocyanine 540; Merocyanine oxazolone
absorbance changes in, 213, 221, 224, 232
in cell membrane potential studies, 217
Merocyanine 540, *see also* Merocyanine dyes
absorption spectra of, 225, 226
in bilayer lipid membranes, 229
in cell membrane potential studies, 217, 219
dye-extinction change in, 237
fractional-fluorescence changes in, 234
molecular structure, 218
in phosphatidylcholine-vesicle membranes, 225–229
signal-to-noise ratio, in fluorescence changes in, 223
Merocyanine oxazolone, in SGA studies, 217
Methane, molecular structure of, 5–6
Micelle stability, hydrophobic bonding and, 6
Microlenses, in BLM, 257
Microorganisms, and membrane field fluctuation, 38, 43–46, 57
Mitochondria, membrane potential of, 217
Model(s)
in aqueous-solution studies, 3–7
see Binary-state conduction model
see Cable theory
see Continuous-state conduction model
of electrical properties, of cells, 342–344
fluctuation-dissipation, of noise, 391, 392
see Hodgkin-Huxley model
linear electric, 342, 352
see Shell model
see Spherical models
Moffitt parameter, in helix-coil transition, 133
Molecular dynamics, computations for, 3–5, 20
Molecular electrostatics, finite element method in, 22–36
Molecular relaxation, in electrical field fluctuations, 39–41
Molecular weight
aggregate structure and, 126–129
in conformational changes, 137
dielectric absorbtion/relaxation and, 123–124, 135, 173
dielectric effects and, in polypeptides, 123, 126, 127
frequency range and, 76, 78
light-scattering measurement of, 127
Monazomycin, in channel formation, 282
Monomer-dimer mechanism, in dye-aggregation changes, 226–229

Monolein membranes, 268–276
Monte Carlo calculations, in liquid water studies, 3–5
Multiplex advantage, 90–91
Muscle fibers, skeletal
dielectrical properties of, frog, 199–205, 217, 344–350
fluorescence changes in, 234
Myxicola axons, gating currint in, 325, 327

Nerve fibers, *see also* Nodes of Ranvier; Squid giant axon (SGA); *and other specific types*
dye-absorption changes in, 224–225
electric-membrane potentials in, 217
optical detection of membrane potential in, 217; *see also under* Dyes
Nerve impulses, biochemical conformation of, 156
Nodes of Ranvier, frog
asymmetrical currents in, 393
flicker ($1/f$) noise component of, 393
gating current measurement in, 325, 327, 333, 336
noise measurement in, 399
perinodal space noise in, 395
permeability versus conductance in, 289
procaine effects on, 334
Noise, *see: under* Bilayer lipid membrane(s) (BLM); *at* Flicker noise; Noise, cell membrane; Noise excitation methods; Noise measurement; Noise (Power) spectra
Noise, cell membrane
in binary-state conduction model evaluation, 416–417
components of, 390
continuous-state conduction models and, 392, 415
criteria for identification of, 403
current-voltage relationships in, 382, 384
fluctuation-dissipation model of, 391–39
gating currents and, 393–395
Langevin equation for, 391
master-equation approach, 391
see Noise measurement
nonequilibrium-cycle model of, 386
peaking response in, 366, 368–378,
PRBS analysis of, 355
relaxation in, 390, 392, 395
sources of, 390–391
Noise-excitation method(s) (Pulse techniques)
in bilayer-lipid membranes, 268–272
characterization of, 92
in chromatin analysis, 171
in conduction-process inquiry, 399–401
in DNA analysis, 171
in gating-current measurement, 323–326, 330–333
in linear-systems analysis, 90, 92–94
see Pseudorandom-binary signal (PRBS) method
Noise measurement
in bilayer-lipid membranes, 274–277
in cell membranes, 399–402
in ion-conduction channels, *see* Potassium ion conduction; Sodium ion conduction
in large membrane areas, 400–402, 413
membrane-area selection for, 400–403
potassium current blockage and, 400
in squid and frog preparations, compared, 426–427
small, patch-area technique, 401–403
Noise (Power) spectra
in cell membranes, 361–362, 403–406, 416
computation of, 385–386
density of, in *m*-current mechanisms, 394
in squid giant axons, 403–408
Nonactin, conductive noise of, 277
Nonpolar solutes, 5, 6, 12
Nonspherical dielectrics, 29–30
Nuclear magnetic resonance (NMR) spectrometry, 90
Nucleic acid complexes, dye interactions of, 175–182
Nucleosomes
helical arrangements of 183–186
negative anisotropy of, 185–186
Null methods, of low-current measurement, 69, 70
Nyquist frequency, 96
Nyquist's theorem, 274
Nystatin, as channel former, 281

Onsager equation, in dipole applications, 59–62
Optical-anisotropy factor, 183, 195
Optical-rotary dispersion (ORD) measurement, 133
Optical technique, in cell membrane potential studies, 217
Organelle membranes, 217
Ornstein-Zernike equation, 61
Overlap volume, 61
Oxonol dyes, 217–219, 224–225

PBLG, *see* Poly(benzyl-L-glutamate)
PCBL, *see* Poly(carbobenzoxy-L-lysine)
PDLPA, *see* Poly(D,L-phenylalanine)

Pearl-chain formation, 208
Peaking
  in noise (power) spectra, 382–387, 424–425
  in nonequilibrium reactions, 386, 424
  periaxonal space effects in, 396–397
  of relaxation noise components, 390, 395
PE membranes (phosphotidyl ethanolamine), conductance characteristics of, 283, 286
Percus-Yevick equation, 61–62
Periaxonal space, 395–397
Periodic noise, generation of, 93; *see also* Pseudorandom noise
Permittivity
  aggregation and, in PBLG, 104–105
  of aqueous solutions, 90
  concentration dependence of, 76–77, 79
  frequency dependence of, 75–77
  gamma dispersion in LDL and, 145
  linear-systems analysis of, 90–94
  of low-density lipoproteins, 142–143, 145
  low-frequency, steady-state, 70
  molecular-weight dependence of, 75
  of PBLG, with PND spectometry, 104
  of PCBL, 100–104
  of polar liquids, 59
  of proteins, in solution, 150–153
  relative, for human LDL, 150–153
Phase transition
  gas-liquid systems, 14
  liquid state, 15–16
Phosphatidyl ethanolamine (PE), 283
Plateau-Gibbs border concept, 256
Poisson equation, 22–24
Polarizability
  of biomolecules, 22, 173
  of polyions, 107–119
Polarizability density, 23, 36
Polarization
  chemical-field effect and, 194
  molecule-segment length and, 174
  two-speed mechanisms in, 173–174
Polyalkylisocynates, solution properties of, 131–132
Poly(benzyl-L-aspartate)
  aggregation of, 124
  birefringence of, 195
  helix-coil transition of, 134–135
Poly(benzyl-L-glutamate) (PBLG)
  aggregation of, 125–126
  disaggregation of, 129, 132–134
  electrical-field effects in, 193, 195
  permittivity plotting of, 104–105
  thermal transition in, 133
Poly(carbobenzoxy-L-lysine) (PCBL)
  helix-coil transition of, 100–104, 135
  permittivity plotting in, 100–102
Polyelectrolytes
  aqueous solutions of linear, 74–86
  characterization of
  concentration dependence of, 75–78
  dielectric dispersion process in, 107–120
  dielectric properties of, 74–79
  flexible, two-phase model of, 108, 118–119
  at infinite dilution, 83–84
  see Rigid-molecular model
  simpler electrolytes compared with, 80–81
  van der Touw-Mandel model of, 81–86
Poly(hexylisocyanate) (PHIC), aggregate structure of, 131–132
Polyions, *see also* Counterions
  characterization of, 107
  fixed charges on, 81
at infinite dilution, 83–84
  longitudinal diffusion of, 111, 116, 120
  perpendicular diffusion of, 116–119
  polarizability of, 117–119
rigid molecular model of, 81
  van der Touw-Mandel model of, 81–84
Polymers, *see also* Polypeptides
  aggregation patterns of, 124
  chain length and transition temperature of, 139
Polymerization, *see* Aggregation; Conformational changes
Polypeptides, *see also* Biomolecules
  absorption, dielectric, 123–124, 135
  aggregation of, 124–131
  in bilayer-lipid membranes, 244
  conformational changes in, 129–134
  dielectric analysis of structure, 123–124
  disaggregation of, 132
Poly(pheny-L-glutamate) (PPLG), 125–126
Poly(D,L-phenylalanine) (PDLPA)
  aggregation of, 124–125
  disaggregation of, 134–135
Polysaccharides, conformational changes in, 135–140
Polystyrene, antiparallel aggregation in, 131 132
Pore(s), *see* Channels
Porphyrin chromophore, 223
Potassium ion conduction
  binary-state conduction model, 416, 417, 424
  blockage of, 311–314, 318, 406–410
  H-H model of, 389, 417–418
  membrane potential fluctuation and, 44
  noise generated in, 369–370, 390, 400, 406–411
  periaxonal space and noise in, 395–397
  power spectrum equation of, 392
  single channel values in, 424
  suppressed conductance of, 374

Potential, cell membrane
dye-absorption changes in, 219–222
in grating current inactivation, 329–330
in microorganisms, 44–45
optical detection of, 217–240
resting value of, 311
Potential energy
computation of, 5
in dipole interaction, 60
PPLG, *see* Poly(phenyl-L-glutamate)
PRBS, *see* Pseudorandom-binary signal
Procaine, in intramembrane-charge blockage, 334
Proflavine, intercalation of, in DNA, 176, 179
Protein(s)
in bilayer lipid membranes, 224, 247
dielectric properties of, 142
human serum albumin as, 52–55
hydrophobic, in BLM, 243–244
permittivity of, in solution, 150
see Protein molecule
water of hydration in, 2, 150
Protein-bound water, dielectric properties of, 200–202
Protein-ligand binding, kinetics of, 47–55
Protein molecule(s), *see also* Biomolecules
dipole molecule compared with, 57
dispersion properties of globular, 50
electric field behavior of, 47
Pseudorandom binary signal (PRBS)
characteristics of, 359, 362, 380
discrete Fourier transform in, 359–360, 375
generation of, 95–97
in impedance/admittance analysis, 357–365, 374–378
magnitude spectrum of, 366
prewhitening of, 375–376
problems of method in, 377–378
utilization of, 359–362
Pseudorandom noise (PN)
binary, *see* Pseudorandom binary signal (PRBS)
characterization of, 93, 95
Fourier synthesis generation of, 97
generation of, 95–100
in impedance/admittance measurement, 357–358, 359, 365
spectometry of, *see* Pseudorandom noise dielectric (PDN) spectometry
Pseudorandom noise dielectric (PDN) spectometry
applications of, 90–91
equations for, 97–98
hardware for, 97–98
permittivity measurement by, 104–105
principles and methods of, 91–95
software flow-chart, 98–100
Pulse, *see* Noise
Purine, 175
Pyrimidine, 175

Random gel model
characterization of, 14–15
hydrophobic effect and, 17–18
Refractive indexes, anisotropy in, 64
Relaxation, dielectric, *see also* Relaxation time
cell membrane conductivity and, 316–317
chemical induction of, 194
in conformational changes, of polymers, 123–124, 131–132
in dye intercalation, of DNA-like molecules, 170, 174, 179, 182
frequency-response stimulations of, 50, 51
of induced dipole polarization, 82
high and low-frequency types, compared, 172–173, 146–148, 194
low-frequency dipole, 59, 67–73
in muscle tissue, 199–200
orientation, of rigid biomolecules, 82
of permittivity, linear analysis of, 90–91, 94
in protein-ligand binding, 47–55
single-pulse measurement of, 47–49
time-response simulations of, 50–51
Relaxation time
of birefringence, in DNA, 161–166
concentration dependence and, 78–79, 82–83
in counterion equilibrium distribution, 82, 83
of DNA segmentation, 174
in field fluctuation, of ionic solutions, 39, 41
in muscle cell membrane, 199
in polyion polarizability, 111–119
persistence length and, 214–215
in superhelical structures, 214
Resistor-capacitor (R-C) circuit
bilayer lipid membrane analogy with, 269, 272, 280
cell membrane analogy with, 307, 344, 368
Resting potential, of cells, 342
Rhodamine B, 213
Rigid-molecular model(s), of polyelectrolytes
dielectric behavior of, 74, 108–109, 194
Fourier transform of counterion fluctuation and, 108
perpendicular diffusion and, 118
polarization and segmentation in, 174

relaxation time and length of, 215
static field polarizability of, 215
Rotational diffusion, dipole, 65–67; *see also* Kerr effect
Rouse-Zimm model, of polymers, 119

Sarcomere, 345
Sarcoplasmic reticulum, 345, 346, 350
Scatchard equation, in bound dye calculations, 162
Schwann cell layer
reactance due to, 299, 304
series resistance and, 307, 308
Scorpion venom, 340
Sea water, artificial (ASW)
cell-membrane admittance in, 367–369
recipe for, 369, 403*n*
SGA, *see* Squid giant axon
Shell model, of hydration in proteins, 2
Shot-noise level, equation for, 235
Side-chain polarity, in polypeptide aggregation, 124–126
Signal-to-noise ratio
in cell membrane area selection, 401–403, 413
in fluorescence changes, 233
in fluorescence measurement, 235–237
Single-state conduction model, *see* Continuous-state conduction model
Single-pulse response method, 47–58
Skeletal muscles, *see* Muscle fibers
Sodium ion conduction
blockage of, 310–314, 318, 333–335, 407
chemical blockage of, 333–335
depolarization inactivation, 331–333
gating current in, 334–335, 393
H-H model of, 322–323, 329, 389
inactivation of, 329, 335
membrane-potential inactivation of, 329
noise in, *see* Sodium-conduction noise
power-spectrum equation in, 392
protein-receptor analogy of, 340
ultraviolet-light inactivation of, 333
voltage-dependent capacitance in, 320
zinc ion blockage of, 334
Sodium conduction noise
in cell membrane admittance, 370
cesium fluoride perfusion and, 413
characteristics of, 390, 411–413
fluctuations in, 392–393, 400
measurement of, 400, 411–415
potassium ion effects in, 411
potassium ion replacement and, 411–413, 415
spectra of, 413–415
TEA in measurement of, 410, 413
Sonication, of DNA, 160, 170, 172–173, 182
Spectroscopy, computerized application of, 90
Spherical models
of dipole interactions, 61–62
eye lens as, 351
of Toupin and Lax, 60
of Wertheim, 61–62
Squid giant axons (SGA)
BLM ion translocation and, 272
cell-membrane capacitance in, 306–320
dye-absorbance changes in, 221, 225
dye-fluorescence changes in, 232, 234, 236
fluctuations, spontaneous, in ion movement probe of, 399–420
gating currents in, 272, 325–336, 393
impedance in, 298–304
ion-conduction measurement in, 355–358
noise measurement in, 400–401
optical-change mechanisms and, 225–229
periaxonal-space noise in, 395
resistivity in, 308
series resistance in, 299, 307
zinc ion grating current effects in, 334–335
Stark effect, 222
Superhelical structures, relaxation time in, 214
Suzukacillin, channel formation by, 282
Svedberg units
Synaptosomal ghosts, potential changes and, 219

Tail current, in periaxonal K ions, 395
TEA, *see* Tetraethylammonium ion
Tetraethylammonium (TEA) ions
noise induction by, 410–411
in potassium current blockage, 370, 407, 409–410
sodium conduction noise and, 410–413
Tetramethyl ammonium fluoride (TMAF), sodium conduction noise and, 413
Tetraphenylborate (TFB) ions
conductive effects in BLM and, 266–278
thermal current fluctuations due to, 274–275
Tetrodotoxin (TTX)
in cell membrane admittance, 369, 372, 376
in intramembrane charge movement, 323, 336
sodium noise reduction and, 407, 412, 413, 415
TFB, *see* Tetraphenylborate
Thermal contraction, elasticity and, 19

Thermal noise
of hydrophobic ions, in BLM, 274–277
of permittivity, 94, 95
Time-domain refractometer (TDR), 90, 94
Tissue water, dielectric properties of, 200–202
Tocopherol-sphingomyelin, as membrane, 295–297
Triethyldecylammonium (TEDA) ion, noise of, in SGA, 409–410
TTX, *see* Tetrodotoxin
Tubular system, muscle fiber
characterization of, 346–350
network of, 347–348
propogation constant in, 349
Two-state conduction model(s), *see* Binary-state conduction model(s)

Urea, in hydrophobic bonding, 6

Valinomycin
conductive noise of, in BLM, 276
in nonaqueous media membranes, 297
Van der Touw-Mandel model, of charged polyelectrolyte chains, 81–86, 173–174
Vapor-condensation process, 15–16
Venom, black-widow spider, 282
Vesicles, dyes in membrane potential studies of, 219
Viscosity
in dye binding, of DNA, 182
near glassy temperature, 11
pressure of, and thermal expansion, 12
Voltage change, in membrane-channel formation, 281–285
Voltage clamp (potential control)
application of, 355
in current noise of SGA, 410–411
in impedance/admittance measurement, 357, 380
H-H model parameters and, 389
periaxonal space effects and, 397
Voltage dependence, *see also* Potential, cell membrane
in bilayer-lipid membranes, 267
of cell-membrane capacitance, in SGA, 306–320
intramembrane-charge movement and, 323
in symmetrical membranes, 267
Volume effects, in hydrocarbon-water systems, 6

Water, *see also* Aqueous solution; Liquid state
anomalous physical characteristics of, 10–19
bonded coordination structure of, 18
bond reshuffling in, 21
dielectric properties of, 200–201
dielectric role, in BLM, 292–293
empirical pair potential (EPEN), 3–5
flicker cluster model of, 20
freezing point of, 10
of hydration, 149–153, 156–157
hydrophobic effects and, 6–7, 10, 12–13, 17–18, 21
molecular dynamics of, 3–5, 20
permittivity of pure, 149, 150
see Protein-bound water
see Random-gel model
structure of, 10–18, 20–21
thermal expansion of, 10–11
tissue and free water compared, 200–203
Wertheim spherical model, 61–62
White noise, 95–96, 357–358, 395
Wiener concept, 356, 365

Zinc ions, gating current blockage by, 334